DUCTILE SHEAR ZONES

Ductile Shear Zones
From Micro- to Macro-scales

Edited by

Soumyajit Mukherjee

Department of Earth Sciences, Indian Institute of Technology Bombay, Powai, Mumbai 400076,
Maharashtra, India

Kieran F. Mulchrone

Department of Applied Mathematics, University College, Cork, Ireland

WILEY Blackwell

This edition first published 2016 © 2016 by John Wiley & Sons, Ltd

Registered Office
John Wiley & Sons, Ltd, The Atrium, Southern Gate, Chichester, West Sussex, PO19 8SQ, UK

Editorial Offices
9600 Garsington Road, Oxford, OX4 2DQ, UK
The Atrium, Southern Gate, Chichester, West Sussex, PO19 8SQ, UK
111 River Street, Hoboken, NJ 07030-5774, USA

For details of our global editorial offices, for customer services and for information about how to apply for permission to reuse the copyright material in this book please see our website at www.wiley.com/wiley-blackwell.

Library of Congress Cataloging-in-Publication Data

Ductile shear zones: from micro- to macro-scales / edited by Soumyajit Mukherjee and Kieran F. Mulchrone.
    pages   cm
  Includes bibliographical references and index.
  ISBN 978-1-118-84496-0 (cloth)
1. Shear zones (Geology)   2. Geology, Structural.   I. Mukherjee, Soumyajit, editor.   II. Mulchrone, Kieran F., editor.
  QE606.D82 2016
  551.8'72–dc23
                                    2015024896

A catalogue record for this book is available from the British Library.

Wiley also publishes its books in a variety of electronic formats. Some content that appears in print may not be available in electronic books.

Cover image: A sigmoid-shaped muscovite fish with top-to-right ductile shear sense. Photo length: 5 mm. Under cross polarized light. Location: Karakoram range. Reproduced from fig. 3b of Mukherjee (2011). Ref: Mukherjee S. (2011) Mineral fish: their morphological classification, usefulness as shear sense indicators and genesis. International Journal of Earth Sciences 100, 1303–1314.

Set in 9.5/11.5 pt Melior by SPi Global, Pondicherry, India
Printed and bound in Malaysia by Vivar Printing Sdn Bhd

1   2016

Contents

Contributors

Sukumar Baruah
Department of Applied Geology, Dibrugarh University, Dibrugarh 786004, Assam, India

Subhadip Bhadra
Department of Earth Sciences, Pondicherry University, R.V. Nagar, Kalapet, Puducherry 605014, India

Rakesh Biswas
Geodata Processing and Interpretation Centre, Oil and Natural Gas Corporation Limited, Dehradun, India

Santanu Bose
Department of Geology, University of Calcutta, Kolkata, 700019, India

David Alexandre Boutelier
The University of Newcastle, School of Environmental and Life Sciences, University Drive, Callaghan, NSW, 2308, Australia

Emmanuelle Boutonnet
Institute of Geosciences, Johannes Gutenberg University Mainz, J.-J.-Becher-Weg 21, D-55128 Mainz, Germany

Laboratoire de Géologie de Lyon – Terre, Planètes, Environnement UMR CNRS 5276, UCB Lyon1 – ENS Lyon, 2 rue Raphael Dubois, 69622 Villeurbanne, France

Fernando Calamita
Dipartimento di Ingegneria e Geologia, Università degli Studi "G. D'Annunzio" di Chieti-Pescara,Via dei Vestini 31, 66013, Chieti Scalo (CH), Italy

Paulo Castro
Laboratório Nacional de Energia e Geologia, Rua da Amieira, Apartado 1089, 4466 901 S. Mamede de Infesta, Portugal

Sadhana M. Chatterjee
Department of Geological Sciences, Jadavpur University, Kolkata 700032, India

Nandini Chattopadhyay
Department of Geological Sciences, Jadavpur University, Kolkata 700032, India

Carlos Coke
Departamento de Geologia, Universidade de Trás-os-Montes e Alto Douro, Apartado 1013, 5001-801 Vila Real, Portugal

Sujoy Dasgupta
Department of Geological Sciences, Jadavpur University, Kolkata, 700032, India

Sergio Delpino
INGEOSUR (CONICET-UNS), Departamento de Geología (UNS), San Juan 670 (B8000ICN), Bahía Blanca, Argentina

Departamento de Geología, (Universidad Nacional del Sur), Bahía Blanca, Argentina

Narciso Ferreira
Laboratório Nacional de Energia e Geologia, Rua da Amieira, Apartado 1089, 4466 901 S. Mamede de Infesta, Portugal

Tapos Kumar Goswami
Department of Applied Geology, Dibrugarh University, Dibrugarh 786004, Assam, India

Bernhard Grasemann
Department of Geodynamics and Sedimentology, Structural Processes Group, University of Vienna, Austria

Saibal Gupta
Department of Geology and Geophysics, Indian Institute of Technology Kharagpur, Kharagpur 721302, West Midnapore, West Bengal, India

Pitsanupong Kanjanapayont
Department of Geology, Faculty of Science, Chulalongkorn University, Bangkok 10330, Thailand

Ali Karrech
The University of Western Australia, School of Civil, Environmental and Mining Engineering, 35 Stirling Highway, Crawley, 6009, WA, Australia

Matthew J. Kohn
Department of Geosciences, Boise State University, 1910 University Drive, Boise, ID 83725, USA

Phillipe-Hervé Leloup
Laboratoire de Géologie de Lyon – Terre, Planètes, Environnement UMR CNRS 5276, UCB Lyon1 – ENS Lyon, 2 rue Raphael Dubois, 69622 Villeurbanne, France

Sergio Llana-Fúnez
Departamento de Geología, Universidad de Oviedo, Arias de Velasco s/n, 33005 Oviedo, Spain

Tanushree Mahadani
Department of Earth Sciences, Indian Institute of Technology Bombay, Powai, Mumbai 400076, Maharashtra, India

Nibir Mandal
Department of Geological Sciences, Jadavpur University, Kolkata, 700032, India

Stefano Mazzoli
Dipartimento di Scienze della Terra, dell'Ambiente e delle Risorse (DiSTAR), Università degli studi di Napoli 'Federico II', Largo San Marcellino 10, 80138 Napoli, Italy

Dave J. McCarthy
Department of Geology, University College, Cork, Ireland

Patrick A. Meere
Department of Geology, University College, Cork, Ireland

Rupsa Mukherjee
Department of Earth Sciences, Indian Institute of Technology Bombay, Powai, Mumbai 400076, Maharashtra, India

Soumyajit Mukherjee
Department of Earth Sciences, Indian Institute of Technology Bombay, Powai, Mumbai 400076, Maharashtra, India

Kieran F. Mulchrone
Department of Applied Mathematics, University College, Cork, Ireland

Clyde J. Northrup
Department of Geosciences, Boise State University, 1910 University Drive, Boise, ID 83725, USA

Paolo Pace
Dipartimento di Ingegneria e Geologia, Università degli Studi "G. D'Annunzio" di Chieti-Pescara,Via dei Vestini 31, 66013, Chieti Scalo (CH), Italy

Jorge Pamplona
Institute of Earth Sciences (ICT)/University of Minho Pole, Universidade do Minho, Campus de Gualtar, 4710-057 Braga, Portugal

Eurico Pereira
Laboratório Nacional de Energia e Geologia, Rua da Amieira, Apartado 1089, 4466 901 S. Mamede de Infesta, Portugal

Jahnavi Narayan Punekar
Department of Geosciences, Princeton University, Princeton, NJ, USA

Sayan Ray
Department of Geological Sciences, Jadavpur University, Kolkata 700032, India

Klaus Regenauer-Lieb
The University of Western Australia, School of Earth and Environment, 35 Stirling Highway, Crawley, 6009, WA, Australia

The University of New South Wales, School of Petroleum Engineering, Tyree Energy Technologies Building, H6 Anzac Parade, Sydney, NSW, 2052, Australia

Benedito C. Rodrigues
Institute of Earth Sciences (ICT)/University of Minho Pole, Universidade do Minho, Campus de Gualtar, 4710-057 Braga, Portugal

José Rodrigues
Laboratório Nacional de Energia e Geologia, Rua da Amieira, Apartado 1089, 4466 901 S. Mamede de Infesta, Portugal

Marina Rueda
Departamento de Geología, (Universidad Nacional del Sur), Bahía Blanca, Argentina

Sanjoy Sanyal
Department of Geological Sciences, Jadavpur University, Kolkata 700032, India

Christoph Eckart Schrank
Queensland University of Technology, School of Earth, Environmental and Biological Sciences, Brisbane, 4001, QLD, Australia

The University of Western Australia, School of Earth and Environment, 35 Stirling Highway, Crawley, 6009, WA, Australia

Pulak Sengupta
Department of Geological Sciences, Jadavpur University, Kolkata 700032, India

Sudipta Sengupta
Department of Geological Sciences, Jadavpur University, Kolkata 700032, India

Pedro Pimenta Simões
Institute of Earth Sciences (ICT)/University of Minho Pole, Universidade do Minho, Campus de Gualtar, 4710-057 Braga, Portugal

Yutaka Takahashi
Geological Survey of Japan, AIST, 1-1-1 Higashi, Tsukuba, Ibaraki 305-8567, Japan

Enrico Tavarnelli
Dipartimento di Scienze Fisiche, della Terra e dell'Ambiente, Università degli Studi di Siena, Via Laterina 8, 53100, Siena, Italy

Ivana Urraza
Departamento de Geología, (Universidad Nacional del Sur), Bahía Blanca, Argentina

Stefano Vitale
Dipartimento di Scienze della Terra, dell'Ambiente e delle Risorse (DiSTAR), Università degli studi di Napoli 'Federico II', Largo San Marcellino 10, 80138 Napoli, Italy

Andrea M. Wolfowicz
Department of Geosciences, Boise State University, 1910 University Drive, Boise, ID 83725, USA

Shell International Exploration and Production, Shell Woodcreek Complex, 200 N Dairy Ashford Rd, Houston, TX 77079, USA

Acknowledgments

Thanks to **Ian Francis, Delia Sandford**, and **Kelvin Matthews** (Wiley Blackwell) for handling this edited volume. Comments by the three anonymous reviewers on the book proposal helped us to remain cautious in editing. We thank all the authors and reviewers for participation.

Thanks to **Beth Dufour** for looking at permission issues and to **Alison Woodhouse** for assistance in copyediting, and **Kavitha Chandrasekar, Sukanya Shalisha Sam** and **Janyne Ste Mary** for numerous help for all the chapters in this book.

Introduction

Kinematics of ductile shear zones is a fundamental aspect of structural geology (e.g. Ramsay 1980; Regenauer-Lieb and Yuen 2003; Mandal et al. 2004; Carreras et al. 2005; Passchier and Trouw 2005; Mukherjee 2011, 2012, 2013, 2014a; Koyi et al. 2013; Mukherjee and Mulchrone 2013; Mukherjee and Biswas 2014; and many others). This edited volume compiles a total of 17 research papers related to various aspects of ductile shear zones.

In the first section "Theoretical Advances and New Methods", **Vitale and Mazzoli** (2016) describe an inverse method to deduce the incremental strain path in heterogeneous ductile shear zones, and have applied the method in a wrench zone hosted in pre-Alpine batholiths. Field studies, analog and numerical modeling presented by **Dasgupta et al.** (2016) strongly indicate that volume reduction augments transpression in ductile shear zones. In addition, two parameters that control shortening perpendicular to shear zones are defined. **Mulchrone et al.** (2016) analytically model steady state and oblique foliation development in shear zones and present the possibility of estimating: (i) the relative strength of foliation destroying processes, (ii) the relative competency of the grains, and (iii) the kinematic vorticity number. **Schrank et al.** (2016) analytically model the deformation of inclusions with a hyperelastoviscoplastic rheology under ductile lithosphere conditions, and predict the evolution of the shape of the inclusion. **Mukherjee and Biswas** (2016) presented kinematics of layered curved simple shear zones. Considering Newtonian viscous rheology of the litho-layers, they explain how aspect ratios of inactive initially circular markers keep changing.

In the second section "Examples from Regional Aspects", **Boutonnet and Leloup** (2016) discuss Quartz strain-rate-metry from shear zones at Ailao Shan–Red River (China) and Karakoram (India) and strain rate variation within these zones. They decipher high slip rates of the order of cm per year from both these zones. Applying Titanium-in-quartz thermobarometry on Scandinavian Caledonides, **Wolfowicz et al.** (2016) deduce a geothermal gradient and support a critical taper mechanism of deformation. **Pace et al.** (2016) study brittle-ductile shear zones from the Central–Northern Apennines related to frontal and oblique ramps and deduced structural inheritance of extensional faults. **Sengupta and Chatterjee** (2016) deduce lower amphibolites facies metamorphism in the Phulad Shear Zone (India). Antithetically oriented clasts indicate a general shear deformation; however, the method of vorticity analysis applied was found unsuitable since the deformation was heterogeneous and the shear zone contains a number of phases of folds.

Chattopadhyay et al. (2016) describe how ductile shear altered the mineralogy, chemistry and texture of rocks from the South Purulia Shear Zone (India). Using phase diagrams, they also constrain the temperature of metasomatism. In a study of ductile shear zones in the Khariar basin (India), **Bhadra and Gupta** (2016) decipher two movement phases along the Terrane Boundary Shear Zone. **Mukherjee et al.** (2016) review morphology and genesis of intrafolial folds and deduce their Class 1C and Class 2 morphologies from Zanskar Shear Zone from Kashmir Himalaya. **Pamplona et al.** (2016) classify the Malpica Lamego Ductile Shear Zone into sectors of different deformation patterns, such as sinistral or dextral shear and flattening. Pseudosection studies by **Delpino et al.** (2016) for ductile shear zone in the Pringles Metamorphic Complex (Argentina) yield thermal curves. **Kanjanapayont** (2016) reviews ductile shear zones in Thailand, presents structural details, and constrains and correlates when they were active. **Takahashi** (2016) studies the Nihonkoku Mylonite Zone (Japan) and found that its mylonitization during 55–60 Ma correlates with deformation in the Tanagura Tectonic Line. Flanking structures (Passchier 2001; Mukherjee and Koyi 2009; Mukherjee 2014b, etc.) has recently been of great interest in the context of ductile shear zones. **Goswami et al.** (2016) describe the geometry of flanking structures from Arunachal Pradesh, Higher Himalaya, India and use contractional flanking structures to deduce shear sense.

Soumyajit Mukherjee
Kieran F. Mulchrone

REFERENCES

Bhadra S, Gupta S. 2016. Reworking of a basement–cover interface during Terrane Boundary shearing: An example from the Khariar basin, Bastar craton, India. In Ductile Shear Zones: From Micro- to Macro-scales, edited by S. Mukherjee and K.F. Mulchrone, John Wiley & Sons, Chichester.

Boutonnet E, Leloup P-H. 2016. Quartz-strain-rate-metry (QSR), an efficient tool to quantify strain localization in the continental crust. In Ductile Shear Zones: From Micro- to Macro-scales, edited by S. Mukherjee and K.F. Mulchrone, John Wiley & Sons, Chichester.

Carreras J, Druguet E, Griera A. 2005. Shear zone-related folds. Journal of Structural Geology 27, 1229–1251.

Chattopadhyay N. et al. 2016. Mineralogical, textural, and chemical reconstitution of granitic rock in ductile shear zones: A study from a part of the South Purulia Shear Zone, West Bengal, India. In Ductile Shear Zones: From Micro- to Macro-scales, edited by S. Mukherjee and K.F. Mulchrone, John Wiley & Sons, Chichester.

Dasgupta S, Mandal N, Bose S. 2016. How far does a ductile shear zone permit transpression?. In Ductile Shear Zones: From Micro- to Macro-scales, edited by S. Mukherjee and K.F. Mulchrone, John Wiley & Sons, Chichester.

Delpino S. et al. 2016. Microstructural development in ductile deformed metapelitic–metapsamitic rocks: A case study from the Greenschist to Granulite facies megashear zone of the Pringles Metamorphic Complex, Argentina. In Ductile Shear Zones: From Micro- to Macro-scales, edited by S. Mukherjee and K.F. Mulchrone, John Wiley & Sons, Chichester.

Goswami TK, Baruah S. 2016. Flanking structures as shear sense indicators in the higher Himalayan gneisses near Tato, West Siang District, Arunachal Pradesh, India. In Ductile Shear Zones: From Micro- to Macro-scales, edited by S. Mukherjee and K.F. Mulchrone, John Wiley & Sons, Chichester.

Kanjanapayont P. 2016. Strike–slip ductile shear zones in Thailand. In Ductile Shear Zones: From Micro- to Macro-scales, edited by S. Mukherjee and K.F. Mulchrone, John Wiley & Sons, Chichester.

Koyi H, Schmeling H, Burchardt S, Talbot C, Mukherjee S, Sjöström H, Chemia Z. 2013. Shear zones between rock units with no relative movement. Journal of Structural Geology 50, 82–90.

Mandal N, Samanta SK, Chakraborty C. 2004. Problem of folding in ductile shear zones: a theoretical and experimental investigation. Journal of Structural Geology 26, 475–489.

Mukherjee S. 2011. Mineral fish: their morphological classification, usefulness as shear sense indicators and genesis. International Journal of Earth Sciences 100, 1303–1314.

Mukherjee S. 2012. Simple shear is not so simple! Kinematics and shear senses in Newtonian viscous simple shear zones. Geological Magazine 149, 819–826.

Mukherjee S. 2013. Deformation Microstructures in Rocks. Springer, Berlin.

Mukherjee S. 2014a. Atlas of shear zone structures in meso-scale. Springer, Berlin.

Mukherjee S. 2014b. Review of flanking structures in meso-and micro-scales. Geological Magazine, 151, 957–974.

Mukherjee S, Biswas R. 2014. Kinematics of horizontal simple shear zones of concentric arcs (Taylor–Couette flow) with incompressible Newtonian rheology. International Journal of Earth Sciences 103, 597–602.

Mukherjee S, Biswas R. 2016. Biviscous horizontal simple shear zones of concentric arcs (Taylor–Couette Flow) with incompressible Newtonian rheology. In Ductile Shear Zones: From Micro- to Macro-scales, edited by S. Mukherjee and K.F. Mulchrone, John Wiley & Sons, Chichester.

Mukherjee S, Koyi HA. 2009. Flanking microstructures. Geological Magazine 146, 517–526.

Mukherjee S, Mulchrone KF. 2013. Viscous dissipation pattern in incompressible Newtonian simple shear zones: an analytical model. International Journal of Earth Sciences 102, 1165–1170.

Mukherjee S. et al. 2016. Intrafolial Folds: Review and examples from the Western Indian Higher Himalaya. In Ductile Shear Zones: From Micro- to Macro-scales, edited by S. Mukherjee and K.F. Mulchrone, John Wiley & Sons, Chichester.

Mulchrone KF, Meere PA, McCarthy DJ. 2016. 2D model for development of steady-state and oblique foliations in simple shear and more general deformations. In Ductile Shear Zones: From Micro- to Macro-scales, edited by S. Mukherjee and K.F. Mulchrone, John Wiley & Sons, Chichester.

Pace P, Calamita F, Tavarnelli E. 2016 Brittle-ductile shear zones along inversion-related frontal and oblique thrust ramps: Insights from the Central–Northern Apennines curved thrust system (Italy). In Ductile Shear Zones: From Micro- to Macro-scales, edited by S. Mukherjee and K.F. Mulchrone, John Wiley & Sons, Chichester.

Passchier C. 2001. Flanking structures, Journal of Structural Geology 23, 951–962.

Passchier CW, Trouw RAJ. 2005. Microtectonics, second edition. Springer, Berlin.

Pamplona J. et al. 2016. Structure and Variscan evolution of Malpica–Lamego ductile shear zone (NW of Iberian Peninsula). In Ductile Shear Zones: From Micro- to Macro-scales, edited by S. Mukherjee and K.F. Mulchrone, John Wiley & Sons, Chichester.

Ramsay JG. 1980. Shear zone geometry: a review. Journal of Structural Geology 2, 83–99.

Regenauer-Lieb K, Yuen DA. 2003. Modeling shear zones in geological and planetary sciences: solid-and fluid-thermal-mechanical approaches. Earth Science Reviews 63, 295–349.

Schrank, CE. et al. 2016. Ductile deformation of single inclusions in simple shear with a finite-strain hyperelastoviscoplastic rheology. In Ductile Shear Zones: From Micro- to Macro-scales, edited by S. Mukherjee and K.F. Mulchrone, John Wiley & Sons, Chichester.

Sengupta S, Chatterjee SM. 2016. Microstructural variations in quartzofeldspathic mylonites and the problem of vorticity analysis using rotating porphyroclasts in the Phulad Shear Zone, Rajasthan, India. In Ductile Shear Zones: From Micro- to Macro-scales, edited by S. Mukherjee and K.F. Mulchrone, John Wiley & Sons, Chichester.

Takahashi Y. 2016. Geotectonic evolution of the Nihonkoku Mylonite Zone of north central Japan based on geology, geochemistry, and radiometric ages of the Nihonkoku Mylonites: Implications for Cretaceous to Paleogene tectonics of the Japanese Islands. In Ductile Shear Zones: From Micro- to Macro-scales, edited by S. Mukherjee and K.F. Mulchrone, John Wiley & Sons, Chichester.

Vitale S, Mazzoli S. 2016. From finite to incremental strain: Insights into heterogeneous shear zone evolution. In Ductile Shear Zones: From Micro- to Macro-scales, edited by S. Mukherjee and K.F. Mulchrone, John Wiley & Sons, Chichester.

Wolfowicz AM, Kohn MJ, Northrup CJ. 2016. Thermal structure of shear zones from Ti-in-quartz thermometry of mylonites: Methods and example from the basal shear zone, northern Scandinavian Caledonides. In Ductile Shear Zones: From Micro- to Macro-scales, edited by S. Mukherjee and K.F. Mulchrone, John Wiley & Sons, Chichester.

PART I
Theoretical Advances and New Methods

Chapter 1

From finite to incremental strain: Insights into heterogeneous shear zone evolution

STEFANO VITALE and STEFANO MAZZOLI

Dipartimento di Scienze della Terra, dell'Ambiente e delle Risorse (DiSTAR), Università degli studi di Napoli 'Federico II', Largo San Marcellino 10, 80138, Napoli, Italy

1.1 INTRODUCTION

Heterogeneous ductile shear zones are very common in the Earth's lithosphere and are particularly well exposed in mountain belts (e.g. Iannace and Vitale 2004; Yonkee 2005; Vitale et al. 2007a,b; Okudaira and Beppu 2008; Alsleben et al. 2008; Sarkarinejad et al. 2010; Kuiper et al. 2011; Dasgupta et al. 2012; Zhang et al. 2013; Samani 2013; Mukherjee 2013, 2014; also see Chapter 9), where they provide useful tools for a better understanding of the processes and parameters controlling strain localization, type of deformation, and rock rheology. The occurrence of strain markers such as fossils, ooids and ellipsoidal clasts in sedimentary rocks, or equant minerals, deflected veins and dykes in igneous rocks, allows one to quantify the finite strain by means of various methods (e.g. Dunnet 1969; Fry 1979; Lisle 1985; Erslev 1988; Vitale and Mazzoli 2005, 2010).

Finite strains are all quantities, directly measured or derived, related to the final state of deformation. These finite quantities, such as strain ratio, effective shear strain (*sensu* Fossen and Tikoff 1993), and angle θ' between the shear plane and oblique foliation in heterogeneous ductile shear zones, cannot furnish unequivocal information about the temporal strain evolution (i.e. strain path; Flinn 1962). This is because there are several combinations of deformation types such as simple shear, pure shear and volume change, that can act synchronously or at different times, leading to the same final strain configuration (Tikoff and Fossen 1993; Fossen and Tikoff 1993; Vitale and Mazzoli 2008, 2009; Davis and Titus 2011). Appropriate constraints are needed to obtain a unique solution – or at least reduce the underdetermination. This also implies introducing some assumptions in the definition of the strain model. The strain path may be envisaged as a temporal accumulation of small strain increments, and the final strain arrangement as the total addition (Ramsay 1967). A possible relationship between final strain configuration and temporal evolution (i.e. incremental strains) was suggested by different authors, such as Hull (1988), Mitra (1991) and Means (1995). The latter author envisaged strain softening/hardening as the main rheological control on shear zone evolution: shear zones characterized by a thickness decreasing with time (Type II) result from strain softening, whereas shear zones characterized by increasing thickness (Type I) are produced by strain hardening (Means 1995). Based on this view, each part of a heterogeneous ductile shear zone is the result of a different strain evolution, and taken all together, the various shear zone sectors may be able to record the whole strain history.

During the last few years, several papers dealt with the possibility of calculating the incremental strain knowing the temporal and spatial evolution of the deformation. Provost et al. (2004) reconstruct the deformation history by means of the n times iteration of the transforming equation characterizing the incremental strain, where n is the number of deformation stages. Horsman and Tikoff (2007) focus the opportunity of separating, by previous method, the strain related to the shear zone margins, where according to Ramsay (1980) the deformation is weak, and that associated with the more deformed shear zone sectors. The authors consider heterogeneous shear zones as consisting of sub-zones, each characterized by a roughly homogeneous deformation. Based on their temporal and spatial evolution, shear zones are then classified into three main groups: (i) constant-volume deformation, (ii) localizing, and (iii) delocalizing deformation. In the first case, the shear zone boundaries remain fixed (Type III shear zone of Hull (1988); Mitra (1991); Means (1995)), whereas for the latter two groups the shear zone boundaries migrate with time, leading to decreasing (group ii) or increasing (group iii) thickness of the actively deforming zone (respectively Type II and Type I shear zones of Hull (1988); Mitra (1991); Means (1995).

Following Means (1995), Vitale and Mazzoli (2008) provide a mathematical forward model of strain accumulation within an ideal heterogeneous shear zone by subdividing it into n homogenously deformed layers, each one

Ductile Shear Zones: From Micro- to Macro-scales, First Edition. Edited by Soumyajit Mukherjee and Kieran F. Mulchrone.

bound by shear planes (C-planes: Passchier and Trouw 2005) and characterized by a specific evolution, this being related to that of adjacent layers within the framework of a defined temporal succession. In the case of strain hardening, the strain evolution starts with a homogeneous deformation affecting originally a specific volume of rock (being represented by a single layer in the model). As the original "single layer" is able to accumulate only a specific amount of strain (due to strain hardening), further shearing involves new material located along the shear zone margins (Mazzoli and Di Bucci 2003; Mazzoli et al. 2004), thereby increasing the active shear zone volume (delocalizing zone of Horsman and Tikoff 2007). On the contrary, in the strain softening case, the deformation – originally homogeneous and affecting an ideal "multilayer" – progressively abandons the layers located at the shear zone margins due to easier strain accumulation in the central sector (localizing zone of Horsman and Tikoff 2007). The difference between the approach of Horsman and Tikoff (2007) and that of Vitale and Mazzoli (2008) is that the latter authors relate the temporal and spatial evolution of the shear zone to strain softening/hardening, whereas the former authors avoid any genetic implication.

Building on the results obtained by Vitale and Mazzoli (2008), and following the mathematical approach of Provost et al. (2004) and Horsman and Tikoff (2007), a technique is proposed in this paper, which is able to provide information on the incremental strain path based on measured finite strains. The incremental strain analysis is then applied to a heterogeneous wrench zone

characterized by no stretches along the shear direction and no volume change.

1.2 INCREMENTAL STRAIN

To obtain a mathematical relationship between incremental and finite strain, consider the general case of deformation being localized within a heterogeneous ductile shear zone with synchronous deformation in the host rock. The shear zone is composed of n deformed layers, each characterized by homogeneous strain. The strain evolution is illustrated in Fig. 1.1 when strain softens (localizing shear zone) and hardens (delocalizing shear zone), where matrices $\mathbf{B}_i$ and $\mathbf{C}^{\text{fin}}$ represent finite strain within the shear zone and in the host rock in the last configuration, respectively, whereas matrices $\mathbf{A}_i$ and $\mathbf{C}$ are related to incremental strain.

In the case of a localizing shear zone, indicating with $\mathbf{B}_i$ the finite strain matrix of the i-th layer in the last configuration (n), the finite strain matrix is related to the incremental matrices by the following relationships (with i ranging between 2 to $n-1$):

$$\mathbf{B}_1 = \mathbf{C}^{n-1}\mathbf{A}_1 \rightarrow \left(\mathbf{C}^{n-1}\right)^{-1}\mathbf{B}_1 = \left(\mathbf{C}^{n-1}\right)^{-1}\left(\mathbf{C}^{n-1}\right)\mathbf{A}_1 \rightarrow$$
$$\mathbf{A}_1 = \left(\mathbf{C}^{n-1}\right)^{-1}\mathbf{B}_1,$$
$$\mathbf{B}_2 = \mathbf{C}^{n-2}\mathbf{A}_2\mathbf{A}_1 \rightarrow \mathbf{B}_2 = \mathbf{C}^{n-2}\mathbf{A}_2\left(\mathbf{C}^{n-1}\right)^{-1}\mathbf{B}_1 \rightarrow \qquad (1)$$
$$\mathbf{A}_2 = \left(\mathbf{C}^{n-2}\right)^{-1}\mathbf{B}_2\left(\mathbf{B}_1\right)^{-1}\mathbf{C}^{n-1},\ldots\ldots$$
$$\mathbf{A}_i = \left(\mathbf{C}^{n-i}\right)^{-1}\mathbf{B}_i\left(\mathbf{B}_{i-1}\right)^{-1}\mathbf{C}^{n-i+1}.$$

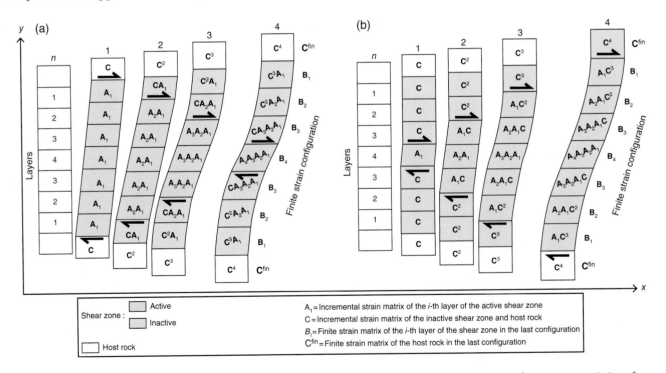

Fig. 1.1. Examples of strain evolution along the *xy* plane of the coordinate reference frame for heterogeneous shear zones consisting of seven homogeneously deformed layers and characterized by transtension in the active shear zone and synchronous pure shear elsewhere. (a) Localizing shear zone. (b) Delocalizing shear zone.

In the case of a delocalizing shear zone the relationship is:

$$\mathbf{A}_i = \mathbf{B}_i \left(\mathbf{C}^{n-i} \right)^{-1} \mathbf{C}^{n-i+1} \left(\mathbf{B}_{i-1} \right)^{-1} = \mathbf{B}_i \mathbf{C} \left(\mathbf{B}_{i-1} \right)^{-1}. \quad (2)$$

1.2.1   Special case of no deformation in the host rock

In the case of $\mathbf{C} = 1$ (identity matrix), i.e. no deformation in the host rock, Equations (1) and (2) are always substituted by the equation:

$$\mathbf{A}_i = \mathbf{B}_i \left(\mathbf{B}_{i-1} \right)^{-1}. \quad (3)$$

In this circumstance, the relationships between incremental and finite strain quantities (stretches and effective shear strains) may be directly obtained by rewriting Equation 3 in an explicit form. Consider a wrench zone in which each i-th layer is characterized by finite strain represented by the matrix (Tikoff and Fossen 1993):

$$\mathbf{B}_i = \begin{bmatrix} \left(k_1^{fin} \right)_i & \Gamma_i^{fin} & 0 \\ 0 & \left(k_2^{fin} \right)_i & 0 \\ 0 & 0 & \left(k_3^{fin} \right)_i \end{bmatrix}, \quad (4)$$

where

$$\Gamma_i^{fin} = \gamma_i^{fin} \left(\frac{\left(k_1^{fin} \right)_i - \left(k_2^{fin} \right)_i}{\ln \left(\frac{\left(k_1^{fin} \right)_i}{\left(k_2^{fin} \right)_i} \right)} \right) \quad (5)$$

is the i-th finite effective shear strain.

Applying Equation 3 this yields:

$$\mathbf{A}_i = \mathbf{B}_i \mathbf{B}_{i-1}^{-1} = \begin{bmatrix} \dfrac{\left(k_1^{fin} \right)_i}{\left(k_1^{fin} \right)_{i-1}} & \dfrac{\Gamma_i^{fin}}{\left(k_2^{fin} \right)_{i-1}} - \dfrac{\left(k_1^{fin} \right)_i}{\left(k_1^{fin} \right)_{i-1}} \dfrac{\Gamma_{i-1}^{fin}}{\left(k_2^{fin} \right)_{i-1}} & 0 \\ 0 & \dfrac{\left(k_2^{fin} \right)_i}{\left(k_2^{fin} \right)_{i-1}} & 0 \\ 0 & 0 & \dfrac{\left(k_3^{fin} \right)_i}{\left(k_3^{fin} \right)_{i-1}} \end{bmatrix}, \quad (6)$$

where $\mathbf{A}_i$ is the incremental matrix referred to the i-th step.

The relationships between incremental and finite quantities are obtained:

$$\left(k_1^{incr} \right)_i = \frac{\left(k_1^{fin} \right)_i}{\left(k_1^{fin} \right)_{i-1}}; \left(k_2^{incr} \right)_i = \frac{\left(k_2^{fin} \right)_i}{\left(k_2^{fin} \right)_{i-1}}; \left(k_3^{incr} \right)_i = \frac{\left(k_3^{fin} \right)_i}{\left(k_3^{fin} \right)_{i-1}} \quad (7)$$

and

$$\Gamma_i^{incr} = \frac{\Gamma_i^{fin}}{\left(k_2^{fin} \right)_{i-1}} - \frac{\left(k_1^{fin} \right)_i}{\left(k_1^{fin} \right)_{i-1}} \frac{\Gamma_{i-1}^{fin}}{\left(k_2^{fin} \right)_{i-1}}. \quad (8)$$

In the case of a transpressional/transtensional wrench zone with $k_1 = 1$ and $k_2 k_3 = 1$, equation (8) becomes:

$$\Gamma_i^{incr} = \frac{\Gamma_i^{fin}}{\left(k_2^{fin} \right)_{i-1}} - \frac{\Gamma_{i-1}^{fin}}{\left(k_2^{fin} \right)_{i-1}} = \left(\Gamma_i^{fin} - \Gamma_{i-1}^{fin} \right) \left(k_3^{fin} \right)_{i-1}. \quad (9)$$

Note that, for a deformation characterized by simple shear only ($k_1 = k_2 = k_3 = 1$), the incremental shear strain corresponds to the finite shear strain increment.

In the case of no synchronous pure shear deformation in the host rock (i.e. $\mathbf{C} = 1$), the final strain configurations of localizing and delocalizing shear zones (Fig. 1.1) are indistinguishable. However, according to Means (1995), useful information may be obtained by the analysis of the shear strain across the shear zone. In the case of strain softening, which is assumed to control the development of localizing shear zones, the finite shear strain profile displays a peaked shape. On the contrary, in the case of strain hardening (i.e. in the case of delocalizing shear zones) the profile is flat-shaped. Therefore an accurate study of the shear strain gradient across the shear zone may effectively unravel the type of rheological behaviour (strain hardening/softening) characterizing the analysed structure.

1.3   FINITE STRAIN

It is generally very difficult to obtain all four parameters of the finite matrix (Equation 4) by analyzing naturally deformed shear zones. However, in some cases, it is possible to determine all derived strain parameters starting from measured ones. Among these, the case of a shear zone characterized by no volume variation and nor stretch along the x direction ($k_1 = 1$).

Consider a wrench zone (Fig. 1.2) where each i-th layer is characterized by synchronous simple and pure shear represented by the finite strain matrix (in order to simplify the formulae, the label i is omitted and all quantities have to be considered as finite values):

$$\mathbf{B} = \begin{bmatrix} 1 & \Gamma & 0 \\ 0 & k_2 & 0 \\ 0 & 0 & k_3 \end{bmatrix}. \quad (10)$$

Generally one can directly measure only the (R, θ') and/or (Γ, θ') values, where R is the aspect ratio of the finite strain ellipse (e.g. Ramsay and Huber 1983) and θ' is the angle between the finite strain ellipsoid XY plane and the shear plane xy (e.g. Vitale and Mazzoli 2008, 2010). R and Γ can be obtained using the R_f/ϕ method

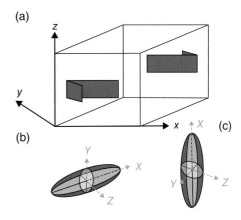

Fig. 1.2. (a) Cartoon showing a homogeneous wrench zone. (b) Finite strain ellipsoid geometry in case the x-axis lies in the xy plane of the coordinate reference frame. (c) Finite strain ellipsoid geometry in case the x-axis is parallel to the z-axis of the coordinate reference frame.

(Dunnet 1969) and the cotangent rule (Ramsay and Huber 1983; Vitale and Mazzoli 2010), respectively. Starting from the values of R and θ', let us try to find other strain parameters such as shear strain and stretches. To obtain a relationship between k and Γ, let us calculate the mathematical expression of the principal strain ratios R_{XZ}, R_{YZ} and R_{XY} starting from the magnitudes of the strain ellipsoid axes ($\lambda_{1,2,3}$), corresponding to the eigenvalues of the matrix $\mathbf{BB^T}$ (where $\mathbf{B^T}$ is the transposed matrix of $\mathbf{B}$):

$$\left(\mathbf{BB^T} - \lambda\mathbf{I}\right)\mathbf{e} = \left(\mathbf{BB^T} - \lambda\mathbf{I}\right)\begin{bmatrix} x \\ y \\ z \end{bmatrix} = \begin{bmatrix} \left(1+\Gamma^2-\lambda\right)x + \Gamma k_3 z \\ \left(k_2^2-\lambda\right)y \\ \Gamma k_3 x + \left(k_3^2-\lambda\right)z \end{bmatrix} = 0,$$

(11)

where $\mathbf{e}$ is the eigenvector and $\mathbf{I}$ is the identity matrix.

The solution is:

$$\lambda_{1,2,3} = \begin{cases} k_2^2 \\ \frac{1}{2}\left(k_3^2+\Gamma^2+1+\sqrt{k_3^4+2k_3^2\Gamma^2-2k_3^2+\Gamma^4+2\Gamma^2+1}\right). \\ \frac{1}{2}\left(k_3^2+\Gamma^2+1-\sqrt{k_3^4+2k_3^2\Gamma^2-2k_3^2+\Gamma^4+2\Gamma^2+1}\right) \end{cases}$$

(12)

Indicating with λ_1, λ_2 and λ_3 the maximum, intermediate and minimum value, respectively (i.e. $\lambda_1 > \lambda_2 > \lambda_3$), the strain ratios of the principal finite strain ellipses are:

$$R_{XZ} = \sqrt{\frac{\lambda_1}{\lambda_3}}; R_{YZ} = \sqrt{\frac{\lambda_2}{\lambda_3}}; R_{XY} = \sqrt{\frac{\lambda_1}{\lambda_2}}.$$

(13)

In order to find the angle θ' between the xz plane of the reference coordinate system (parallel to the shear plane) and the XY plane of the finite strain ellipsoid (parallel to

the foliation) from the first equation of the system (Equation 11) and choosing the appropriate value of λ^*, the angle θ' is obtained as follows:

$$-\frac{1+\Gamma^2-\lambda^*}{\Gamma k_3} = \frac{z}{x} = \tan\theta'.$$

(14)

The X-axis of the strain ellipsoid (the maximum stretching) can lie in the xz plane (Fig. 1.2b) or be parallel to the y-axis (Fig. 1.2c) of the reference coordinate system. In both cases the value of λ^* in the Equation 14 is

$$\lambda^* = \frac{1}{2}\left(k_3^2+\Gamma^2+1+\sqrt{k_3^4+2k_3^2\Gamma^2-2k_3^2+\Gamma^4+2\Gamma^2+1}\right);$$

(15)

substituting the value of λ^* of Equation 15 in the formula (Equation 14), the resulting equation is

$$(\tan\theta')k_3\Gamma+\Gamma^2+1-$$
$$\frac{1}{2}\left(k_3^2+\Gamma^2+1+\sqrt{k_3^4+2k_3^2\Gamma^2-2k_3^2+\Gamma^4+2\Gamma^2+1}\right)=0,$$

which simplifies to

$$(\tan\theta')k_3^2-\Gamma\left((\tan\theta')^2-1\right)k_3-(\tan\theta')(\Gamma^2+1)=0.$$

(16)

Solving for k_3 yields

$$k_3 = \frac{\Gamma(\tan\theta')^2-\Gamma+\sqrt{\Gamma^2(\tan\theta')^4+2\Gamma^2(\tan\theta')^2+\Gamma^2+4(\tan\theta')^2}}{2\tan\theta'}$$

(17)

and

$$k_3 = \frac{\Gamma(\tan\theta')^2-\Gamma-\sqrt{\Gamma^2(\tan\theta')^4+2\Gamma^2(\tan\theta')^2+\Gamma^2+4(\tan\theta')^2}}{2\tan\theta'}.$$

(18)

The solution for k_3 in Equation 18 provides negative values (k_3 must be ≥ 0), and hence has to be eliminated.

In order to find a suitable equation to join with Equation 17 in the variables k_3 and Γ, let us consider the strain ratio relationship

$$R-\sqrt{\frac{k_3^2+\Gamma^2+1+\sqrt{k_3^4+2k_3^2\Gamma^2-2k_3^2+\Gamma^4+2\Gamma^2+1}}{k_3^2+\Gamma^2+1-\sqrt{k_3^4+2k_3^2\Gamma^2-2k_3^2+\Gamma^4+2\Gamma^2+1}}}=0.$$

(19)

If the direction of the maximum lengthening (x-axis of the strain ellipsoid) lies in the xy plane of the reference coordinate system (Fig. 1.2a) than the second value of λ in the Equation 12 is the maximum one and $R = R_{XZ}$, else if the X-axis is parallel to the z-axis (Fig. 1.2b), the second value of λ is the intermediate one and $R = R_{YZ}$. Solving for k_3 yields

$$k_3 = \frac{\pm\left(-1-R^2\right)\pm\sqrt{1-R^2+R^4-4R^2\Gamma^2}}{2R} \qquad (20)$$

In order to obtain real solutions the argument of the square root must be zero or positive, hence

$$1-R^2+R^4-4R^2\Gamma^2 \geq 0,$$

with $R \geq 1$ and $\Gamma \geq 0$. These inequalities hold only when $R \geq \Gamma + \sqrt{\Gamma^2+1}$. Under this condition, the solutions that furnish positive values are

$$k_3 = \frac{1+R^2 \pm \sqrt{1-R^2+R^4-4R^2\Gamma^2}}{2R}. \qquad (21)$$

Combining Equations 21 and 17 yields

$$\frac{\Gamma \tan\theta'^2 - \Gamma + \sqrt{\Gamma^2\tan\theta'^4 + 2\Gamma^2\tan\theta'^2 + \Gamma^2 + 4\tan\theta'^2}}{2\tan\theta'}$$
$$= \frac{1+R^2 \pm \sqrt{1-R^2+R^4-4R^2\Gamma^2}}{2R}. \qquad (22)$$

Solving for Γ gives the only positive and real solution

$$\Gamma = \frac{R^2-1}{R}\frac{\tan\theta'}{1+\tan\theta'^2}. \qquad (23)$$

To find k_3 we can substitute the formula (Equation 23) into Equation 17. Furthermore $k_2 = k_3^{-1}$ and for the others strain quantities, such as shear strain and kinematic vorticity number, we can use the following equations, respectively:

$$\gamma = \frac{\Gamma \log(k_3)}{(1-k_2)} \qquad (24)$$

and

$$W_k = \frac{\gamma}{\sqrt{2\left(\log(k_2)^2+\log(k_3)^2\right)+\gamma^2}} = \frac{\gamma}{\sqrt{4log(k_3)^2+\gamma^2}}. \qquad (25)$$

Summarizing, in the special case of $k_1 = k_2k_3 = 1$, starting from the strain ratio (R_{XZ} if the strain ellipsoid X-axis lies in the xz plane, or R_{YZ} if the strain ellipsoid x-axis is parallel to the y-axis) and the angle θ' that the foliation forms with the shear plane, it is possible to obtain the values of k_3 (and k_2), γ and W_k by means of Equations 16, 23, 24, and 25.

1.4   PRACTICAL APPLICATION OF INCREMENTAL AND FINITE STRAIN ANALYSES

The technique proposed in this study is applied to a heterogeneous ductile wrench zone (Fig. 1.3a), previously analyzed by Vitale and Mazzoli (2010), exposed in a low-strain domain of an elsewhere extensively deformed and mylonitized pre-Alpine intrusive granitoid body included within the amphibolite facies *Zentralgneiss* (Penninic units exposed within the Tauern tectonic window, Eastern Alps; Fig. 1.3c; Mancktelow and Pennacchioni 2005; Pennacchioni and Mancktelow 2007). Shear zone nucleation was controlled by the presence of precursor joints, and occurred by a widespread reactivation process that characterizes solid-state deformation of granitoid plutons also elsewhere (e.g. Pennacchioni 2005; Mazzoli et al. 2009). Wrench zones are characterized by a well-developed foliation and deformed quartz veins that are intersected by the shear zone themselves (Fig. 1.3a). Wrench zones are subvertical and characterized by sub-horizontal slip vectors, and sinistral and dextral sense of shear. Geochemical analyses of major and trace elements of deformed and undeformed rocks (Pennacchioni 2005), indicate no geochemical changes occurred during deformation and hence suggesting that the deformation involved no volume variation (Grant 1986).

1.4.1   Finite strain

The analyzed wrench zone is characterized by localized synchronous simple shear and pure shear. The main finite strain parameters of the shear zone were evaluated by analyzing deformed planar markers (Vitale and Mazzoli 2010). In this case, the shear zone was divided into layers characterized by a roughly homogeneous internal deformation (Fig. 1.3b). For each layer, finite quantities of θ' and Γ were measured and plotted in a scatter diagram (Fig. 1.4). The latter also includes a k_2–γ grid that was constructed considering the known conditions of no volume change ($\Delta = 0$) and assuming $k_1 = 1$ (i.e. transpressional/transtensional deformation *sensu* Sanderson and Marchini 1984) by varying the stretch k_2 and the shear strain γ in the equations:

$$\Gamma = \frac{\gamma(1-k_2)}{\log(k_3)} \qquad (26)$$

and

$$\theta' = \tan^{-1}\left(-\frac{\Gamma^2+1-\lambda_{max}}{k_2\Gamma}\right) \qquad (27)$$

The obtained data plot along a general path involving increasing shear strain γ and decreasing values of the stretch k_2 moving from the margin toward the shear zone centre. Using Equations 24 and 27 one can obtain the exact value of the stretch k_3 (and hence $k_2 = k_3^{-1}$) and of the shear strain γ. The finite values of the strain ratios R_{XZ}, R_{YZ} and R_{XY} and the kinematic vorticity number W_k are obtained by applying Equations 19 and 25.

In order to smooth out the data in the incremental strain analysis that will be carried out in the following

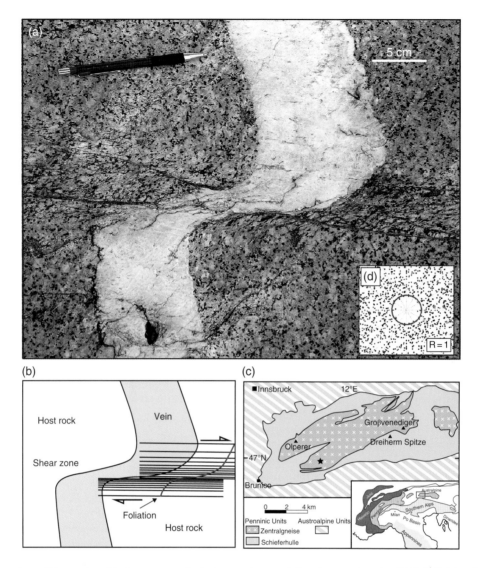

Fig. 1.3. (a) Outcrop view of the analyzed heterogeneous dextral wrench zone deforming quartz vein. (b) Subdivision into homogeneously deformed layers. (c) Geological sketch map of part of the Eastern Alps, showing site of the structural analysis (star) (modified after Pennacchioni and Mancktelow 2007). (d) Normalized Fry plot showing finite strain measured in the host rock.

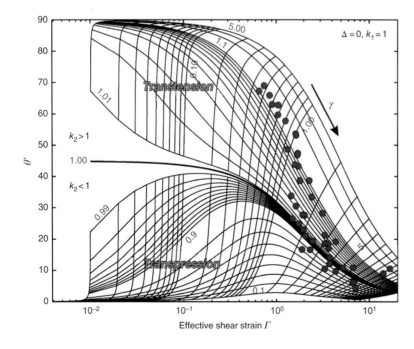

Fig. 1.4. $\Gamma - \theta'$ scatter diagram. The $k_2 - \gamma$ grid was constructed for a transpressional/transtensional deformation ($\Delta = 0$ and $k_1 = 1$).

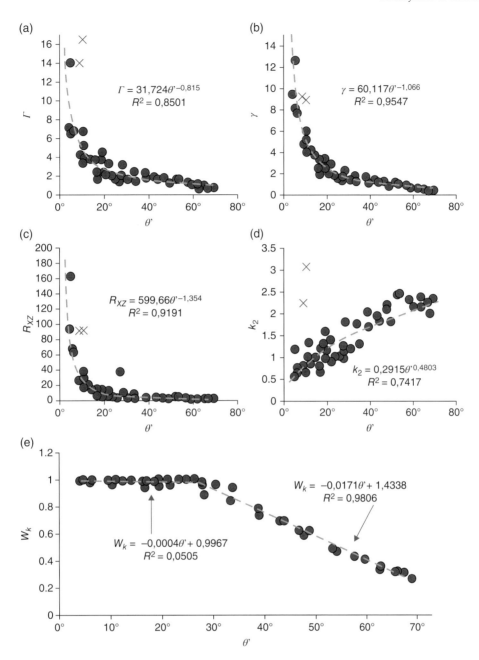

Fig. 1.5. Scatter diagrams of (a) finite effective shear strain Γ, (b) finite shear strain γ, (c) finite strain ratio R_{XZ}, (d) finite elongation k_2, and (e) finite kinematic vorticity number W_k versus finite angle θ'. Best-fit curves are also shown with associated equation and coedfficient of determination R^2.

section, best-fit power-law curves are determined for the finite values of effective shear strain Γ (Fig. 1.5a), shear strain γ (Fig. 1.5b), strain ratio R_{XZ} (Fig. 1.5c), and stretch k_2 (Fig. 1.5d) as a function of the angle θ'. A power-law curve is used to fit the scatter data because it provides the best values of the coefficient of determination R^2 (which is a measure of how well the data fit the adopted statistical model).

The finite kinematic vorticity number W_k increases linearly for θ', ranging from the maximum observed value (about 70°) to about 30°, becoming constant (and close to

unity) for angles between about 30° and the minimum observed θ' values (<5°; Fig. 1.5e). Note that if the increase in effective shear strain, shear strain and strain ratio with decreasing θ' are somewhat expected, much less obvious is the decrease in elongation k_2. To evaluate finite strain in the host rock, the normalized-Fry method (Fry 1979; Erslev 1988) was applied to the rock areas surrounding the structure. The obtained values of ellipticity on the XZ plane (R_{XZ}^{host}) are of about 1, indicating no strain (Fig. 1.3d).

Summarizing, the ductile wrench zone is characterized by: (i) non-constant finite values of the stretches

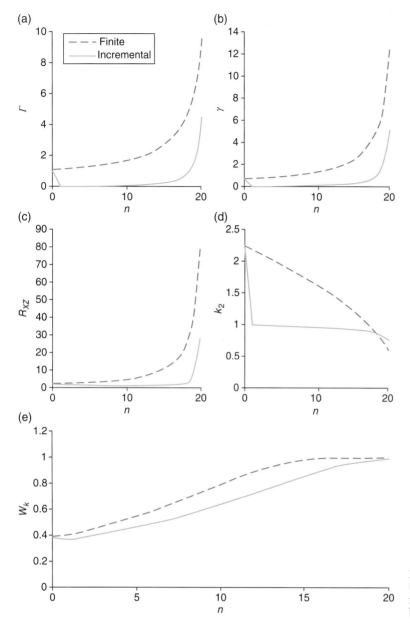

Fig. 1.6. Profiles of finite and incremental values of: (a) effective shear strain Γ, (b) shear strain γ, (c) strain ratio R_{XZ}, (d) stretch k_2, and (e) kinematic vorticity number W_k versus layer number n.

k_2 and k_3, with k_2 values mostly larger than unity (having assumed $k_1 = 1$); (ii) localized deformation occurring within the shear zone only (i.e. undeformed host rock); and (iii) no volume variation. These features point out the occurrence of material flow along the z-axis of the reference frame (i.e. in the vertical direction). This effect is dominant in the low-strain parts of the shear zone (i.e. for high values of the angle θ') characterized by k_2 values above unity (transtensional deformation), becoming negligible for the simple shear-dominated ($W_k \approx 1$) central sector of localized high strain.

1.4.2 Incremental strain

The incremental strain analysis of the studied shear zone was carried out considering a total number of $n = 20$ steps forthe application of the inverse method. For each

step, discrete values of k_2 and Γ were obtained from the best-fit equations (Fig. 1.5). Equation 3 has been used to calculate the incremental strain because this shear zone displays no deformation outside of it (Fig. 1.3a, d).

Figure 1.6 displays the profiles of finite and incremental values for effective shear strain Γ, shear strain γ, stretch k_2, strain ratio R_{XZ} and kinematic vorticity number W_k versus number of steps, whereas in Fig. 1.7(a,b) finite and incremental strain paths are plotted on: (i) $\Gamma - \theta'$ and (ii) logarithmic Flinn diagrams (Flinn 1962; Ramsay 1967), respectively. It must be stressed that the first incremental strain step corresponds to the first step of finite strain (transtension); on the contrary, the subsequent incremental strain values (from 2 to 20) indicate a transpressional deformation, although the finite strain is of dominantly transtensional type. For example, the first incremental value of k_2 is 2.23 (transtension), whereas

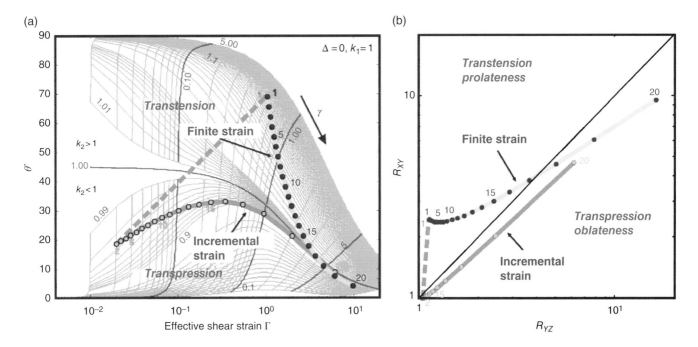

Fig. 1.7. (a) $\Gamma - \theta'$ and (b) logarithmic Flinn diagrams showing incremental and finite strain paths.

the following incremental k_2 path (from 2 to 20 steps, Fig. 1.7a) is generally characterized by values lower than 1, decreasing toward the center of the shear zone.

Finite and incremental profiles of effective shear strain Γ, shear strain γ and strain ratio R_{XZ} show similar peaked shapes, pointing out a non-linear increase during shear zone evolution (Fig. 1.6a–c). The peaked shape of the curve confirms the interpretation, already suggested, based on finite strain analysis, that shear zone rheology was characterized by strain softening, involving both simple shear and the pure shear components of the deformation. However, both finite and incremental kinematic vorticity numbers increase, up to unity, for increasing deformation (Fig. 1.6e). Therefore, the partitioning between simple shear and pure shear changed during shear zone evolution from prevailing pure shear ($W_k \approx 0.4$) to dominant simple shear, eventually reaching conditions of simple shear alone ($W_k \approx 1$) for the latest stages of highly localized deformation in the softened central sector of the shear zone.

The incremental strain paths plotted on the $\Gamma - \theta'$ diagram (Fig. 1.7a) and in the logarithmic Flinn diagram (Fig. 1.7b) confirm that the incremental strain from steps 2 to 20 consistently lies in the transpressional oblate field. Also, note that none of the incremental strain parameters holds a constant value during the temporal evolution; therefore the incremental deformation evolution is non-steady state.

As previously mentioned, the first increment points out a transtensional deformation whereas the subsequent strain increments are of transpressional type. In the case of no synchronous deformation in the host rock, the final strain configuration does not provide information about the temporal evolution of the shear zone. However, the suggested strain evolution is more consistent with a localizing shear zone (Fig. 1.1a, with **C** = 1) characterized by: (i) an early homogeneous transtensional deformation affecting the whole shear zone; and (ii) a following localization of the transpressional strain in the central part, probably driven by strain softening processes, with respect to a delocalizing shear zone (Fig. 1.1b, with **C** = 1) where an initial transtension affects the central part and subsequently migrates outward, with synchronous transpression affecting the inner sectors.

1.5   CONCLUSIONS

According to the Means hypothesis (Means 1975) and as suggested by Provost et al. (2004) and Horsman and Tikoff (2007), information on the incremental strain history may be obtained from the analysis of the final strain configuration in a heterogeneous shear zone. In particular, in case the structural evolution was characterized by strain softening or hardening – which are assumed to control the development of localizing and delocalizing shear zones, respectively – one can unravel the relationship between finite strain across the shear zone and incremental strains. Based on this assumption, an inverse method was derived which is able to evaluate the incremental strain matrices starting from measured finite strain quantities in a heterogeneous ductile shear zone. The proposed technique uses the finite values of the effective shear strain Γ and the finite angle θ' (angle between the foliation, i.e. the xy plane of the finite strain ellipsoid, and the shear plane) obtained for n layers, in which a shear zone may be subdivided according to homogeneity criteria. Shear zone deformation may then be described in terms of *n* finite strain matrices, each representing homogeneous deformation of the related layer. Starting from these matrices, it is possible to derive the incremental strain matrices. For simple cases, the proposed method furnishes symbolic formulae relating finite and incremental values of the main strain parameters, such as shear strain and stretches. The inverse method requires knowing the strain parameters k_1, k_2, k_3, and Γ. In particular instances such as that analyzed in this paper, it is possible to derive all of the required strain parameters from simple formulae. For the analyzed wrench zone, characterized by no stretches along the *x* direction and no volume change, the incremental strain path suggests a localizing shear zone evolution characterized by an initial homogeneous transtensional deformation in the whole shear zone and a subsequent incremental transpression driven by strain softening processes affecting progressively inner sectors.

ACKNOWLEDGMENTS

We thank Soumyajit Mukherjee for editorial assistance and useful suggestions and the two anonymous reviewers for the precious comments and corrections. Thanks to Kelvin Matthews, Delia Sandford, and Ian Francis (Wiley Blackwell) for support.

REFERENCES

Alsleben H, Wetmore PH, Schmidt KL, Paterson SR. 2008. Complex deformation as a result of arc-continent collision: Quantifying finite strain in the Alisitos arc, Peninsular Ranges, Baja California. Journal of Structural Geology 30, 220–236.

Dasgupta N, Mukhopadhyay D, Bhattacharyya T. 2012. Analysis of superposed strain: A case study from Barr Conglomerate in the South Delhi Fold Belt, Rajasthan, India. Journal of Structural Geology 34, 30–42.

Davis JR, Titus SJ. 2011. Homogeneous steady deformation: A review of computational techniques. Journal of Structural Geology 33, 1046–1062.

Dunnet D. 1969. A technique of finite strain analysis using elliptical particles. Tectonophysics 7, 117–136.

Erslev EA. 1988. Normalized center-to-center strain analysis of packed aggregates. Journal of Structural Geology 10, 201–209.

Flinn D. 1962. On folding during three-dimensional progressive deformation. Quarterly Journal of the Geological Society of London 118, 385–428.

Fossen H, Tikoff B. 1993. The deformation matrix for simultaneous simple shearing, pure shearing and volume change, and its application to transpression–transtension tectonics. Journal of Structural Geology 15, 413–422.

Fry N. 1979. Random point distributions and strain measurement in rocks. Tectonophysics 60, 89–105.

Grant JA. 1986. The isocon diagram – a simple solution to Gresens' equation for metasomatic alteration. Economic Geology 81, 1976–1982.

Horsman E, Tikoff B. 2007. Constraints on deformation path from finite strain gradients. Journal of Structural Geology 29, 256–272.

Hull J. 1988. Thickness-displacement relationships for deformation zones. Journal of Structural Geology 10, 431–435.

Iannace A, Vitale S. 2004. Ductile shear zones on carbonates: the *calcaires plaquettés* of Northern Calabria (Italy). Comptes Rendues Geosciences 336, 227–234.

Kuiper YD, Lin S, Jiang D. 2011. Deformation partitioning in transpressional shear zones with an along-strike stretch component: An example from the Superior Boundary Zone, Manitoba, Canada. Journal of Structural Geology 33, 192–202.

Lisle RJ. 1985. Geological Strain Analysis: A Manual for the Rf/φ Method. Pergamon Press, Oxford.

Mancktelow NS, Pennacchioni G. 2005. The control of precursor brittle fracture and fluid-rock interaction on the development of single and paired ductile shear zones. Journal of Structural Geology 27, 645–661.

Mazzoli S, Di Bucci D. 2003. Critical displacement for normal fault nucleation from en-échelon vein arrays in limestones: a case study from the southern Apennines (Italy). Journal of Structural Geology 25, 1011–1020.

Mazzoli S, Invernizzi C, Marchegiani L, Mattioni L, Cello G. 2004. Brittle-ductile shear zone evolution and fault initiation in limestones, Monte Cugnone (Lucania), southern Apennines, Italy. In Transport and Flow Processes in Shear Zones, edited by I. Alsop, and R.E. Holdsworth, Geological Society, London, Special Publications, 224, pp. 353–373.

Mazzoli S, Vitale S, Delmonaco G, Guerriero V, Margottini C, Spizzichino D, 2009. "Diffuse faulting" in the Machu Picchu granitoid pluton, Eastern Cordillera, Peru. Journal of Structural Geology 31, 1395–1408.

Means WD. 1995. Shear zones and rock history. Tectonophysics 247, 157–160.

Mitra G. 1991. Deformation of granitic basement rocks along fault zones at shallow to intermediate crustal levels. In Structural Geology of Fold and Thrust Belts, edited by S. Mitra, and G. W. Fisher, Johns Hopkins University Press, Baltimore, pp. 123–144.

Mukherjee S. 2013. Deformation Microstructures in Rocks. Springer, Heidelberg.

Mukherjee S. 2014. Atlas of shear zone structures in Meso-scale. Springer International Publishing, Cham.

Okudaira T, Beppu Y. 2008. Inhomogeneous deformation of metamorphic tectonites of contrasting lithologies: Strain analysis of metapelite and metachert from the Ryoke metamorphic belt, SW Japan. Journal of Structural Geology 30, 39–49.

Passchier C, Trouw R. 2005. Microtectonics. Springer Verlag, Berlin.

Provost A, Buisson C, Merle O. 2004. From progressive to finite deformation and back. Journal of Geophysical Research: Solid Earth and Planets 109, B02405.

Pennacchioni G. 2005. Control of the geometry of precursor brittle structures on the type of ductile shear zone in the Adamello tonalites, Southern Alps (Italy). Journal of Structural Geology 27, 627–644.

Pennacchioni G, Mancktelow NS. 2007. Nucleation and initial growth of a shear zone network within compositionally and structurally heterogeneous granitoids under amphibolite facies conditions. Journal of Structural Geology 29, 1757–1780.

Ramsay JG. 1967. Folding and Fracturing of Rocks. McGraw-Hill, New York.

Ramsay JG. 1980. Shear zone geometry: a review. Journal of Structural Geology 2, 83–99.

Ramsay JG, Huber M. 1983. The Techniques of Modern Structural Geology. Volume I: Strain Analysis, Academic Press, London.

Samani B. 2013. Quartz c-axis evidence for deformation characteristics in the Sanandaj–Sirjan metamorphic belt, Iran. Journal of African Earth Sciences 81, 28–34.

Sanderson D, Marchini RD. 1984. Transpression. Journal of Structural Geology 6, 449–458.

Sarkarinejad K, Samani B, Faghih A, Grasemann B, Moradipoor M. 2010. Implications of strain and vorticity of flow analyses to interpret the kinematics of an oblique convergence event (Zagros Mountains, Iran). Journal of Asian Earth Sciences 38, 34–43.

Tikoff B, Fossen H. 1993. Simultaneous pure and simple shear: the unified deformation matrix. Tectonophysics 217, 267–283.

Vitale S, Mazzoli S. 2005. Influence of object concentration on finite strain and effective viscosity contrast: Insights from naturally deformed packstones. Journal of Structural Geology 27, 2135–2149.

Vitale S, Mazzoli S. 2008. Heterogeneous shear zone evolution: the role of shear strain hardening/softening. Journal of Structural Geology 30, 1363–1395.

Vitale S, Mazzoli S. 2009. Finite strain analysis of a natural ductile shear zone in limestones: insights into 3-D coaxial vs. non-coaxial deformation partitioning. Journal of Structural Geology 31, 104–113.

Vitale S, Mazzoli S. 2010. Strain analysis of heterogeneous ductile shear zones based on the attitude of planar markers. Journal of Structural Geology 32, 321–329.

Vitale S, Iannace A, Mazzoli S. 2007a. Strain variations within a major carbonate thrust sheet of the Apennine collisional belt, northern Calabria, southern Italy. In Deformation of the Continental Crust: The Legacy of Mike Coward, edited by A.C. Ries, R.W.H. Butler, and R.H. Graham, Geological Society, London, Special Publications 272, pp. 145–156.

Vitale S, White JC, Iannace A, Mazzoli S. 2007b. Ductile strain partitioning in micritic limestones, Calabria, Italy: the roles and mechanisms of intracrystalline and intercrystalline deformation. Canadian Journal of Earth Sciences 44, 1587–1602.

Yonkee A. 2005. Strain patterns within part of the Willard thrust sheet, Idaho–Utah–Wyoming thrust belt. Journal of Structural Geology 27, 1315–1343.

Zhang Q, Giorgis S, Teyssier C. 2013. Finite strain analysis of the Zhangbaling metamorphic belt, SE China e Crustal thinning in transpression. Journal of Structural Geology 49, 13–22.

Chapter 2
How far does a ductile shear zone permit transpression?

SUJOY DASGUPTA[1], NIBIR MANDAL[1], and SANTANU BOSE[2]

[1] *Department of Geological Sciences, Jadavpur University, Kolkata, 700032, India*
[2] *Department of Geology, University of Calcutta, Kolkata, 700019, India*

2.1 INTRODUCTION

Understanding the shear zone kinematics has enormous implications in interpreting a wide variety of geological processes, ranging from the exhumation of deep crustal rocks to the formation of sedimentary basins. Kinematically, ductile shear zones are defined as regions marked by localization of intense non-coaxial deformations. Considering a homogeneous strain model, Ramberg (1975) first provided a theoretical analysis of the general non-coaxial deformations by combining pure shear and simple shear flows. Based on the kinematic vorticity number, expressed as: $W_k = \dfrac{W}{\left[2\left(\dot{\varepsilon}_1^2 + \dot{\varepsilon}_2^2 + \dot{\varepsilon}_3^2\right)\right]^{\frac{1}{2}}}$ (Truesdell 1954), where $\dot{\varepsilon}_i$ is the principal longitudinal strain rate and W is the magnitude of the vorticity vector, Ramberg (1975) has shown characteristic flow patterns in ductile shear zones. His analysis derives W_k as a function of the ratio (S_r) of pure and simple shear rates, and the orientation of the principal axes of pure shear with respect to the simple shear frame. For $W_k = 1$, shear zone deformations are characterized by shear-parallel flow paths, implying simple shear kinematics. On the other end, non-coaxial deformations in shear zones with $W_k < 1$ develop open hyperbolic particle paths, which transform into closed paths as $W_k > 1$. However, deformations with $W_k > 1$ described as a pulsating type, have been rarely reported from natural shear zones. Ramsay and his co-workers included volume strain in the kinematic analysis of ductile shear zones (Ramsay and Graham 1970; Ramsay 1980; Ramsay and Huber 1987). A range of natural structures (at both micro- and mesoscopic scales), for example, anastomose mylonitic fabrics (Gapais et al. 1987), porphyroclast tail patterns (Ghosh and Ramberg 1976; Simpson and De Paor 1997; Passchier and Simpson 1986; Mandal et al. 2000; Kurz and Northrup 2008), and instantaneous strain axis (ISA) (Passchier and Urai 1988; Tikoff and Fossen 1993; Xypolias 2010, and references therein) have been used to explain these structures and to demonstrate the effects of pure and simple shear kinematics in ductile shear zones. Similarly, a parallel line of studies has dealt with structural (e.g. stylolites; Tondi et al. 2006) and chemical criteria (e.g. enrichment of immobile elements; O'Hara and Blackburn 1989; Mohanty and Ramsay 1994; Srivastava et al. 1995; Fagereng 2013) to determine the volume loss in shear zones.

In many tectonic settings shear zones developed under general non-coaxial deformation show components of shear and shortening parallel and orthogonal to the shear zone boundaries, respectively. From a kinematic point of view, these shear zones have been described as transpression zones, a term first coined by Harland (1971) to describe the zone of oblique convergence between two crustal blocks. Sanderson and Marchini (1984) modeled a vertical transpression zone sandwiched between two laterally moving rigid blocks, which gave rise to a strike–slip motion coupled with shortening across the zone. Their model imposes a slip boundary condition to allow the material extrusion in the vertical direction, that is, perpendicular to the shear direction. It was pointed out later that a slip boundary condition had a mechanical limitation to produce shearing motion in a transpression zone (Robin and Cruden 1994; Dutton 1997). To meet this shortcoming, they advanced the transpression model with a non-slip boundary condition, and explained the structural variations in terms of the heterogeneous flow developed in their non-slip model. A series of kinematic models have been proposed afterwards; all of them, in principle, pivot on the same theoretical formulation of Ramberg (1975) that combines the homogeneous simple shear and pure shear fields. As a consequence, the transpression can show either a monoclinic or a triclinic kinematic symmetry, depending upon the orientation of pure shear axes with respect to the simple shear (Dewey et al. 1998; Jones and Holdsworth 1998; Ghosh 2001; Fernández and Díaz-Azpiroz 2009).

Despite an enormous volume of research work over the last several decades, there have been so far few attempts to analyze the mechanics and to estimate the degree of transpression physically possible in a ductile

Ductile Shear Zones: From Micro- to Macro-scales, First Edition. Edited by Soumyajit Mukherjee and Kieran F. Mulchrone.

shear zone. Flow modeling for a viscous layer sandwiched between two rigid walls (Mandal et al. 2001) show that the viscous extrusion is much more energetically expensive, as compared to the gliding motion of the rigid blocks. According to Mandal et al. (2001), this contrasting energetics thus leads to dominantly simple shear kinematics even when the bulk compression is at a high angle to the shear zone. This theoretical prediction is in good agreement with the dominantly simple shear kinematics of deep-crustal ductile shear zones reported from high-grade metamorphic terrains (Puelles et al. 2005). Furthermore, our own field observations of centimeter to meter scale shear zones from the Chotanagpur Granite Gneiss Complex point to limited flattening. In contrast, many field studies have documented a range of structural evidences in support of the transpressional kinematics (Ghosh and Sengupta 1987; Sengupta and Ghosh 2004; Žák et al. 2005; Massey and Moecher 2013). A revisit to the problem of transpressional movement in ductile shear zones is needed to address such apparently contradictory field observations. Using analog experiments the present study aims to physically evaluate how much flattening is possible in a coherent ductile shear zone hosted in an undeformable medium. Our experimental results confirm earlier theoretical predictions that shear zones with large aspect ratios are unlikely to undergo significant flattening. We also demonstrate from the laboratory experiments that appreciable flattening can occur under specific geometrical conditions. Finite element model simulations were performed to substantiate these findings.

Section 2.2 presents a set of field examples of small scale shear zones along with a qualitative evaluation of their flattening. Section 2.3 deals with analog modeling, giving a quantitative estimation of flattening as a function of shear zone geometry. In Section 2.4, we discuss simulated ductile shear zones in finite element (FE) models, considering physical properties approximated to the crustal rheology. In Section 2.5 we discuss the experimental findings in the context of actual field observations, and propose syn-shearing volume changes as a mechanically potential mechanism for transpression in ductile shear zones. Section 2.6 concludes the principal outcomes of this work.

2.2 FIELD OBSERVATIONS: GEOMETRIC ANALYSIS

We investigated outcrop-scale ductile shear zones in the Chotanagpur Granite Gneiss Complex (CGGC) terrain, lying north of the Singhbhum Proterozoic mobile belt (Fig. 2.1). The dominant mineral constituents of the CGGC rocks are quartz, feldspars, micas and garnet with secondary phases, such as tourmaline and amphibole. Petrological studies suggest that the terrain has undergone a peak metamorphism ($P \sim 6$ kbar, $T \sim 850°C$) in granulite facies (Maji et al. 2008), and regional deformations under N–S

convergence, giving rise to an overall E–W structural trend (Fig. 2.1). The CGGC in the study area recorded four major ductile deformation episodes (Mahmoud, 2008). In the course of these ongoing events, the terrain has experienced regionally a brittle–ductile transition, producing vertical shear zones on centimeter to meter scales. In the western part of Purulia district the gneissic rocks display excellent outcrops of such shear zones, broadly in two sets, trending NNE–SSW and NE–SW, showing both dextral and sinistral strike–slip motion, as revealed from the deflection of dominant gneissic foliations or markers, like quartz veins.

The deflection patterns of mylonitic foliation produced by the motion of the shear zones were used to undertake a qualitative analysis of the shear zone kinematics. We assumed that the distortion pattern of transverse markers is a function of the ratio (S_r) between pure and simple shear rates (Mandal et al. 2001; Puelles et al. 2005). The distortion of a passive marker at an instant can be reconstructed from the following velocity equations (after Jaeger 1969) in a Cartesian space (x, y):

$$u = 3\dot{\varepsilon}_b x \left[\frac{1}{2} - \left(\frac{y}{t} \right)^2 \right] + \dot{\gamma}_b x \qquad (1)$$

$$v = \frac{1}{2} \dot{\varepsilon}_b y \left[\left(\frac{y}{t} \right)^2 - 3 \right], \qquad (2)$$

Where, $\dot{\varepsilon}_b$ and $\dot{\gamma}_b$ are the bulk flattening and shear rates, u and v are the components of velocity along the x and y directions, respectively and t is the instantaneous shear zone thickness. Using Equations 1 and 2, it can be demonstrated that a line parallel to the y-axis will be distorted with outward convexity, the symmetry of which depends on the ratio of pure and simple shear rates $\left(S_r = \dfrac{\dot{\varepsilon}_b}{\dot{\gamma}_b} \right)$ and the initial orientation of markers with the shear direction. To investigate the distortion pattern, we chose a set of lines across the shear zone (Fig. 2.2) described by the following equation:

$$y = m(x - C) \qquad (3)$$

where, $m = \tan\theta$; θ and C are the initial angles of lines with respect to the shear direction and their distance from the shear zone center along the shear direction, respectively (Fig. 2.2). The lines were deformed by incremental displacements by using the velocity function in Equations 1 and 2. All the parameters, such as shear zone thickness t were modified at each step of the iterations.

Using this theoretical approach we obtained the distortion patterns of a set of parallel passive markers as a function of S_r and θ, and compared them with those observed in natural shear zones. For $S_r = 0$, the markers show angular deflections without any distortion, irrespective of their initial orientation (θ). The degree of outward convexity decreases with decreasing values of

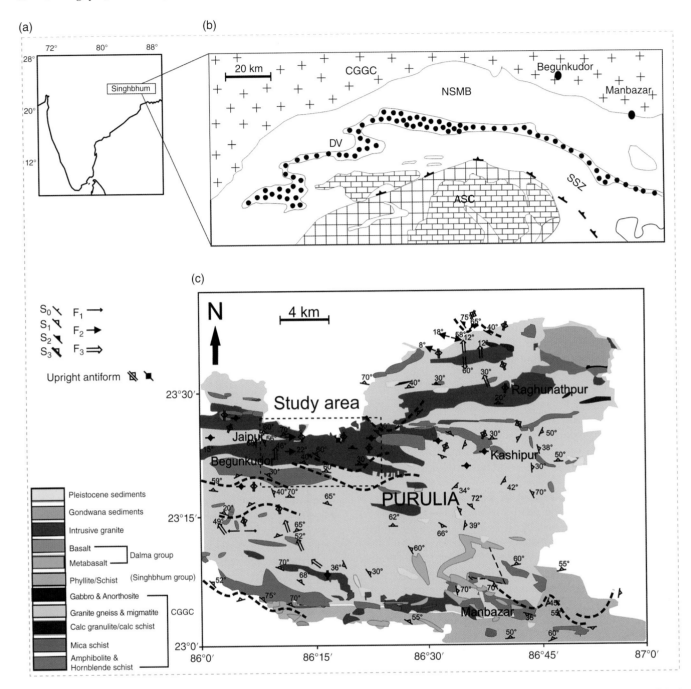

Fig. 2.1. A generalized geological map of Singhbhum Craton, modified after Saha (1994). (a) Position of the Singhbhum region. (b) Simplified geological map showing the distribution of lithological units in Singhbhum craton (after Saha 1994). Abbreviated units: CGGC: Chhotanagpur Granite Gneiss Complex; ASC: Archean Singhbhum craton; NSMB: North Singhbhum mobile belt; SSZ: Singhbhum shear zone, DV: Dalmaviolcanics. (c) Geological map of Purulia district (modified after map published by Geological Survey of India, 2001), West Bengal within the eastern part of CGGC. Black dashed box is the study area. Symbols in the legend indicate lithology and structural fabrics and fold axis.

both S_r and θ (Fig. 2.2). The graphical plots indicate that markers oriented initially normal to the shear zone boundaries ($\theta = 90°$) do not show any appreciable outward convexity for S_r values less than 0.05.

Figure 2.3(a) shows an isolated shear zone at a right angle to the foliation, with an aspect ratio of 10.8. The foliation marker has been deflected in a sinistral sense by

an angle of about 70°. A close inspection of the distorted pattern of foliation markers within the shear zone reveals that they maintain their planar geometry along the entire length of the shear zone. Comparing the distortion patterns with the corresponding theoretical one (Fig. 2.2), the shear zone is predicted to have undergone little flattening, and evolved dominantly by simple shear (i.e. $S_r \sim 0$).

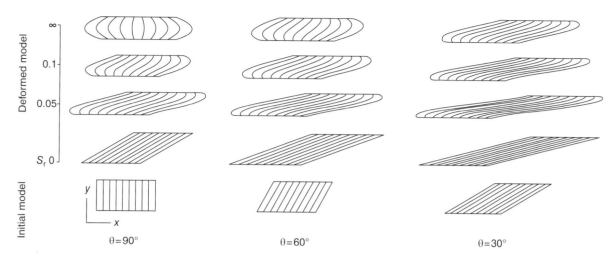

Fig. 2.2. Distortion patterns of passive markers in transpression zones for different S_r (ratio of pure and simple shear rates) values. Markers were initially oriented at an angle of θ with shear directions. The bottom row shows the initial orientation of markers.

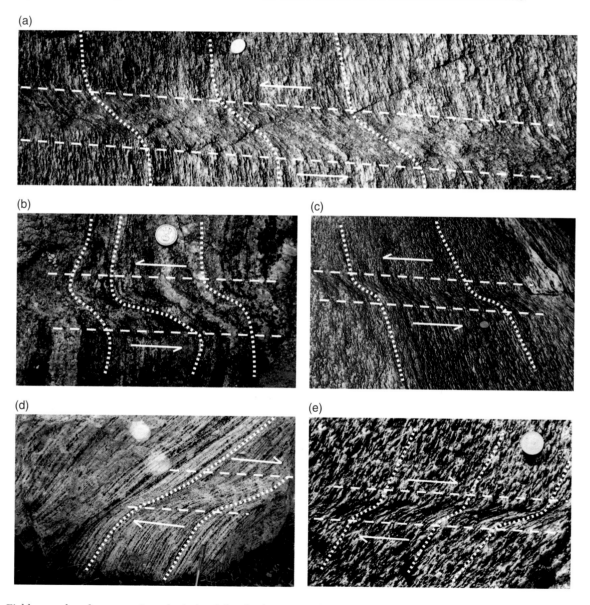

Fig. 2.3. Field examples of mesoscopic-scale, isolated ductile shear zones from the CGGC (20 km west of Purulia town). Note that the foliation markers passing through the shear zone at a high angle have been uniformly deflected sinistrally within the shear zones (marked in white line). The aspect ratios of the shear zones were (a) 10.3, (b) 4.4 and (c) 11.7. Small scale dextral shear zones are also presented here showing deflections of the foliation markers oriented t an angle of: 61° (d) and 34° (e) with the shear direction respectively. The aspect ratios of the shear zones were measured to be 7.7 in (d) and 7.2 in (e). See text for detailed descriptions. Straight dashed lines mark the shear zone boundary, while curved dashed lines indicate deflection of foliations.

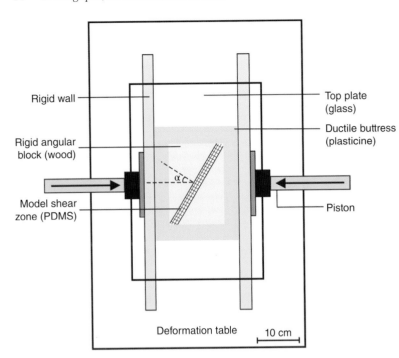

Fig. 2.4. A schematic plan view of the experimental set-up used for analog experiments of the transpressional kinematics in planar ductile shear zones under plane strain conditions. α: inclination of shear zone normal to the bulk compression direction. Black arrows indicate the direction of moving pistons.

Considering this simple shear kinematics a maximum finite shear of 1.65 is estimated from the foliation deflections. Figure 2.3(b) shows a spectacular shear zone of low aspect ratio (~4). It is marked by deflection of a gneissic foliation at a right angle to the shear zone trend. In the central zone the foliations show a sharp sinistral deflection by an angle of nearly 60°. The deflections become progressively diffused towards the flank regions, and the foliations describe a slightly curvilinear geometry with outward convexity. Based on qualitative matching with the theoretically predicted patterns, S_r is found to be in the order of 0.1. Figure 2.3(c) presents a similar shear zone with a large aspect ratio (~11), displaying a distortion pattern closely resembling that shown in Fig. 2.3(a). Figure 2.3(d) shows a narrow shear zone (aspect ratio ~7.5) with the foliation marker at an angle of 31° to the shear zone. The absence of any appreciable outward convexity in the deflected markers within the shear zone indicates little flattening in the shear zone. Comparing the overall distorted pattern with those obtained from the theoretical calculations (Fig. 2.2), S_r appears be very small (<0.05). We present another example of an isolated dextral shear zone with an aspect ratio of 7.7 (Fig. 2.3e). The foliation marker has developed a sigmoidal pattern with a uniformly right-handed vergence. This distortion geometry is, however, unlike those produced in shear zones with transpression (Fig. 2.2), implying dominantly simple shear kinematics.

The field examples presented above suggest that shear zones of large aspect ratios are unlikely to undergo large flattening strain, as also inferred from theory (Mandal et al. 2001). However, it still awaits an experimental verification of the degree of transpression mechanically possible in a ductile shear zone. To meet this gap, we performed physical experiments, the results of which have been substantiated by FE models.

2.3   SHEAR ZONE KINEMATICS: ANALOG EXPERIMENTS

2.3.1   Experimental approach

We simulated ductile shear zones in analog models consisting of a vertical viscous slab sandwiched between two angular rigid blocks. The experimental set-up is schematically shown in Fig. 2.4. PDMS (Polydimethyl-xiloxane; Dow Corning – SGM 36) was used to simulate the shear zone material (viscosity in the order of 10^5 Pa s), which has been used by several workers for modeling structures developed in a ductile regime (Koyi 1991; Exner 2005; Godin et al. 2011; Mukherjee et al. 2012). The angular rigid blocks were made of wood. The model was encased within ductile walls of modeling clay. The main purpose of using the modeling clay walls was to decouple the rigid blocks from the piston plate, and allow them to move freely under the bulk horizontal shortening. Model experiments were run on a hydraulically driven deformation table, equipped with a pair of horizontal pistons. These two pistons approach each other at equal speed in order to shorten the model placed at the center of the deformation table. We controlled the piston speed of 3 cm/min to maintain the rate of bulk strain in the order of 10^{-4}/s. The model was arrested by a horizontal glass plate at its top to restrict extension in the vertical direction (i.e. $\lambda_2 = 1$), and keep the model deformation under plane strain condition (i.e. λ_1–λ_3 on horizontal plane).

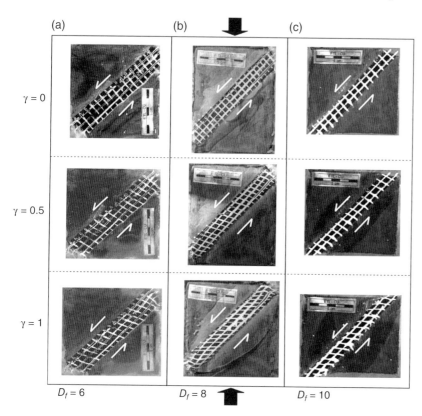

Fig. 2.5. Development of ductile shear zones in analogue models with $\alpha = 45°$. Bold arrow indicates the direction of maximum compression. D_f: Shear zone aspect ratio. The square grids were stamped in the initial models. Scale bar: 5 cm.

The PDMS slab was placed at an angle (α) to the bulk horizontal shortening (Fig. 2.4).

Consequently, the movement of rigid blocks caused shear localization in the PDMS layer, giving rise to a planar ductile shear zone with strike–slip motion. The top surface of the model was gridded by two orthogonal sets of lines with respect to the PDMS–wood boundary. The passive markers across the shear zone boundary enabled us to determine the amount of bulk shearing in the ductile shear zone, and at the same time to evaluate the relative flattening strain from their distortion patterns, as discussed in Section 2.2.

We performed these experiments in three different sets, placing the viscous slab within the wooden blocks at 45°, 60°, and 70° to the bulk shortening direction. In other words, the inclination (α) of the shear zone normal to the principal bulk compression axis (σ_3) was 45°, 30°, and 20° respectively. These experiments aimed to corroborate the theoretical evaluations suggested by Mandal et al. (2001) and to verify our field observations and the theoretical results presented in the previous section. In each set of experiments for a specific α value, we varied the initial aspect ratio (D_i) of shear zones to 5, 6, 7, 8, 9, and 10, considering the shear zone geometry observed in the field. Thus, our experimental models in combination simulated 18 different geometric settings (Fig. 2.5).

During experimental runs photographs were taken at regular intervals from the top, keeping the camera axis vertical. These snap shots were used for post-experimental data processing using the available image processing software. We evaluated the following shear zone parameters

at any instant of an experimental run: (1) aspect ratio (D_f); (2) bulk finite flattening strain (ε_b), and finite shear (γ_b); (3) the ratio of bulk pure and simple shear rates $\left(S_r = \dfrac{\dot{\varepsilon}_b}{\dot{\gamma}_b} \right)$.

2.3.2   Model results

Models with $\alpha = 45°$ developed shear zones with dominantly simple shear kinematics. Figure 2.5(a, b and c) presents successive stages of the model runs with $D_f = 6$, 8 and 10. Low-aspect ratio ($D_f = 6$) shear zones showed angular deflections of the markers, but without showing any appreciable outward curvatures (Fig. 2.5a). This implies that the shear zone has evolved dominantly under simple shear movement, as discussed in the preceding section. We measured the shear zone thickness in incremental steps of the progressive deformation, and estimated the amount of bulk shortening normal to the shear zone. For a finite shear of 1, the transpression is limited to about 5%. With increasing D_f, the magnitude of transpression further drops to <1% when $D_f = 10$ (Fig. 2.5b, c and Fig. 2.6a). The shear zone movement thus took place essentially by simple shear kinematics, as observed in our field studies (Fig. 2.3) as well as predicted in earlier studies (Mandal et al. 2001; Puelles et al. 2005). We analyzed the S_r parameter as a function of D_f. A ductile shear zone with $D_f = 6$ (Fig. 2.5a) shows a S_r value of 0.4, which decreases non-linearly with increasing D_f (Fig. 2.6b). The experimental results indicate that

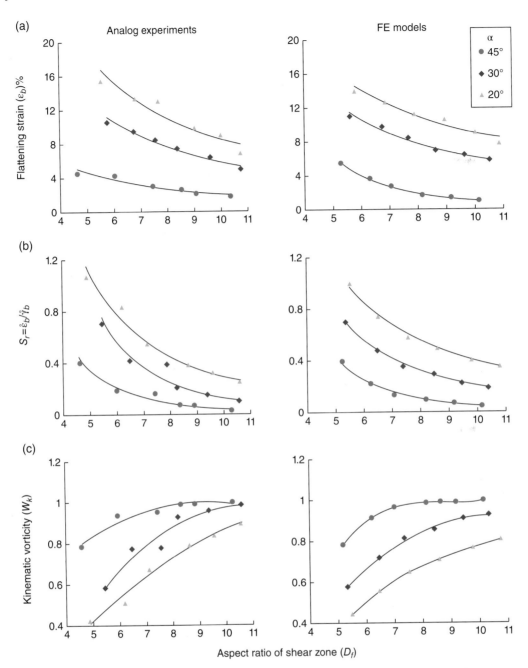

Fig. 2.6. Graphical presentations of the results obtained from analog experiments (left) and FE simulations (right). (a) Finite bulk shortening across the shear zones (ε_b), (b) ratio of bulk pure and simple shear rates (S_r) and (c) bulk vorticity (W_k) versus shear zone aspect ratio (D_f). Symbols (on the top right) indicate the inclination (in degrees) of shear zone normal with the bulk compression direction (α).

shear zones of low aspect ratios can undergo a small amount of transpression at the initial stage, but their kinematics subsequently transforms into simple shear. Using the following equation (Ghosh 1987), we evaluated the bulk kinematic vorticity as a measure of transpression:

$$W_K = \frac{1}{\left(1 + 4 S_r^2\right)^{\frac{1}{2}}} \qquad (4)$$

$W_k = 1$ for simple shear kinematics, which tends to zero with increasing flattening strain. W_k corresponding to $S_r = 0.4$ is nearly 0.8 (Fig. 2.6c), which increases to nearly 1 when the bulk shear is 1. The analysis indicates that low-aspect ratio shear zones can undergo some flattening in the beginning, but the simple shear dominates in the advanced stages ($\gamma_b > 1$) of a shear zone (Section 2.4.2). With increasing aspect ratio W_k approaches 1, implying that narrow shear zones cannot undergo large transpression (Fig. 2.6c), as predicted in earlier theoretical studies.

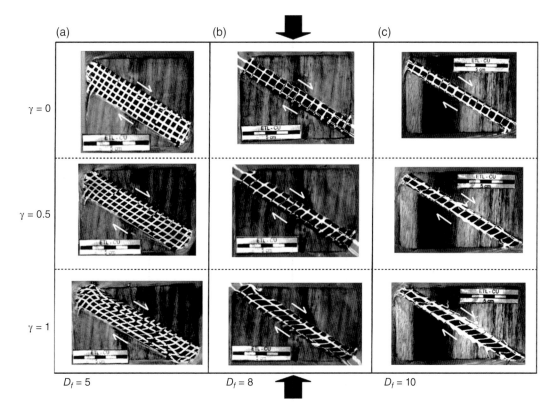

Fig. 2.7. Progressive development of ductile shear zones in analog models with $\alpha = 30°$ for different aspect ratio (D_f). Bold arrow indicates the direction of maximum compression. (a) $D_f = 5$; (b) $D_f = 8$, and (c) $D_f = 10$. Scale bar: 5 cm.

The experimental findings corroborate our field observations of shear zones with $D_f = 10$, displaying the foliation markers without any outward convexity (Fig. 2.3a, c, e).

The effect of pure shear component of deformation was somewhat noticeable in experiments with $\alpha = 30°$ (Fig. 2.7). The results for $D_f = 5$, 8, and 10 are presented in Fig. 2.7. Low aspect-ratio ($D_f = 5$) shear zones develop discernible outward convexity of the grid markers stamped initially at right angles to the shear zone boundaries (Fig. 2.7a). Their overall distortion patterns imply relatively larger amounts of bulk flattening strain, as compared to the previous models. The curvatures of distorted lines tend to die out with increasing aspect ratios (Fig. 2.7b, c). For example, it is almost zero for $D_f = 10$. This variation is consistent with the S_r and W_k analysis. $S_r = 0.8$ for $D_f = 5$, which decreases nonlinearly with increasing D_f (Fig. 2.6b). Similarly, W_k approaches 1 as D_f becomes large (Fig. 2.6c). With progressive deformation, these shear zones, irrespective of their initial aspect ratios, show a decreasing S_r trend with increasing finite shear, similar to the models with $\alpha = 45°$. The S_r value drops non-linearly to a low value when γ becomes large (the details of these variations are discussed later). Both S_r and W_k analyses indicate that, in $\alpha = 30°$ models the shear zones with large aspect ratios ($D_f > 5$) undergo dominantly simple shear (Fig. 2.6b, c). They show outward convexity of the markers, indicating little flattening only in the initial stages of their evolution (Fig. 2.7a).

Experiments performed with $\alpha = 20°$ (Fig. 2.8) developed shear zones with relatively larger pure shear component, as compared to models with $\alpha = 45°$ and $30°$ (Figs 2.5 and 2.7). Low aspect ratio ($D_f = 5$) shear zones show distortion patterns of the markers with discernible outward convexity (Fig. 2.8a). For a bulk shear of 1, the amount of flattening (ε_b) was around 14%. ε_b is inversely related to D_f, and becomes insignificant (<7%) when $D_f = 10$ (Fig. 2.6a). We analyzed S_r and W_k as a function of D_f (Fig. 2.6b, c). S_r is large (~1.1) for $D_f = 5$, but decreases to ~0.4 when $D_f = 10$. Similarly, W_k tends to 1. Both S_r and W_k analyses suggest that shear zones even at high angle to the bulk compression direction (>60°) undergo movement virtually in simple shear if their aspect ratios are greater than 10. Furthermore, the progressive movement in shear zones involved an exponential decrease of S_r values with increasing bulk shear, approaching the simple shear kinematics as the bulk shear exceeds 1 (see Section 2.4.2).

2.4 FINITE ELEMENT MODELING

2.4.1 Method

Physical experiments presented in Section 2.3 suggest that a ductile shear zone can undergo appreciable transpression only when its length to thickness ratio is

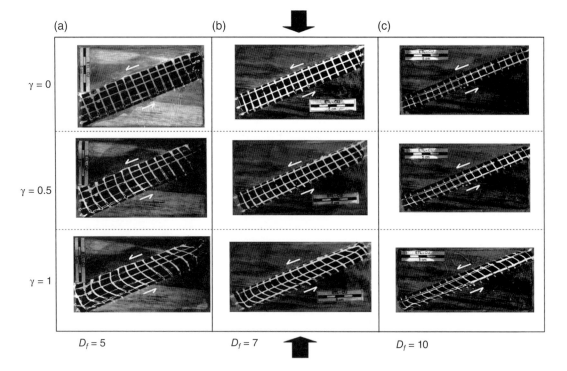

Fig. 2.8. Progressive development of ductile shear zones in analog models with $\alpha = 20°$ for different aspect ratio (D_f). Bold arrow indicates the direction of maximum compression. (a) $D_f = 5$; (b) $D_f = 7$, and (c) $D_f = 10$. Scale bar: 5 cm.

very low (less than 10) and oriented at a high angle to the compression direction (>60°). Otherwise, they are likely to evolve in simple shear kinematics, as observed in CGGC (Fig. 2.3). To substantiate our experimental findings we simulated similar shear zones in FE models, and examined the possible magnitudes of transpression as a function of shear zone geometry, considering the material properties applicable to real rocks. The crustal rock rheology has been approximated with Maxwell viscoelastic rheology, as used in several earlier studies (Passchier and Druguet 2002; Mandal et al. 2007). The constitutive equation for this rheology is as follows:

$$\dot{\varepsilon} = \frac{1}{2\mu_m}\frac{\mathrm{d}\sigma}{\mathrm{d}t} + \frac{\sigma}{2\eta_m} \quad (5)$$

where $\dot{\varepsilon}$ and σ are the strain rate and stress respectively, and μ_m and η_m represent the Maxwell shear modulus and the Maxwell coefficient of viscosity, respectively. Geological evidence suggests that, over a prolonged period of time the crustal materials in active tectonic belts behave closely like a viscoelastic substance with the viscosity in the order of 10^{17}–10^{23} Pa s (Copley and McKenzie 2007; Mukherjee 2013). Based on these available data for crustal viscosity, we chose the value of η_m in the order of 10^{21} Pa s, keeping the relaxation time at 10^{11} s (i.e., $\mu_m \sim 10^{10}$ Pa), as estimated from the post-glacial continental uplift (Larsen et al. 2005) and evidence of continental break-up (Clift et al. 2002). The details of rheological and material parameters and element geometry

Table 2.1. Material properties and model parameters used in the finite element modeling

Material properties used in numerical modeling				
Model domains	Density (kg/m³)	Shear modulus (Pa)	Bulk modulus (Pa)	Viscosity (Pa s)
Wall rock	2800	3×10^{10}	7×10^{10}	3×10^{21}
Shear Zone	2700	3×10^{8}	7×10^{8}	3×10^{18}
Finite element consideration				
Type of elements	Visco88*, Tetrahedral viscoelastic, 2 degrees of freedom at each node			

*Elements are ANSYS defined.

employed in our FE modeling are provided in Table 2.1. Using the FE code of solid mechanics for Maxwell rheology, we used the ANSYS code (ANSYS version 11, 2007) to model ductile shear zones.

FE models were developed in a two-dimensional Cartesian space to represent the deformation plane containing the principal directions of strain (Fig. 2.9). The models were designed with centimeter- to meter-scale dimensions to replicate outcrop-scale shear zones, as described in Section 2.2. To simulate a planar shear zone in the model we introduced a narrow zone with a relatively lower viscosity at an angle to the model length. Natural shear zones are characterized by mechanical

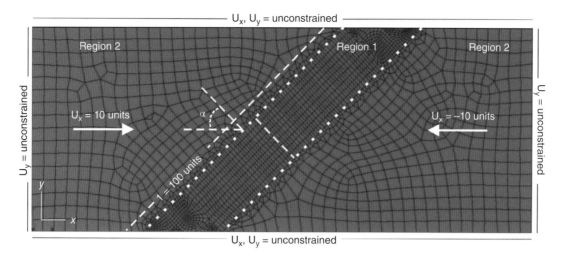

Fig. 2.9. Initial FE models showing the boundary conditions for simulation of a ductile shear zone at an angle α with the bulk compression direction (bold arrows). Both the inclinations (α) and the length (l) to thickness (t) ratio of the shear zones were varied in FE simulations. Regions 1 and 2 represent the shear zone and wall rock, respectively with a non-slip boundary (white dotted lines) condition. U_x and U_y represent the amount of displacement along x and y directions respectively.

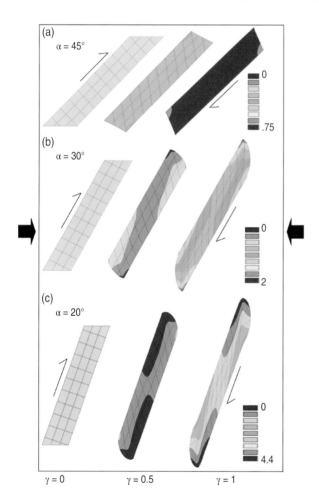

Fig. 2.10. Progressive deformation in ductile shear zones simulated in FE models withinitial $D_f = 5$. (a) $\alpha = 45°$, (b) $\alpha = 30°$, and (c) $\alpha = 20°$. Note that the deformed grid patterns closely resemble those in equivalent physical experiments (Figs 2.5, 2.7, and 2.8). The color contours corresponding to each model indicates the value of the first principal strain. Bold arrow indicates the direction of maximum compression.

softening, resulting in significant decrease in viscosity with respect to the country rock (Brun and Cobbold 1980; Gapais 1989; Rutter 1999). The viscosity ratio between the country and shear zone rocks in Earth's crust is generally ~10^3 (Marsh 1982). Based on this available information, we considered the viscosity of model shear zones in the order of 10^{18} Pa s, which gave rise to a viscosity ratio of 10^3 (Table 2.1). For this viscosity contrast, the region outside the model shear zone underwent little deformation, and behaved almost like rigid blocks similar to those in the physical experiments. The interface between the shear zone and the walls was ascribed to a non-slip condition. All other mechanical conditions imposed on FE models are depicted in Fig. 2.9. Model runs were performed in incremental steps of shortening on a real time scale, allowing viscous flow in the models. We designed the model geometry, similar to analog models, to simulate ductile shear zones with $\alpha = 45°$, $30°$ and $20°$ and $D_f = 5$, 6, 7, 8, 9, and 10, as used in the physical experiments.

2.4.2   Model simulations and their results

Under bulk shortening the deformation localize in the model preferentially along the low-viscosity layer, retaining its walls as rigid blocks, sliding past each other. Consequently, the model developed a ductile shear zone (Fig. 2.10 and Fig. 2.11), as in the physical experiments (Figs 2.5, 2.7, and 2.8). Using the same analytical approach discussed in the preceding section we evaluated the degree of transpression as a function of shear zone geometry. For a given value of $\alpha = 45°$ (Fig. 2.10a), we estimated a flattening strain (ε_b) of 5% across a model shear zone for $D_f = 5$, when $\gamma_b \sim 1.1$ (Fig. 2.6a). For the same finite shear, ε_b decreases non-linearly with increasing D_f, and drops to less than 1% as $D_f = 10$. This ε_b versus D_f regression is consistent with the experimental data. The degree of

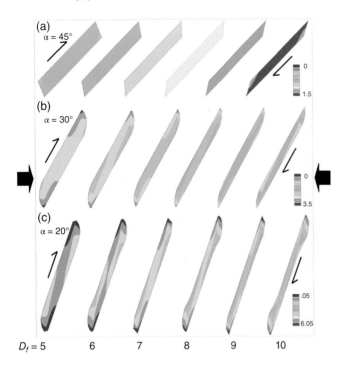

Fig. 2.11. Ductile shear zones in FE models showing shear zone normal flattening as a function of α and D_f (details in the text). Color contours indicate the value of the first principal strain. Bold arrow indicates the direction of maximum compression.

flattening in model shear zones is found to be sensitive to their initial inclination (α) to the bulk shortening direction (λ_3) (Fig. 2.10b, c). For a given aspect ratio of shear zones (e.g. 5), the flattening is around 5% when α = 45°, which becomes large (14%) when α = 20° (Fig. 2.6a). The amount of flattening is evidently a function of the finite bulk shear. However, our model results show that the flattening tends to attain a stable value in course of progressive deformation (Fig. 2.12a).

The S_r analysis of model simulations confirms that the pure to simple shear rates drops nonlinearly with increasing finite shear in a shear zone, as observed in the physical experiments. To demonstrate, we present two sets of data corresponding to a shear zone with D_f = 6 and 8 for α = 30° from physical experiments and model simulations (Fig. 2.12b). In both cases S_r = 0.4 at an initial stage, and decreases exponentially to a small value (~0.1) as the finite bulk shear exceeds 1. We compiled all the data sets obtained from model simulations run with varying α and D_f, which are shown in Fig. 2.13. The plots show an inverse relation of S_r with D_f, whereas there is a positive relation with α. At the initial stage S_r shows a wide variation, ranging from almost zero when α = 45° and D_f = 10 to as high as 1 when α = 20° and D_f = 5. However, the variation narrows down with progressively increasing shear, and S_r tends to zero,

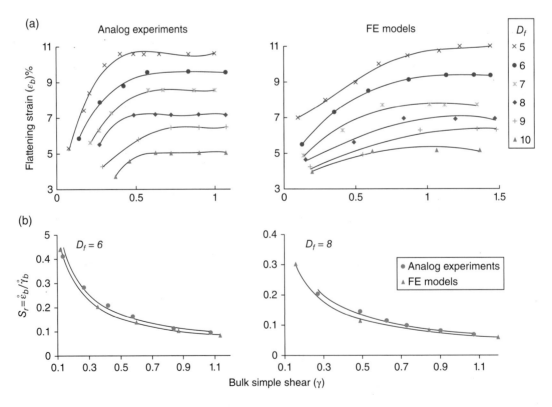

Fig. 2.12. Variations of (a) the bulk flattening strain (ε_b), (b) the ratio of pure and simple shear rates with progressive shear in physical experiments and FE models. α = 30°. Note an excellent match of the results of physical experiments with those obtained from FE models in both (a) and (b). In Fig. 2.2(a), symbols indicate the values of aspect ratio (D_f).

(a)

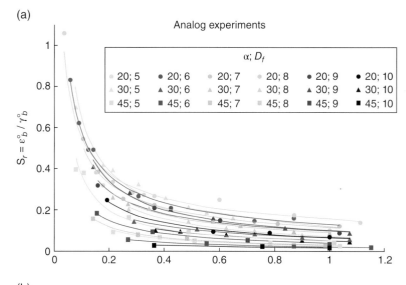

(b)

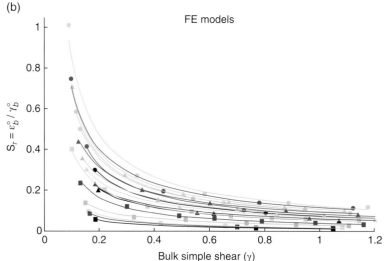

Bulk simple shear (γ)

Fig. 2.13. Graphical plots of the variations of S_r with progressive shearing (γ) in ductile shear zones for their different inclinations (α) and aspect ratios (D_f). (a) Physical experiments. (b) FE simulations. The first and the second numerical values corresponding to each symbol denote α and D_f values, respectively.

irrespective of their initial geometry. We find, overall, a spectacular match between the model simulations and experimental S_r values.

Using Equation 4, we evaluated the bulk kinematic vorticity number (W_k) to analyze the degree of transpression in ductile shear zones as a function of the geometric parameters: D_f and α. Figure 2.6c presents the variations of W_k with D_f for $\alpha = 45°$, $30°$ and $20°$. For a given value of α, W_k is somewhat lower than 1 when the aspect ratio is low, but asymptotically tends to be 1, implying simple shear kinematics for large aspect ratios. As an example, for $\alpha = 45°$, $W_k = 0.8$ when $D_f \sim 5$ and approaches 1 as $D_f > 8$. For a given D_f, W_k drops further below 1 for smaller values of α. As an example, when $D_f = 5.5$, $W_k = 0.8$ for $\alpha = 45°$, which decreases to 0.4 for $\alpha = 20°$, implying that shear zones can undergo some flattening. These variations obtained from FE model simulations show an excellent match with the experimental data (Fig. 2.6c).

2.5   DISCUSSION

2.5.1   Effects of shear zone geometry on transpression

The experimental results presented in Section 2.4 confirm earlier theoretical predictions that an isolated ductile shear zone can undergo a limited amount of bulk flattening strain. This finding confirms our inference of the simple shear dominated kinematics of small scale shear zones observed in the CGGC terrain. Earlier workers (Passchier and Coelho 2006) have also demonstrated simple shear kinematics in deep-crustal shear zones from a granulite terrain. Our study, therefore, refines the understanding of shear zone kinematics from a mechanical perspective. Our experiments show that under a constant volume condition the flattening deformation must involve material extrusion against enormous viscous resistance exerted by the rigid walls. Consequently, the gliding movement of rigid walls along the shear zone becomes a mechanically much easier process to accommodate the bulk shortening. It is the resistive force to viscous extrusion that fundamentally determines the degree of transpression possible in a shear zone. This viscous resistance depends on the shear zone geometry. Shear zones with large length to thickness ratios involve high resistance to viscous extrusion, leading to little transpression, as observed in both physical and FE experiments. According to our experimental findings, shear zones with aspect ratios ≥10 undergo little or no transpression, irrespective of their initial orientations with respect to the principal axis of bulk compression. The kinematic vorticity would be virtually 1 at any location within the shear zone. Lower aspect ratios can give rise to some transpression when the shear zones are at an angle >45° to the bulk compression direction. The bulk flattening, however, does not increase steadily with progressive deformation, but asymptotically attains a stable value. Our experiments suggests a maximum of 15–16% shortening for shear zones with an aspect ratio of 5, and oriented at angle of 70° to the bulk compression direction. The shortening drops down to around 5% when the shear zone inclination is 45°. The initial kinematic states of such low-aspect shear zones are characterized by $S_r = 0.4$ and 1, and $W_k = 0.8$ and 0.4, respectively for shear zone inclinations 45° and 70°, respectively. Experiments performed on different materials, such as rocks (Rudnicki and Rice 1975), metals (Anand and Spitzig 1982) and polymers (Bowden and Raha 1974) have shown that ductile shear bands generally develop at angles around 45° to the bulk compression direction. It then follows from our experimental results that natural shear zones can undergo bulk shortening, but not exceeding 5%.

Both the physical and FE modeling considered in this study developed ductile shear zones in a static state. However, natural shear zones generally involve dynamic rheological transformation of the wall rocks, leading to migration of the shear zone boundaries (Burg and Laurent 1978). The geometric factors that control the kinematics must change in an evolving shear zone. To discuss the shear kinematics in such a dynamic setting, we need to consider the possible growth behavior of a ductile shear zone. Experimental observations suggest that shear zones grow lengthwise at much faster rate, as compared to the widening rate. The ratio of growth rates along and across the length of the shear zone and thickness appears to be in order of 5 to 10 (Misra and Mandal 2007; Misra et al. 2009). Dynamic shear zones thus increase their aspect ratios during their growth. The flattening can thus be important at an initial stage, but it must exponentially drop as the shear zones grow in dimensions with rapidly increasing aspect ratios. We will advance our present study to predict the progressively changing kinematics in dynamically growing shear zones.

There are several other factors to modify the geometry of shear zones during their growth. Discrete shear zones generally nucleate at several locations, and propagate lengthwise, as discussed above, and coalesce with one another with progressive deformations (Shen et al. 1995). Such coalescence processes abruptly increase their length to thickness ratios. Considering a simple case, the aspect ratio will multiply by an order of 2 or more, depending upon the number of shear zones coalescing with one another along their length at an instant. Our study shows a non-linear decrease of flattening with aspect ratio. The ongoing coalescence process in a shear zone system would dramatically reduce the degree of flattening following a coalescence event.

2.5.2 Shear zone kinematics with volume change

Ramsay and Graham (1970) in their classic work dealt with the kinematics of shear zone flattening entirely attributed to volume reduction in the shear zones. According to their kinematic model, a shear zone can undergo flattening to any extent depending upon the degree of syn-shearing volume change. Despite a significant accretion of literature on the phenomenaof transpression over the last few decades, it is yet to validate Ramsay and Graham's model in the context of mechanical constraint for shear zone flattening, as discussed in Section 2.4. A large number of workers have reported transpressional type of shear zones from many tectonic belts (Harland 1971; Sanderson and Marchini 1984; Jones and Tanner 1995). However, our present experimental verifications point to a narrow possibility of transpression in natural shear zones with aspect ratios in the order of 10^1 to 10^2. Volume reduction, as predicted by Ramsay and Graham seems to be a potential model to account for large transpression. However, this model is essentially developed upon a kinematic consideration, and thus demands a revisit from a mechanical point of view. To verify the issue, we ran a few FE model experiments permitting compressibility in the shear zone material, as presented in the following discussion.

FE simulations were performed with a constant aspect ratio ($D_f = 10$), but varying inclinations (Fig. 2.14). We ran these simulations in two sets with a difference in the bulk modulus (K) of shear zone material by a factor of 10^{-1}, keeping all other rheological parameters as in the previous FE models. The model results show that increasing the angle of the shear zones with the bulk compression direction promotes the amount of volume loss, which in turn enhances the degree of flattening (Fig. 2.14a–c). For $\alpha = 30°$, the flattening increases from 6% to 33%, attributed to the volume loss by 19% (Fig. 2.14b). The degree of flattening increases further to nearly 50% as $\alpha = 20°$, and the shear zone involves volume loss by about 30% (Fig. 2.14c). Our results have led to infer that a shear zone with an aspect ratio ≥10 can undergo transpression by more than 30% when the shear zone involves syn-shearing volume loss, as documented from chemical analyses by many workers (O'Hara 1988; Mohanty and Ramsay 1994; Srivastava et al. 1995).

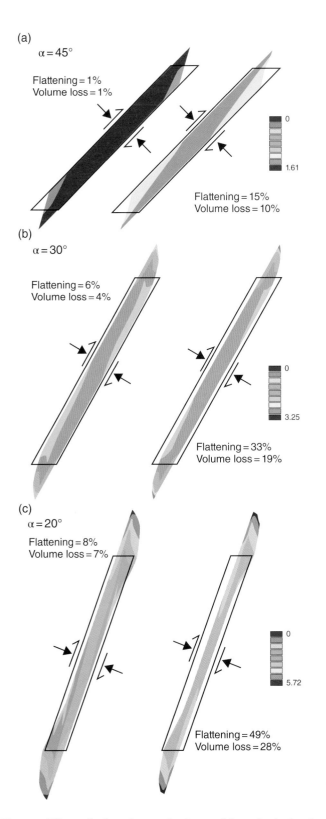

Fig. 2.14. Effects of volume loss on the degree of flattening in ductile shear zones of varying orientations: α = (a) 45°, (b) 30° and (c) 20°. Models on the right column had the bulk modulus of the shear zone 10^{-1} times that for models on the left column. Note that the amount of flattening is much greater in shear zones undergoing relatively larger volume loss (shown on the right column). Color contours indicate the value of the first principal strain. Solid arrows and half arrows indicate the directions of flattening and simple shear respectively.

2.5.3   Limitations

This study has some limitations, which need to be discussed here. (1) Our physical experiments were performed, keeping the extrusion along the shear direction. This experimental setup simulates a specific tectonic condition, as considered in many earlier kinematic models (Ramberg 1975; Ghosh and Ramberg 1976). However, several workers reported transpression zones with bulk extension perpendicular to the shear direction (Sanderson and Marchini 1984). (2) Both the physical and numerical models presented here are applicable to plane strain conditions, where the intermediate principal quadratic elongation (λ_2) was always 1 at any point within the shear zone. Field studies documented that natural shear can undergo shearing with $\lambda_2 >$ or <1 (Dewey et al. 1998; Czeck and Hudleston 2003; Ghosh and Sengupta 1987). (3) Due to a mechanical limitation of our experimental setup we had to run our experiments for a limited range of finite shear (not exceeding 1.5). On the other hand, natural shear zones often show large finite shear, exceeding 5 (Ramsay and Graham 1970; Lacassin et al. 1993; Law et al. 2013). However, the flattening ceases to take place at much lower shear (~1.2). (4) This study does not take into account the effects of wall rock deformations. The model results are applicable only to ductile shear zones hosted in a rigid country rock.

2.6   CONCLUSIONS

The main conclusions of this study are: (1) Small-scale strike–slip ductile shear zones with aspect ratios more than 6 in the Chotanagpur Granite Gneiss Complex show little or no flattening. (2) Ductile shear zones with aspect ratios <10 can undergo flattening by about 15% if they are oriented at angles larger than 60° to the bulk compression direction. In contrast, shear zones with aspect ratios ≥10 experience virtually no flattening, irrespective of their initial orientations. (3) The flattening ceases to occur in course of progressive shearing, and attain a stable value. (4) The ratio of pure and simple shear rates shows a strongly non-linear, inverse relation with the aspect ratio. (5) The bulk kinematic vorticity approaches 1, implying simple shear kinematics for shear zones either with large aspect ratios or oriented at 45° to the bulk compression direction. (6) Syn-shearing volume reduction appears to be the most effective mechanism for transpressional deformations in shear zones.

ACKNOWLEDGMENTS

We thank Amiya Baruah and Dr. Manas Kumar Roy for their insightful discussions at different stages of this study. The work has been supported by SERB, Department of Science and Technology, Govt. of India (N.M. and S.B.). S.D. acknowledges CSIR, India for providing him the research fellowship (Award no 09/096(0628)/2009-EMR-I).

REFERENCES

Anand L, Spitzig WA. 1982. Shear-band orientations in plane strain. Acta Metallurgica 30(2), 553–561.

ANSYS® Academic Research, (2007) Release 11.0. ANSYS, Inc., Canonsburg, PA, USA.

Bowden PB, Raha S. 1974. A molecular model for yield and flow in amorphous glassy polymers making use of a dislocation analogue. Philosophical Magazine 29(1), 149–166.

Brun JP, Cobbold PR. 1980. Strain heating and thermal softening in continental shear zones: a review. Journal of Structural Geology 2(1), 149–158.

Burg JP, Laurent P. 1978. Strain analysis of a shear zone in a granodiorite. Tectonophysics 47(1), 15–42.

Clift P, Lin J, Barckhausen U. 2002. Evidence of low flexural rigidity and low viscosity lower continental crust during continental break-up in the South China Sea. Marine and Petroleum Geology 19(8), 951–970.

Copley A, McKenzie D. 2007. Models of crustal flow in the India–Asia collision zone. Geophysical Journal International 169(2), 683–698.

Czeck DM, Hudleston PJ. 2003. Testing models for obliquely plunging lineations in transpression: a natural example and theoretical discussion. Journal of Structural Geology 25(6), 959–982.

Dewey JF, Holdsworth RE, Strachan RA. 1998 Transpression and transtension zones. Geological Society, London, Special Publications 135(1), 1–14.

Dutton BJ. 1997. Finite strains in transpression zones with no boundary slip. Journal of Structural Geology 19(9), 1189–1200.

Exner U. 2005. Analog modelling of flanking structures. Unpublished PhD thesis, ETH, Zurich, pp. 65–78.

Fagereng Å. 2013. On stress and strain in a continuous-discontinuous shear zone undergoing simple shear and volume loss. Journal of Structural Geology 50, 44–53.

Fernández C, Díaz-Azpiroz M. 2009. Triclinic transpression zones with inclined extrusion. Journal of Structural Geology 31(10), 1255–1269.

Gapais D. 1989. Shear structures within deformed granites: mechanical and thermal indicators. Geology 17(12), 1144–1147.

Gapais D, Bale P, Choukroune P, Cobbold P, Mahjoub Y, Marquer D. 1987. Bulk kinematics from shear zone patterns: some field examples. Journal of Structural Geology 9(5), 635–646.

Geological Survey of India. 2001. District Resource Map of Purulia, West Bengal. 1:125,000. India.

Godin L, Yakymchuk C, Harris LB. 2011. Himalayan hinterland verging superstructure folds related to foreland-directed infrastructure ductile flow: insights from centrifuge analogue modeling. Journal of Structural Geology 33, 329–342.

Ghosh SK. 1987. Measure of non-coaxiality. Journal of Structural Geology 9, 111–113.

Ghosh SK. 2001. Types of transpressional and transtensional deformation. Memoirs-Geological Society of America, 1–20.

Ghosh SK, Ramberg H. 1976. Reorientation of inclusions by combination of pure shear and simple shear. Tectonophysics 34(1), 1–70.

Ghosh SK, Sengupta S. 1987. Progressive development of structures in a ductile shear zone. Journal of Structural Geology 9(3), 277–287.

Harland WB. 1971. Tectonic transpression in Caledonian Spitsbergen. Geological Magazine 108, 27–42.

Jaeger JC. 1969. Elasticity, fracture and flow Chapman & Hall, London, p. 268.

Jones RR, Holdsworth RE. 1998. Oblique simple shear in transpression zones. Geological Society, London, Special Publications 135(1), 35–40.

Jones RR, Tanner PWG. 1995. Strain partitioning in transpression zones. Journal of Structural Geology 17(6), 793–802.

Koyi H. 1991. Mushroom diapirs penetrating into high viscous overburden. Geology 19, 1229–1232.

Kurz GA, Northrup CJ. 2008. Structural analysis of mylonitic rocks in the Cougar Creek Complex, Oregon–Idaho using the porphyroclast hyperbolic distribution method, and potential use of SC′-type extensional shear bands as quantitative vorticity indicators. Journal of Structural Geology 30(8), 1005–1012.

Lacassin R, Leloup PH, Tapponnier P. 1993. Bounds on strain in large Tertiary shear zones of SE Asia from boudinage restoration. Journal of Structural Geology 15(6), 677–692.

Law RD, Stahr III DW, Francsis MK, Ashley KT, Grasemann B, Ahmad T. 2013. Deformation temperatures and flow vorticities near the base of the Greater Himalayan Series, Sutlej Valley and Shimla Klippe, NW India. Journal of Structural Geology, 54, 21–53.

Larsen CF, Motyka RJ, Freymueller JT, Echelmeyer KA, Ivins ER. 2005. Rapid viscoelastic uplift in southeast Alaska caused by post-Little Ice Age glacial retreat. Earth and Planetary Science Letters 237(3), 548–560.

Maji AK, Goon S, Bhattacharya A, Mishra B, Mahato S, Bernhardt HJ. 2008. Proterozoic polyphase metamorphism in the Chhotanagpur Gneissic Complex (India), and implication for trans-continental Gondwanaland correlation. Precambrian Research 162(3), 385–402.

Mandal N, Samanta SK, Chakraborty C. 2000. Progressive development of mantle structures around elongate porphyroclasts: insights from numerical models. Journal of Structural Geology 22(7), 993–1008.

Mandal N, Chakraborty C, Samanta SK 2001. Flattening in shear zones under constant volume: a theoretical evaluation. Journal of Structural Geology 23(11), 1771–1780.

Mandal N, Dhar R, Misra S, Chakraborty C. 2007. Use of boudinaged rigid objects as a strain gauge: Insights from analogue and numerical models. Journal of Structural Geology 29(5), 759–773.

Mahmoud M. 2008. Structural evolution in the Chotanagpur Granite Gneiss Complex Purulia, W. Bengal: a kinematic study. Unpublished thesis, Jadavpur University, India.

Marsh BD. 1982. On the mechanics of igneous diapirism, stoping, and zone melting. American Journal of Science 282(6), 808–855.

Massey MA, Moecher DP. 2013. Transpression, extrusion, partitioning, and lateral escape in the middle crust: Significance of structures, fabrics, and kinematics in the Bronson Hill zone, southern New England, USA. Journal of Structural Geology 55, 62–78.

Misra S, Mandal N. 2007. Localization of plastic zones in rocks around rigid inclusions: Insights from experimental and theoretical models. Journal of Geophysical Research 112, B09206.

Misra S, Manda, N, Dhar R, Chakraborty C. 2009. Mechanisms of deformation localization at the tips of shear fractures: Findings from analogue experiments and field evidence. Journal of Geophysical Research 114, B04204.

Mohanty S, Ramsay JG. 1994. Strain partitioning in ductile shear zones: an example from a Lower Penninenappe of Switzerland. Journal of Structural Geology 16(5), 663–676.

Mukherjee S. 2013. Channel flow extrusion model to constrain dynamic viscosity and Prandtl number of the Higher Himalayan Shear Zone. International Journal of Earth Sciences 102, 1811–1835.

Mukherjee S, Koyi HA, Talbot CJ. 2012. Implications of channel flow analogue models for extrusion of the Higher Himalayan Shear Zone with special reference to the out-of-sequence thrusting. International Journal of Earth Sciences 101, 253–272.

O'Hara K. 1988. Fluid flow and volume loss during mylonitization: an origin for phyllonite in an overthrust setting, North Carolina USA. Tectonophysics 156(1), 21–36.

O'Hara K, Blackburn WH. 1989. Volume-loss model for trace-element enrichments in mylonites. Geology 17(6), 524–527.

Passchier CW, Coelho S. 2006. An outline of shear-sense analysis in high-grade rocks. Gondwana Research 10(1), 66–76.

Passchier CW, Druguet E. 2002. Numerical modelling of asymmetric boudinage. Journal of Structural Geology 24(11), 1789–1803.

Passchier C, Simpson C. 1986. Porphyroclast systems as kinematic indicators. Journal of Structural Geology 8(8), 831–843.

Passchier CW, Urai JL. 1988. Vorticity and strain analysis using Mohr diagrams. Journal of Structural Geology 10(7), 755–763.

Puelles P, Mulchrone KF, Ábalos B, Ibarguchi JI. 2005. Structural analysis of high-pressure shear zones (Bacariza Formation, Cabo Ortegal, NW Spain). Journal of Structural Geology 27(6), 1046–1060.

Ramberg H. 1975. Particle paths, displacement and progressive strain applicable to rocks. Tectonophysics 28(1), 1–37.

Ramsay JG. 1980. Shear zone geometry: a review. Journal of Structural Geology 2(1), 83–99.

Ramsay JG, Graham RH. 1970. Strain variation in shear belts. Canadian Journal of Earth Sciences 7(3), 786–813.

Ramsay JG, Huber MI. 1987. The Techniques of Modern Structural Geology, volume 2: Folds and fractures. Academic Press, London.

Robin PYF, Cruden AR. 1994. Strain and vorticity patterns in ideally ductile transpression zones. Journal of Structural Geology 16(4), 447–466.

Rudnicki JW, Rice JR. 1975. Conditions for the localization of deformation in pressure-sensitive dilatant materials. Journal of the Mechanics and Physics of Solids 23(6), 371–394.

Rutter EH. 1999. On the relationship between the formation of shear zones and the form of the flow law for rocks undergoing dynamic recrystallization. Tectonophysics 303(1), 147–158.

Saha AK. 1994. Crustal evolution of Singhbhum-North Orissa, Eastern India. Memoir Geological Society of India 27, 341.

Sanderson DJ, Marchini WRD. 1984. Transpression. Journal of Structural Geology 6, 449–458.

Sengupta S, Ghosh SK. 2004. Analysis of transpressional deformation from geometrical evolution of mesoscopic structures from Phulad shear zone, Rajasthan, India. Journal of Structural Geology 26(11), 1961–1976.

Shen B, Stephansson O, Einstein HH, Ghahreman B. 1995. Coalescence of fractures under shear stresses in experiments. Journal of Geophysical Research: Solid Earth 100(B4), 5975–5990.

Simpson C, De Paor DG. 1997. Practical analysis of general shear zones using the porphyroclast hyperbolic distribution method: an example from the Scandinavian Caledonides. In Evolution of Geological Structures in Micro- to Macro-scales, edited by S. Sengupta, Chapman & Hall, London, pp. 169–184.

Srivastava HB, Hudleston P, Earley III D. 1995. Strain and possible volume loss in a high-grade ductile shear zone. Journal of Structural Geology 17(9), 1217–1231.

Tikoff B, Fossen H. 1993. Simultaneous pure and simple shear: the unifying deformation matrix. Tectonophysics 217(3), 267–283.

Tondi E, Antonellini M, Aydin A, Marchegiani L, Cello G. 2006. The role of deformation bands, stylolites and sheared stylolites in fault development in carbonate grainstones of Majella Mountain, Italy. Journal of Structural Geology 28(3), 376–391.

Truesdell C. 1954. The Kinematics of Vorticity. Indiana University Press, Bloomington, IN.

Xypolias P. 2010. Vorticity analysis in shear zones: A review of methods and applications. Journal of Structural Geology 32(12), 2072–2092.

Žák J, Schulmann K, Hrouda F. 2005. Multiple magmatic fabrics in the Sázava pluton (Bohemian Massif, Czech Republic): a result of superposition of wrench-dominated regional transpression on final emplacement. Journal of Structural Geology 27(5), 805–822.

Chapter 3

2D model for development of steady-state and oblique foliations in simple shear and more general deformations

KIERAN F. MULCHRONE[1], PATRICK A. MEERE[2], and DAVE J. McCARTHY[2]

[1] Department of Applied Mathematics, University College, Cork, Ireland
[2] Department of Geology, University College, Cork, Ireland

3.1 INTRODUCTION

The use of grain shape foliations within shear zones as a means to quantify non-coaxial progressive deformation and to determine sense of shear is now well established in the literature (see Passchier and Trouw 2005, and references therein). The development of foliations in shear zones is generally categorized into strain sensitive fabrics that record the complete finite strain history of a shear zone, and those strain insensitive fabrics that do not record the full strain history. One of the most common strain insensitive fabrics observed in low to medium grade mylonitic shear zones is a microscopic oblique foliation that is typically preserved in aggregates of small dynamically recrystallized grains (Means 1980, 1981, Lister and Snoke 1984). Typically this grain shape preferred orientation (GSPO) fabric is defined by aligned sub-grains in monomineralic aggregates or layers within mylonitic shear zones. Oblique foliations are typically developed in quartz (Law et al. 1990; Lister and Snoke 1984; Dell Angelo and Tullis 1989; Mukherjee and Koyi 2010; Mukherjee, 2013) and calcite aggregates (de Bresser 1989) but examples of oblique fabrics in olivine peridotites have also been recorded (Van der Wal et al. 1992). The angle between this fabric and the plane of shear (fabric attractor) typically ranges from 20° to 40° but can be as high as 60° and lower than 5° (Passchier and Trouw 2005). Oblique foliations are thought to represent fabrics that have reached a steady state as a consequence of two competing sets of processes; on one hand those processes that relate to grain elongation and shape fabric development that are expected in a non-coaxial strain environment competing against those dynamic recrystallization processes such as grain boundary migration that tend to counteract the development of the expected strain sensitive fabric (Means 1980, 1981; Ree 1991). Oblique fabrics are often assumed to be a product of multiple progressive deformation cycles, whereby a fabric is being continually created following a finite strain path and destroyed by subsequent or concomitant dynamic recrystallization. Thus, oblique fabrics are thought to represent a stable steady state, where the orientation and intensity of the foliation, once established, do not significantly change over the strain history of a given shear zone. As such, the orientation of these fabrics in the sense of shear (SOS) plane of a shear zone will therefore stabilize and lie in an intermediate position somewhere between the orientation of instantaneous stretching axis (ISA) of the instantaneous strain ellipse and the finite stretching axis (FSA) of the finite strain ellipse. The implication drawn from this is that the orientation of an oblique fabric with respect to the shear plane is a function of where the clock stopped in the last progressive deformation cycle of shear strain and dynamic recrystallization. The conditions of deformation, especially temperature and overall strain rate, play a significant role in controlling the relative rates of competing processes that operate in a progressive deformation cycle which will in turn indirectly determine the final orientation of oblique fabrics preserved in shear zones.

A key requirement for the development of steady-state foliations and oblique foliations is that there is a foliation forming process that is balanced by some foliation-destroying process (Means 1981; Hanmer 1984). Although Means' (1981) insight was somewhat theoretical at the time, convincing natural examples have been reported since. For example, the occurrence of a strain-insensitive distribution of porphyroblasts in the Mont Mary mylonites (Pennacchioni et al. 2001) and numerous reports of oblique fabrics in shear zones (e.g. Ebert et al. 2007; Menegon et al. 2008; Gottardi and Teyssier 2013). There is abundant evidence that the key foliation-forming process in natural rocks is deformation, whereby inclusions and grains tend to elongate and align along a particular direction with fabric intensity increasing with finite strain. Grains and inclusions may behave passively, or alternatively, may be less competent or more competent than the enclosing material. Foliation-destroying

Ductile Shear Zones: From Micro- to Macro-scales, First Edition. Edited by Soumyajit Mukherjee and Kieran F. Mulchrone.

processes are less well understood. However, recrystallization processes such as grain boundary migration (Passchier and Trouw 2005, pp. 36–40) may cause grains to become more equant rather than elongate. For inclusions, Means (1981) suggests many different processes which may contribute to reducing clast elongation, such as vorticity effects, rigidity, clast on clast interference, interaction with micro shear zones, interaction with obstacles, clast breakage, etc.

In this chapter the deformational context is described and a mathematical model for describing the development of steady-state and oblique foliations is presented. It is shown how the aspect ratio and orientation of a steady state foliation can be used to determine both the deformation type (i.e. kinematic vorticity number) and relative strength of foliation destroying processes. If the type of deformation can be independently assumed or measured then both the relative strength of the foliation destroying processes and the competency of the grains may be estimated. Finally, the methods are applied to experimental and natural data.

3.2 MODELING STEADY-STATE AND OBLIQUE FOLIATION DEVELOPMENT

Although natural inclusions and grains in rocks occur in all shapes and sizes (Mukherjee 2013, 2014), as a first step all components are idealized as ellipses. This abstraction simplifies theoretical developments and, in addition, a best-fit ellipse can be found for typical shapes found in rocks (Mulchrone and Choudury 2004). In the development that follows the modification of elliptical linear viscous inclusions during a general deformation in two dimensions (2D) is combined with an expression encapsulating the foliation destroying process.

Homogeneous deformation may be described by a constant coefficient velocity gradient tensor ($\mathbf{L}$), which in a suitably chosen Cartesian coordinate system assigns to each position vector ($\mathbf{x}$) a velocity vector ($\mathbf{v}$) (Mulchrone 2013), that is $\mathbf{v} = \mathbf{L}\mathbf{x}$. In this paper a general isochoric 2D deformation is considered with the following form:

$$\mathbf{L} = \begin{bmatrix} L_{xx} & L_{xy} \\ 0 & -L_{xx} \end{bmatrix} \quad (1)$$

Note that in order for area to be constant (isochoric) the diagonal elements of $\mathbf{L}$ must sum to 0; a condition automatically satisfied by Equation 1. To further simplify the mathematics the deformation is scaled by a factor of L_{xy} that is, $L_{xx} = \beta L_{xy}$ so that:

$$\mathbf{L} = L_{xy} \begin{bmatrix} \beta & 1 \\ 0 & -\beta \end{bmatrix} \quad (2)$$

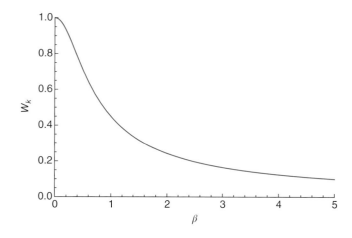

Fig. 3.1. The relationship between β and W_k (Eq. 4).

The following scaled velocity gradient tensor is used throughout the paper:

$$\mathbf{L}_s = \begin{bmatrix} \beta & 1 \\ 0 & -\beta \end{bmatrix} \quad (3)$$

This means that if $\mathbf{L}_s$ is used to calculate a velocity vector then the result must first be multiplied by the scaling factor L_{xy} to determine the actual velocity. The kinematic vorticity number (Ghosh 1987) associated with $\mathbf{L}_s$ is:

$$W_k = \sqrt{\frac{1}{4\beta^2 + 1}} \quad (4)$$

and the relationship between β and W_k is illustrated in Fig. 3.1. If $\beta = 0$ then $W_k = 1$ and the deformation is simple shear whereas for large β then W_k approaches 0 and pure shear dominates. Varying β between 0 and 10 covers most typical deformation scenarios (Ramberg 1975). This approach simplifies the analysis and a range deformations may be considered by varying a single parameter (β). Additionally, it emphasizes that in most geological situations the actual values of the parameters of the velocity gradient tensor are unknown but ratios may be estimated or assumed. In Fig. 3.2 the flow field associated with $\mathbf{L}_s$ is illustrated for typical values of β and W_k. The general deformation described by Equation 3 relates to common structural geological situations and may be decomposed as follows:

$$\mathbf{L}_s = \begin{bmatrix} \beta & 0 \\ 0 & -\beta \end{bmatrix} + \begin{bmatrix} 0 & 1 \\ 0 & 0 \end{bmatrix} \quad (5)$$

which represent the components of pure shear and simple shear. Hence, the general deformation in Equation 3 also represents simple shear combined with pure shear, known as general shear or sub-simple shear (i.e. transtension and transpression), and may be readily related to structures observed in the field. If β is positive then

Flow field $\beta = 0.0$, $W_k = 1.00$

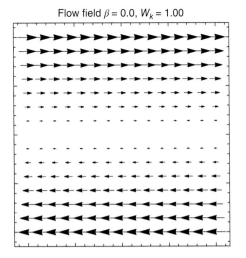

Flow field $\beta = 2.0$, $W_k = 0.24$

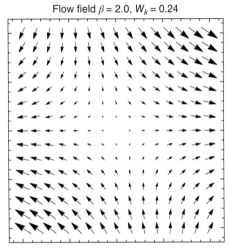

Flow field $\beta = 4.0$, $W_k = 0.12$

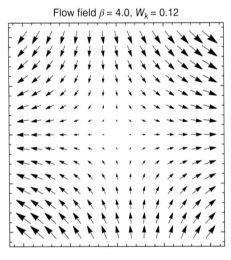

Fig. 3.2. Typical distributions of velocity corresponding to different values of β and W_k. From top to bottom: dextral simple shear, intermediate dextral shear and pure shear dominated.

deformation is transpressional and if β is negative then deformation is transtensional. In all cases the shear strain is dextral; however, by changing the sign of the L_{xy} component the sinistral case can also be studied. In the

absence of foliation-destroying processes, foliations typically develop due to deformation. Common models for the evolution of grains and inclusions during deformation include passive behavior (Fossen 2010, p. 62) and rigid behavior (Jeffery 1922). A model is adopted here which encompasses passive, rigid, competent and incompetent inclusion behaviors (Mulchrone and Walsh 2006) and builds on the approach of Jeffery (1922). A commonly used alternative follows Bilby and Kolbuszewski (1977), who adapted the method of Eshelby (1957) using the well-known analogy between slow incompressible linear viscous flow and linear elasticity. Mulchrone and Walsh (2006) derived non-linear ordinary differential equations describing the time evolution of both the long axis orientation (ϕ) and axial ratio (R) of a deformable inclusion immersed in a Newtonian viscous flow. As well as the velocity gradient tensor describing the deformation in the surrounding material a key parameter is the ratio of viscosity (μ_r) between the materials outside (μ_e) and inside (μ_i) the inclusion i.e. $\mu_r = \mu_e/\mu_i$. If $\mu_r = 0$, the inclusion is rigid, if $0 < \mu_r < 1$, the inclusion is incompetent, if $\mu_r = 1$, the inclusion is passive, and finally if $\mu_r > 1$, the inclusion is incompetent. Substituting for the scaled deformation (Equation 3) the governing equations are:

$$\frac{d\phi}{dt} = \frac{1}{2}\left(-1 + \frac{(R+1)(1+R(R+2\mu_r-2))(\cos 2\phi - 2\beta \sin 2\phi)}{(R-1)(1+2\mu_r R + R^2)} \right)$$

(6)

$$\frac{dR}{dt} = \frac{\mu_r R (R+1)^2 (2\beta \cos 2\phi + \sin 2\phi)}{2R + \mu_r (R^2+1)}$$

(7)

In summary, Equations 6 and 7 model the evolution of the shape and orientation of grains and inclusions in a general 2D isochoric deformation for any chosen viscosity ratio. The model in Equations 6 and 7 must be extended to include foliation-destroying processes. However, given that these processes are an eclectic mix (Means 1981; Passchier and Trouw 2005, pp. 36–40), the mathematical model is constrained to be heuristic in nature. In developing the model the following assumptions are made:

1 Foliation-destroying processes tend to reduce the elongation of a grain or inclusion i.e. R decreases.
2 If $R = 1$ then the inclusion is equant and the foliation-destroying process no longer operates.
3 Foliation-destroying processes are independent of inclusion orientation.
4 The intensity of the foliation-destroying process increases the more elongate an inclusion becomes.

The first assumption is supported by the common observation that in the absence of other processes, metamorphic rocks tend towards an equilibrium texture of coarse-grained equant grains and the second assumption simply enforces this by preventing an equant grain from becoming more elongate. The third assumption is borne out of consideration of processes such as grain boundary migration, where the boundary moves in the direction of highest dislocation density. In effect, this assumption implies that dislocation densities are concentrated near the long axis of a grain or inclusion hence the boundary tends to move to reduce the axial ratio. This process is thus independent of orientation. Finally, the last assumption arises from the need to balance foliation-enhancing processes (i.e. deformation) with foliation-destroying processes. If such a balance cannot be achieved then steady-state and oblique foliations cannot form. Supposing that the foliation-destroying process increases with elongation admits the possibility of such a balance occurring.

An expression for the rate of change of R due to foliation-destroying processes which encapsulates the above assumptions is:

$$\frac{dR}{dt} = -\alpha\left(R^2 - 1\right) \qquad (8)$$

where α is a positive rate constant that controls the size of the effect of the process. Given that $R \geq 1$, it is clear that $dR/dt < = 0$ in Equation 8, which satisfies assumption 1. Furthermore, as R approaches 1.0, dR/dt approaches zero, meaning that the more equant the shape the lesser it is affected by the process thus satisfying assumption 2. Equation 8 is independent of ϕ, which satisfies the third assumption and finally as R gets large the effect of the process increases quadratically.

Combining the foliation-forming process of Equations 6 and 7 with the foliation-destroying process of Equation 8 gives the model explored in this contribution:

$$\frac{d\phi}{dt} = \frac{1}{2}\left(-1 + \frac{(R+1)\left(1 + R(R + 2\mu_r - 2)\right)(\cos 2\phi - 2\beta \sin 2\phi)}{(R-1)\left(1 + 2\mu_r R + R^2\right)}\right) \qquad (9)$$

$$\frac{dR}{dt} = \frac{\mu_r R(R+1)^2\left(2\beta \cos 2\phi + \sin 2\phi\right)}{2R + \mu_r\left(R^2 + 1\right)} - \alpha\left(R^2 - 1\right) \qquad (10)$$

The foliation-forming process due to deformation is modelled by a physically based derivation and the foliation-destroying process is based on heuristic reasoning. However, as the results of analysis of the model demonstrate, it provides insight into the dynamics of systems producing oblique and steady-state foliations.

3.3 ANALYSIS OF THE MODEL

3.3.1 Introduction

In this section the model is explored and analyzed principally using the tools of non-linear dynamical systems (see Strogatz 1994). The model comprises a

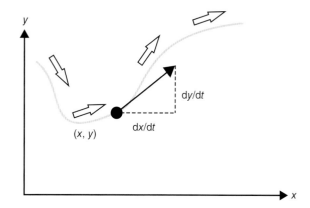

Fig. 3.3. Given a point (x, y) in the xy-plane we can construct a corresponding velocity vector. The trajectory of the motion or flow of the point must be tangential to the velocity vector. If this is calculated for many points, then the trajectory (gray line direction given by white arrows) can be found.

system of non-linear ordinary differential equations, which in general, cannot be solved analytically. However, it is possible to gain much insight into the system by qualitative and graphical analysis. Furthermore, a key concern is whether or not there exists values of R and ϕ (which are denoted as R^* and ϕ^*), which are stable and which the system ought to approach, given enough time. The passive case is considered first and the non-passive case is considered second. Considerable use is made of phase portraits (Strogatz 1994, p. 145–148) in this section and they are now briefly introduced.

Phase portraits are ideal for getting a qualitative understanding of coupled systems of ordinary differential equations such as those considered here (Equations 9 and 10). First consider the physical setting in Fig. 3.3. A point (x, y) in the plane is shown, and suppose we also know its velocity as a function of x and y (i.e. its velocity in the x-direction is dx/dt and in the y-direction it is dy/dt). Then, it can be described as follows:

$$dx/dt = f(x, y)$$
$$dy/dt = g(x, y)$$

where f and g are some known functions. Given any point (x, y) in the plane, we can draw the corresponding velocity vector at that point, and hence at that position we know the direction of motion or flow. By considering many points and corresponding velocity vectors we can trace out the trajectory followed points in the plane. This is the phase portrait. Hence, in the case of our equations we can draw the trajectories of (ϕ, R) because we know the form of $\left(\dfrac{d\phi}{dt}, \dfrac{dR}{dt}\right)$.

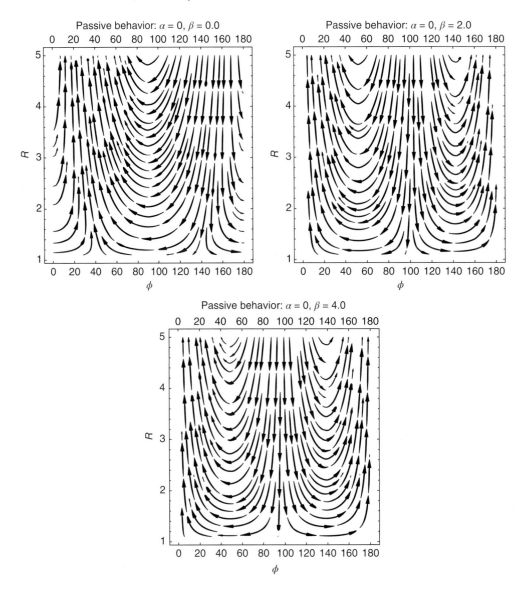

Fig. 3.4. Phase portrait for passive deformation and no foliation destruction process ($\alpha = 0$). ϕ varies from $0°$ (direction of positive horizontal axis in Figure 3.2) to $180°$ (direction of negative horizontal axis in Figure 3.2).

3.3.2   Passive deformation response and foliation destruction

In this case the viscosity of the grains and the surrounding material is equal and $\mu_r = 1$, which after simplification is governed by the following equations:

$$\frac{d\phi}{dt} = \frac{1}{2}\left(-1 + \frac{(R^2 + 1)(\cos 2\phi - 2\beta \sin 2\phi)}{(R^2 - 1)}\right) \qquad (11)$$

$$\frac{dR}{dt} = \frac{1}{2}R(2\beta \cos 2\phi + \sin 2\phi) - \alpha(R^2 - 1) \qquad (12)$$

For comparison, consider the phase portraits in Fig. 3.4 in the case of passive deformation and no foliation destruction process ($\alpha = 0$). Under simple shear, objects

making angles greater than $90°$ with the horizontal tend to become more equant; however, as they rotate clockwise to lower values of ϕ then R tends to increase without bound (Mulchrone and Walsh 2006). A steady-state foliation is not possible. Although it may appear counterintuitive, objects making small angles with the horizontal rotate opposite to that of the simple shear (i.e. sinistral rotation in a dextral simple shear). This is because elliptical inclusions behave differently to material lines where such behavior is not possible. Ultimately, all elliptical objects end uporiented, such that R becomes extremely large. Similar behavior is demonstrated for intermediate- and pure-shear dominated situations.

Figure 3.5 presents phase portraits where the foliation destroying parameter (α) is set to 0.75, a reasonable value based on analysis of experimental and natural data presented below. The pictures are significantly different to

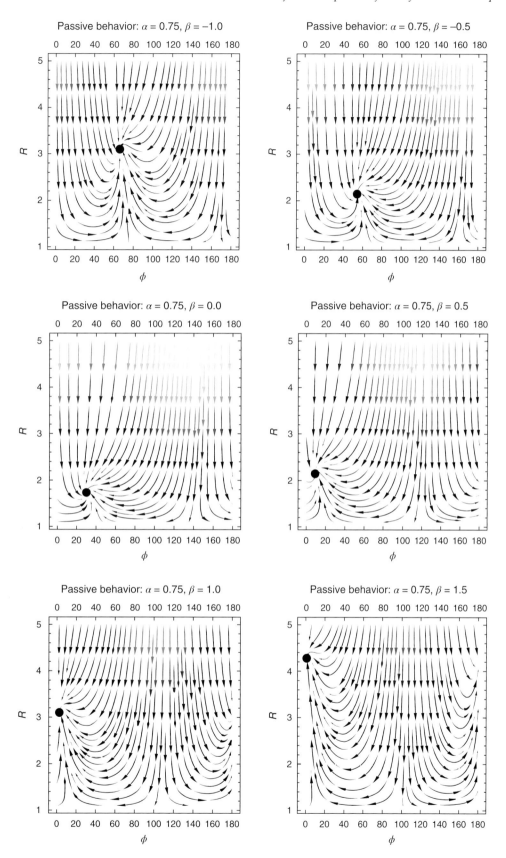

Fig. 3.5. Phase portrait for passive deformation coupled with the foliation destruction process ($\alpha = 0.75$). Stable fixed points are indicated as filled circles.

those in Fig. 3.4 and most notable is the presence of a fixed point (i.e. a point where neither R nor ϕ change), that is, a steady state where a balance between the deformation and foliation-destroying processes is attained. Furthermore it is a stable fixed point (which may be confirmed by linear stability analysis) and thus all trajectories are attracted to the steady state aspect ratio and orientation given enough time. Stable fixed points are denoted by filled black circles in Fig. 3.5.

It is clear that all trajectories will eventually approach the fixed point. It is also worth noticing that the position of the fixed point (i.e. both aspect ratio and orientation) varies systematically with β, the type of deformation. This implies that under the assumptions of the model, including passive deformation, a steady-state foliation readily develops and its characteristics can be used to determine the relative importance of the foliation-destroying process (α), the type of deformation (β) and kinematic vorticity.

Fixed points can be found by setting the right-hand side of both Equations 11 and 12 to zero and finding the values of R and ϕ for which these equations are satisfied. Let such a solution be denoted by R^* and ϕ^*. It is possible to write out analytical solutions, but they are lengthy and not insightful. In Fig. 3.6 the values of R^* and ϕ^* are contoured to illustrate the dependence of the stable fixed point position on the parameters α and β. In the case of transtensional regimes, steady-state foliations are predicted to make an angle of the order of 60° with the shear zone boundary whereas for simple shear it is between 20° and 40° (Fig. 3.6a). Under transpression the steady state foliations become subparallel to the shear zone boundary. Aspect ratios tend to be lower in simple shear and become higher in both transtensional and transpressional regimes (Fig. 3.6b). Solving for R^* and ϕ^* is algebraically involved; however, given a measured value for R^* and ϕ^* from field observations or microscope work, estimation of α and β is relatively straightforward:

$$\alpha = \frac{R^*\left(\cot\phi^* + R^{*2}\tan\phi^*\right)}{R^{*4} - 1} \quad (13)$$

$$\beta = \frac{\cot\phi^* - R^{*2}\tan\phi^*}{2\left(R^{*2} + 1\right)} \quad (14)$$

Hence, in situations where a steady-state foliation develops, a simple measurement of mean grain aspect ratio and mean orientation with respect to the shear zone boundary enables estimation of deformation type and strength of the foliation destruction process.

3.3.3 Non-passive deformation response and foliation destruction

In this section the situation where $\mu_r \neq 1$ is considered and the system is governed by the most general equations (Equations 9 and 10). When $\mu_r < 1$ grains behave in a

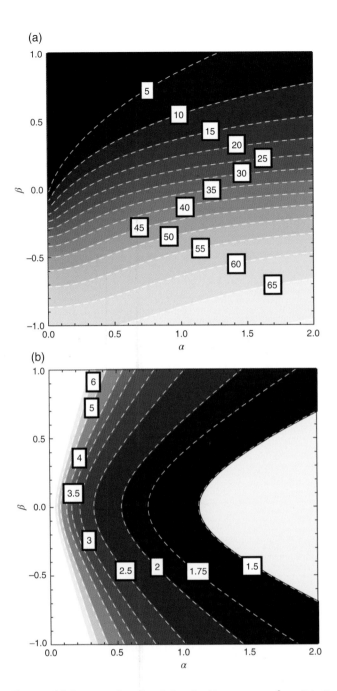

Fig. 3.6. (a) Contour plot of variation in ϕ^* as α ranges from 0 to 2 and β goes from −1 (transtension) to 1 (transpression). (b) Contour plot of variation in R^* as α ranges from 0 to 2 and β goes from −1 (transtension) to 1 (transpression).

competent manner (i.e. are more viscous than the surrounding materials) and typical phase portraits are illustrated in Fig. 3.7. On the other hand if $\mu_r > 1$ grains behave incompetently (i.e. are less viscous than the surrounding materials) and Fig. 3.8 shows some typical phase portraits. Comparison of phase portraits for different viscosity ratios indicates that steady state foliation angles (ϕ^*) are only slightly affected by these variations whereas the effect on steady state axial ratios (R^*) is more pronounced. In the competent case axial ratios are lower,

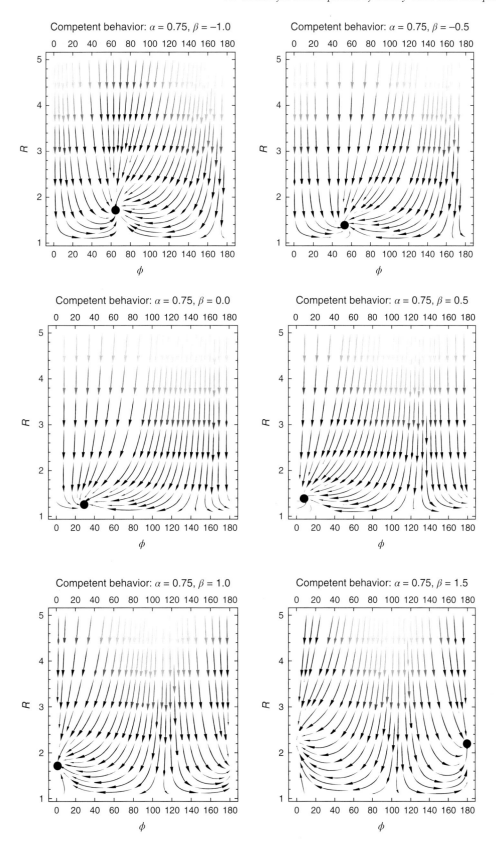

Fig. 3.7. Phase portrait for competent grain behavior ($\mu_r = 0.25$) coupled with the foliation destruction process ($\alpha = 0.75$). Stable fixed points are indicated as filled circles.

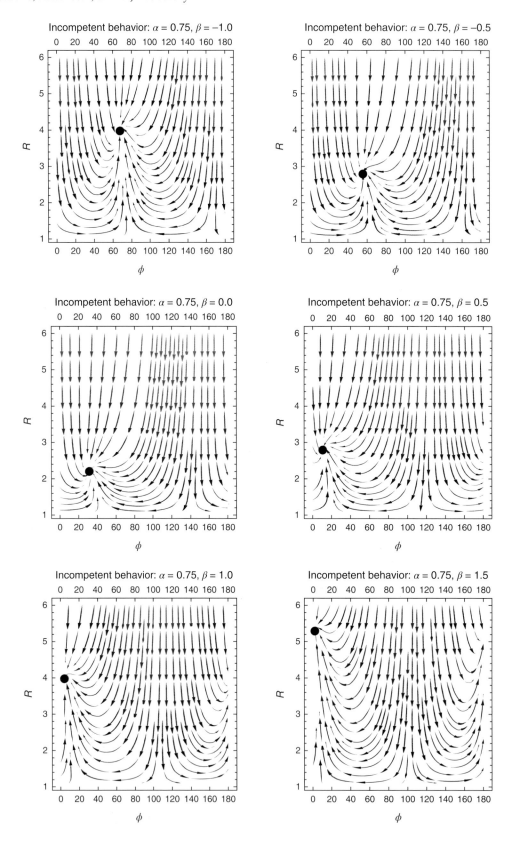

Fig. 3.8. Phase portrait for incompetent grain behavior ($\mu_r = 4$) coupled with the foliation destruction process ($\alpha = 0.75$). Stable fixed points are indicated as filled circles. Note R ranges from 1 to 6 instead of 1 to 5 (as is the case for Figs 3.4, 3.5, and 3.7).

whereas they are higher in the less competent case, as might be expected. The example in Fig. 3.9 illustrates this point for a chosen value of $\alpha = 0{:}75$. There is little difference between the curves for ϕ^* (Fig. 3.9a) when μ_r takes then values 0.2 (five times more competent, lower thick dashed curve), 1.0 (passive, thick solid curve) and 5.0 (five times less competent, thin dashed line). By contrast, there is a significant difference between the curves for R^* (Fig. 3.9b) clearly indicating that lower R^* values

are anticipated for competent grains and higher values for less competent grains.

Fixed points can be found for this system by setting the right-hand side of both Equations 9 and 10 to zero and finding the values of R and ϕ (R^* and ϕ^*) for which these equations are satisfied. Analytical solutions are not possible; however, given measured values for R^* and ϕ^*, the corresponding values of α and β can be estimated from:

$$\alpha = \frac{\left\{R^*\mu_r\left[\left(R^*+1\right)\left(R^{*2}+2\left(\mu_r-1\right)R^*+1\right)-\cos 2\phi^*\left(R^*-1\right)\left(R^{*2}+2\mu_r R^*+1\right)\right]\right\}}{\sin 2\phi^*\left(R^*-1\right)\left[\mu_r R^{*4}+2\left(\mu_r^2-\mu_r+1\right)\left(R^{*3}+R^*\right)+\left(6\mu_r-4\right)R^{*2}+\mu_r\right]} \tag{15}$$

$$\beta = \frac{\left[\cos 2\phi^*\left(R^*+1\right)\left(R^{*2}+2\left(\mu_r-1\right)R^*+1\right)-\left(R^*-1\right)\left(R^{*2}+2\mu_r R^*+1\right)\right]}{2\sin 2\phi^*\left[\left(R^*+1\right)\left(R^{*2}+2\left(\mu_r-1\right)R^*+1\right)\right]} \tag{16}$$

These equations are only useful if an independent estimate for μ_r is available; for example, in Section 3.3.2 the case of passive behavior (assuming $\mu_r = 1$) was analyzed. It may be easier to assume the nature of the deformation (e.g. simple shear) on the basis of other

structures and vorticity indicators in particular (Xypolias and Koukouvelas 2001; Xypolias 2010). If the value of β can be independently assumed, then estimates for α and μ_r are obtained as follows:

$$\alpha = \frac{R^*\left(2\beta\cos 2\phi^*+\sin 2\phi\right)\left[R^{*2}+1+\left(R^{*2}-1\right)\left(2\beta\sin 2\phi^*-\cos 2\phi\right)\right]}{\left\{R^*\left(R^{*3}-2R^{*2}+2\right)-1+\left(\cos 2\phi^*-2\beta\sin 2\phi^*\right)\left[R^*\left(2R^{*2}+2R^*+2-R^{*3}\right)-1\right]\right\}} \tag{17}$$

$$\mu_r = -\frac{\left(R^*-1\right)^2}{2R^*}+\frac{R^*-1}{1-R^*+\left(R^*+1\right)\left(\cos 2\phi^*-2\beta\sin 2\phi^*\right)} \tag{18}$$

3.4 COMPARISON WITH NATURAL AND EXPERIMENTAL DATA

The existence of oblique steady state fabrics has been established experimentally (Ree 1991) and is interpreted to occur in many natural examples (Bestmann et al. 2000; Passchier and Trouw 2005; Trouw et al. 2010; Mukherjee 2011). In this section a suite of thirteen examples of steady state and oblique foliations in naturally and experimentally deformed samples are analyzed in detail. The results strongly support the validity of the mathematical model developed above. Grain boundaries were traced from microphotographs and image analysis software was used to extract the corresponding best fit ellipses for the grains (Mulchrone et al. 2013). The average grain axial ratio and orientation was estimated using the mean radial length method (MRL, Mulchrone et al. 2003). Although, typically, MRL is used to estimate finite strain – it can

also serve to effectively estimate average geometric properties of a population of ellipses. The basic premise of the MRL method is that the MRL of a population of ellipses with an isotropic distribution is a constant and forms a circle; hence, if the population deviates from this configuration, the circle becomes an ellipse. The characteristics of the resulting ellipse represent the anisotropy of the population. The calculated mean aspect ratio and orientation of this ellipse are taken as estimates for R^* and ϕ^*, respectively.

Ree (1991) analyzed foliation formation processes in octachloropropane (OCP), an analog material with similar optical properties as quartz, but with a lower melting temperature. Samples were deformed in a shear rig apparatus under a dextral simple shear regime at 80°C, (i.e. $\beta = 0$, $W_k = 1.0$). The experiment allowed the sample to be imaged at different stages of deformation. A series of eight microphotographs documenting the range of deformation states from pre-deformation (Fig. 3.10a), through simple shearing (Fig. 3.10b), to post-deformation recovery (Fig. 3.10c), were analyzed (original data presented in Fig. 2 of Ree 1991). Our analysis complements the findings of Ree (1991). At the beginning of the experiment, the grains were

(a)

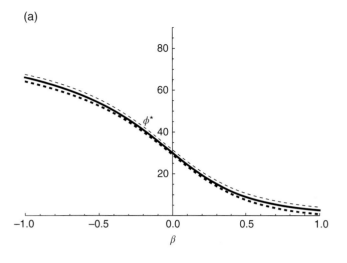

(b)

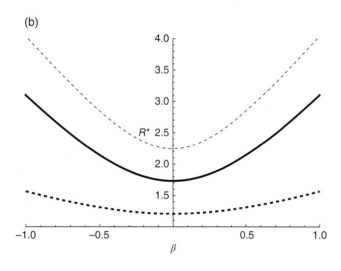

Fig. 3.9. (a) Variation of ϕ^* with β for $\alpha = 0.75$. Solid curve $\mu_r = 1$, passive behavior, lower bold dashed curve $\mu_r = 0.2$, competent behavior and finally upper dashed line $\mu_r = 5$, incompetent behavior. (b) Variation of R^* with β for $\alpha = 0.75$. Curve interpretation as for part (a).

largely equant with a low aspect ratio and a strong preferred orientation is absent (Fig. 3.10a). As deformation progresses our analysis suggests that a foliation did not become established until a shear strain of 0.4 had accumulated (Fig. 3.10d). After this point the aspect ratio and orientation becomes relatively constant with $R^* \approx 1.45–1.55$ and $\phi^* \approx 25–35°$. Clearly, the foliation is oblique to the shear plane and attains a strain-insensitive intensity and orientation (Fig. 3.10e).

Values for α and β were calculated for each of the samples (using Equations 13 and 14, see Table 3.1) assuming passive deformation coupled with a foliation destroying process. Furthermore, values for α and μ_r were also calculated under the assumption of simple shear (note that this particular value of α is in the column headed α' in Table 3.1 for clarity). This makes

particular sense for the data of Ree (1991) because the deformation was experimentally constrained to be simple shear.

Samples A and B are taken from the earliest stages of deformation and are interpreted to be from a transitional phase prior to the establishment of steady state foliation. Hence, the estimated values of α and β are invalid because the fixed point of the system has not yet been reached. For samples C to H equilibrium (the fixed point of the system) has been reached and calculated values of β are close to the expected value of 0.0 for simple shear. Typical values of α are between 1.05 and 1.5 and most are close to 1.2. This is OCP's material property rather than a general value for the strength of foliation-destroying processes. On the other hand, if simple shear is assumed ($\beta = 0$) then Equations 17 and 18 are used to estimate α and μ_r. This suggests that OCP grains behave as if they were slightly incompetent ($\mu_r \approx 0.5–0.7$) along with lower estimated values for α (0.4–1.2). Figure 3.11(a) illustrates the positions of (ϕ^*; R^*) calculated from experimental data on the phase portrait for passive behavior and $\alpha = 1:2$. The match is reasonable and it is interesting that the path from the initial phase (A, B) to the steady foliation phase (C–H) lies along the theoretical trajectories. Under the assumption of simple shear along with the estimated values of $\alpha = 0:7$ and $\mu_r = 0:5$ the calculated (ϕ^*; R^*) trajectory fits the theoretical phase portrait closely (Fig. 3.11b). The development of oblique foliations is also commonly observed in naturally deformed mylonites. To establish consistency with the model and to calculate representative parameters, microphotographs of mylonites from Trouw et al. (2010) and Bestmann et al. (2000) were analyzed. The images from Trouw et al. (2010, Figs 9.5.10–9.5.13) are of a medium-grade mica-quartz mylonite from Conceição do Rio Verde, SE Brazil. This mylonite is used as a classic example of mica fish, but also features a recrystallized quartz matrix defining a clear foliation lying oblique to the shear plane (Paaschier and Trouw 1996; Trouw et al. 2010). The quartz grains appear to have deformed by sub-grain recrystallization and grain boundary migration. The foliation is consistently oblique to the shear planes between 16° and 28° with a moderate intensity (R values: 1.5–1.8; Fig. 12). Analyzing these samples under the assumption of passive behavior the deformation (Table 3.1) is found to be close to simple shear with a tendency towards transpression. Associated values of α lie between 0.24 and 0.67 but tend towards 0.3. On the other hand by assuming simple shear, the grains appear to have behaved competently ($\mu_r \approx 0.3–0.5$) with associated values of $\alpha \approx 0.25–0.7$. The sample from Bestmann et al. (2000, Fig. 8) is from a fine-grained calcitic ultramylonite from Thassos Island of Northern Greece. Similar to the samples from Trouw et al. (2010) it exhibits an oblique foliation to the shear plane (21°) but has a slightly lower foliation intensity ($R = 1.3$; Fig. 3.12e). The

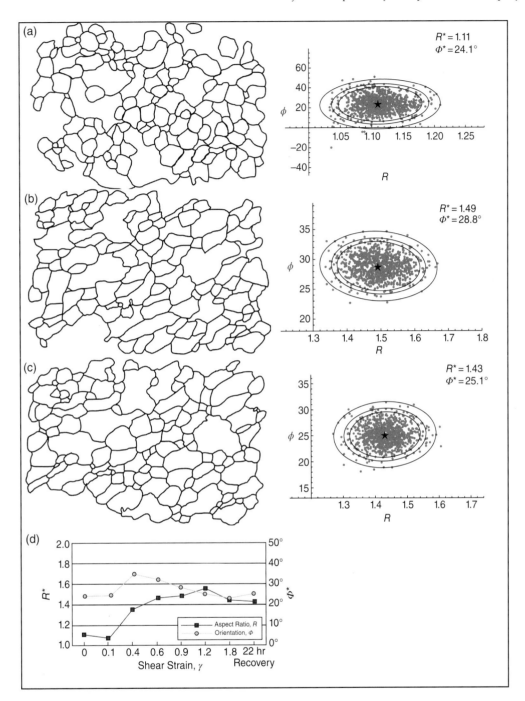

Fig. 3.10. Examples of analyzed grain boundary maps and results from microphotographs in Fig. 2 of Ree (1991). All grain boundary maps are of an area of 2 mm width. (a) Grain boundary map of the sample prior to the experimental deformation, most of the grains are largely equant and the MRL analysis does not detect a grain preferred orientation (GPO). (b) Grain boundary map of the sample after a moderate dextral shear strain ($\gamma = 0:9$), the shear direction is horizontal in this case. There is a well-defined GPO at an oblique angle to the direction of shear which MRL analysis clearly identifies. (c) Grain boundary map of the sample 22 hours after the experiment has finished, a total shear strain of = 1:8 has accumulated. Despite some minor alterations to the GPO due to recovery mechanisms the oblique foliation still persists and is clearly detected by MRL. (d) Plot of R^* and ϕ^* against shear strain (γ). After initial rapid changes in R^* and ϕ^* at the onset of deformation, both values stabilize as the experiment continues.

matrix has undergone complete recrystallization (calcite grains exhibit no twinning) by grain boundary migration. Calculated parameters (Table 3.1) under the assumption of passive behavior give $\alpha = 2.29$, which is out of line with other samples; however, this may be

due to the difference in mineralogy. Associated deformation is transpressive. By assuming simple shear α takes the value 0.38, which is similar to other samples and $\mu_r = 0.18$, which may indicate significantly competent behavior.

Table 3.1. Summary of data obtained from analysis of microphotographs

Sample (Source)	R*	ϕ*	γ	α	β	W_k	α'	μ_r
A (Ree 1991)	1.11	24.0	0.0	5.99	0.38	0.63	0.52	0.08
B (Ree 1991)	1.08	24.4	0.1	8.19	0.39	0.63	0.55	0.06
C (Ree 1991)	1.37	34.5	0.4	1.49	0.03	1.00	1.23	0.72
D (Ree 1991)	1.47	32.1	0.6	1.18	0.04	0.99	0.95	0.70
E (Ree 1991)	1.49	28.8	0.9	1.15	0.09	0.97	0.70	0.50
F (Ree 1991)	1.55	25.6	1.2	1.05	0.14	0.93	0.53	0.43
G (Ree 1991)	1.44	23.4	1.8	1.40	0.23	0.83	0.45	0.29
H (Ree 1991)	1.43	25.2	rec	1.39	0.19	0.87	0.32	0.52
9.5.10 (Trouw et al. 2010)	1.70	16.7	n/a	0.97	0.32	0.71	0.24	0.31
9.5.11 (Trouw et al. 2010)	1.82	20.7	n/a	0.71	0.16	0.91	0.33	0.45
9.5.12 (Trouw et al. 2010)	1.52	28.4	n/a	1.09	0.09	0.97	0.67	0.52
9.5.13 (Trouw et al. 2010)	1.52	18.4	n/a	1.32	0.34	0.69	0.29	0.26
Fig. 8a (Bestmann et al. 2000)	1.30	20.9	n/a	2.29	0.37	0.65	0.38	0.18

Note that sample H is taken 22 hours after the simple shear ceased, hence the entry for simple shear is rec meaning recovered. The simple shear associated with the natural examples is unknown. The columns α, β and W_k are calculated assuming passive behavior whereas the columns α' and μ_r are calculated assuming simple shear ($\beta = 0$; $W_k = 1$).

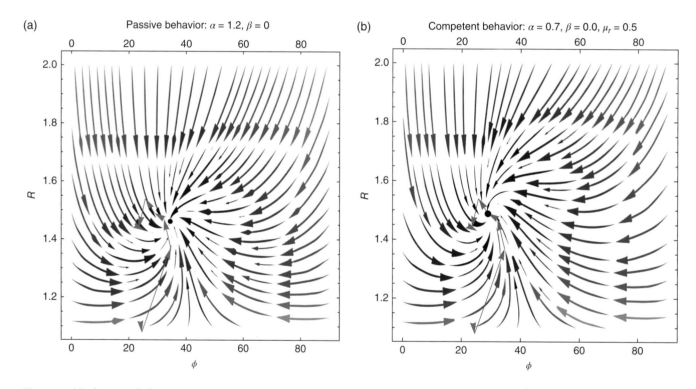

Fig. 3.11. (a) Theoretical phase portrait for passive behavior under simple shear deformation and $\alpha = 1.2$. Arrows are constructed from the analyzed data of Ree (1991) and point in the direction of increasing shear strain. Notice that the direction of movement for the experimental data in the phase portrait is consistent with theoretical trajectories. (b) Same experimental data represented by arrows as in (a) and the phase portrait is constructed for competent behavior ($\mu_r = 0.5$) under simple shear with $\alpha = 0.7$. Experimental movement directions and theoretical trajectories are consistent.

3.5 DISCUSSION AND CONCLUSION

A mathematical model for the development of steady state and oblique foliations is presented in this paper. Previously this development has been couched in terms of successive cycles of foliation formation followed by foliation destruction (Ree 1991), which tends to lead to a discrete model of separate processes for fabric development. The approach here is continuous and supposes that foliation-forming and foliation-destroying processes operate concurrently. In other words, deformation tends to continuously (i.e. operates at all times) elongate grains and at the same time foliation-destroying processes tend to make grains more equant. This leads to a

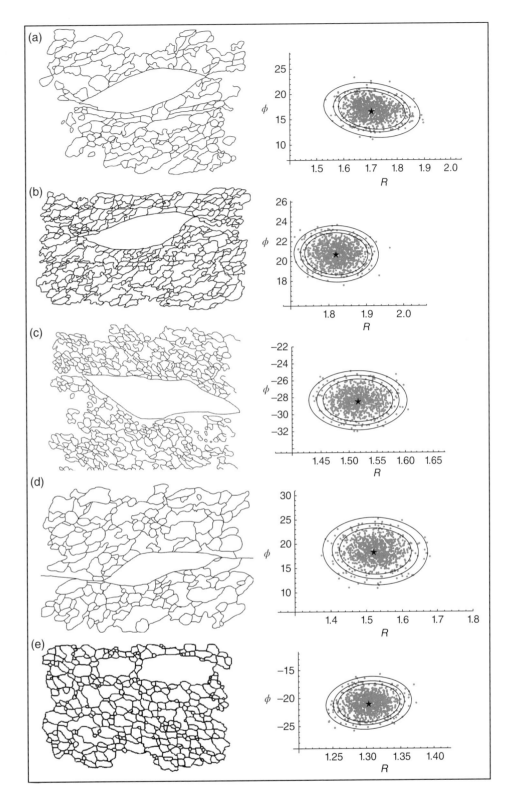

Fig. 3.12. Examples of analyzed grain boundary maps and MRL results from naturally deformed rocks from Trouw et al. (2010) and Bestmann et al. (2000). All images are from microphotographs of a thin sections cut perpendicular to the shear plane and the shear direction is horizontal. In the case of images (a)–(d) the mica fish were excluded from the analysis, similarly in image (e) the two large grains in the top right of the image were also excluded. (a) Grain boundary map and MRL results from Fig. 9.5.10 of Trouw et al. (2010). The image is from a moderate grade quartz-rich mylonite with dextral shear sense, similar to the images from Ree (1991); there is a very strong GPO at an oblique orientation to the shear plane. (b) Grain boundary map and MRL results from Fig. 9.5.11 of Trouw et al. (2010). This image is from the same locality as image (a) and shows a similarly strong oblique GPO to the shear plane. (c) Grain boundary map and MRL results from Fig. 9.5.12 of Trouw et al. (2010). This image is from the same locality as image (a), but interestingly has a sinistral sense of shear, yet it still features a strong oblique foliation to the shear direction. (d) Grain boundary map and MRL results from Fig. 9.5 of Trouw et al. (2010). This image is from the same locality as image (a) and again has a strong oblique GPO to the sense of shear. (e) Grain boundary map and MRL results from Fig. 8 of Bestmann et al. (2000). The sample is from a calcitic mylonite with a sinistral shear sense and similar to other examples it has an oblique GPO to the shear direction, but at a much lower $R^\star$ value. This could be due to the differing recrystallization (i.e. foliation destruction) mechanisms of calcite compared to quartz.

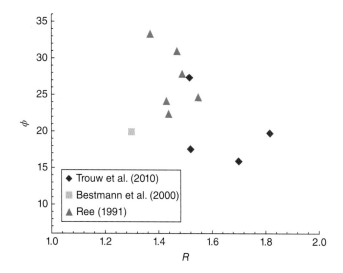

Fig. 3.13. Graph of R^*/ϕ^* data for samples examined in this study.

formulation in terms of ordinary differential equations (Equations 9 and 10), which can be analyzed quantitatively in terms of phase portraits (Figs 3.4, 3.5, 3.7, 3.8, and 3.11) and quantitatively assuming that system equilibrium (or stable fixed point) has been attained (Equations 13, 14, 15, 16, 17, and 18). Importantly, this analysis allows calculation of the relative importance of foliation destroying processes (the parameter α) and also the type of deformation (β) or kinematic vorticity number (W_k), assuming a passive response by grains to deformation.

On the other hand, if the kinematics of deformation can be confidently assumed, it is possible to estimate the relative strength of the deformation (α) and the competency (μ_r) of the inclusions.

The method and analysis presented in this chapter leads naturally to many practical applications. For example, in a tectonic terrain containing a suite of suitable shear zones occurring at different structural levels, and assuming simple shear, then it may be possible to compare the viscosity ratio of host to clast material and the strength of foliation-destroying processes between shear zones. Another possibility is that, in cases where passive behavior can be confidently assumed, oblique foliations may be used to estimate kinematic vorticity number and strength of foliation-destroying processes.

Analysis of experimental and natural examples are consistent with the developed models. In particular the measured trajectory of steady state aspect ratio and orientation in experimental data follows theoretical trajectories for estimated parameters.

ACKNOWLEDGMENTS

We thank Dr Soumyajit Mukerjee for editorial assistance. Support from Delia Sandford and Ian Francis (Wiley Blackwell) is gratefully acknowledged.

REFERENCES

Bestmann M, Karsten K, Matthews A. 2000. Evolution of a calcite marble shear zone complex on Thassos Island, Greece: microstructural and textural fabrics and their kinematic significance. Journal of Structural Geology 22, 1789–1807.

Bilby BA, Kolbuszewski ML. 1975. The finite deformation of an inhomogeneity in two-dimensional slow viscous incompressible flow. Proceedings of the Royal Society of London, Series A 355, 335–353.

de Bresser JPH. 1989. Calcite c-axis textures along the Gavervie thrust zone, central Pyrenees. Geologie en Mijnbouw 68, 367–376.

Dell Angelo LN, Tullis J. 1989. Fabric development in experimentally sheared quartzites. Tectonophysics 169, 1.21.

Ebert A, Herwegh M, Evans B, Piffner A, Austin N, Vennemann T. 2007. Microfabrics in carbonate mylonites along a large-scale shear zone (Helvetic Alps). Tectonophysics 444(1–4), 1–26.

Eshelby JD. 1957. The determination of the elastic field of an ellipsoidal inclusion, and related problems. Proceedings of the Royal Society of London, Series A 241, 376–396.

Fossen H. 2010. Structural Geology. Cambridge University Press, Cambridge, UK.

Ghosh SK. 1987. Measure of non-coaxiality. Journal of Structural Geology 9, 111–113.

Gottardi R, Teyssier C. 2013. Thermomechanics of an extensional shear zone, Raft River metamorphic core complex, NW Utah. Journal of Structural Geology 53, 54–69.

Hanmer S. 1984. Strain-insensitive foliations in polymineralic rocks. Canadian Journal of Earth Sciences 21(12), 1410–1414.

Jeffery GB. 1922. The motion of ellipsoidal particles immersed in a viscous fluid. Proceedings of the Royal Society of London A 102, 201–211.

Law RD, Schmid SM, Wheeler J. 1990. Simple shear deformation and quartz crystallographic fabrics: a possible natural example from the Torridon area of NW Scotland. Journal of Structural Geology 12, 29–45.

Lister GS, Snoke AW. 1984. S-C mylonites. Journal of Structural Geology 6, 617–638.

Means WD. 1980. High temperature simple shearing fabrics: A new experimental approach. Journal of Structural Geology 2, 197–202.

Means WD. 1981. The concept of steady-state foliation. Tectonophysics 78(1–4), 179–199.

Menegon L, Pennacchioni G, Heilbronner R, Pittarello L. 2008. Evolution of quartz microstructure and c-axis crystallographic preferred orientation within ductilely deformed granitoids (Arolla unit, Western Alps). Journal of Structural Geology 30(11), 1332–1347.

Mukherjee S. 2011. Mineral fish: their morphological classification, usefulness as shear sense indicators and genesis. International Journal of Earth Sciences 100, 1303–1314.

Mukherjee S. 2013. Deformation Microstructures of Rocks. Springer, Berlin.

Mukherjee S. 2014. Mica inclusions inside host mica grains-examples from Sutlej section of the Higher Himalayan Shear Zone, India. Acta Geologica Sinica 88, 1729–1741.

Mukherjee S, Koyi HA. 2010. Higher Himalayan Shear Zone, Zanskar Indian Himalaya: microstructural studies and extrusion mechanism by a combination of simple shear and channel flow. International Journal of Earth Sciences 99, 1083–1110.

Mulchrone KF. 2013. Qualitative and quantitative analysis of vorticity, strain and area change in general non-isochoric 2D deformation. Journal of Structural Geology 55(0), 114–126.

Mulchrone KF, Choudhury KR. 2004. Fitting an ellipse to an arbitrary shape: implications for strain analysis. Journal of Structural Geology 26(1), 143–153.

Mulchrone KF, Walsh K. 2006. The motion of a non-rigid ellipse in a general 2D deformation. Journal of Structural Geology 28(3), 392–407.

Mulchrone KF, O. Sullivan F, Meere PA. 2003. Finite strain estimation using the mean radial length of elliptical objects with bootstrap confidence intervals. Journal of Structural Geology 25(4), 529–539.

Mulchrone KF, McCarthy DJ, Meere PA. 2013. Mathematica code for image analysis, semi-automatic parameter extraction and strain analysis. Computers & Geosciences 61(0), 64–70.

Passchier CW, Trouw RAJ. 2005. Microtectonics, 2nd edition. Springer, Berlin.

Pennacchioni G, Di Toro G, Mancktelow NS. 2001. Strain-insensitive preferred orientation of porphyroclasts in Mont Mary mylonites. Journal of Structural Geology 23(8), 1281–1298.

Ramberg H. 1975. Particle paths, displacement and progressive strain applicable to rocks. Tectonophysics 28, 1–37.

Ree JH. 1991. An experimental steady state foliation. Journal of Structural Geology 13, 1001–1011.

Strogatz SH. 1994. Nonlinear Dynamics and Chaos with Applications to Physics, Biology, Chemistry, and Engineering. Perseus Books, Reading, MA.

Trouw RAJ, Passchier CW, Siersma D. 2010. Atlas of Mylonites – and related microstructures. Springer-Verlag, Berlin, Heidelberg.

Van der Wal D, Vissers RMD, Drury, MR. 1992. Oblique fabrics in porphyroclastic Alpine peridotites: a shear sense indicator for upper mantle flow. Journal of Structural Geology 14, 839–846.

Xypolias P. 2010. Vorticity analysis in shear zones: A review of methods and applications. Journal of Structural Geology 32(12), 2072–2092.

Xypolias P, Koukouvelas IK. 2001. Kinematic vorticity and strain rate patterns associated with ductile extrusion in the Chelmos Shear Zone (External Hellenides, Greece). Tectonophysics 338(1), 59–77.

Chapter 4

Ductile deformation of single inclusions in simple shear with a finite-strain hyperelastoviscoplastic rheology

CHRISTOPH ECKART SCHRANK[1,2], ALI KARRECH[3], DAVID ALEXANDRE BOUTELIER[4], and KLAUS REGENAUER-LIEB[2,5]

[1] Queensland University of Technology, School of Earth, Environmental and Biological Sciences, Brisbane, 4001, QLD, Australia

[2] The University of Western Australia, School of Earth and Environment, 35 Stirling Highway, Crawley, 6009, WA, Australia

[3] The University of Western Australia, School of Civil, Environmental and Mining Engineering, 35 Stirling Highway, Crawley, 6009, WA, Australia

[4] The University of Newcastle, School of Environmental and Life Sciences, University Drive, Callaghan, NSW, 2308, Australia

[5] The University of New South Wales, School of Petroleum Engineering, Tyree Energy Technologies Building, H6 Anzac Parade, Sydney, NSW, 2052, Australia

4.1 INTRODUCTION

The behavior of deformable and rigid inclusions in shear flows poses a fundamental geological problem and has thus attracted attention in the respective literature for almost a century (for a recent review, see Marques et al. 2014). As highlighted in the literature summary tables of Jessell et al. (2009) and Griera et al. (2013), the majority of work has focused on rigid or deformable purely elastic or viscous inclusions in a purely elastic or viscous matrix (e.g. Jeffery 1922; Gay 1968; Bilby and Kolbuszewski 1977; Schmid and Podladchikov 2003). However, it seems generally accepted that the long-term, inelastic deformation of the lithosphere is not only viscous but sensitive to elastic and plastic contributions to rheology (Moresi et al. 2002; Kaus and Podladchikov 2006; Regenauer-Lieb et al. 2006, 2011; Schmalholz et al. 2009; Schrank et al. 2012). Therefore, the question arises if and how the addition of elasticity and plasticity to rheology affects the inelastic deformation behavior of inclusions in shear. This work aims to provide a systematic reference study of the large deformation of single, initially round, fully bonded, deformable inclusions in isothermal two-dimensional (2D) simple shear with Dirichlet boundary conditions (constant velocity) and a hyperelastoviscoplastic rheology (Karrech et al. 2011b, c).

The prefix "hyper-" indicates that the stresses are derived from the strain energy potential and not simply assumed to be a single-valued function of strain (e.g. chapter 2 of Houlsby and Puzrin 2007). The consideration of elasticity at large strains poses a particular challenge because the mathematical treatment of large transformations requires an objective formulation of the stress rate considering both advective and corotational terms (e.g. Mühlhaus and Regenauer-Lieb 2005; Beuchert and Podladchikov 2010). An additional complexity arises due to the constraint to balance the rate of elastic energy stored during the finite deformation process. This is where classical stress rate formulations fail, and a new energetically consistent formulation has been proposed (Xiao et al. 1997; Bruhns et al. 1999; Karrech et al. 2011c) and used here. Considering weak and strong inclusions, we explored the wide parameter space spanned by the large variability of effective viscosity, yield stress, and strain rate in the lithosphere and find significant differences in the deformation of single inclusions compared to the Newtonian viscous case (Bilby and Kolbuszewski 1977).

4.2 METHODS

4.2.1 Hyperelastoviscoplasticity at finite strain

The derivation of our finite-strain theory (FST) and its numerical implementation are described in detail by Karrech et al. (2011c), and will not be repeated here. We only summarize its basic properties in the following. Our FST employs rate-dependent hyperelastoviscoplasticity with a multiplicative decomposition of the deformation gradient and a frame-indifferent corotational stress rate $\hat{t}$ with logarithmic spin:

$$\hat{t} = \dot{t} + t\Omega - \Omega t \qquad (1)$$

Ductile Shear Zones: From Micro- to Macro-scales, First Edition. Edited by Soumyajit Mukherjee and Kieran F. Mulchrone.

t is the Kirchoff stress tensor, the over-dot denotes the time derivative, and Ω is the logarithmic spin tensor of Xiao et al. (1997, 1998). This logarithmic spin includes the commonly used Jaumann spin (Zaremba 1903; Jaumann 1911), augmented by a correction term, which accounts for the stretching of the eigenprojections, not only their rotation. Equation 1 measures the stress seen by an observer who corotates with the infinitesimal volume element under consideration (Mühlhaus and Regenauer-Lieb 2005). Strain is measured logarithmically by the Hencky tensor, a natural measure of strain (Nadai 1937):

$$\boldsymbol{h} = \frac{1}{2}\ln\boldsymbol{b} \qquad (2)$$

where $\boldsymbol{b}$ is the Finger deformation tensor (Karrech et al. 2011c; Zhilin et al. 2013). The constitutive model is derived from the Helmholtz free-energy potential and a general dissipation function using the principle of maximum-entropy production (MEP, e.g. Dewar (2005)), thus accounting for dissipation without arbitrary assumptions regarding the rheology of the material (Poulet et al. 2009; Regenauer-Lieb et al. 2010; Karrech et al. 2011b). Hence, in this formulation, the flow rules and yield function emerge from fundamental thermodynamic principles encapsulated in the dissipation function, the free-energy potential, and the optimization procedure associated with MEP. In summary, the FST provides a proper large-strain measure and a self-consistent (in the sense of Hill's work conjugacy relation (Hill 1968)), integrable constitutive model, ensuring that the energy conservation and the entropy inequality (first and second law of thermodynamics) are satisfied (Houlsby and Puzrin 2007; Karrech et al. 2011c).

Elasticity is considered to be isotropic for simplicity. As done commonly for ductile materials (e.g. Hobbs et al. 1990), inelastic deformation is assumed to be controlled by deviatoric stress alone, which is expressed by the equivalent stress with the von Mises norm:

$$\sigma_{eq} = \sqrt{\frac{3}{2}\boldsymbol{t}'\boldsymbol{t}'} \qquad (3)$$

where $\boldsymbol{t}' = \boldsymbol{t} - \dfrac{tr(\boldsymbol{t})}{3}$ is the deviatoric stress tensor. Beyond the elastic limit, an overstress formulation is used (Karrech et al. 2011a, b; Veveakis and Regenauer-Lieb 2014). In the overstress model, the equivalent stress is the sum of the yield stress σ_y and the viscous stress σ_V:

$$\sigma_{eq} = \sigma_y + \sigma_V. \qquad (4)$$

While the full formulation of the FST is capable of simulating temperature- and pressure-dependent non-linear viscosities, due to a combination of diffusion or dislocation creep (Karrech et al. 2011b), we use a constant Newtonian viscosity η to calculate σ_V:

$$\sigma_V = \eta\dot{\gamma}_{eq} \qquad (5)$$

where $\dot{\gamma}_{eq}$ is the equivalent strain rate (see section 3 of Karrech et al. 2011a). In addition, the FST permits a pressure-dependent elastic limit (i.e. yield envelope; with strain hardening, if needed). However, for simplicity, we restrict ourselves to a constant yield stress (von Mises yield envelope) in this contribution.

4.2.2 Set-up and numerical solution

We consider a unit square in isothermal, plane-strain, quasi-static simple shear with Dirichlet boundary conditions, which contains a circular, fully bonded, deformable inclusion in its center (Fig. 4.1a). Body forces and inertia terms are neglected. The inclusion has a diameter of 5% of the length of the surrounding square in order to minimize possible edge effects. Taborda et al. (2004) examined flow around rigid elliptical inclusions in 2D plane-strain simple shear and found that their results are not affected by shear zone width if this width is more than 10 times larger than the short axis of the inclusion. This ratio always exceeds 10 in our experiments. Hence, we assume that boundary effects are negligible. We apply the kinematic boundary conditions:

$$\begin{aligned}
&v_x\left(y = 0\right) = 0 \\
&v_x\left(y = 1\right) = const. > 0 \\
&v_y\left(y\right) = 0 \\
&\frac{\partial v_x}{\partial y}\left(0 > y > 1\right) = const. > 0
\end{aligned} \qquad (6)$$

where v denotes velocity, x and y are the horizontal and vertical direction, respectively, and the subscripts indicate the directions of the velocity vector components. The FST is implemented via a Fortran user material subroutine (UMAT) in the finite-element code ABAQUS Standard (ABAQUS/Standard 2009), a Lagrangian code with implicit integration and the option of adaptive time-stepping. We adjusted time stepping such that the transient viscoelastic response is captured. Briefly, over the first three relaxation times (Equation 7), the time increments are non-linearly increased from <1% of the relaxation time to 50%. We use 4-node quadrilateral bilinear plane-strain elements (Fig. 4.1b) and a Newton solver. Remeshing was not employed here to avoid the problem of artificial numerical diffusion (i.e. smoothing of stresses and displacement fields due to interpolation). Monitoring energy, strains, and stresses within and near the inclusion, we found that our numerical solutions are acceptable up to a bulk shear strain $\gamma = 3$. γ is defined as $\tan(\psi)$ of the matrix square (Fig. 4.1a). This strain range proved sufficient to test our hypothesis.

4.2.3 Parameter space and scaling of experiments

As outlined above, the rheology of the inclusion and the matrix square is hyperelastoviscoplastic. For simplicity, elastic stiffness and yield stress are kept identical in

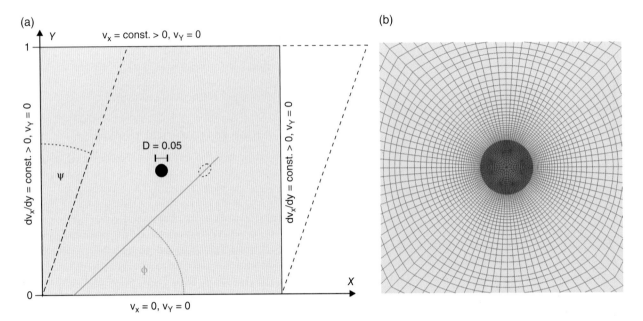

Fig. 4.1. (a) Coordinate system and set-up of our numerical experiment. The gray unit square with the circular black inclusion of diameter D denotes the initial state. The boundary conditions are noted next to the relevant sides of the square. The dashed parallelogram and dashed ellipsoidal inclusion are an example of a deformed state. The long axis orientation of the inclusion is indicated by the green line and quantified by the angle ϕ relative to the shear plane. ψ is the finite shear angle. (b) Magnified view of the finite-element mesh in the model center (the inclusion is green).

inclusion and matrix. We only vary the Newtonian viscosity of the inclusion relative to the matrix over four orders of magnitude, considering both weak and strong inclusions. For an overview of classical geological examples, the reader is referred to the comprehensive review of Marques et al. (2014) and the references therein. The effects of differences in elastic and plastic properties will be examined elsewhere. In the following, we outline our strategy for exploring the geologically relevant parameter space efficiently for the vast range of rock viscosities, yield stresses, and strain rates encountered in nature.

4.2.3.1   Elastic properties

For most minerals, the shear modulus G varies over one order of magnitude, from 10^{10} to 10^{11} Pa (e.g. Bass 1995). Since the variation of viscous and plastic properties extends over several orders of magnitude, we regard the variation in G negligible and therefore choose a fixed value of $G = 10^{11}$ Pa in the following. Poisson's ratio is fixed at 0.25, a typical value for felsic continental crust (Zandt and Ammon 1995). However, the model is isochoric, by definition of simple shear.

4.2.3.2   Viscosity

Laboratory experiments and geophysical observations suggest (for a recent review, see Bürgmann and Dresen 2008) that the effective viscosity of the lithosphere during long-term deformation processes is in the range of 10^{18} to 10^{27} Pa s. This large variability is attributed to the strong sensitivity of viscosity to temperature, water fugacity, pressure (in the high-pressure regime), strain rate, rock composition, and grain size.

4.2.3.3   Time dependence

Strain rates for geological long-term deformation are assumed to vary between 10^{-17} s^{-1} to 10^{-10} s^{-1}. When viscous and elastic rheologies are combined, it depends on the timescale of deformation if the elastic or viscous response dominates (e.g. Bailey 2006; Kaus and Becker 2007). A common timescale for the transition between these regimes is the Maxwell relaxation time t_R:

$$t_R = \frac{\eta}{G}. \qquad (7)$$

Loading processes on timescales well below t_R are mainly elastic, while those on larger timescales are controlled by the viscous material properties. Using the previously mentioned values, relaxation times in lithospheric materials range from 10^7 s (~0.3 a) to 10^{16} s (~317 Ma). A convenient dimensionless number for describing the degree of non-linearity of the rheological response of a viscoelastic material is the Weissenberg number W (e.g. Dealy 2010):

$$W = t_R \dot{\gamma} \qquad (8)$$

where $\dot{\gamma}$ is the shear strain rate. This concept is illustrated in Fig. 4.2, which shows a plot of normalized shear stress over shear strain for homogeneous simple

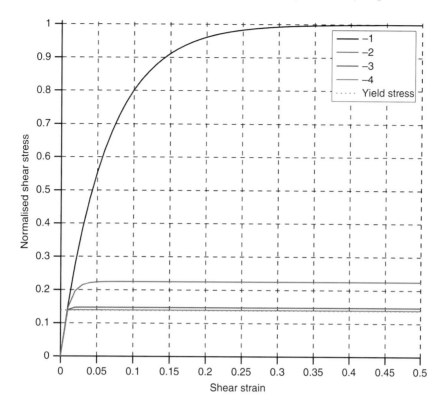

Fig. 4.2. Plot of normalized shear stress over shear strain γ for homogeneous simple shear with all four values of *W*. The legend indicates log10(*W*) for the respective curve colors. All stresses are normalized to the steady-state value of the curve for *W* = 10⁻¹. The dashed green line marks the normalized yield stress, which here assumes the maximum value considered in this study (one tenth of Frenkel's maximum theoretical shear strength (Frenkel 1926), see Section 4.2.3.4 and Equations 10 and 11). The vertical difference between the individual plateau values of the shear stress curves and the yield stress corresponds to the (normalized) steady-state viscous stress. Hence, the curve for *W* = 10⁻⁴ resembles a perfectly elastoplastic rheology, while the one for *W* = 10⁻¹ is very similar to a viscoelastic rheology. Note that the latter curve is strongly non-linear over the strain range of this diagram because five relaxation times correspond to γ = 0.5. The graph for *W* = 10⁻² has reached five relaxation times at γ = 0.05 and hence remains flat afterwards.

shear of the matrix without inclusion for four different *W*. A shear strain of 0.5 corresponds to five relaxation times for the model with *W* = 10⁻¹. As a result, it shows a non-linear viscoelastic stress-strain curve for most of this range. In contrast, a model with *W* = 10⁻⁴ has experienced five relaxation times at a shear strain of 5 × 10⁻⁴ and thus exhibits an effectively constant linear steady-state stress after yield. Hence, the stress–strain curve resembles that of a classical linear elastoplastic material. Considering the range of relaxation times and strain rates above, *W* varies between 10⁻¹⁰ and 10⁶ in nature. In thermal convection simulations for the upper mantle with a Maxwell rheology (Beuchert and Podladchikov 2010), the model lithosphere exhibits a *W* range of ~10⁻⁴ to 10¹. We restrict ourselves to a *W* range of 10⁻⁴ to 10⁻¹ because it denotes the most common loading regime expected for the ductile lithosphere. This helps to limit the parameter space we explore below. In our simple-shear problem, the steady-state viscous stress can hence be calculated as:

$$\sigma_V = \frac{WG}{\sqrt{3}} \tag{9}$$

and thus ranges from ~5.7 × 10⁶ Pa to 5.7 × 10⁹ Pa for the parameters adopted here.

4.2.3.4   Yield strength

Frenkel (1926) estimated the maximum theoretical yield strength of undamaged homogeneous solids as:

$$\hat{\sigma}_y = \frac{G}{2\pi} \tag{10}$$

Since natural materials are internally heterogeneous and damaged, their yield strength is a fraction of Frenkel's theoretical maximum, usually between a tenth and a ten-thousandth (Renshaw and Schulson 2004, 2007; Braeck and Podladchikov 2007). Therefore, the yield strength of rock should be in the range from ~1.6 × 10⁶ Pa to 1.6 × 10⁹ Pa. We thus parameterize the plastic yield strength in our model logarithmically as multiples of the Frenkel's maximum strength:

$$\sigma_y = K\hat{\sigma}_y$$
$$K \in \{10^{-1}, 10^{-2}, 10^{-3}, 10^{-4}\} \tag{11}$$

4.2.3.5   Stress ratio in the inelastic regime

In our rate-dependent overstress model, the equivalent stress is the sum of the yield and viscous stress (Equation 4, Fig. 4.2). Therefore, it is useful to consider the ratio of yield over viscous stress, σ_R, to identify parameter combinations, in which the inelastic stress is mainly plastic ($\sigma_R \gg 1$) or viscous ($\sigma_R \ll 1$). Equation 12 demonstrates that σ_R can be expressed as a function of K and W:

$$\sigma_R = \frac{\sigma_y}{\sigma_V} = \frac{K\sqrt{3}}{2\pi W} \Leftrightarrow \sigma_y = \frac{K\eta\dot{\gamma}}{2\pi W} \quad (12)$$

Using the ranges for K and W discussed previously, σ_R varies between ~2.8×10^{-4} and 2.8×10^{2}. Figure 4.3 shows the range of σ_R covered in this study as a function of K and W.

4.2.3.6   Choice of model parameters

The geologically relevant parameter space for our purely mechanical model can be explored in terms of the dimensionless numbers K and W. For computational convenience, we scale our models down. They will retain dynamical similarity with nature as long as σ_R is identical in model and nature (Buckingham 1914). We keep the model shear modulus fixed at 1000 Pa, and apply a constant velocity of 10^{-5} m s^{-1} to the upper boundary of the square. Hence, η is controlled by the choice of W (Equations 5 and 9). Since we fix the choices for K as well, σ_Y is given by Equation 12. Therefore, we can cover the geologically relevant parameter space for lithospheric deformation with 16 experiments, varying both K and W logarithmically over four orders of magnitude. We also

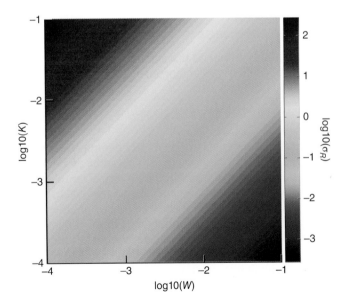

Fig. 4.3. The range σ_R of in log10(K–W) space. Color contours give the decadic logarithm of σ_R.

vary the viscosity ratio $\Delta\eta$ logarithmically over four orders of magnitude:

$$\Delta\eta = \frac{\eta_{inc}}{\eta_{mat}} \in \left\{10^{-2}, 10^{-1}, 10^{1}, 10^{2}\right\} \quad (13)$$

where the subscripts inc and mat denote inclusion and matrix, respectively. For smaller or larger viscosity ratios, the inclusion can be regarded as a hole or rigid, respectively. We focus on deformable inclusions. Hence, we conducted a total of 64 experiments, the results of which are summarized in Section 4.3.

4.2.4   Geometrical analysis of inclusion

We present two commonly employed parameters to delineate the deformation history of the inclusion: its aspect ratio (AR), defined as the quotient of the length of its long axis over the length of its short axis, and the inclination of the inclusion long axis with respect to the shear plane, measured by the angle ϕ (reported in degrees here, see Fig. 4.1a). To measure these properties, images of the deformed inclusion were exported for each time step for each experiment. The images were analyzed with the image processing toolbox of the commercial software package MATLAB® (The MathWorks 2011). We used the "regionprops" command to calculate the best-fit ellipse for each inclusion, which permits straightforward calculation of AR and ϕ.

4.3   RESULTS

4.3.1   Inclusion aspect ratio

Let us consider the AR evolution of weak inclusions first (Fig. 4.4). The AR evolution for both viscosity ratios (10^{-1} and 10^{-2}) is quite similar in the strain range studied here (solid black and red curves). At $\gamma = 3$, their difference is at maximum (for any given K–W combination) and of order 10%, with weaker inclusions ($\Delta\eta = 10^{-2}$) deforming more readily. However, the AR–γ curves are non-linear, and hence it can be expected that the difference increase for larger strains, at least for cases with $\sigma_R \leq 1$. For $W \leq 10^{-2}$, AR evolution noticeably depends on K. For $K = 10^{-1}$ and $W = 10^{-4}$, the inclusion AR evolution is effectively identical to that of the strain ellipse for homogeneous simple shear (see also Fig. 4.5). A decrease in K results in enhanced deformation of the inclusion, relative to the matrix. The strongest influence of yield stress is seen for the lowest W, where the final AR varies by a factor of ~2 over the full K range. For $W = 10^{-1}$, the choice of K has a negligible influence on AR. AR evolution also depends on W, most notably for $K \geq 10^{-2}$. For higher W, the inclusion deforms faster. Final AR increases by up to a factor of ~2 with increasing W for $K = 10^{-1}$. At $K = 10^{-3}$, this increase is restricted to a maximum factor of ~1.4. For $K = 10^{-4}$, the choice of W has no appreciable effect on AR

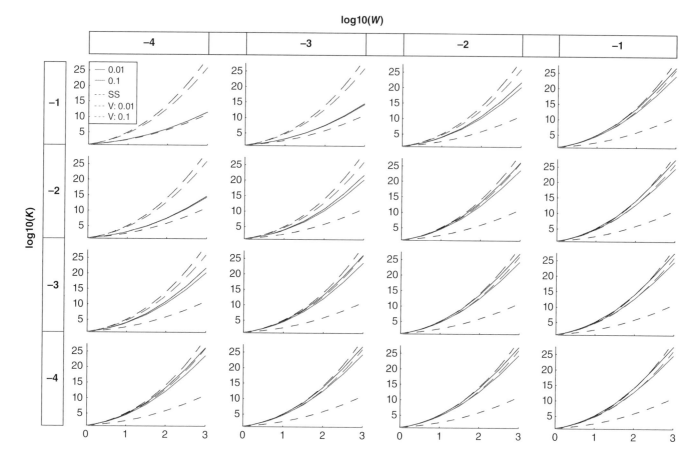

Fig. 4.4. Plots of aspect ratio (AR, vertical axes) of inclusion over γ (horizontal axes) for weak inclusions (solid lines) in log10(K–W) space. The respective viscosity ratios are color-coded in the legend. The dashed blue lines represent the aspect ratio of the finite-strain ellipsoid for homogeneous simple shear of the matrix. The remaining dashed lines display the AR evolution for a Newtonian rheology with the respective viscosity ratios, calculated numerically with the instantaneous analytical solutions of Bilby and Kolbuszewski (1977).

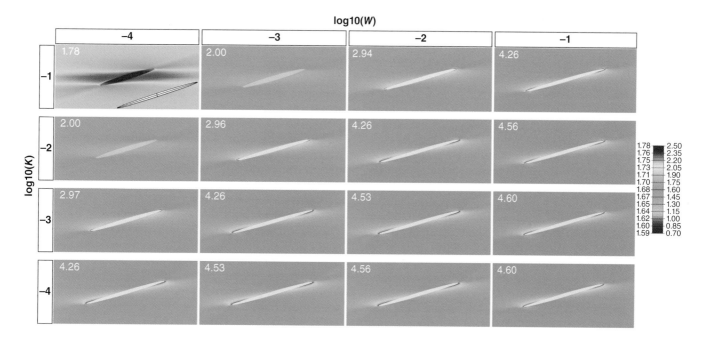

Fig. 4.5. Maps of finite equivalent viscoplastic strain of inclusions with $\Delta\eta = 10^{-1}$ at γ = 3 in log10(K–W) space. The color bar on the right has two numeric scales. The left one corresponds to the map in the upper-left corner only. This map has its own scale because it hardly deviates from homogeneous strain (maximum relative difference less than 5%). It also shows a plot of the finite strain ellipse for the corresponding purely viscous case for comparison (Bilby and Kolbuszewski 1977, size of ellipse not to scale). White colors not shown in the color bar denote the maximum plastic strain attained in the weak inclusion. Its respective numeric value is given in each map (white font in upper-left corner).

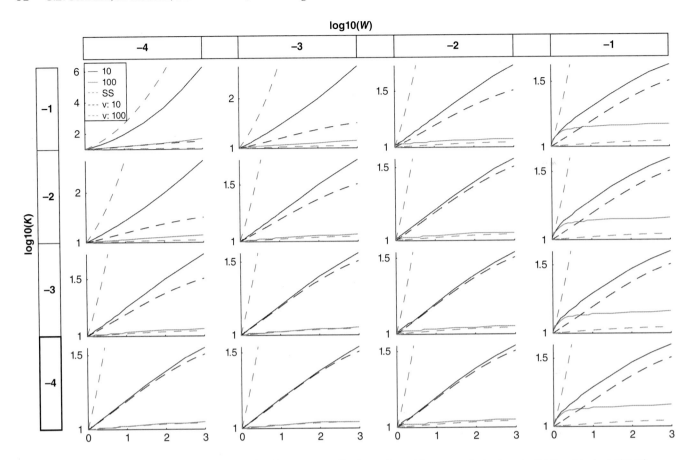

Fig. 4.6. Plots of aspect ratio (AR, vertical axes) of inclusion over γ (horizontal axes) for strong inclusions (solid lines) in log10(K–W) space. The respective viscosity ratios are color-coded in the legend. The dashed lines again denote homogeneous simple shear (blue) and the evolution of viscous inclusions in a viscous matrix (Bilby and Kolbuszewski 1977).

evolution. The inclusions generally deform more slowly than in the purely viscous case (dashed black and red curves), most notably so at $\sigma_R \geq 1$ (see also Fig. 4.5). For $W < 10^{-1}$, σ_R governs the AR evolution. This can be seen by comparing plots in the experimental matrix of Fig. 4.4, which lie along diagonals from the lower left to the upper right. In doing so, one moves along lines of constant σ_R (for example, the case $K = 10^{-2}$ with $W = 10^{-4}$ has the same σ_R as $K = 10^{-1}$ and $W = 10^{-3}$). Results with common σ_R appear almost identical. This is perhaps illustrated more clearly in Fig. 4.5. It displays the finite inelastic strain field for the case of $\Delta\eta = 10^{-1}$ at $\gamma = 3$ for the entire K–W range. Maps with equal σ_R show almost identical strain fields and strain maxima. The strain maximum decreases with increasing σ_R. Highly localized matrix strains are restricted to fairly small areas around the tips of the inclusions and decay to the value of the background strain over short distances.

Let us now consider the strong inclusions (Fig. 4.6, solid black and red curves). As expected, they deform less than the matrix (dashed blue curves), but generally more than in the purely viscous case (dashed black and red curves). Comparing the results for both viscosity ratios, one notices clear differences. The inclusion with $\Delta\eta = 10$ shows a final AR larger than that of the stiffer

one by a factor of ~1.3 for $W = 10^{-1}$ up to a factor of ~4 at $W = 10^{-4}$ and $K = 10^{-1}$. For the latter parameter combination, both inclusions attain their maximum AR: ~1.7 for $\Delta\eta = 100$, and ~6.3 for $\Delta\eta = 10$. For all other cases, the strongest inclusion shows a final AR < 1.2 while the weaker one displays final AR between 1.5 and 3. The difference in AR compared to the respective viscous cases increases with increasing σ_R, with the exception of all cases with $W = 10^{-1}$: there, the choice of K does not affect the results greatly, and an interesting non-linearity in the AR-γ curves is observed at $\gamma < 1$, in particular for the strongest inclusions. They exhibit a larger rate of deformation in this strain range, which decreases noticeably at $\gamma > 0.3$. The curves for the strongest inclusions exhibit slight undulations. Since AR is quite small for $\Delta\eta = 100$, it is possible that these undulations are a result of the imperfect discretization of a spherical object with quadrilateral finite elements. Moreover, plots with equal σ_R are almost identical for $W < 10^{-1}$, as in the case of weak inclusions. However, maps of finite strain, for the inclusions with $\Delta\eta = 10$ as a representative example (Fig. 4.7), exhibit additional differences to the weak case. Strain maxima are now located in the matrix and increase with growing σ_R. In addition, quite pronounced zones of highly localized strains form. Especially in cases with

log10(*W*)

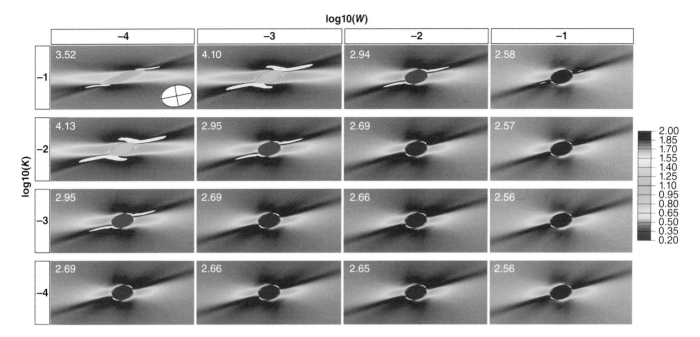

Fig. 4.7. Maps of finite equivalent viscoplastic strain of inclusions with $\Delta\eta = 10$ at $\gamma = 3$ in log10(K–W) space. The map in the upper-left corner also shows a plot of the finite strain ellipse for the corresponding purely viscous case (Bilby and Kolbuszewski 1977, size of ellipse not to scale). White colors denote areas where the plastic strain in the matrix exceeded 2, the maximum of the color bar. The maximum strain value attained in the matrix is given in each map (white font in upper-left corner).

$\sigma_R \geq 1$, these high-strain zones perhaps resemble the tails of sigmoidal clasts (e.g., Trouw et al. 2010; Mukherjee 2011). Low-strain zones associated with the median plane of the inclusion also develop, the extent of which increases with growing σ_R. In summary, strong inclusions display the opposite behavior than weak ones: they deform more for increasing σ_R, and they exhibit quite pronounced strain localization.

4.3.2 Inclusion inclination

The evolution for inclusion inclination angle ϕ is presented in Figs 4.8 and 4.9. For the entire parameter space, the evolution of the weak inclusions (Fig. 4.8, solid black and red curves) hardly deviates from that of the finite strain ellipse for homogeneous simple shear (dashed blue curves) and the purely viscous cases (dashed black and red curves). The case of $W = 10^{-4}$ and $K = 10^{-1}$ matches homogeneous shear almost perfectly, while all other cases show slight differences of order 2% or less. The subtle divergence of the curves commences at $\gamma \approx 1$, with the inclusions exhibiting a smaller ϕ than the homogeneous and viscous solutions. Since the curves are non-linear, this difference may become more relevant at significantly larger strains (e.g. Treagus and Treagus 2001).

Let us consider the strong inclusions (Fig. 4.9). The inclination angle ϕ of the inclusions is generally larger than that of purely viscous ones at $\gamma > 1$. This difference in ϕ is most pronounced for inclusions with $\Delta\eta = 100$, where the maximum difference exceeds a factor of 10, compared to the corresponding viscous solution. Except

for $W = 10^{-1}$, this difference generally increases with increasing K. For $W = 10^{-1}$, K does not exert a noticeable effect on the ϕ evolution. For a given K, the evolution of ϕ depends on the choice of W, most obviously so in the case of $\Delta\eta = 100$. For $K \leq 10^{-3}$, ϕ at $\gamma = 3$ increases with increasing W, by a factor of order 10 for the strongest inclusions, and by a factor of ~1.3 for the weaker ones. For $K \geq 10^{-2}$, ϕ at $\gamma = 3$ decreases from $W = 10^{-4}$ to $W = 10^{-3}$ and then increases again for larger W.

For $W \leq 10^{-3}$ and $K \leq 10^{-2}$, the curves exhibit a small non-linearity for $\gamma \leq 0.5$. We note that this is an artifact of the image analysis routine employed to calculate ϕ. In the finite-element mesh, the initially circular inclusion is of course not a perfect circle, despite the high mesh resolution used here. In addition, the image analysis routine uses bitmaps of the deformed inclusions, which introduce another small discretization error due to their pixel nature. For the strongest inclusions, the deformation is so small in the low-strain domain that this error becomes noticeable. It becomes irrelevant at $\gamma \geq 0.5$.

4.4 DISCUSSION

4.4.1 Weak inclusions

The weak inclusions show a significant dependence of AR evolution on K for $W \leq 10^{-2}$. For $K = 10^{-1}$ and $W = 10^{-4}$, the inclusion AR evolution resembles that of the strain ellipse for homogeneous simple shear. This is because the contribution of the viscous stress to the total deviatoric

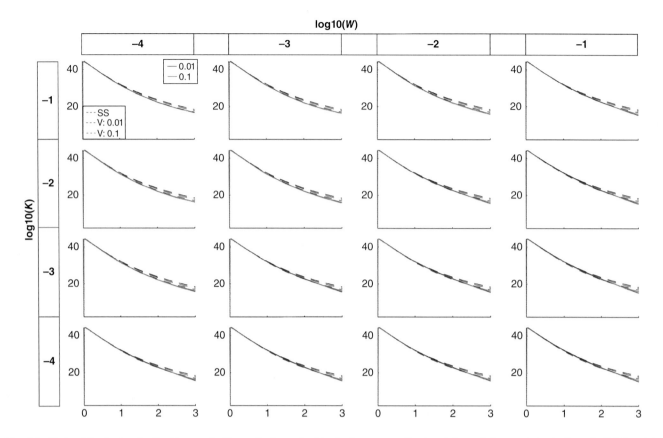

Fig. 4.8. Plots of inclusion inclination angle (ϕ, vertical axes) of inclusion over γ (horizontal axes) for weak inclusions (solid lines) in log10(K–W) space. The respective viscosity ratios are color-coded in the legend. The dashed lines represent homogeneous simple shear (blue) and the evolution of viscous inclusions in a viscous matrix, after Bilby and Kolbuszewski (1977). At $\gamma = 0$, $\phi = 0$ for the numerical inclusions. This is because the inclusion is circular at this stage, and the image analysis routine employed to analyze inclusion deformation assigns an arbitrary long-axis inclination of 0° for circular objects. This is also the case in Fig. 4.9.

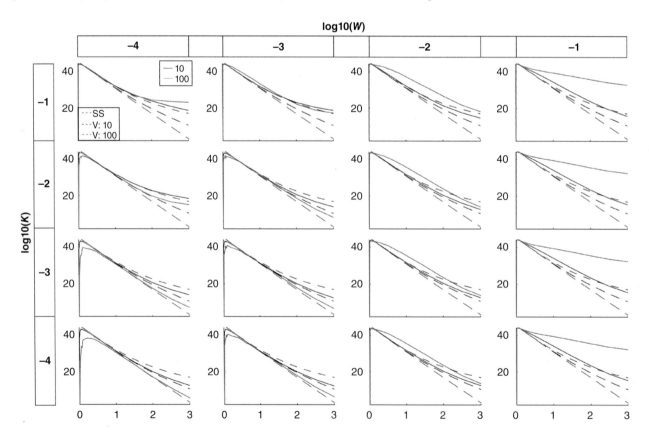

Fig. 4.9. Plots of inclusion inclination angle (ϕ, vertical axes) of inclusion over γ (horizontal axes) for strong inclusions (solid lines) in log10(K–W) space. The respective viscosity ratios are color-coded in the legend. The dashed lines represent homogeneous simple shear (blue) and the evolution of viscous inclusions in a viscous matrix, after Bilby and Kolbuszewski (1977).

stress is very small, i.e., $\sigma_R \approx 80$ in the matrix and one or two orders of magnitude larger in the inclusions (Fig. 4.2). Since elastic and plastic properties of matrix and inclusion are identical, there is no effective rheological difference between the two for this parameter combination. A decreasing K causes a growing increase of the viscous stress contribution, and hence the rheological difference between matrix and inclusion is amplified, as is the deformation of the inclusion relative to a homogeneous matrix. For $W = 10^{-1}$, the choice of K has a negligible influence on AR because $\sigma_R < 0.08$ (Fig. 4.3). The dependence of AR evolution on W for $K \geq 10^{-2}$ can be explained by similar arguments. For these high K values, increases in W result in a significant decrease of σ_R, which in turn leads to more deformable inclusions because the effective rheological difference with respect to the matrix grows. Hence, the shape evolution of weak inclusions can be summarized as follows: for decreasing σ_R, weak inclusions deform more readily; if hyperelasticity and -plasticity are considered, weak inclusions deform less than purely viscous ones – if the related elastoplastic properties are equal in inclusion and matrix. For large W and small K, the latter effect becomes less important over the strain range studied here because viscous stresses are most dominant in the post-yield domain. The evolution of orientation angle ϕ, however, differs only slightly from homogeneous and purely viscous simple shear. As mentioned earlier, for large σ_R, the weak inclusions exhibit no rheological difference compared to the matrix, which explains this similarity. For smaller σ_R, weak inclusions behave increasingly similar to purely viscous ones because elastic and plastic stress contributions become negligible. Mancktelow (2011) pointed out that inclusions tend to rotate like passive marker lines regardless of rheology once their aspect ratios become large. Our results reflect this notion.

4.4.2 Strong inclusions

In terms of aspect ratio, strong inclusions behave in the opposite way to weak ones. Since for lower σ_R (i.e. higher W and lower K) the contrast in viscosity becomes more apparent mechanically, the inclusions behave more stiffly and deform less in this parameter range than the homogeneous matrix. However, the inclusions generally deform faster than their purely viscous counterparts, in particular for lower W. This can also be explained with the fact that an increase in σ_R suppresses the rheological contrast and thus amplifies inclusion deformation compared to the viscous case. The effect of transient viscoelasticity probably explains the initial non-linearity of AR curves at $W = 10^{-1}$ (Fig. 4.6). The matrix deforming at this high W reaches its steady-state viscous response only after $\gamma = 0.5$ (Fig. 4.2). Inside the inclusions, W is even larger. Hence, elastic effects are important up to at least $\gamma = 0.5$. Since there is no contrast in elastic moduli between inclusions and matrix, and the viscous steady-state stress is not yet reached in this strain range, the effective rheological con-

trast is smaller and transient. Therefore, the inclusions deform more in tune with the homogeneous matrix and hence faster than purely viscous ones. However, once more than five matrix relaxation times have passed, the deformation rate slows down as the viscous contrast takes full effect. In addition, in the strongest inclusions, the viscous steady-state regime is never reached in the strain range covered here because their relaxation times exceed that of the matrix by one or two orders of magnitude (Fig. 4.10a, b). In summary, it seems an interesting, perhaps counterintuitive, result that higher K and very high W serve to amplify inclusion deformation rate relative to the purely viscous case by masking the rheological contrast. This is of course not the general case in nature because one can expect differences in elastic moduli and yield stress between inclusions and matrix as well. Nevertheless, this observation highlights that contributions of elastic and plastic stresses are likely quite important for inclusion deformation history in geological materials, in particular if their differences are of the same order as the differences in viscous stresses. This notion is also supported by the observed differences in inclusion inclination angle ϕ. For our rheology, forward rotation is always slower than in the purely viscous case, most notably so for the strongest inclusions and at the highest W and K, the former resulting in transient viscoelastic behavior, the latter suppressing the apparent rheological contrast (Figs 4.9 and 4.10). In addition, quite significant strain localization effects occur at high σ_R (Figs 4.7 and 4.10). Grain-scale models of stiff porphyroclasts (Griera et al. 2011) indicate that strain localization can inhibit inclusion rotation, and this effect may contribute to explaining the higher inclusion inclination angle observed for the strong inclusions at high K (Fig. 4.10c).

4.4.3 Geological relevance of covered parameter space

Given the range of material properties discussed in Section 4.2.3, one notices that equivalent stresses are of the order 10^9 to 10^{10} Pa for experiments with $K = 10^{-1}$ and/or $W = 10^{-1}$. Yet differential-stress estimates in the lithosphere are typically of order 10^6 to 10^8 Pa (e.g. England and McKenzie 1982; Townend and Zoback 2000; Lithgow-Bertelloni and Guynn 2004). Hence, although our chosen limits for material properties are based on quite well established theoretical and experimental results, it appears that the models applying the upper limits of K and W do not or, at best rarely, occur in nature. However, a growing body of literature on the concept of tectonic overpressure (Mancktelow 1993, 2008; Moulas et al. 2013; Lechmann et al. 2014; Schmalholz et al. 2014) suggests that equivalent stresses $\geq 10^9$ Pa might indeed be reached in contractional orogens. Similar evidence comes from grain-scale modeling of thermal-elastic deviatoric internal stresses (Schrank et al. 2012), suggesting that very high deviatoric stresses may be common at the grain scale. Hence, it cannot be ruled out that the high stress

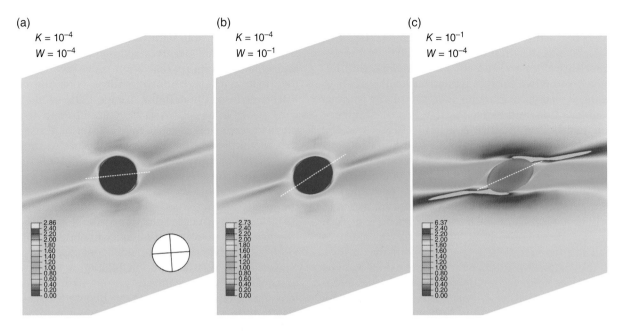

Fig. 4.10. Maps of finite equivalent viscoplastic strain of inclusions with $\Delta\eta = 100$ at $\gamma = 3$. The respective K and W values are given in the maps. The dashed white line in the center of each inclusion indicates its long-axis orientation. The white ellipse in the lower-right corner of map (a) is the result of the corresponding purely viscous case. Map (a) resembles the viscous case most closely because of the low K and W values. Map (b) denotes the result with the highest W (and lowest σ_R) and represents an effectively viscoelastic rheology, for which the inclusion remains in the transient non-linear regime (compare Fig. 4.2). The strain patterns for (a) and (b) are quite similar. Hence, plastic localization can be ruled out as explanation for the difference in inclusion deformation. Map (c) corresponds to a case with an effectively elastoplastic rheology (σ_R is at its maximum). Quite significant high-strain zones form.

magnitudes of models with $K = W = 10^{-1}$ occur in nature. Even if one disregarded cases with $K = 10^{-1}$ and $W = 10^{-1}$, the remaining experiments deviate from the purely viscous case for most of the covered parameter space. In addition, our results demonstrate that the dimensionless viscoplastic stress ratio σ_R mainly controls the deformation history of the inclusions for $W < 10^{-1}$ (Figs 4.4 to 4.9). Hence, experiments with the high plastic yield stress corresponding to $K = 10^{-1}$ can be used for extrapolation to parameter combinations not shown here. For example, the case of $K = 10^{-1}$ and $W = 10^{-4}$ is equivalent to $K = 10^{-2}$ and $W = 10^{-5}$.

4.4.4   Geological implications

Our simple-shear experiments demonstrate that hyperelasticity and -plasticity have significant effects on inclusion shape evolution for weak and strong inclusions over most of the parameter space explored here. For weak inclusions, orientation evolution is only affected weakly. However, the non-linearity of the curves implies that the observed differences may become more significant at larger strains. Deformed inclusions are commonly employed for determining finite strain, deformation path and/or viscosity contrasts from deformed rocks, in most cases using Newtonian or power-law viscous rheologies (e.g. Gay 1968; Dunnet 1969; Freeman 1987; Treagus 2002, 2003; Treagus and Treagus 2002; Fletcher 2004; Mancktelow 2011; Jiang 2013). Our results suggest that

the assumption of a purely viscous system should be assessed carefully for each natural case study to avoid potentially erroneous measures of finite strain and viscosity ratio. Therefore, it seems desirable to extend studies of deforming inclusions to more general rheologies. Our experiments also show that stiff inclusions deform in the transient viscoelastic regime for quite a large range of shear strains at high Weissenberg numbers ($W \geq 10^{-2}$). This implies that purely viscous models of strong clasts in lithospheric shear zones (e.g. Bons et al. 1997; Jessell et al. 2009) may neglect an important elastic contribution to the force and energy balance. Inclusion-rich shear zones may store more energy than previously thought.

We have presented an application of a new energetically consistent finite-strain approach for hyperelastoviscoplastic materials (Karrech et al. 2011c) and highlighted the important elastic and highly non-linear viscoplastic effects for the deformation of natural rock samples. The highly variable first-order effect of this rheology on the deformation behavior of inclusions may be used in the future to devise more comprehensive tools for the derivation of material properties from field observations. Inversely, if material properties are known, our findings can be used as a robust tool for kinematic interpretation, giving further insights into the fundamental behavior of inclusions in shear zones. In order to achieve these goals, we focus, in ongoing work, on the effects of varying elastic moduli and yield stress, the impact of a strain-softening rheology with damage mechanics, multi-inclusion systems (e.g. Dabrowski et al. 2012), different boundary conditions (pure and/or general shear (e.g. Treagus and Lan 2000, 2004), Neumann conditions), inclusion–matrix interface properties (Schmid and Podladchikov 2003), and rheologically relevant feedbacks such as shear heating (Fleitout and Froidevaux 1980; Regenauer-Lieb and Yuen 1998). Moreover, it will be worthwhile to explore the parameter space for viscosity ratios <10 more closely because Bilby and Kolbuszewski (1977) noted three characteristic regimes in AR–ϕ space for simple shear of linear viscous systems. Studies on natural (ten Grotenhuis et al. 2003; Mukherjee 2011) and modeled (ten Grotenhuis et al. 2002; Treagus and Lan 2003) mineral fish demonstrate the importance of (non-circular) shape for inclusion deformation, which highlights another research avenue.

ACKNOWLEDGMENTS

The authors gratefully acknowledge funding by the Australian Research Council, through Discovery project DP140103015. C.S. thanks Oliver Gaede for discussions. Soumyajit Mukherjee is thanked for his editorial handling and detailed review of this contribution. We gratefully acknowledge Albert Griera and Susanta Kumar Samanta for their insightful reviews. Support from Delia Sandford and Ian Francis (Wiley Blackwell) is appreciated.

REFERENCES

ABAQUS/Standard (2009). User's Manual, Version 6.9. (Hibbit, Karlsson, and Sorensen Inc., Pawtucket, Rhode Island, 2000).

Bailey RC. 2006. Large time step numerical modelling of the flow of Maxwell materials. Geophysical Journal International 164(2), 460–466.

Bass JD. 1995. Elasticity of Minerals, Glasses, and Melts. Mineral physics and crystallography: a handbook of physical constants. T.J. Ahrens, American Geophysical Union. AGU Reference shelf 2, pp. 45–63.

Beuchert MJ, Podladchikov YY. 2010. Viscoelastic mantle convection and lithospheric stresses. Geophysical Journal International 183(1), 35–63.

Bilby BA, Kolbuszewski ML. 1977. The finite deformation of an inhomogeneity in two-dimensional slow viscous incompressible flow. Proceedings of the Royal Society of London. A. Mathematical and Physical Sciences 355(1682), 335–353.

Bons PD, Barr TD, ten Brink CE. 1997. The development of δ-clasts in non-linear viscous materials: a numerical approach. Tectonophysics270(1–2): 29–41.

Braeck S, Podladchikov YY. 2007. Spontaneous thermal runaway as an ultimate failure mechanism of materials. Physical Review Letters 98(9), 095504.

Bruhns OT, Xiao H, Mayers, A. 1999. Self-consistent Eulerian rate type elasto-plasticity models based upon the logarithmic stress rate. International Journal of Plasticity 15(5), 479–520.

Buckingham E. 1914. On physically similar systems; illustrations of the use of dimensional equations. Physical Review 4(4), 345.

Bürgmann R, Dresen G. 2008. Rheology of the lower crust and upper mantle: Evidence from rock mechanics, geodesy, and field observations. Annual Review of Earth and Planetary Sciences 36, 531–567.

Dabrowski M, Schmid DW, Podladtchikov I. 2012. A two-phase composite in simple shear: effective mechanical anisotropy development and localization potential. Journal of Geophysical Research: Solid Earth 117(B8), B08406.

Dealy JM. 2010. Weissenberg and Deborah Numbers – their definition and use. Rheology Bulletin 79(2), 28.

Dewar RC. 2005. 4 Maximum entropy production and non-equilibrium statistical mechanics. In Non-equilibrium Thermodynamics and the Production of Entropy, edited by A. Kleidon and R. Lorenz, Springer, Berlin, Heidelberg, pp. 41–55.

Dunnet D. 1969. A technique of finite strain analysis using elliptical particles. Tectonophysics 7(2), 117–136.

England P, McKenzie D. 1982. A thin viscous sheet model for continental deformation. Geophysical Journal of the Royal Astronomical Society 70(2), 295–321.

Fleitout L, Froidevaux C. 1980. Thermal and mechanical evolution of shear zones. Journal of Structural Geology 2(1–2), 159–164.

Fletcher RC. 2004. Anisotropic viscosity of a dispersion of aligned elliptical cylindrical clasts in viscous matrix. Journal of Structural Geology 26(11), 1977–1987.

Freeman B. 1987. The behaviour of deformable ellipsoidal particles in three-dimensional slow flows: implications for geological strain analysis. Tectonophysics 132(4), 297–309.

Frenkel J. 1926. Zur Theorie der Elastizitätsgrenze und der Festigkeit kristallinischer Körper. Zeitschrift für Physik A Hadrons and Nuclei 37(7), 572–609.

Gay NC. 1968. Pure shear and simple shear deformation of inhomogeneous viscous fluids. 1. Theory. Tectonophysics 5(3), 211–234.

Griera A, Bons PD, Jessell MW, Lebensohn RA, Evans L, Gomez-Rias E. 2011. Strain localization and porphyroclast rotation. Geology 39(3), 275–278.

Griera A, Llorens M-G, Gomez-Rivas E, Bons PD, Jessell MW, Evans LA. 2013. Numerical modelling of porphyroclast and porphyroblast rotation in anisotropic rocks. Tectonophysics 587(0), 4–29.

Hill R. 1968. On constitutive inequalities for simple materials—I. Journal of the Mechanics and Physics of Solids 16(4), 229–242.

Hobbs BE, Mühlhaus HB, Ord A. 1990. Instability, softening and localization of deformation. In Deformation Mechanisms, Rheology and Tectonics, edited by R.J. Knipe and E.A. Rutter, The Geological Society of London, London, pp. 143–165.

Houlsby GT, Puzrin AM. 2007. Principles of Hyperplasticity, An Approach to Plasticity Theory Based on Thermodynamic Principles. Springer, London.

Jaumann G. 1911. Geschlossenes System physikalischer und chemischer Differential-Gesetze. Sitzberichte der Akademie der Wissenschaften Wien, Abteilung IIa 120, 385–530.

Jeffery GB. 1922. The motion of ellipsoidal particles immersed in a viscous fluid. Proceedings of the Royal Society of London. Series A 102(715), 161–179.

Jessell MW, Bons PD, et al. 2009. A tale of two viscosities. Journal of Structural Geology 31(7), 719–736.

Jiang D. 2013. The motion of deformable ellipsoids in power-law viscous materials: Formulation and numerical implementation of a micromechanical approach applicable to flow partitioning and heterogeneous deformation in Earth's lithosphere. Journal of Structural Geology 50(0), 22–34.

Karrech A, Regenauer-Lieb K, Poulet T. 2011a. Continuum damage mechanics for the lithosphere. Journal of Geophysical Research 116(B4), B04205.

Karrech A, Regenauer-Lieb K, Poulet T. 2011b. A damaged visco-plasticity model for pressure and temperature sensitive geomaterials. International Journal of Engineering Science 49(10), 1141–1150.

Karrech A, Regenauer-Lieb K, Poulet T. 2011c. Frame indifferent elastoplasticity of frictional materials at finite strain. International Journal of Solids and Structures 48(3–4), 397–407.

Kaus BJP, Becker TW. 2007. Effects of elasticity on the Rayleigh–Taylor instability: implications for large-scale geodynamics. Geophysical Journal International 168(2), 843–862.

Kaus BJP, Podladchikov YY. 2006. Initiation of localized shear zones in viscoelastoplastic rocks. Journal of Geophysical Research 111, B4, B04412.

Lechmann SM, Schmalholz SM, Hetényi G, May DA, Kaus BJP. 2014. Quantifying the impact of mechanical layering and underthrusting on the dynamics of the modern India-Asia collisional system with 3-D numerical models. Journal of Geophysical Research: Solid Earth 119(1), 616–644.

Lithgow-Bertelloni C, Guynn JH. 2004. Origin of the lithospheric stress field. Journal of Geophysical Research: Solid Earth 109(B1), B01408.

Mancktelow NS. 1993. Tectonic overpressure in competent mafic layers and the development of isolated eclogites. Journal of Metamorphic Geology 11(6), 801–812.

Mancktelow NS. 2008. Tectonic pressure: Theoretical concepts and modelled examples. Lithos 103(1–2), 149–177.

Mancktelow NS. 2011. Deformation of an elliptical inclusion in two-dimensional incompressible power-law viscous flow. Journal of Structural Geology 33(9), 1378–1393.

Marques FO, Mandal N, Taborda R, Antunes J, Bose S. 2014. The behaviour of deformable and non-deformable inclusions in viscous flow. Earth-Science Reviews 134(0), 16–69.

Moresi L, Dufour F, Mühlhaus H. 2002. Mantle convection modeling with viscoelastic/brittle lithosphere: numerical methodology and plate tectonic modeling. Pure and Applied Geophysics 159(10), 2335–2356.

Moulas E, Podladchikov YY, Aranovich LY, Kostopolous D. 2013. The problem of depth in geology: When pressure does not translate into depth. Petrology 21(6), 527–538.

Mühlhaus H-B, Regenauer-Lieb K. 2005. Towards a self-consistent plate mantle model that includes elasticity: simple benchmarks and application to basic modes of convection. Geophysical Journal International 163(2), 788–800.

Mukherjee S. 2011. Mineral fish: their morphological classification, usefulness as shear sense indicators and genesis. International Journal of Earth Sciences : Geologische Rundschau 100(6), 1303–1314.

Nadai A. 1937. Plastic behavior of metals in the strain-hardening range. Part I. Journal of Applied Physics 8(3), 205.

Poulet T, Regenauer-Lieb K, Karrech A. 2009. A unified multi-scale thermodynamical framework for coupling geomechanical and chemical simulations. Tectonophysics 483(1–2), 178–189.

Regenauer-Lieb K, Karrech A, Chua HT, Horowitz FG, Yuen D. 2010. Time-dependent, irreversible entropy production and geodynamics. Philosophical Transactions of the Royal Society A: Mathematical, Physical and Engineering Sciences 368(1910), 285–300.

Regenauer-Lieb K, Weinberg RF, Rosenbaum G. 2006. The effect of energy feedbacks on continental strength. Nature 442(7098), 67–70.

Regenauer-Lieb K, Weinberg RF, Rosenbaum G. 2011. The role of elastic stored energy in controlling the long term rheological behaviour of the lithosphere. Journal of Geodynamics 55, 66–75.

Regenauer-Lieb K, Yuen DA. 1998. Rapid conversion of elastic energy into plastic shear heating during incipient necking of the lithosphere. Geophysical Research Letters 25(14), 2737–2740.

Renshaw CE, Schulson EM. 2004. Plastic faulting: Brittle-like failure under high confinement. Journal of Geophysical Research 109(B9), B09207.

Renshaw CE, Schulson EM. 2007. Limits on rock strength under high confinement. Earth and Planetary Science Letters 258(1–2), 307–314.

Schmalholz SM, Kaus BJP, Burg JP. 2009. Stress-strength relationship in the lithosphere during continental collision. Geology 37(9), 775–778.

Schmalholz SM, Medvedev S, Lechmann SM, Podladchikov Y. 2014. Relationship between tectonic overpressure, deviatoric stress, driving force, isostasy and gravitational potential energy. Geophysical Journal International 197(2), 680–696.

Schmid DW, Podladchikov YY. 2003. Analytical solutions for deformable elliptical inclusions in general shear. Geophysical Journal International 155(1), 269–288.

Schrank CE, Fusseis F, Karrech A, Regenauer-Lieb K. 2012. Thermal-elastic stresses and the criticality of the continental crust. Geochemistry, Geophysics, and Geosystems, 13, Q09005.

Taborda R, Antunes J, Marques FO. 2004. 2-D rotation behavior of a rigid ellipse in confined viscous simple shear: numerical experiments using FEM. Tectonophysics 379(1–4), 127–137.

ten Grotenhuis SM, Saskia M, Passchier CW, Bons PD. 2002. The influence of strain localisation on the rotation behaviour of rigid objects in experimental shear zones. Journal of Structural Geology 24(3), 485–499.

ten Grotenhuis SM, Trouw RAJ, Passchier CW. 2003. Evolution of mica fish in mylonitic rocks. Tectonophysics 372(1–2): 1–21.

The MathWorks. 2011. MATLAB 7.12 and Image Processing Toolbox. The MathWorks, Inc., Natick, MA.

Townend J, Zoback MD. 2000. How faulting keeps the crust strong. Geology 28(5), 399–402.

Treagus SH. 2002. Modelling the bulk viscosity of two-phase mixtures in terms of clast shape. Journal of Structural Geology 24(1), 57–76.

Treagus SH. 2003. Viscous anisotropy of two-phase composites, and applications to rocks and structures. Tectonophysics 372(3–4), 121–133.

Treagus SH, Lan L. 2000. Pure shear deformation of square objects, and applications to geological strain analysis. Journal of Structural Geology 22(1), 105–122.

Treagus SH, Lan L. 2003. Simple shear of deformable square objects. Journal of Structural Geology 25(12), 1993–2003.

Treagus SH, Lan L. 2004. Deformation of square objects and boudins. Journal of Structural Geology 26(8), 1361–1376.

Treagus SH, Treagus JE. 2001. Effects of object ellipticity on strain, and implications for clast–matrix rocks. Journal of Structural Geology 23(4), 601–608.

Treagus SH, Treagus JE. 2002. Studies of strain and rheology of conglomerates. Journal of Structural Geology 24(10), 1541–1567.

Trouw RA, Passchier CW, Siersma D. 2010. Atlas of Mylonites – and related microstructures, Springer, Berlin.

Veveakis E, Regenauer-Lieb K. 2014. The fluid dynamics of solid mechanical shear zones. Pure and Applied Geophysics, 171(11), 3159–3174.

Xiao H, Bruhns O, Meyers A. 1997. Hypo-elasticity model based upon the logarithmic stress rate. Journal of Elasticity 47(1), 51–68.

Xiao H, Bruhns OT, Meyers A. 1998. On objective corotational rates and their defining spin tensors. International Journal of Solids and Structures 35(30), 4001–4014.

Zandt G, Ammon CJ. 1995. Continental crust composition constrained by measurements of crustal Poisson's ratio. Nature 374(6518), 152–154.

Zaremba S. 1903. Sur une forme perfectionnée de la théorie de la relaxation. Bulletin International de l'Académie des Sciences de Cracovie, Classe des Sciences Mathématiques et Naturelles, 594–614.

Zhilin PA, Altenbach H, Ivanova EA, Krivtsov AM. 2013. Material strain tensor. generalized continua as models for materials. In Mechanics of Generalized Continua, edited by H. Altenbach, S. Forest, and A. Krivtsov, Springer, Berlin Heidelberg, vol 22, pp. 321–331.

Chapter 5

Biviscous horizontal simple shear zones of concentric arcs (Taylor–Couette flow) with incompressible Newtonian rheology

SOUMYAJIT MUKHERJEE[1] and RAKESH BISWAS[2]

[1] *Department of Earth Sciences, Indian Institute of Technology Bombay, Powai, Mumbai 400076, Maharashtra, India*
[2] *Geodata Processing and Interpretation Centre, Oil and Natural Gas Corporation Limited, Dehradun, India*

5.1 INTRODUCTION

Fluid caught between rotating cylinders has been intriguing physicists for over 300 years...

R.J. Donnelly (1991)

Ductile shear zones have so far been modeled mainly as zones of single lithology and with straight parallel and rigid boundaries (Ramsay 1980). Following this, thermal models of ductile shear zones were also provided (Fleitout and Froideavaux 1980). However, (i) natural shear zones can have curved boundaries in regional-scale, and (ii) may consist of more than one lithology. For example, crustal cross-sections of collisional orogens deduced from geophysical studies reveal shear zones with curved boundaries (Beaumont et al. 2001 and references therein). On the other hand, pronounced ductile shear segregates specific mineral assemblages for polymineralic rocks into zones with their interfaces parallel to the shear zone boundaries (Druguet et al. 2009). Layered shear zones have been reported/studied in granulite facies rocks (Ji et al. 1997), in models with ice (Wilson et al. 2003), from collisional terrains (Mukherjee and Koyi 2010), and in granular materials (Börzsönyi et al. 2009), besides most common cases of micaceous minerals alternating with quartzofeldspathic minerals in mylonites (Lister and Snoke 1984). Those two natural cases (i) and (ii) have recently been modeled individually (Mukherjee and Biswas 2014; Mulchrone and Mukherjee, in press) to deduce velocity profiles and shear senses. This work considers the two cases together to deduce and interpret velocity profiles of biviscous curved ductile simple shear zones. We do not address here shear zone related folds (see Mukherjee et al. 2016, Chapter 12).

5.2 THE MODEL

We use the Taylor–Couette flow model (Taylor 1923) to explain the kinematics of biviscous curved shear zone, as follows. Consider a ductile shear zone with concentric circular boundaries of radii R_1 and R_2 $(R_1 > R_2)$ with two immiscible incompressible Newtonian viscous fluids within: an outer layer of fluid A with a viscosity μ_a, and an inner fluid layer B with a viscosity μ_b $(\mu_a > \mu_b)$. Their interface is a circle of radius R_b. The inner boundary rotates clockwise with an angular velocity ω and the outer boundary remains static. Such flow in fluid mechanics has been known as Taylor–Couette flow/circular Couette shear, etc. for a long time (Donnelly, 1991), both for rotation of two boundaries and one of the boundaries, and for single and two fluids (Schulz et al. 2003). Even if one considers the two fluids (in geology, "ductile lithologies") were mixed, upon circular shear they segregate with lighter fluid near the core and the denser fluid near the periphery (Baier 1999; Vedantam et al. 2006). Taylor–Couette flow has already been classified in fluid mechanics into three types: (i) homogeneous dispersion, (ii) banded dispersion, and (iii) segregated/stratified flow. We discuss here a kind of stratified flow.

The velocity distributions in both the layers obey the velocity equation:

$$v_\theta = (C_1/2)r + C_2(1/r) \tag{1}$$

(from eqn 15.38 of Williams and Elder 1989)

Velocity equation for
fluid A : $v_\theta^a = (C_1^a/2)r + C_2^a(1/r)$ (2)

That for fluid B is : $v_b^\theta = (C_1^b/2)r + C_2^b(1/r)$ (3)

Here v_θ is azimuthal velocity; θ is meridional angle (Fig. 5.1a); C_1^a, C_2^a, C_1^b, and C_2^b are integration constants.

$$\text{At } r = R_1, v_\theta^a = 0 \tag{4}$$

And at

$$r = R_2, v_\theta^b = R_2\omega \tag{5}$$

Ductile Shear Zones: From Micro- to Macro-scales, First Edition. Edited by Soumyajit Mukherjee and Kieran F. Mulchrone.

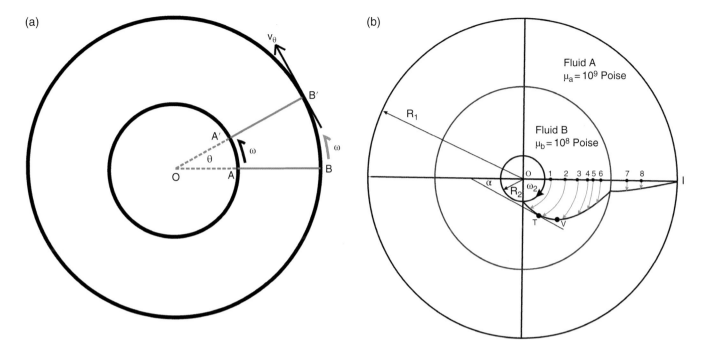

Fig. 5.1. (a) Angular velocity ω acts on the two concentric circular boundaries of a curved horizontal shear zone. A marker AB turns A′B′. The meridional angle (θ) and the azimuthal velocity (v_θ) are shown. Source: Mukherjee & Biswas, 2014. Reproduced with permission from Springer Science + Business Media. (b) Velocity profiles for Taylor–Couette flow with two Newtonian fluids "A" and "B" within two concentric circular boundaries. The red circle marks the interface between the two fluids. The inner boundary rotates clockwise. The outer boundary is static. Here R_1 = 100 cm., R_2 = 50 cm, ω = 2° hr^{-1}, μ_a = 10⁹ Poise and μ_b = 10⁸ Poise.

At the interface, $r = R_b$, the two additional conditions are as follows. (i) The two fluids stick together:

$$v_\theta^a = v_\theta^b \tag{6}$$

(ii) The momentum transfer through the interface is continuous:

$$\tau_{r\theta}^a = \tau_{r\theta}^b \text{ at } r = R_b \tag{7}$$

Or,

$$\mu_a r \frac{\mathrm{d}}{\mathrm{d}r}\left(\frac{v_\theta^a}{r}\right) = \mu_b r \frac{\mathrm{d}}{\mathrm{d}r}\left(\frac{v_\theta^b}{r}\right) \tag{8}$$

Using Equations 2 to 8 and after doing some algebra, the velocity equations for fluids A and B are:

$$v_\theta^a = \left\{\left(\mu_b \omega R_b^2 R_2^2 R_1^2\right) / \left[\mu_b\left(R_1^2 R_2^2 - R_b^2 R_2^2\right) - \mu_a\left(R_1^2 R_2^2 - R_b^2 R_1^2\right)\right]\right\} \times \left(\frac{1}{r} - \frac{r}{R_1^2}\right) \tag{9}$$

$$v_\theta^b = \omega r + \left\{\left(\mu_a \omega R_b^2 R_2^2 R_1^2\right) / \left[\mu_b\left(R_1^2 R_2^2 - R_b^2 R_2^2\right) - \mu_a\left(R_1^2 R_2^2 - R_b^2 R_1^2\right)\right]\right\} \times \left(\frac{1}{r} - \frac{r}{R_2^2}\right) \tag{10}$$

Notice that the velocity equations depend on the viscosity of both the fluids. Starting from line OI, velocity profiles developed in the two fluids are shown in Fig. 5.1b. Flow paths of both the fluids are segments of circles that are concentric with the circular boundaries of the shear zone. Angular shear at some particular moment can be measured at any point on the profile by drawing a tangent at that point and finding the angle between that tangent and the line OI (Fig. 5.1b). The point of highest curvature on the velocity profile is shown as 'V' in Fig. 1b, which is also the point of highest speed induced by ductile shear of the curved inner boundary. Reverse ductile shear senses develop simultaneously across point 'V'. In detail: from the outer boundary of the circular shear zone up to point 'V', a shear sense same as that produced by the rotating inner boundary is produced. From 'V' up to the inner boundary, an opposite ductile shear sense develops. The point of intersection between the velocity profile and the line OI, point 'I', is called the "neutral point". It is the unique static point inside the shear zone. A circle concentric with the shear zone boundaries and passing through the neutral point is called the "neutral curve" (Mukherjee and Biswas 2014). Material points on the neutral curve remain stationary during ductile shear. In the present case, the neutral curve coincides with the static outer boundary of the shear zone. Note that the term "neutral curve" has been used here in a different context than that used by Fossen and Rykkelid (1992), Peng and Zhu (2010), and Ovchinnikova (2012). Had there been rotation of the outer curved boundary of the shear zone in a direction opposite to that of the inner boundary (that is, anticlockwise), the neutral point would have plotted inside the shear zone. Note that we can at best decipher relative shear movements in shear zones, and not the absolute movements of its boundaries. Therefore, locating neutral point in real shear zones seems not possible even though it is discussed here. Taylor–Couette flow apparatus has been in use in structural geological analogue models to simulate high-strain ductile shear (e.g. Bons and Jessell 1999).

Shear strain (tanα) at any point 'T' on the profile (Fig. 5.1b) can be obtained from the angle α between the tangent at 'T' on the profile and the line AI. Figure 5.2 shows how shear strain varies inside the model shear zone measured for eight points with initial positions '1' to '8' shown in Fig. 5.1b. Shear strain is minimum at 'V' and increases away from it in both directions. Higher shear strain attains within Fluid B- at few locations than in Fluid A layer: compare the plots in Fig. 5.2 for points '7' and '8' with points '1', '2', '4', '5', and '6'.

Five circular markers of equal radius before deformation (Fig. 5.3), considered in both the fluid layers on ductile shear, become irregular shaped, indicating their non-homogeneous deformation. Figure 5.4 shows the temporal evolution of aspect ratios of these markers at four instances. In general, aspect ratios increase temporally. It can also decrease since points 'y' and 'z' (inset in Fig. 5.3), that were increasing distance between them during deformation, can also start decreasing. This can be understood from the green dots showing evolution of marker 'd' in Fig. 5.4. The inset in Fig. 5.4 defines the aspect ratio tentatively from irregular objects. Marker 'c'

was positioned deliberately partly in one fluid layer and partly in the other. Fluid 'B' undergoes more shear than that of fluid 'A', as can be visually appreciated more from marker 'c'. The reasons are, first: only the (inner) boundary of the curved zone shears. Second, fluid B is less viscous than fluid A. This is also corroborated from shear strains within these layers (Fig. 5.2). Biviscous Taylor–Couette flows may develop instability at the contact between the two fluid layers (Gelfgat et al. 2004). This manifest as warping of the interface. The present study did not consider development of such an instability. Andereck et al. (1986) pointed out that Taylor–Couette flow kinematics depends on aspect ratio and radius ratio of the region where the fluid is kept, and on the Reynold's number of the fluid. Taylor–Couette flow has been studied in fluid mechanics for a single rotating cylinder and for rotation of both cylinders (White 2005). Layered Taylor–Couette flow of non-Newtonian fluids for eccentric/non-axisymmetric cylinder are already available in fluid mechanics, such as Escudier et al. (2002). We are working to adopt them in ductile shear zone studies in structural geology.

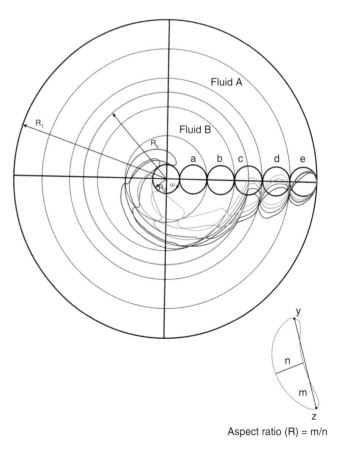

Fig. 5.3. Five circular markers, "a" to "e", inside a concentric circular shear zone. Caption of Fig. 5.1 presents detail of the shear zone. Sky blue markers: markers after 60 hr, green markers: after 90 hr, deep blue markers: after 120 hr, and black markers: after 150 hr. $\omega = 2°$ hr^{-1}

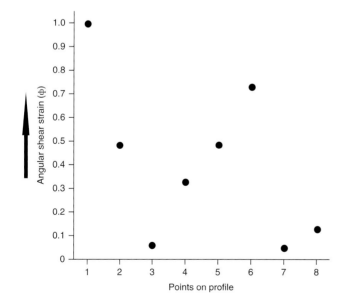

Fig. 5.2. Magnitudes of shear strain variation in biviscous shear zone, detailed in the caption of Fig. 5.1, at one particular instant. Locations of points "1" to "8" on the marker before shear are shown in Fig. 5.1b.

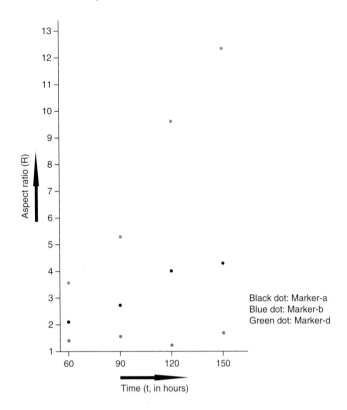

Fig. 5.4. Temporal variation of aspect ratios of three markers, "a", "b", and "d".

ACKNOWLEDGMENTS

Department of Science and Technology's (New Delhi) grant number: SR/FTP/ES-117/2009 supported SM. RB received IIT Bombay's fellowship. Gretchen Baier (The Dow Chemical Company) and Christoph Schrank (Queensland University of Technology) are thanked for comments.

REFERENCES

Andereck CD, Liu SS, Swinney HL. 1986. Flow regimes in a circular-Couette system with independently rotating cylinders. Journal of Fluid Mechanics 164, 155–183.

Baier G. 1999. Liquid-liquid extraction based on a new flow pattern: Two fluid Taylor Couette flow. PhD thesis. University of Wisconsin – Madison. pp. 1–231.

Baier G, Graham MD. 2000. Two-fluid Taylor–Couette flow with countercurrent axial flow: Linear theory for immiscible liquids between corotating cylinders. Physics of Fluids 12, 294–303.

Beaumont C, Jamieson RA, Nguyen MH, Lee B. 2001. Himalayan tectonics explained by extrusion of a low-viscosity crustal channel coupled to focused surface denudation. Nature 414:738–742.

Bons PD, Jessell MW, 1999. Micro-shear zones in experimentally deformed octachloropropane. Journal of Structural Geology 21, 323–334.

Börzsönyi T, Unger T, Szabó1 B. 2009. Shear zone refraction and deflection in layered granular materials. Physical Review E 80, 060302(R).

Donnelly RJ. 1991. Taylor Couette flow: The early days. Physics Today 32, 36–39.

Druguet E, Alsop GI, Carreras J. 2009. Coeval brittle and ductile structures associated with extreme deformation partitioning in a multilayer sequence. Journal of Structural Geology, 31, 498–511.

Escudier MP, Oliveira PJ, Pinho FT (2002) Fully developed laminar flow of purely viscous non-Newtonian liquid through annuli, including the effects of eccentricity and inner cylinder rotation. International Journal of Heat and Fluid Flow 23, 52–73.

Fleitout L, Froideavaux C. 1980. Thermal and mechanical evolution of shear zones. Journal of Structural Geology, 2, 159–164.

Fossen, H., Rykkelid, E. The interaction between oblique and layer-parallel shear in high-strain zones: Observations and experiments. Tectonophysics 207, 331–343.

Gelfgat AY, Yarin A, Bar-Yoseph PZ, Graham MD, Bai G. 2004. Numerical modeling of two-fluid Taylor–Couette flow with deformable capillary liquid–liquid interface. Physics of Fluids 16, 4066–4047.

Ji S, Long C, Martignole J, Salisbury M. 1997. Seismic reflectivity of a finely layered, granulite-facies ductile shear zone in the southern Grenville Province (Quebec). Tectonophysics 279, 113–133.

Lister GS, Snoke AW. 1984. S-C Mylonites. Journal of Structural Geology 6, 617–638.

Masuda T, Mizuno N, Kobayashi M, Nam TN. 1995. Stress and strain estimates for Newtonian and non-Newtonian materials in a rotating shear zone. Journal of Structural Geology 17, 451–454.

Mukherjee S, Biswas R. 2014. Kinematics of horizontal simple shear zones of concentric arcs (Taylor–Couette flow) with incompressible Newtonian rheology. International Journal of Earth Sciences 103, 597–602.

Mukherjee S, Koyi HA, 2010. Higher Himalayan Shear Zone, Sutlej section: structural geology and extrusion mechanism by various combinations of simple shear, pure shear and channel flow in shifting modes. International Journal of Earth Sciences 99, 1267–1303.

Mukherjee S, Punekar JN, Mahadani T, Mukherjee R. 2016. Intrafolial folds: Review and examples from the western Indian Higher Himalaya. In Ductile Shear Zones: From Micro- to Macro-scales, edited by S. Mukherjee and K.F. Mulchrone, John Wiley & Sons, Chichester.

Mulchrone KF, Mukherjee S. in press. Shear senses and viscous dissipation of layered ductile simple shear zones. Pure and Applied Geophysics, doi: 10.1007/s00024-015-1035-8.

Ovchinnikova SN. 2012. Oscillatory instability of the Couette flow between the two unidirectionally rotating cylinders. Fluid Dynamics 47, 454–464.

Peng J, Zhu K-Q. 2010. Linear instability of two-fluid Taylor–Couette flow in the presence of surfactant. Journal of Fluid Mechanics 651, 357–385.

Ramsay JG, 1980. Shear zone geometry: a review. Journal of Structural Geology 334 83–99.

Wilson CJL, Russell-Head D.S., Sim HM. 2003. The application of an automated fabric analyzer system to the textural evolution of folded ice layers in shear zones. Annals of Glaciology 37, 7–17.

Sathe MJ, Deshmukh SS, Joshi JB, Koganti SB, 2010. Computational fluid dynamics simulation and experimental investigation: study of two-phase liquid–liquid flow in a vertical Taylor–Couette contactor. Industrial & Engineering Chemistry Research 49, 14–28.

Schulz A, Pfister G, Tavener SJ, 2003. The effect of outer cylinder rotation on Taylor–Couette flow at small aspect ratio. Physics of Fluids 15, 417–425.

Taylor GI, 1923. Stability of a viscous liquid contained between rotating cylinders. Philosophical Transactions of the Royal Society of London A 223, 289–343.

Vedantam S, Joshi JB, Koganti SB, 2006. Three-dimensional CFD simulation of stratified two-fluid Taylor–Couette flow. The Canadian Journal of Chemical Engineering 84, 279–288.

Williams J, Elder SA, 1989. Fluid Physics for Oceanographers and Physicists. Pergamon Press, Oxford, pp. 253–255.

White FM. 2005. Viscous Fluid Flow. Tata McGraw-Hill Education, New York.

PART II

Examples from Regional Aspects

Chapter 6

Quartz-strain-rate-metry (QSR), an efficient tool to quantify strain localization in the continental crust

EMMANUELLE BOUTONNET[1,2] and PHILLIPE-HERVÉ LELOUP[2]

[1] *Institute of Geosciences, Johannes Gutenberg University Mainz, J.-J.-Becher-Weg 21, D-55128 Mainz, Germany*
[2] *Laboratoire de Géologie de Lyon – Terre, Planètes, Environnement UMR CNRS 5276, UCB Lyon1 – ENS Lyon, 2 rue Raphael Dubois, 69622 Villeurbanne, France*

6.1 INTRODUCTION

How continental crust and lithosphere absorbs ductile deformations is debated. In particular, how far deformation in the middle and deep crust localizes in narrow shear zones or is broadly distributed is discussed. Some see the continental crust as coherent blocs separated by fault zones where most of the deformation is absorbed (e.g. Tapponnier et al. 2001), while others perceive it as a continuous viscous medium where deformation is widely distributed (e.g. Beaumont et al. 2001; Mukherjee 2012). If GPS studies constrain the short-term repartition of deformation at the surface of the continents, we know less about deeper and longer-term deformations significant for the geological history of continents. This is because even if many theories and descriptions of ductile deformations exist (e.g. Ramsay 1980; Mukherjee 2012, 2013), quantification of their amount and furthermore rate are scarce. Indeed, ductile deformation rates in natural settings have been effectively measured in only three cases (Christensen et al. 1989; Müller et al. 2000; Sassier et al. 2009). However, Boutonnet et al. (2013) proposed recently a method to measure deformation rates in quartz bearing rocks deformed in the dislocation–creep regime, which could be used in numerous ductile shear zones.

This method, called quartz-strain-rate-metry (QSR), relies both on a piezometer, a flow law calibrated for quartz dislocation-creep recrystallization, and precise measurements of the temperature of deformation (Boutonnet et al. 2013). Such a method was formalized from laboratory experiments that quantitatively describe the properties of quartz at millimeter scales and at deformation rates of $\sim 10^{-6}$ s^{-1}. A first set of experiments established piezometer relationships linking the size of recrystallized grains to the applied stress (e.g. Twiss 1977; Stipp and Tullis 2003) while a second set established power flow laws linking the stress to the temperature and the deformation rate (e.g. Hirth et al. 2001; Gleason and Tullis 1995; Paterson and Luan 1990; Luan and Paterson 1992). However, extrapolat-

ing from the scale of the experiment to the scale of the natural shear zones is a considerable leap across 8–10 orders of magnitude for the deformation rate, in order to reach the natural values of $\sim 10^{-14}$ s^{-1}. Furthermore, for a given crystal size and a given temperature, results of the QSR vary by five orders of magnitude, depending on the piezometer and power flow law that are chosen (Jerabek et al. 2007). In resolving that problem, Boutonnet et al. (2013) performed an empiric calibration of the QSR method using quartz ribbons sampled in an outcrop where the local strain rate had been previously estimated by an independent method (Sassier et al. 2009) and testing different flow laws and piezometers calibrated experimentally. They conclude that the combinations of (1) Hirth et al.'s (2001) flow law and Shimizu's (2008) piezometer and (2) Paterson and Luan's (1990) flow law and Stipp and Tullis's (2003) piezometer led to correct strain rate measurements in the conditions of deformation of the Ailao Shan–Red River (ASRR) shear zone (China).

One important prerequisite for the QSR method to yield accurate results is to characterize the mechanism of quartz recrystallization and the precise temperature of deformation. Boutonnet et al. (2013) used the TitaniQ thermobarometer (Wark and Watson 2006; Thomas et al. (2010) combined this with a fluid inclusions study and a previously determined local pressure–temperature (P–T) path, as well as crystallographic preferred orientation (CPO) of quartz to constrain the P–T conditions of recrystallization. However, many shear zones exhume during shearing and this complicates deciphering the precise P–T conditions. Furthermore, the ability of the TitaniQ and CPO to accurately constrain the temperature of deformation have been recently challenged (e.g. Grujic et al. 2011; Kidder et al. 2013). In this study, we re-investigated the two shear zones studied in Boutonnet et al. (2013) in order to focus on the quartz recrystallization processes, and discuss the way to accurately constrain the deformation temperature in shear zones undergoing exhumation through time. Finally,

Ductile Shear Zones: From Micro- to Macro-scales, First Edition. Edited by Soumyajit Mukherjee and Kieran F. Mulchrone.
© 2016 John Wiley & Sons, Ltd. Published 2016 by John Wiley & Sons, Ltd.

we discuss the accuracy of the QSR-metry, a cheap and fast method allowing generalizing measurements of local strain rates in the continental crust.

6.2   METHODS

Experimental studies show a close relationship (called a piezometer) between the average size D of quartz crystals recrystallized during dislocation creep, at medium to high temperature, and differential stress σ (e.g. Shimizu 2008; Stipp and Tullis 2003; Twiss 1977):

$$\sigma = K D^{-p} \tag{1}$$

where p and K are determined either experimentally or theoretically. The QSR method combines this equation with the ductile rheological law under the same thermodynamic conditions, that links the strain rate $\dot{\varepsilon}$, the differential stress σ, the temperature T (e.g. Gleason and Tullis 1995; Paterson and Luan 1990; Luan and Paterson 1992), and in some studies the water fugacity f_{H2O} (Hirth et al. 2001; Rutter and Brodie 2004):

$$\dot{\varepsilon} = d\varepsilon / dt = A\sigma^n f_{H2O}{}^m e^{-Q/RT} \tag{2}$$

where the activation energy Q, the prefactor A, and the exponents n and m are determined experimentally, and R is the ideal gas constant. Combining Equations 1 and 2 yields the strain rate $\dot{\varepsilon}$ from the grain size D when the deformation temperature T is known (e.g. Stipp et al. 2002b).

In most geological contexts, rocks vary in pressure and temperature through time, due to burial/exhumation. It is the case for the main strike–slip shear zones of the India–Asia collision zone, where the centers of the shear zones were exhumed during lateral shearing (Leloup et al. 1995, 2001; Boutonnet et al. 2012). The tricky part of the QSR method is to correlate the measured thermodynamic conditions (T, P and f_{H2O}) with the corresponding size (D) of the quartz grains that recrystallized during shearing. Quartz microstructures have therefore to be studied carefully.

6.2.1   Quartz microstructures

All samples are pure quartz ribbons, with as little other mineral as possible. Indeed, it has been shown that polymineralic assemblages (quartz + feldspar or micas) approximately follow diffusion creep (Kilian et al. 2011a). However, the QSR method is based on the use of dislocation creep flow laws, and one has to avoid quartz grains recrystallized in the neighborhood of foreign minerals. The quartz microstructure was investigated by optical methods in ~30 μm thin sections cut parallel to the lineation (X axis) and orthogonal to the foliation (XZ plane).

The dominant recrystallization mechanisms of quartz is inferred from the type of microstructures, including bulge nucleation (BLG), subgrain rotation (SGR), and grain boundary migration (GBM). These mechanisms often occur in tandem (Hirth and Tullis 1991; Stipp and Kunze 2008; Stipp et al. 2002b), but the relative contribution of each mechanism to the bulk microstructure has been demonstrated to vary with stress and/or temperature by laboratory experiments (Hirth and Tullis 1991) and by observing natural microstructures (Stipp et al. 2002a, b). Consequently, the recrystallization regimes have long been used to infer temperatures of deformation. The GBM mechanism, characterized by Hirth and Tullis (1991) as regime 3, activate at >510°C and low stress conditions (Stipp et al. 2002a, b), when there is enough heat energy to allow fast growth of the low-energy grains, those without defects, at the expense of the deformed grains. SGR, described as regime 2 by Hirth and Tullis (1991) is a nucleation regime in which free-energy at the grain boundaries is not high enough to accommodate recovery by boundary migration. Dislocation creep and defect accumulations along crystal planes develop individual subgrains, and then nucleate new grains. This process is activated at medium temperatures of 400 to 510°C, and medium stress conditions (Stipp et al. 2002a, b). Bulging (BLG) or local grain boundary migration, described as regime 1 by Hirth and Tullis (1991), is the nucleation regime dominating at low temperature (below 400°C) and high stress conditions (Hirth and Tullis 1991; Stipp and Kunze 2008). In a recent compilation of natural quartz recrystallization mechanisms, Stipp et al. (2010) also suggested that the transitions amongst these three mechanisms occur at characteristic grain sizes. Finally, the transition between ductile and brittle deformation places the lowest boundary for quartz recrystallization at ~250°C.

6.2.2   Grain sizes and stress

Quartz boundaries were mapped using quartz lattice preferred orientations, measured by Fabric Analyzer (LGGE Grenoble, France). This method identifies the grains by building a map from the <c>-axis (optical axis) orientations. Contiguous pixels with orientation <10–15° misorientation are interpreted as belonging to the same grain. Orientation maps with pixel size of 6.8 μm are analyzed using ImageJ analysis software (NIH image) and macros developed by the LGGE (Grenoble, France). Grain boundaries are detected and corrected visually to avoid foreign minerals and artefacts. For each grain, we estimate a surface (S, μm²) and we deduce an equivalent diameter (D, μm) considering each grain as a circle (Stipp et al. 2002a; Stipp and Tullis 2003): $D = 2(S/\pi)^{-2}$. The grain sizes are plotted as frequency histograms, and a Kernell density estimation is calculated. For a single recrystallization event, the size frequency histograms have usually log-normal distributions (Slotemaker and Bresser 2006; Shimizu 2008). For samples with composite microstructures, several log-normal distributions overlap. To distinguish the different grain generations, it is generally accepted that newly recrystallized grains are relatively strain free, whereas older host grains are more internally deformed, and contain subgrains.

Distributions are analyzed with the Past mixture analysis tool (Hammer et al. 2001), which indicates what combination of log-normal distributions produces the best fit of the histogram. The mode(s) of the Gaussian curve(s) is/are the mean grain size. The standard deviation depends on the mean grain size: the higher the mode, the larger the width of the Gaussian curve (Gueydan et al. 2005). Our size error calculation takes into account the fabric analyser error and a small correction due to the thickness of the grain boundaries, and is approximately on the order of one pixel (6.8 μm).

Paleopiezometry is a relationship linking the dynamically recrystallized grain size formed during dislocation creep and the differential stress (Twiss 1977). As most of the piezometers are calibrated for recrystallization regime 2 (Twiss 1977; Hirth and Tullis 1991; Stipp and Tullis 2003; Shimizu 2008), we select for each sample the quartz grains that correspond to the SGR recrystallization event, based on both the grain size (Stipp et al. 2010) and the microstructure. The grain size analysis is performed on two-dimensional (2D) thin sections, implying that the apparent sizes are different than the actual three-dimensional (3D) grain size. The relationship between the real grain size (D_3) and the apparent 2D-size (D_2), for spherical grains, can be shown to be: $D_3 = 4/\pi \times D_2$. In our study the experimental piezometers (Stipp and Tullis 2003) are calibrated with the 2D value of the grain size, whereas the theoretical piezometers (Twiss 1977; Shimizu 2008) are calculated with the 3D value.

6.2.3 Crystallographic preferred orientation

Quartz crystal CPO describes the orientation of <c>-axis, and sometimes of <a>-axes, of the quartz grains within the ribbon. The CPO indicates the active glide system(s) during deformation and has long been used to infer the type and the temperatures of deformation. For deformation close to simple shear, the activation of the basal plane along the <a> direction, leading to the <c> axis concentrated near the maximum shortening axis (Z axis), is supposed to occur at low temperatures: <400°C (Gapais and Barbarin 1986; Stipp et al. 2002a; Passchier and Trouw 1998). At higher temperatures, the subordinate activation of the romb-<a> (or rhomb-<a+c>) slip system develops a girdle in the CPO plots (Menegon et al. 2008; Peternell et al. 2010). According to Stipp et al. (2002a), the transition from combined basal, rhomb, and prism <a> slip to dominantly prism <a> slip, and therefore from a YZ girdle to a dominant single Y maximum in the <c>-axis pole figures, is rather abrupt and occurs at about 500°C. The temperature range of dominantly prism-<a> slip is between 500°C and ~600–650°C. This former temperature is that of the onset of dominant prism-<c> slip (Mainprice et al. 1986).

The Fabric Analyzer method (Type G50, LGGE, Grenoble), used to measure the CPOs, is a cheap alternative to electron back scattered diffraction (EBSD), but allows only <c>-axis orientation measurements. We followed a method described by Peternell et al. (2010), based on a stack of eight microphotographs taken with different orientations of the cross-polarized light. The spatial step is 6.8 μm for all samples. The data (colatitudes, azimuth and quality factor) are extracted and their analyses are performed using the package G50c Investigator and personal Matlab programs. The plots are performed using Stereo32. For equigranular quartz ribbons, the <c>-axis repartition is plotted with one point per pixel. However, they are plotted with one point per grain if the quartz sample displays different families of grain sizes.

6.2.4 The TitaniQ method

The TitaniQ geothermobarometer (Wark and Watson 2006; Thomas et al. 2010; Huang and Audedat 2012) is based on the dependence of the chemical substitution between Si^{4+} and Ti^{4+} in quartz with pressure and temperature. At high temperatures and high pressures, the number of substituted sites increases (Thomas et al. 2010), leading to the calibration of a thermobarometer specific to quartz (Wark and Watson 2006; Thomas et al. 2010; Huang and Audedat 2012). It has been proposed that Ti concentrations in quartz can re-equilibrate during dynamic recrystallization at low temperatures (Kohn and Northrup 2009), despite the very low diffusion rates (Cherniak et al. 2007). Ti concentrations in quartz are determined by ICP-MS (Element XR) coupled to a laser ablation system (Microlas platform and Excimer CompEx Laser, spot diameters of 33 μm and repetition rates of 10 Hz) at the Geosciences Montpellier (France) and at IUEM Brest (France). Two or three of the Ti isotopes were analyzed: [47]Ti (7.3% of total Ti), [49]Ti (5.5%) and [48]Ti (73.8%). The total Ti content of the sample is calculated averaging Ti contents estimated from each isotope. The alignment of the instrument and mass calibration is performed before every analytical session using the NIST 612 reference glass. US Geological Survey (USGS) basalt glass reference materials BCR and BIR are used during experiment as standards. Masses of isotopes are analyzed over 20 cycles for each analysis. Isotopes of [27]Al, [29]Si, [43]Ca, and [7]Li are used to monitor the quartz ablation, and [85]Rb, [86]Sr, and [137]Ba to control if other mineral inclusions are also ablated.

Although calibrated for quartz crystallized in the presence of rutile, the thermobarometer can also be applied to rutile-absent systems if TiO_2 activity is constrained. A Ti activity of ≥0.6 is appropriate for most continental rocks containing a Ti-rich phase (rutile, ilmenite, sphene, biotite) (Wark and Watson 2006; Ghent and Stout 1984). The measurements have been made both in areas where small recrystallized grains are frequent and inside the larger older grains. The Ti-contents of both the centre and the borders of the quartz ribbons were measured in order to check the influence of the matrix, source of titanium (Fig. 6.1).

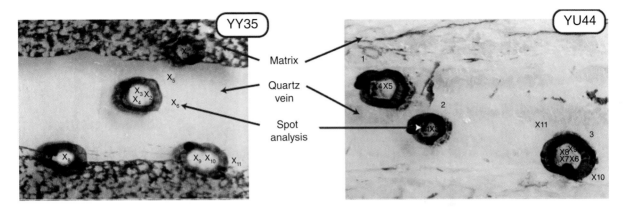

Fig. 6.1. Location of the laser ablation spots for samples YY35 and YU44: point X_n corresponds to data dd_n of Fig. 6.8. The Ti-contents of both the center and the borders of the quartz ribbons were measured in order to check the influence of the matrix (Ti-source).

6.3   GEOLOGICAL SETTING

6.3.1   The Ailao Shan–Red River shear zone

6.3.1.1   Geological context

The Red River zone is a major physiographic and geological discontinuity in continental East Asia. It stretches for more than 1000 km from eastern Tibet to the Tonkin Gulf, separating the South China and Indochina blocks (Fig. 6.2a), (e.g. BGMRY 1983; Helmcke 1985) and possibly extends into the mantle (Huang et al. 2007). This discontinuity corresponds to at least two different structures: the Red River fault zone (RRF) and the ASRR. The most recent one is the RRF that shows morphological evidences for recent right-lateral/normal motion (Tapponnier and Molnar 1977; Allen et al. 1984; Leloup et al. 1995; Wang et al. 1998; Replumaz et al. 2001). The RRF straddles four 10–20 km wide high-grade metamorphic ranges (Fig. 6.2b): the XueLong Shan, the Diancang Shan, the Ailao Shan and the Day Nui Con Voi. These massifs are interpreted as the exhumed ductile root of the ASRR, which is an Oligo-Miocene left-lateral ductile shear zone (e.g. Tapponnier et al. 1986, 1990; Leloup and Kienast 1993; Leloup et al. 1995, 2001). The metamorphic rocks display a strong ductile deformation, with a generally steep foliation bearing a horizontal lineation. Both parallel the trend of the gneissic cores. Numerous shear criteria indicate that the gneisses are intensively left-lateral sheared (e.g. Tapponnier et al. 1986, 1990; Leloup and Kienast 1993; Leloup et al. 1995, 2001; Jolivet et al. 2001; Anczkiewicz et al. 2007).

In the Ailao Shan massif, the shear zone crops out as a ~10 km wide belt of high grade mylonitic gneiss framed by slightly deformed Mesozoic sediments to the north and schists to the south (Fig. 6.2). The shear zone rocks include thinly banded, biotite-sillimanite-garnet- bearing paragneisses, orthogneisses, augengneisses with large feldspar porphyroblasts, migmatites, deformed leucocratic veins, and intrusions of anatectic leucogranites and granodiorites. The paragneisses are found along the northeast side of the range and contain large, up to

several tens of meters wide, marbles boudins (Fig. 6.2c). Most rocks are mylonitic, but the deformation is more impressive in the paragneiss, as this could be due to a lack of indicators in the orthogneiss. The RRF bounds the range to the northeast.

Amphibole-rich levels and synkinematic leucocratic dykes are common (Leloup et al. 1995) (Fig. 6.2c). Both of them, within their gneissic country rock, form spectacular boudins trails that have been used to estimate shear strains (Lacassin et al. 1993; Sassier et al. 2009). In most outcrops, shear strains are high and all the dykes transpose parallel to the main foliation, so the ductile deformation is difficult to quantify. However, in the orthogneissic core of the Ailao Shan, about 3 km southwest of YuangJiang, the site C1 (Leloup et al. 1995) exhibits various generations of syntectonic dykes. Sassier et al. (2009) determined the strain rate by measuring independently the shear strain (γ) recorded by the dykes and the emplacement age (t) of the same dykes. The minimum strain rates deduced from this study range from 3 to 4×10^{-14} s^{-1}. This value had been taken as reference to test the different power flow laws and piezometers used by the QSR method (Boutonnet et al. 2013).

The left-lateral shearing occurred under a high geothermic gradient ($\geq 35°$C/km, Leloup and Kienast 1993). The petrologic studies in the Ailao Shan show a metamorphic peak in amphibolite facies conditions (4.5 ± 1.5 kbar and 700 ± 70°C, Leloup and Kienast 1993). Left-lateral deformation continued in retrograde, greenschist facies conditions (<4 kbar and <500°C, Leloup and Kienast 1993; Nam et al. 1998; Jolivet et al. 2001; Leloup et al. 2001). Monazite U–Th/Pb dating from the mylonitic fabric and as inclusion within synkinematic garnets constrains the duration of high-temperature metamorphism from 34 to 21 Ma (Gilley et al. 2003). Felsic and alkaline magmatism, dated from 35 to 22 Ma (Schärer et al. 1994; Zhang and Schärer 1999), was coeval with both metamorphism and deformation. Considering that left lateral shearing was coeval with cooling, the ^{40}Ar/^{39}Ar data constrain the timing of ductile deformation between ~31 and 17 Ma (e.g. Leloup et al. 1995, 2001). Moreover, Briais et al. (1993) proposed

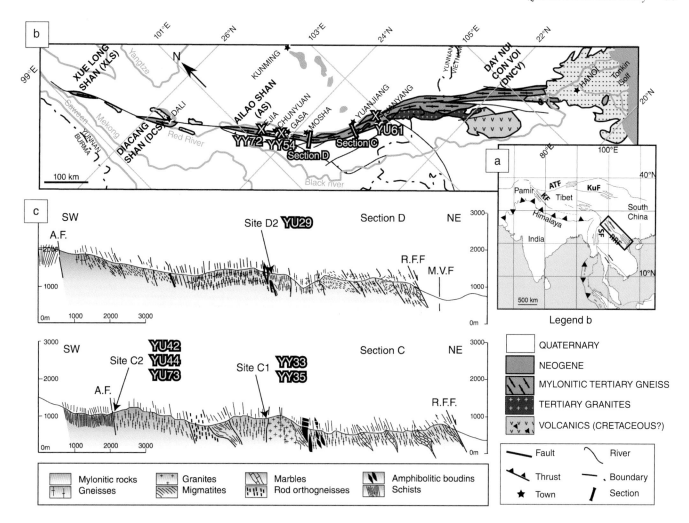

Fig. 6.2. Geological maps and sections of Ailao Shan Red River (ASRR) shear zone. (a) Location of the major strike–slip shear zones in the India–Asia collision zone. RRF: Red River fault, KuF: Kunlun Fault, ATF: Altyn Tagh Fault, KF: Karakorum fault, SF: Sagaing Fault. (b) Simplified geological map of the left-lateral ASRR shear zone. Adapted from Leloup et al. 2001 and from Sassier et al. 2009. (c) Geological cross-sections C and D of Ailao Shan metamorphic range, with sites C1, C2 and D2 location. Modified after Leloup et al. 1995. A.F.: Ailao Shan Fault; R.F.F. = active range front fault (mainly normal slip); M.V.F. = active mid valley fault (purely right-lateral).

that the South China Sea oceanic basin formed between ~32 Ma and ~16 Ma as a pull apart basin at the Southeast termination of the ASRR implying ~540 km of left-lateral motion along the shear zone. These observations and interpretations have been challenged by some authors. For example Searle (2006) considers that all deformed granites within the ASRR are prekinematic, implying that left-lateral shear started only after 21 Ma (see Leloup et al. 2007). Other authors questioned the link between motion on the ASRR and sea-floor spreading in the South China Sea (e.g. Clift et al. 1997; Fyhn et al. 2009), whilst these studies confirm the existence of a fault linking the ASRR to the spreading center and the contemporaneity of the two events.

6.3.1.2 Sample location

We selected nine samples from the ASRR shear zone. Two quartz ribbons have been sampled in the outcrop C1, located in the centre of section C (Leloup et al. 1995,

Fig. 6.2c). YY33 and YY35 both stretch parallel to the main foliation (N120°, vertical, lineation pitch: 18°E). YY35 is a centimeter-wide homogeneous quartz ribbon in a granodioritic gneiss, whereas YY33 displays thin millimeter-wide quartz ribbons in a hornblende-bearing gneiss. These two samples have been used by Boutonnet et al. (2013) to test and calibrate the QSR method. Three other samples come from the site C2 (Section C, Leloup et al. 1995, Fig. 62c), located at the south-western edge of the Ailao Shan massif, bordered by the Ailao Shan fault. YU44 is a large (~5 cm) quartz ribbon, and YU73 and YU42 are thinner (0.5 to 1 cm). They lie into the host gneiss composed mostly of black and white micas, quartz, and feldspar. YU29 is located in section D, in site D2 (Leloup et al. 1995, Fig. 6.2c). This sample displays a quartz-rich matrix, with feldspar eyes showing sinistral asymmetric deformations, and micas defining the foliation. YY72 and YY54 are located in the north-eastern border of the Ailao Shan massif, near Ejia and between Chunyuan and Gasa, respectively (Fig. 6.2b). Thin quartz ribbons compose the foliation of the quartz-rich gneisses, oriented N160°, 67°E for YY54 and N147°, 45°E, with a lineation pitch of 5°S for YY72. Finally, YU61 is a green-hornblende bearing gneiss, with millimetric quartz ribbons, located near YuanYang in the north-eastern border of the south Ailao Shan (Fig. 6.2b). YU42, YU44, YU61 and YU29 quartz fabric, measured using a U-stage microscope, have been studied by Leloup et al. (1995). They showed that YU29 deformed at high temperature, because of the activation of the prismatic <c> glide system, whereas the three other samples deformed at lower temperatures with activation of both the prismatic <a> and basal <a> glide systems.

6.3.2 The Karakorum shear zone (KSZ)

6.3.2.1 Geological context

The NW–SE right-lateral Karakorum fault zone (KFZ) displays a prominent morphological trace from Tash Gurgan in the NW to the Kailash area in the SE (e.g. Weinberg et al. 2000; Lacassin et al. 2004). This is related to Quaternary right-lateral motion of the Karakorum fault, which has deflected the course of the Indus River by 120 km (Gaudemer et al. 1989). However, the rate of recent motion and the active portion(s) of the fault are highly debated, from 1 ± 3 mm yr^{-1} (Wright et al. 2004) to 11 ± 4 mm yr^{-1} (Banerjee and Bürgmann 2002). Ductilely deformed rocks locally outcrop along the KFZ: (1) in the Nubra valley (Ladakh, India), (2) in the Darbuk Tangtse Pangong region (Ladakh, India) (Fig. 6.3), and (3) in the Ayilari Range (China). In all these locations the rocks show mylonitic textures with steep foliations and close to horizontal stretching lineations, with unambiguous right-lateral shear criteria (e.g. Matte et al. 1996; Lacassin et al. 2004; Phillips and Searle 2007; Searle and Phillips 2007; Valli et al. 2007; Rolland et al. 2009; Roy et al. 2010). Because (1) the mylonites are parallel the KFZ for at least 400 km, (2) the mylonites share the same direction and sense of motion as the KFZ, and (3) there is no evidence for major tilting of the mylonites after their formation, these mylonites are interpreted as constituting the KSZ corresponding to the exhumed deep part of the KFZ.

In the Tangtse zone (34°N, 78.2°E), the KFZ splits into two strands, which flank a topographic range, the Pangong Range, in which slightly deformed to mylonitized magmatic, migmatitic and metamorphic rocks outcrop (Fig. 6.3a). They constitute a metamorphic belt which exhibit a foliation trending N131 and plunging 84°SE on average, with a stretching lineation dipping from 20° to the SE to 40° to the NW (15° to the NW on average) as previously described (e.g. Searle et al. 1998; Phillips and Searle 2007; Jain and Singh 2008; Rolland et al. 2009). Shear criteria indicate right-lateral shear. These rocks have been interpreted as the 8 km wide KSZ (e.g. Searle et al. 1998; Rolland et al. 2009). The two mylonitic strands that frame the Pangong range are the Tangtse strand to the SW and the Muglib strand to the NE. The exhumation of granulitic rocks (800°C and 5:5 kbar)

of the Pangong Range (Rolland et al. 2009), and their rapid cooling, has been related to right-lateral transpressive deformation between the two strands (Dunlap et al. 1998; McCarthy and Weinberg 2010; Rolland et al. 2009).

Two valleys perpendicular to the belt give access to the structure of the KSZ: the Darbuk valley to the NW and the Tangtse gorge to the SE (Fig. 6.3a). The series exposed from SW to NE are:

1 The Ladakh batholith.
2 Rocks belonging to the Shyok suture zone including ultramafics and black mudstones containing Jurassic ammonoid fossils and volcanoclastic rocks (Ehiro et al. 2007). These rocks are locally intruded by the 18 Ma South Tangtse granite (Leloup et al. 2011).
3 Mylonites of the Tangtse strand, with dextrally sheared mylonitic ortho- and para-derived gneisses and marbles, and leucocratic dykes parallel to the foliation as well as crosscutting ones. These synkinematic dykes are intruded between 19 and 14 Ma (Phillips et al. 2004; Boutonnet et al. 2012).
4 The Pangong range where the country rocks and the leucocratic dykes appear less deformed. There, synkinematic migmatization affects both a metasedimentary sequence comprising Bt-psammites, calc-silicates and amphibolites, and a calcalkaline granitoid suite comprising Bt-Hbl granodiorites, Bt-granodiorites, and diorites (Reichardt et al. 2010).
5 The Muglib strand with dextrally sheared mylonites, also intruded by synkinematic dykes between 18 and 15 Ma (e.g. Boutonnet et al. 2012).
6 The Karakorum batholith and the Pangong metamorphic complex (PMC) comprising marbles and large ($\leq$10 m) leucogranitic dykes, with foliations trending more easterly than in the KSZ, and locally showing left-lateral shear criteria (McCarthy and Weinberg 2010). Note that the various authors give different names to the geologic formations and that we use the names given in Fig. 6.3.

The cooling histories of the four different structural units of the KSZ in the Tangtse area (South Tangtse granite, Tangtse strand, Pangong Range, and Muglib strand) have been described by Boutonnet et al. (2012), based on new and published U/Pb and Ar/Ar ages. They showed that the KSZ cooled later that the surrounding terrains. The cooling of the SW side of the shear zone was earlier than its NE side, which is compatible with a reverse fault component of the right-lateral deformation suggested by the NW dip of the lineations in that zone (Fig. 6.3). The time range for right lateral ductile deformation is $\geq$18.5 to 14 Ma for the South Tangtse mountain, $\geq$19 to 11 Ma in the Tangtse strand, and $\geq$15 Ma to 7 Ma in the Muglib strand. Combination of the P–T path proposed by Rolland et al. (2009) with the temperature–time (T–t) path of Boutonnet et al. (2012) allows building a P–T–t path for the Pangong range unit, showing that at least 20 km of exhumation occurred during the right-lateral deformation. About 40% of that exhumation

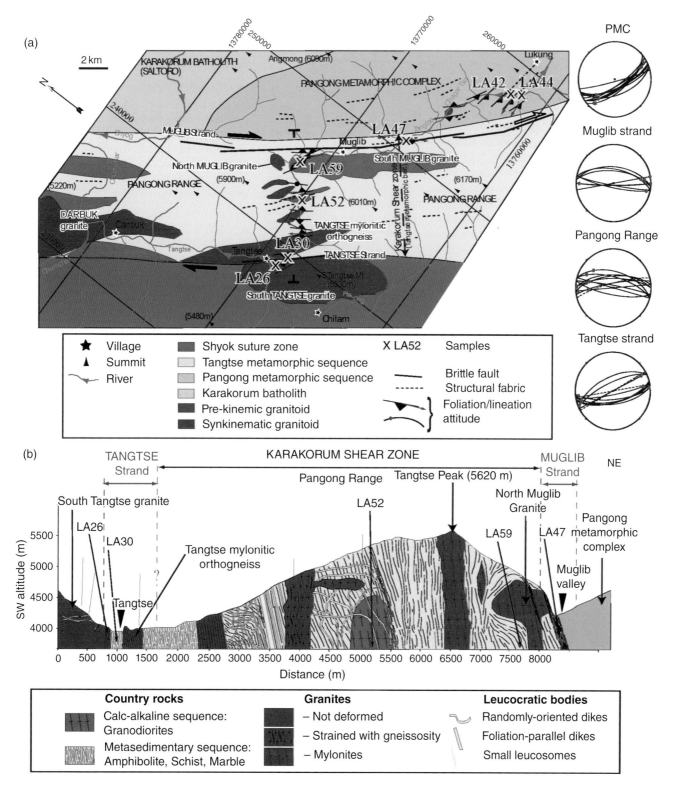

Fig. 6.3. Geological map and section of the Karakorum shear zone (KSZ). The general location of the KSZ is given in Fig. 6.2a. (a) Simplified geological map of the KSZ in the region of Tangtse (India). (b) Corresponding SW–NE directed geological cross-section. All the studied samples are located, as well as the main structures orientations. PMC: Pangong Metamorphic Complex. Adapted from Leloup et al. (2011) and Boutonnet et al. (2012).

occurred before 12 Ma with a mean rate of 1 mm yr^{-1}, while deformation was still ductile. Exhumation of the Pangong range has been attributed to transpressive deformation (Dunlap et al. 1998; Rolland et al. 2009; McCarthy and Weinberg 2010). During this exhumation, the apparent geothermal gradient, assumed as linear in the conductive crust, progressively decreased from <40°C/km to "normal" geothermal conditions prevailing in an unperturbed lithosphere of 30°C/km. The initially high geothermal gradient may have been due to heat advection by rising melts or fluids, as suggested by the abundance of granitic magmatism between ca. 20 and ca. 15 Ma (Boutonnet et al. 2012). The Pangong Metamorphic Complex (PMC), located northeast of the shear zone displays inherited ages from a late Cretaceous metamorphic event at 108 Ma (Streule et al. 2009). Nevertheless, the PMC has been reheated, possibly by viscous dissipation (Mukherjee and Mulchrone 2013) and the samples located close to the Muglib strand show clear Miocene cooling ages at 10.6 Ma (McCarthy and Weinberg 2010).

6.3.2.2 Sample locations

We selected five samples from the KSZ, along the Tangtse gorge, joining the village of Tangtse in the south-western strand of the shear zone to the village of Muglib in the north eastern strand (Fig. 6.3). Sample LA26 is a large quartz ribbon from the deformed part of the South Tangtse granite. This granite, dated at 18.5 ± 0.2 Ma by Leloup et al. (2011), is located slightly outside of the shear zone (Fig. 6.3) and is deformed by the Tangtse strand in its north-eastern part. As for all samples, the quartz ribbon is parallel to the foliation, trending N120, 65°S in this outcrop. Sample LA30 is a millimeter- to centimeter-wide quartz ribbon in a matrix of green schist, located in the centre part of the Tangtse strand, within the mylonites (N120, 65°S). The Pangong range is the less deformed part of the shear zone, and the quartz ribbons, formed by quartz segregation during rock deformations, are rare. LA52 is one of them, taken within the calc-alkaline granodioritic suite. Sample LA59 is a centimeter-wide quartz ribbon parallel to the weakly deformed schists (N110, 55°N), located few tens of meters South-west of the Muglib strand. LA47 is a thin, ~1 mm wide, ribbon of quartz parallel to the main foliation of the Muglib stand mylonites (foliation: N146°, 80°S, lineation: pitch 10°SE), composed of fine light schist with clear dextral criteria in this outcrop. Finally, two samples LA42 and LA44 are located in the Pangong Metamorphic Complex (PMC, Fig. 6.3). In this area, the general foliation turns gradually from ~N110° at ~300 m north-east from the Mublig strand (LA44: foliation N105°, 75°S, pitch lineation 20°E), to ~N90° at ~500 m from it (LA42: foliation N95°, 63°S, pitch lineation 19°W). This is interpreted as the deformation of the Cretaceous PMC (Streule et al. 2009) by the dextral movement of the KSZ.

6.4 STRAIN RATE MEASUREMENTS IN NATURAL SHEAR ZONES

6.4.1 Recrystallization regime and paleo-stresses

Figure 6.4 shows the microstructures of our 15 samples, and Fig. 6.5 displays their grain size frequency diagrams in logarithmic size and their best normal distributions.

All ASRR samples exhibit completely recrystallized quartz grains in cross-polarized microscopy. Samples YU44, LA73, YU61, YU29, YY33, and YY35 show large grains (1 μm) with boundaries showing lobes of amplitude >20 μm, and irregular shapes and sizes (Fig. 6.4). In many cases, some grains that appear separated in the 2D section, have the same crystallographic orientation and actually belong to a common lobed grain in 3D. These microstructures are typical of recrystallization mechanism by grain boundary migration (GBM). In all cases, smaller geometric grains are observed at the triple junctions (Fig. 6.4) and are often associated with zones of undulose extinction and subgrains in the surrounding larger grains. The grains display typically angular shapes, with angles close to 120°. This microstructure is typical of an overprinting quartz recrystallization by subgrain rotation (SGR). As the large (GBM) grains are themselves deformed by processes typical of dislocation migration, we conclude that the SGR event occurred later. In the case of samples YU29 and YY3, the large GBM-grains are more deformed exhibiting undulose extinction and subgrains, and the SGR-grains are more numerous.

This kind of microstructures typically leads to a bimodal grain size repartition, with a population of large grains corresponding to the GBM ones, and a population of smaller grains corresponding to the SGR event. YU44 and YU73 (site C2) display very similar distributions, with the first mode corresponding to the largest grains population centered at 312.9 ± 6.8 and 388.9 ± 6.8 μm and the second mode to the smallest grains population centered at 72.2 ± 6.8 and 78.2 ± 6.8 μm, respectively (Fig. 6.5). For the two samples coming from site C1 (Boutonnet et al. 2013), the population of largest grains is represented by a first mode centered at 150.4 ± 6.8 μm for YY35 and at 127.3 ± 6.8 μm for YY33. The second family of grains, represented by the second mode, display sizes at 55.9 ± 6.8 and 63.3 ± 6.8 μm for these two samples, respectively. The population of large grains of sample YU29 has a mean size of 238.4 ± 6.8 μm, and the small grains have a mean size centered at 52.7 ± 6.8 μm. Finally, the population of large grains of sample YU61 has a mean size of 167.9 ± 6.8 μm, and the small grains have a mean size centered at 61.9 ± 6.8 μm. Quartz grains of sample YU42 are large and rather homogeneous in color, size, and shape. We measured a log-normal distribution of grain sizes, centered at 92.9 ± 6.8 μm. Sample YY54 displays heterogeneous grains, with undulose extinction and subgrains, surrounded by recrystallized grains. This indicates an important recrystallization by SGR. Some larger grains have finely lobed boundaries (~1–10 μm,

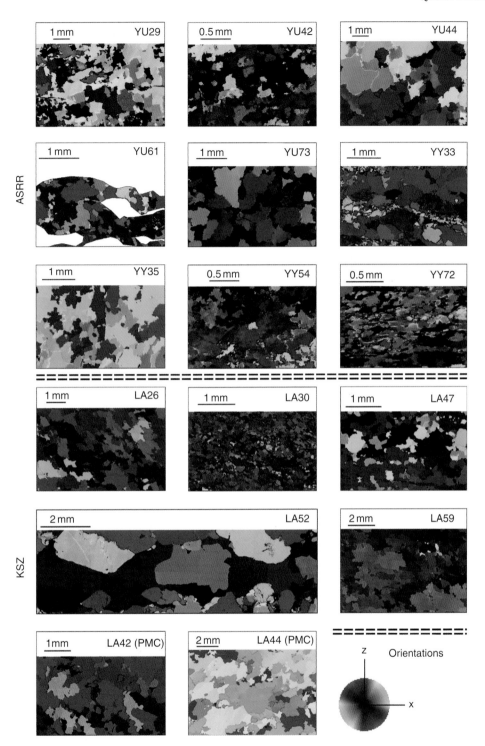

Fig. 6.4. Quartz microstructures. <c>-axis orientations images of 9 samples of the ASRR (YU and YY) and 7 samples of the KSZ (LA) or the Pangong Metamorphic Complex (PMC) obtained by Fabric Analyzer method. The thin sections are cut in the XZ plane, parallel to lineation and normal to foliation (view from above). The color code of the <c>-axis orientation is given in a stereographic plot (lower hemisphere). Large grains with amoeboid shapes, irregular and lobbed boundaries are interpreted as recrystallized by grain boundary migration. Small regular-shaped grains, spatially link to subgrains, are interpreted as subgrain rotation recrystallization mechanism. These pictures are used to map the grain boundaries.

not detected by the Fabric Analyzer). These bulged boundaries are interpreted as a dynamic recrystallization by bulging, for the grains which original orientation did not allow easy dislocation migration (Menegon et al. 2008). The grain size repartition shows one single mode, centered at 61.6 ± 6.8 μm. Finally, sample YY72 displays very elongated and angular quartz grains, typical of a complete recrystallization by subgrain rotation of the rock. The grain size repartition shows one single mode, centered at 36.0 ± 6.8 μm.

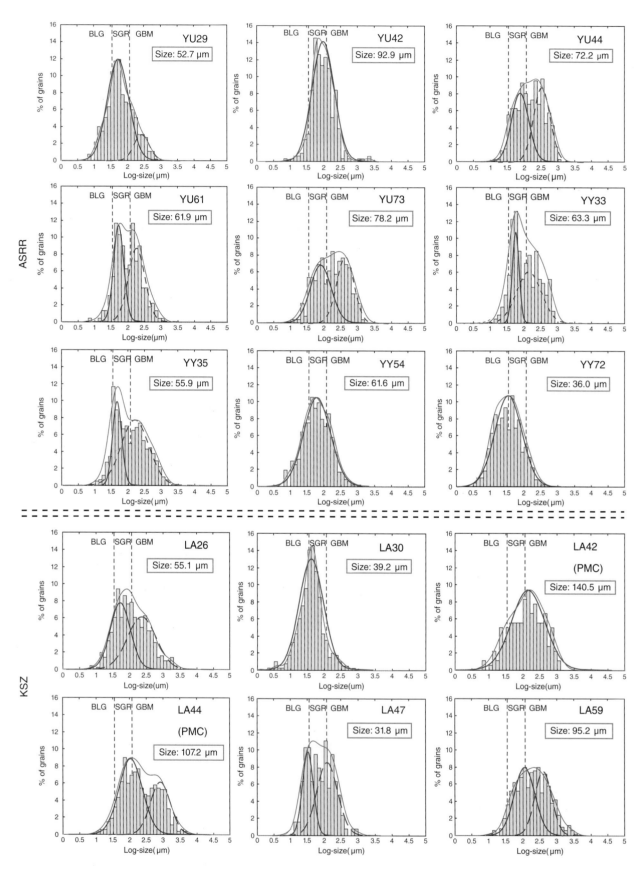

Fig. 6.5. Two-dimensional quartz grain size distribution for 9 ASRR samples (YU and YY) and 6 KSZ samples (LA). The X-scales are in logarithmic scale of the size (μm). The best mixture of normal distributions that could drive to each frequency histogram is calculated. When a bimodal repartition is preferred, the red dashed line represents the large grains, and the red continuous line represents the smaller grains, which size (in μm) is taken for the stress calculation. The results are reported in Table 6.1. BLG: bulging; SGR: subgrain Rotation; GBM: grain boundary migration. The size fields of recrystallization mechanisms are defined by Stipp et al. (2010).

Six of the seven samples of the KSZ/PMC area exhibit completely recrystallized quartz grains in cross-polarized microscope whereas sample LA52 shows large quartz grains without dynamic recrystallization and no mean grain sizes were measured. Samples LA26, LA42, LA44, and LA59 show large lobed grains, typical of grain boundary migration. In the case of LA42, LA44, and LA59, smaller grains are visible, but appear to be lobes of the biggest grains that have been detached to form new grains or could be still attached in 3D. In sample LA44, two grain size populations were identified, one with very large grains (730.3 ± 6.8 μm) and one with smaller grains (107.2 ± 6.8 μm). Sample LA59 also shows two populations of grain sizes, with the smallest one centered around 95.2 ± 6.8 μm. The grain size repartition of sample LA42 is fitted by a Gaussian curve, indicating a single population, with a mean size of 140.5 ± 6.8 μm. The case of sample LA26 is different because the large grains show internal dynamic recrystallization, such as subgrains and undulose extinction, and the small grains, located at triple junctions, are recrystallized by sub-grain rotation (Fig. 6.4). This dichotomy in microstructures is underlined by the clear bimodal repartition of grain sizes. The population of large GBM grains has a mean size of 223.1 ± 6.8 μm and the one of small SGR grains has a mean size of 55.1 ± 6.8 μm. Sample LA47 displays also two populations of quartz grains, but the large GBM grains are more deformed and recrystallized and the small SGR grains are more numerous. The mean size of the large grain population is 119.7 ± 6.8 μm and the mean size of the small grain population is 31.8 ± 6.8 μm. For sample LA30, the SGR recrystallization process is complete. The quartz ribbon displays grains with the same angular shapes with ~120° angles and homogeneous sizes well centered around 39.2 ± 6.8 μm.

In a context of a shear zone exhumation and cooling, one expects a transition between different recrystallization modes. The GBM-type microstructure is overlaid by the SGR one. Gueydan et al. (2005) interpret this as a memory of the different phases of exhumation. At high temperature, the GBM recrystallization mechanism was active and the whole quartz ribbon recrystallized. During the Ailao Shan massif and the Pangong Range massif exhumation, temperature decreased and the SGR recrystallization mechanism was activated below ~500°C, then BLG below ~350°C (Passchier and Trouw 1998). The volume proportion of recrystallized grains decreases with temperature. Stipp et al. (2002b) estimate that only 40 and 10% of the volume recrystallize at 500°C and 400°C, respectively. The GBM-type microstructure can thus be preserved elsewhere. Only the grain populations of the last recrystallization event, generally by SGR, are used to calculate a paleo-stress. The results are given in Table 6.1. For the ASRR, the stresses calculated using Shimizu's (2008) piezometer range between 27.8 ± 5.3 MPa and 46.8 ± 14.6 MPa, and the stresses range between 23.4 ± 3.7 MPa and 55.7 ± 14.8 MPa (Shimizu 2008) for the KSZ/PMC area.

6.4.2 Thermodynamic conditions of recrystallization

6.4.2.1 CPO results

The CPO analysis of samples YU29, YU61, YY33, YY35 (ASRR), LA42 and LA44 (KFZ) indicate a strong concentration of the <c> axis near the maximum shortening axis (Z axis) (Fig. 6.6 and Fig. 6.7). In most cases, the <c> axis fabrics are asymmetrical, and indicates that the basal planes are slightly oblique to the foliation. This geometry is compatible with gliding on the basal plane along the <a> direction under left-lateral simple shear for ASRR and right-lateral simple shear for KSZ (Fig. 6.7). Moreover, the presence of oblique girdles suggests that some crystals show glide along the <a> direction but on the prismatic planes (Fig. 6.7). This CPO is consistent with subordinate romb-<a+ c> slip activation. This romb-<a+ c> slip activation is more important in samples YU29 and YU61 (ASRR). For samples YU42, YU44, YU73 (ASRR), LA26, and LA47 (KFZ), the rhomb-<a+ c> glide is accompanied by a maximum of <c> axis concentrated near the intermediate axis (Y axis). This geometry is compatible with gliding, principally on the prismatic plane, along the <a> direction. The slight asymmetry of the gliding plane is also consistent with the sense of shearing of the considered shear zone. For samples YY54, YY72 (ASRR), LA30, and LA59 (KFZ), the prism-<a> slip is clearly dominant.

The samples for which a second symmetric girdle appears in CPO plots (e.g. YU73, Fig. 6.7) indicate that a minor pure shearing component also occurred (Passchier and Trouw, 1998). In total, therefore, a simple shear-dominated general shear or sub-simple shear is deciphered. We also noticed that the CPOs are the same for the large grains associated to the GBM recrystallization event and the smaller grains associated to the SGR recrystallization event (see two examples in Fig. 6.6). This indicates that all the quartz crystallographic orientations are consistent with the glide system activated during the last recrystallization event, even the CPOs of the large grains inherited from the GBM recrystallization event.

6.4.2.2 Titanium-in-quartz measurements

Most of the samples display low values of titanium-in-quartz (Table 6.2, Fig. 6.8). The lowest value of Ti-content is measured for sample LA42 (KSZ, 1.7 ± 0.3 ppm). Then, values around 5 ppm are measured for two samples of the ASRR, YU44 (5.7 ± 0.9 ppm) and YU73 (4.5 ± 0.9 ppm), and four samples of the KSZ, LA26 (6.3 ± 0.4 ppm), LA30 (4.8 ± 0.5 ppm), LA47 (4.3 ± 0.5 ppm), and LA59 (4.3 ± 0.7 ppm). Applying the thermobarometer

Table 6.1. QSR strain rate measurements in the ASRR and KFZ strike-slip shear zones

Shear zone/ sample	Lat (°D) Long (°D)	Quartz vein size	Recrystal-lization regime[a]	Mean grain size measured (µm) ±error (1σ)	Mean grain size corrected[b] (µm) ±error (1σ)	Stress* (MPa) ±error (1σ)	Stress** (MPa) ±error (1σ)	Method of Temperature determination	Temperature (°C) ±Uncertainty	Pressure (MPa) ±Uncertainty	Hydrostatic pressure (MPa) ±Uncertainty	Strain rate[§] (s⁻¹) Max Min	Strain rate[§§] (s⁻¹) Max Min
ASRR													
YU29	23.767°N 101.710°E	mm	SGR	52.7 ±6.8	67.1 ±10.2	37.3 ±9.4	21.1 ±9.6	Ti-in-Quartz + P-T path	469 ±44	150 ±80	50 ±25	**1.9E-13** 1.6E-12 1.5E-14	**2.6E-13** 2.9E-12 9.9E-15
YU42	23.530°N 101.910°E	mm	SGR	92.9 ±6.8	118.3 ±10.2	27.8 ±5.3	13.4 ±6.0	Ti-in-Quartz + P-T path	402 ±40	120 ±80	32 ±20	**4.2E-15** 3.0E-14 4.3E-16	**7.3E-15** 8.1E-14 2.8E-16
YU44	23.530°N 101.910°E	cm	SGR	72.2 ±6.8	91.9 ±10.2	37.5 ±7.2	16.4 ±7.3	Ti-in-Quartz + P-T path	367 ±40	110 ±80	22 ±15	**2.5E-15** 1.5E-14 3.4E-16	**3.6E-15** 3.3E-14 1.8E-16
YU61	23.244°N 102.781°E	mm	SGR	61.9 ±6.8	78.9 ±10.2	28.1 ±7.5	18.5 ±8.3	Ti-in-Quartz + P-T path	548 ±96	180 ±70	80 ±35	**8.1E-13** 1.7E-11 1.7E-14	**1.4E-12** 3.6E-11 1.6E-14
YU73	23.530°N 101.910°E	cm	SGR	78.2 ±6.8	99.6 ±10.2	36.7 ±6.6	15.4 ±6.8	Ti-in-Quartz + P-T path	352 ±40	100 ±80	18 ±13	**1.1E-15** 6.1E-15 1.4E-16	**1.6E-15** 1.5E-14 8.0E-17
YY33	23.554°N 101.916°E	mm	SGR	63.3 ±6.8	80.6 ±10.2	35.6 ±7.9	18.2 ±8.1	Ti-in-Quartz + P-T path	425 ±40	130 ±80	34 ±25	**2.7E-14** 2.1E-13 2.4E-15	**4.2E-14** 4.6E-13 1.6E-15
YY35	23.554°N 101.916°E	cm	SGR	55.9 ±6.8	71.2 ±10.2	39.4 ±9.1	20.1 ±9.1	Ti-in-Quartz + microthermometry	425 ±38	130 ±80	34 ±25	**4.0E-14** 3.0E-13 3.7E-15	**5.6E-14** 5.9E-13 2.3E-15
YY54	24.277°N 101.378°E	cm	SGR	61.6 ±6.8	78.4 ±10.2	28.4 ±7.6	18.6 ±8.3	Ti-in-Quartz + P-T path	544 ±51	180 ±80	79 ±32	**7.6E-13** 6.3E-12 5.9E-14	**1.3E-12** 1.3E-11 5.6E-14
YY72	24.432°N 101.254°E	cm	SGR	36.0 ±6.8	45.8 ±10.2	46.8 ±14.6	28.5 ±13.7	Ti-in-Quartz + P-T path	507 ±59	160 ±80	72 ±31	**2.0E-12** 2.5E-11 8.2E-14	**1.9E-12** 2.8E-11 4.5E-14
KSZ													
LA26	34.025°N 78.171°E	cm	SGR	55.1 ±6.8	70.2 ±10.2	40.8 ±9.4	20.3 ±9.2	Ti-in-Quartz + P-T path	415 ±40	350 ±80	80 ±37	**8.0E-14** 4.5E-13 1.1E-14	**4.2E-14** 3.3E-13 2.5E-15
LA30	34.023°N 78.175°E	mm	SGR	39.2 ±6.8	49.9 ±10.2	55.7 ±14.8	26.6 ±12.6	Ti-in-Quartz + P-T path	400 ±40	350 ±80	80 ±37	**1.6E-13** 1.0E-12 1.8E-14	**5.6E-14** 4.6E-13 2.9E-15
LA52	34.039°N 78.216°E	mm	N.D.	–	–	–	–	–	–	–	–	**<1.0E-15** –	**<1.0E-15** –
LA59	34.052°N 78.245°E	cm	SGR	95.2 ±6.8	121.2 ±10.2	27.9 ±5.2	13.2 ±5.2	Ti-in-Quartz + P-T path	393 ±40	350 ±80	80 ±37	**8.0E-15** 4.0E-14 1.3E-15	**5.0E-15** 4.0E-14 3.0E-16

Sample	Location	Unit	Recryst. regime					Thermometry				Strain rate	
LA47	34.009°N 78.303°E	mm	SGR	31.8 ±6.8	40.5 ±10.2	67.0 ±20.3	31.4 ±15.6	Ti-in-Quartz + P-T path	394 ±40	350 ±80	80 ±37	**2.7E-13** 1.9E-12 2.4E-14	**7.5E-14** 6.5E-13 3.4E-15
LA44 (PMC)	33.971°N 78.376°E	cm	SGR	107.2 ±6.8	136.5 ±10.2	28.7 ±4.8	12.0 ±5.3	Ti-in-Quartz (LA42) + P-T path	350[c] ±50	350[c] ±80	80 ±37	**1.7E-15** 2.1E-14 8.3E-17	**6.9E-16** 1.5E-14 1.1E-17
LA42 (PMC)	33.971°N 78.376°E	cm	SGR / GBM[c]	140.5 ±6.8	178.8 ±10.2	23.4 ±3.7	9.7 ±4.4	Ti-in-Quartz + P-T path	347 ±40	350 ±80	80 ±37	**6.3E-16** 2.8E-15 1.2E-16	**3.1E-16** 2.4E-15 1.8E-17

Note: ASRR= Ailao Shan Red River; KFZ= Karakorum Fault Zone; Recrystallization regime: SGR= sub-grain rotation; BLG= bulging;; N.D.: No dynamic Recrystallization

Uncertainty calculation takes into account: the experimental measurement errors (LA-ICP-MS, microthermometry, grain size, P-T path, EBSD, Fabric analyser), the errors of equations calibration when available (piezometer, flow law, thermo-barometer) and they are propagated to measure the strain rate.

[a] The recrystallization regime is determined by the shape of the considered grains following criteria of Stipp et al. (2002).

[b] Stereographic correction

[c] Uncertainties in the strain rate value due to the absence of Titanium-in-quartz measurements or defined recrystallization regime

Stress calculated using: *Shimizu (2008) or **Stipp and Tullis (2003) piezometer

Strain rates calculated using: §Hirth et al. (2001)/ Shimizu (2008) or §§Paterson and Luan (1990)/ Stipp and Tullis (2003) combination

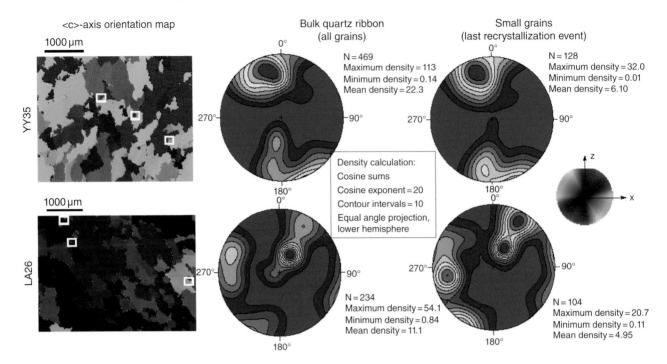

Fig. 6.6. Comparison of the CPOs of bulk quartz ribbon (left) and newly recrystallized quartz grains (right). Up: sample YY35 (ASRR). Down: sample LA26 (KSZ). Small grains belonging to the last recrystallization event are indicated with white squares on CPO maps (left).

calibration of Thomas et al. (2010), these values, combined with a TiO_2 activity of 0.8 ± 0.2, lead to a temperature range of 340–460°C, for pressures ranging between 0 and 6 kbar (Figs 6.9 and 6.10). Intermediate Ti-contents are measured for four quartz ribbons located in the central part of the ASRR: YU42 (10.1 ± 2.1 ppm), YY33 (14.6 ± 1.8 ppm), YY35 (14.3 ± 1.5 ppm), and YU29 (25.9 ± 2.1 ppm), leading to a temperature range of 380–560°C (Thomas et al. 2010) for pressures ranging between 0 and 6 kbar (Figs 6.9 and 6.10). High values of Ti in quartz are measured for the three samples of the north-eastern border of the ASRR: YY72 (40.8 ± 4.0 ppm), YY54 (61.7 ± 2.9 ppm), and YU61 (64.1 ± 8.3 ppm), corresponding to temperatures ranging from 470 to 640°C (Thomas et al. 2010), for pressures ranging between 0 and 6 kbar (Figs 6.9 and 6.10).

Slight differences of Ti-contents are observed for YY35 between the quartz vein core (e.g. dd2 to dd6, Fig. 6.1) and the boundaries (dd8, 9, and 11, Fig. 6.1). However, this difference could be not significant, as it lies within the error bars of the measurements (Fig. 6.8). We observe no difference between core and boundary Ti-contents for the other samples. Nevertheless, for sample YU73, there is a real difference between the Ti-content measured in the quartz ribbon (4.8 ± 0.7 ppm) and in the matrix (12.1 ± 1.9 ppm) (Table 6.2). In most cases, the analyzed points have been made preferentially in the newly recrystallized grains. For control, the Ti-contents of samples YU44 and YU73 quartz veins were investigated for both large and small quartz grains of the quartz vein, and we

found no difference between large relict grains and small newly recrystallized grains.

6.4.2.3 *Inferred thermodynamic conditions*

The *P*–*T* conditions are obtained by combining several thermobarometers. In the initial calibration of the QSR method (YY35 and YY33) (Boutonnet et al. 2013), the authors intersect two independent thermobarometers: the TitaniQ and the fluid inclusions microthermometry. The results for sample YY35 are displayed in Fig. 6.9(a). The conditions of *T* = 425 ± 38°C and *P* = 130 ± 80 MPa are compatible with those of YY33, and are also close to the *P*–*T*-time path previously proposed for the central Ailao Shan (Leloup et al. 2001). The crystallographic preferred orientation of these two samples is consistent with the temperature of deformation. Finally, at 425°C, quartz is supposed to recrystallize by SGR (Stipp et al. 2002b). This suggests that the considered quartz recrystallization event corresponds to the SGR grains and occurred around 23 Ma (Boutonnet et al. 2013). As the fluid inclusions' microthermometry is a tedious method, implying 1 week of work per sample and redundant with the other thermobarometers, in this study we simply combine the titanium-in-quartz thermobarometer with the local *P*–*T* path for the other samples.

Figure 6.9 presents the *P*–*T* diagrams of the ASRR shear zone. The *P*–*T* path of the ASRR has been modified in order to fit exactly the *P*–*T* conditions given by the site

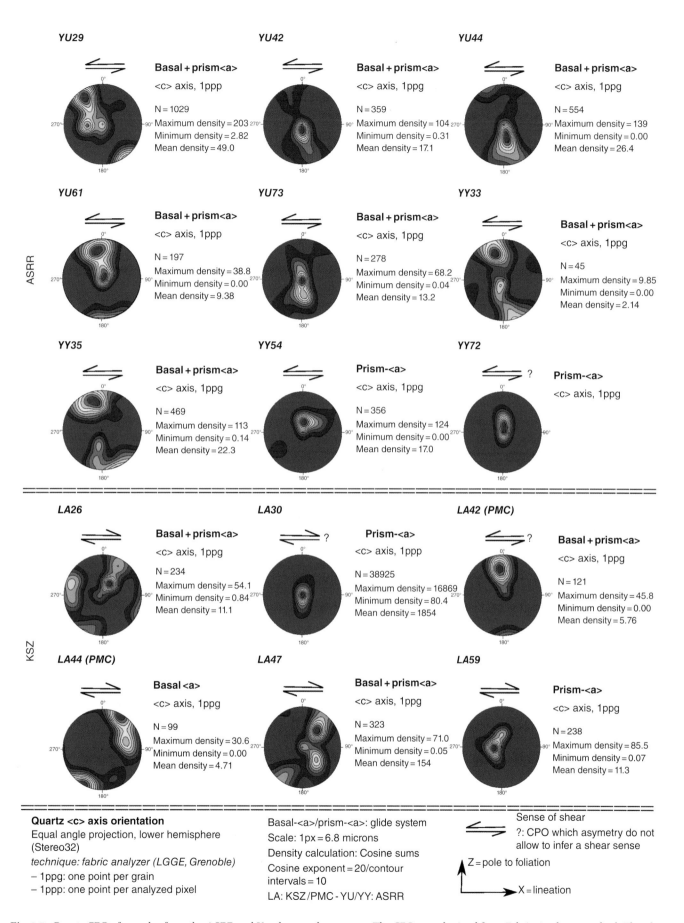

Fig. 6.7. Quartz CPO of samples from the ASRR and Karakorum shear zones. The CPOs are obtained from Fabric Analyzer method. The plots are stereographic lower hemisphere projections, with cosine density contours.

Table 6.2. Titanium-in-quartz measurements

Sample	47Ti[b] (ppm)	48Ti[b] (ppm)	49Ti[b] (ppm)	Ti (ppm)[a] (ppm)	Error (ppm)	TitaniQ thermobarometer[c]			
						a	error	b	Error
ASRR									
YU29	27.4 ± 2.0		24.4 ± 2.1	**25.9**	2.1	4.9E-02	1.2E-03	−35.01	2.22
YU42	10.5 ± 1.7	10.2 ± 2.6	9.5 ± 2.1	**10.1**	2.1	5.4E-02	1.3E-03	−35.01	2.22
YU44	6.3 ± 0.5		5.1 ± 0.3	**5.7**	0.9	5.6E-02	4.2E-04	−35.01	2.22
YU61	64.5 ± 8.3	63.5 ± 7.3	64.2 ± 7.7	**64.1**	8.3	4.5E-02	4.5E-03	−35.01	2.22
YU73 -ribbon	4.8 ± 0.7	4.0 ± 0.4	4.7 ± 0.5	**4.5**	0.9	5.8E-02	4.8E-04	−35.01	2.22
YU73 -matrix	12.1 ± 1.9	12.1 ± 3.7	13.2 ± 3.0	**12.5**	3.7				
YY33	15.0 ± 0.7		14.1 ± 0.9	**14.6**	1.8	5.2E-02	1.1E-03	−35.01	2.22
YY35	14.7 ± 0.4		13.7 ± 0.5	**14.3**	1.5	5.2E-02	3.6E-04	−35.01	2.22
YY54	61.9 ± 1.1	61.5 ± 1.3	61.6 ± 2.9	**61.7**	2.9	4.5E-02	1.1E-03	−35.01	2.22
YY72	40.8 ± 4.0	40.8 ± 3.1	40.7 ± 3.8	**40.8**	4.0	4.7E-02	2.1E-03	−35.01	2.22
KSZ									
LA26	6.4 ± 0.4	5.9 ± 0.4	6.5 ± 0.5	**6.3**	0.4	5.6E-2	4.3E-4	−35.01	2.22
LA30	4.6 ± 0.4	4.0 ± 0.5	4.6 ± 0.7	**4.8**	0.5	5.7E-2	4.6E-4	−35.01	2.22
LA42	1.9 ± 0.4	1.4 ± 0.2	1.9 ± 0.3	**1.7**	0.3	6.2E-2	4.2E-4	−35.01	2.22
LA47	4.4 ± 0.5	3.9 ± 0.7	4.5 ± 0.3	**4.3**	0.5	5.8E-2	4.5E-4	−35.01	2.22
LA59	4.6 ± 0.7	3.6 ± 0.6	4.6 ± 0.8	**4.3**	0.7	5.8E-2	5.6E-4	−35.01	2.22

[a] Ti contents are calculated by combining 47Ti, 48Ti (when available) and 49Ti measurements.
[b] Ti (ppm) calculated with the indicated Ti isotope
[c] thermo-barometer calibrated by Thomas et al. (2010): P = a T + b

C1 samples. These conditions correspond to a slightly lower pressure than was expected from the central Ailao Shan P–T–t path (Leloup et al. 2001) (Fig. 6.9a). This is not surprising, as pressure was not tightly constrained in this part of the path (Leloup et al. 2001). The highest pressure and temperature conditions correspond to the quartz samples located in the north-eastern border of the ASRR: YU61 ($P = 180 ± 70$ MPa and $T = 548 ± 96°C$), YY54 ($P = 180 ± 80$ MPa and $T = 544 ± 51°C$), and YY72 ($P = 160 ± 80$ MPa and $T = 507 ± 59°C$). The samples located close to the center of the ASRR shear zone display intermediate conditions: YY33/YY35 ($P = 130 ± 80$ MPa and $T = 425 ± 40°C$) and YU29 ($P = 150 ± 80$ MPa and $T = 469 ± 44°C$). Finally, the lowest conditions are recorded by the three samples of site C2, located in the south-western side of the ASRR shear zone: YU42 ($P = 120 ± 80$ MPa and $T = 402 ± 40°C$), YU44 ($P = 110 ± 80$ MPa and $T = 367 ± 40°C$), and YU73 ($P = 100 ± 80$ MPa and $T = 352 ± 40°C$).

Figure 6.10 presents the P–T diagrams of the KSZ. The P–T–t path of the Pangong Range is inferred from the P–T exhumation path of its granulitic unit (Rolland et al. 2009), local P–T conditions in the south-western strand (Rutter et al. 2007) and the local Temperature- time constrains of Boutonnet et al. (2012). The P–T–t path of the Pangong Metamorphic Complex is inferred from the studies of Streule et al. (2009) and McCarthy and Weinberg (2010). Although inherited from a late Cretaceous metamorphic event (Streule et al. 2009), the PMC has been reheated during the Miocene (McCarthy and Weinberg 2010), and at least the part of the P–T–t below 500°C is related to the KSZ metamorphic event. All the six samples display rather similar conditions.

The pressures are constant around 350 ± 80 MPa. The temperatures increase slightly from the north-eastern side to the south-western side: 347–350 ± 50°C in the Pangong Metamorphic Complex (LA42/LA44), 393–394 ± 40°C close to the Muglib strand (LA47 and LA59) and 400–415 ± 40°C in the Tangtse strand (LA30 and LA26). P–T conditions of sample LA44 are inferred assuming the same TitaniQ thermobarometer, as sample LA42 located nearby.

6.5 DISCUSSION

6.5.1 Constraining the temperature of deformation

Several methods are commonly used to measure the temperature of deformation. They are all disputed, particularly their ability to re-equilibrate when pressure and temperature vary through time. Next, we discuss the accuracy of three methods, and the correlation between the measured thermodynamic conditions and the paleo-stress.

6.5.1.1 Temperatures deduced from CPOs and recrystallization regimes

The temperature fields inferred from the CPOs and recrystallization regimes have been plotted in Figs 6.9 and 6.10 for all samples, following the ranges of temperatures described in Sections 6.2.3 and 6.2.1. We can notice several inconsistencies. (1) In most cases, the CPO temperatures are higher than the TitaniQ temperatures (e.g. samples LA30, LA59, YU42, YU44, YU73). (2) A more

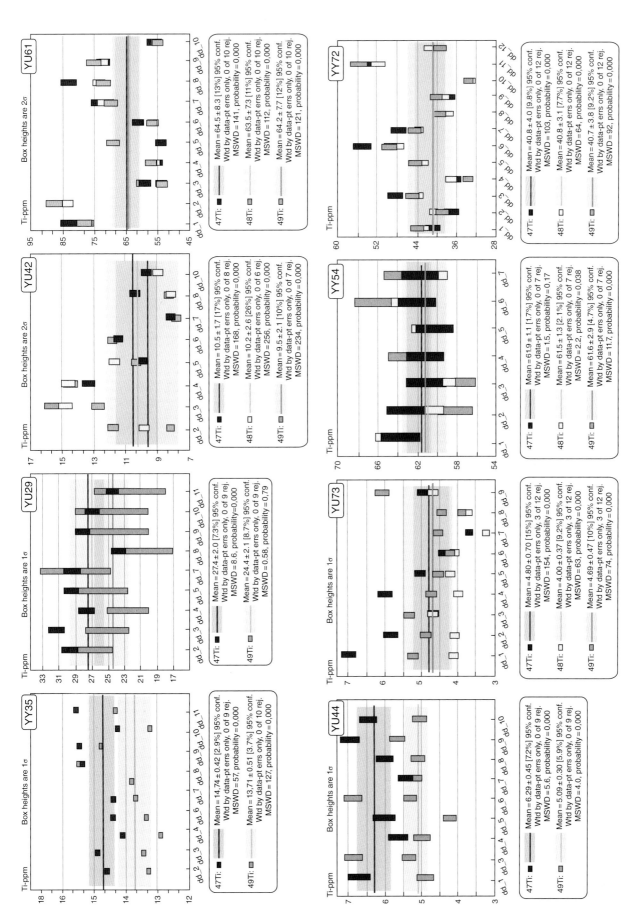

Fig. 6.8. Titanium-in-quartz measurements by La-ICP-MS. ^{47}Ti, ^{49}Ti, and ^{48}Ti (when available) contents of quartz (ppm) for each analysis (dd$_n$), and weighted average for all isotopes. No remarkable difference is noticed between core and rim analysis in sample YY35 and YU44 quartz veins – see Fig. 6.1. For sample YU73, only the quartz vein data are plotted and not the quartz matrix ones. All the data are presented in Table 6.1 and Table 6.2.

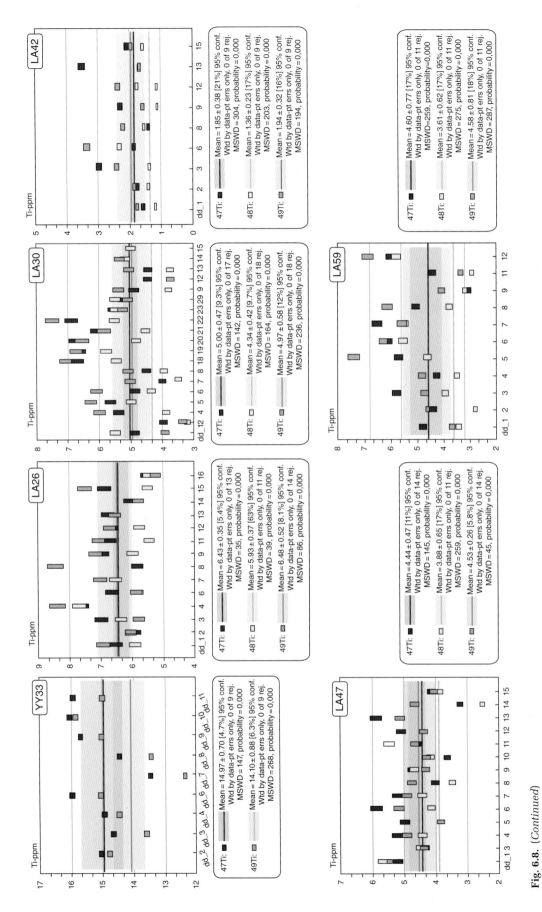

Fig. 6.8. (*Continued*)

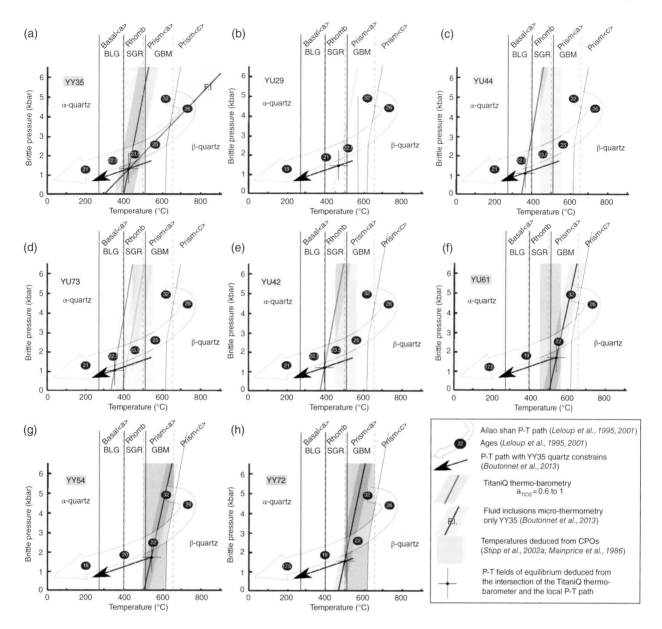

Fig. 6.9. Correlation between *P–T–t* conditions of quartz equilibration and recrystallization events for the ASRR. The temperature conditions are given by the intersection between the TitaniQ thermobarometer for quartz and the local *P–T–t* fields (Leloup et al. 1995, 2001; Harrison et al. 1992, 1996). The ranges of temperatures deduced from microstructures (Stipp et al. 2002a, b) and CPOs (Gapais and Barbarin 1986; Stipp et al. 2002a; Mainprice et al. 1986) are also plotted even if not considered (see Section 6.5.1.1). a) sample YY35; b) sample YU29; c) sample YU44; d) sample YU73; e) sample YU42; f) sample YU61; g) sample YY54; h) sample YY72.

detailed plot of the CPOs for samples showing a clear bimodal grain size repartition indicates that the small grains and the larger grains have the same preferred orientation (Fig. 6.6). (3) Some samples display temperature conditions at which quartz should recrystallize by bulging or GBM, but their sizes (Stipp et al. 2010) and shapes (Stipp et al. 2002b; Passchier and Trouw 1998) indicate that they recrystallized by SGR. Therefore, we conclude that the temperature ranges inferred from CPOs and recrystallization regimes are sometimes inaccurate because these two phenomena do not depend only on temperature.

Dynamic recrystallization is a very complex process depending on temperature but also stress, strain partitioning into a polymineralic rock and former CPOs.

1 The influence of stress on recrystallization regimes has long been observed (Hirth and Tullis 1991; Stipp et al. 2002b) but only recently quantified. The study of Stipp et al. (2010) shows a relationship between the quartz grain size, and thus the stress via a piezometer, and the recrystallization regime. Their results match with our study, in which the populations of grains of the last recrystallization event by SGR have all a

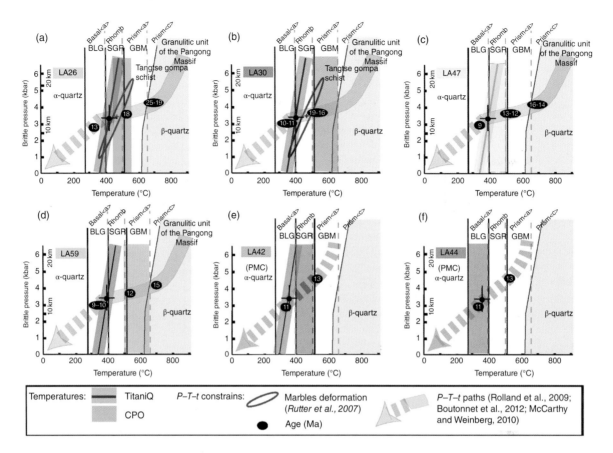

Fig. 6.10. Correlation between *P–T–t* conditions of quartz equilibration and recrystallization events for the KSZ. The temperature conditions are given by the intersection between the TitaniQ thermobarometer for quartz and the local *P–T–t* fields (Boutonnet et al. 2012; Rutter et al. 2007; Rolland et al. 2009; McCarthy and Weinberg 2010). The ranges of temperatures deduced from microstructures (Stipp et al. 2002a, b) and CPOs (Gapais and Barbarin 1986; Stipp et al. 2002a; Mainprice et al. 1986) are also plotted even if not considered (see Section 6.5.1.1). a) sample LA26; b) sample LA30; c) sample LA47; d) sample LA59; e) sample LA42; f) sample LA44.

2D size between ~35 and 120 µm, except sample LA42 (Table 6.1).

2 The usual correlation between temperature and CPOs can be locally inverted if a strong phase, like feldspar porphyroclasts, is present into the quartz ribbon (Peternell et al. 2010). Therefore, within a polymineralic rock where the deformation is partitioned between minerals, quartz microstructures and CPOs may not represent the bulk rock kinematics (Kilian et al. 2011b). This should not be the case for our samples, because we chose quartz ribbons as pure as possible (see also Section 6.5.2.2).

3 Inherited individual crystals orientations create a kind of strain partitioning. When a quartz crystal has a lattice orientation compatible with the present state of stress, the slip system is easily activated and the grain becomes more elongated with increasing strain (Stipp and Kunze 2008). Porphyroclasts with 'hard' orientations (i.e. with an orientation unsuitable for easy slip) are more difficult to deform, and they are selectively removed by dynamic recrystallization (Stipp and Kunze 2008; Pennacchioni et al. 2010). This can explain why both populations of the bimodal samples share the same CPO. The large grains are those that

were previously favorably oriented, whereas the small newly recrystallized grains derive from prophyroclasts with an initial "hard" orientation, which recrystallize by SGR to orient more compatibly.

6.5.1.2 Titanium-in-quartz temperatures

The TitaniQ thermobarometer is one of the most precise ways to determine the temperature of quartz: theoretically ±5°C (Wark and Watson 2006), but practically around ±30°C. This explains why it has been frequently used (e.g. Pennacchioni et al. 2010; Behr and Platt 2011) and is the scope of numerous studies (e.g. Cherniak et al. 2007; Grujic et al. 2011). The principal questions arising from these studies are: (1) How does the chemical system behave during quartz recrystallization? (2) Is the chemical system able to re-equilibrate during exhumation?

The distances of diffusional alteration of Ti concentrations in quartz in 1 Ma are approximately 340 µm at 800°C, 10 µm at 600°C, 1 µm at 500°C, and ~0.2 µm at 400°C (Cherniak et al. 2007). These very slow diffusion rates should greatly limit the applicability of TitaniQ to gauge the deformation temperature, because the duration of plastic shearing along a ductile shear zone is usually not sufficient at $T < 600°C$ to homogeneously reset the Ti concentration in quartz grains >10 μm. Nevertheless, recent work (Grujic et al. 2011) showed that the Ti-diffusion processes may be enhanced at temperatures as low as ~500°C by the fast grain boundary movements during GBM recrystallization in presence of water. Kidder et al. (2013) extended this to temperatures as low as 250–410°C, at which quartz recrystallizes by slow boundary migration through bulging, but they found that the calibration of Huang and Audedat (2012) is more accurate in these conditions. Consequently, in a pure SGR regime, one does not expect an efficient Ti re-equilibration into quartz. Moreover, several authors (Stipp et al. 2002a; Behr and Platt 2011) showed that quartz can keep in memory inherited temperatures; large porphyroclasts generally give much higher Ti concentrations than the surrounding recrystallized grains and this difference is interpreted as low stress/high temperature former conditions kept in memory by the large non-recrystallized grains, and a Ti-re-equilibration for newly recrystallized grains (Kohn and Northrup 2009; Spear and Wark 2009). Thus, during cooling, the TitaniQ temperatures should either be inherited high temperatures ($>500°C$) (Stipp et al. 2002a; Behr and Platt 2011; Grujic et al. 2011) or temperatures close to the ductile–brittle transition ($<400°C$).

However, our temperatures ranging between 350°C and 550°C (Table 6.1), indicate that Ti-contents in quartz re-equilibrated during recrystallization by SGR. Moreover, large grains and small grains display similar Ti-contents. We propose two hypotheses to explain these observations. Hypothesis (1) is related to the fact that most of the studies dealing with the behaviour of the titanium substitution in quartz are made either in steady-state conditions or during prograde metamorphism (Grujic et al. 2011). Negrini et al. (2014) proposed that the retrograde exsolution of titanium is much easier than its prograde incorporation into quartz. This can explain why quartz re-equilibrated at medium temperatures in both the ASRR and the KFZ, in that both underwent a retrograde evolution during shearing. Hypothesis (2) is related to the fact that most of the studies (e.g. Hirth and Tullis 1991) consider that the three recrystallization regimes are independent and activated at different stress/temperature conditions. Another theory associates a nucleation process with a growth process (Derby and Ashby 1987; Shimizu 2008). In their models, the average grain size is derived from a dynamic balance between nucleation, which reduces the grain size, and grain growth. For our samples, the nucleation process is SGR and the growth process is GBM. The relative contribution of each mechanism to bulk microstructure varies with stress and temperature (Hirth and Tullis 1991; Stipp et al. 2002b). Therefore, the small contribution of grain boundary migration at medium temperatures could be sufficient to re-equilibrate the Ti.

Despite all the questions related to the TitaniQ thermobarometer, we consider the measured *P–T* conditions as reliable because they are all consistent with the last recrystallization event, leading to the development of the smallest population of grains. Therefore, this thermobarometer is used to infer the temperature of deformation because it re-equilibrates during retrograde exhumation, as long as quartz recrystallizes, and it closes during/just after the last recrystallization. The other ways to measure the temperature of deformation in quartz, as CPOs and microstructures, are less precise and depend too much on other parameters, such as stress, to be used for the QSR method.

6.5.2 Strain rate measurements

6.5.2.1 Strain rates of the ASRR and KFZ shear zones

The strain rates of some samples have already been measured in a previous study (Boutonnet et al. 2013), but the grain sizes were measured using cross-polarized micrographs. In this study, we used a more robust method, with grain boundary maps based on the <c>-axis orientations. The results are given in Table 6.1 and are close of the previous ones. Boutonnet et al. (2013) showed that, among all piezometers and flow laws calibrated for quartz, only a few combinations of them give accurate strain rates: (1) the experimental piezometer of Stipp and Tullis (2003), corrected for an experimental bias (Holyoke and Kronenberg 2010), yields satisfactory results when associated with Paterson and Luan's (1990) flow law; and (2) Shimizu's (2008) theoretical piezometer gives accurate results when combined with Hirth et al.'s (2001) flow law. The strength of the first combination is the use of the only experimental piezometer calibrated for quartz (Stipp and Tullis 2003). The strength of the second one is the use of the only flow calibrated both on experimental and natural samples (Hirth et al. 2001), and taking into account the water fugacity (f_{H_2O}). The water fugacity can be assimilated to the hydrostatic pressure, calculated knowing the pressure (P, MPa) and a fugacity coefficient, C, determined experimentally, which depends on both pressure and temperature (Tödheide 1972): $f_{H_2O} = P \times C$ (Table 6.1).

The calculated strain rates range between 1.1×10^{-15} s^{-1} and 2.0×10^{-12} s^{-1} for the ASRR shear zone, using the piezometer of Shimizu (2008) and the flow law of Hirth et al. (2001) (Table 6.1, Fig. 6.11). The combination between Stipp and Tullis's (2003) piezometer and Paterson and Luan's (1990) flow law provides very close strain rates, with highest error bars due to the error on the experimental calibration of Stipp and Tullis's (2003) piezometer. The highest values ($>5 \times 10^{-13}$ s^{-1}) correspond to samples located close to the north-eastern border of the shear zone (YU61, YY54, YY72) and the lowest values ($<5 \times 10^{-15}$ s^{-1}) to samples located close to its south-western border (YU42, YU44, YU73). The samples in the centre of the shear zone display intermediate values (YU29, YY33, YY35) (Fig. 6.11a). Plotting the strain rates in linear scale highlights the fact that most of the

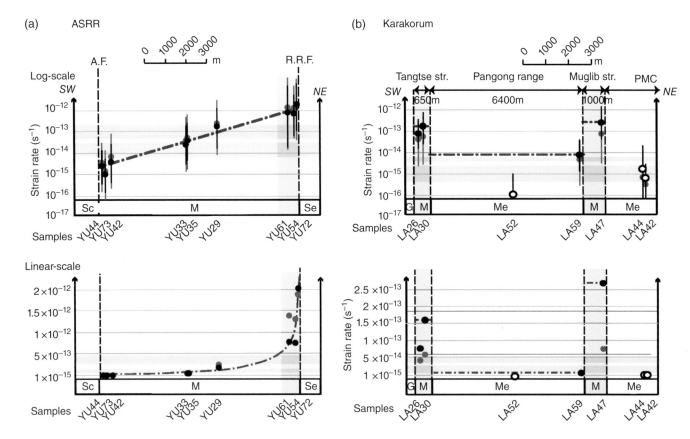

Fig. 6.11. Sections across two major shear zones showing the local strain rates. The quartz-strain-rate-metry (QSR) method used the Shimizu (2008)/Hirth et al. (2001) piezometer rheological/flow law pair (black dots) and the Stipp and Tullis (2003)/Paterson and Luan (1990) piezometer rheological/flow law pair (gray dots). (a) Ailao Shan Red River (ASRR; south-west China) shear zone (see Fig. 6.1b). Bulk strain rates are calculated for a 10-km-wide shear zone, respectively inferring fast fault slip rates between 2.8 and 5.3 cm yr^{-1} (red) or slow ones between 0.5 and 1.4 cm yr^{-1} (blue). (b) Karakorum shear zone (KSZ), India at the latitude of Tangtse village. Bulk strain rates are calculated for a 8-km-wide shear zone, respectively inferring fast fault slip rates between 0.7 and 1.1 cm yr^{-1} (red) or slow ones between 0.1 and 0.5 cm yr^{-1} (blue). Open symbols correspond to samples, for which the strain rates are not well constrained (LA42, LA44 and LA52). Up: log-scale vertical axis; Down: Linear-scale vertical axis. Sc schist, M mylonites, Se sediments, G undeformed granite, Me metamorphic, PMC Pangong Metamorphic Complex, str. strand. Dot-dashed lines indicate shear rate profiles used for the calculation of the integrated shear rates. Adapted from Boutonnet et al. 2013.

deformation was absorbed by the mylonitic strand located close to the Red River fault.

The calculated strain rates range between 6.3×10^{-16} s^{-1} and 2.7×10^{-13} s^{-1} for the KSZ, using the piezometer of Shimizu (2008) and the flow law of Hirth et al. (2001) (Table 6.1, Fig. 6.11). The combination between Stipp and Tullis's (2003) piezometer and Paterson and Luan's (1990) flow law provides lower strain rates, with half an order of magnitude of difference. This difference between the results of the two combinations is higher for the KSZ than for the ASRR, where the initial calibration was made (Boutonnet et al. 2013). This happened since the confining pressure is higher for the KSZ (ca. 350 MPa) than that for the ASRR (ca. 150 MPa). This difference of confining pressure is taken into account by Hirth et al.'s (2001) flow law through the water fugacity parameter, whereas it is ignored by Paterson and Luan's (1990) flow law. For this reason, we prefer to rely on the strain rates provided by the combination of Shimizu's (2008) piezometer and Hirth et al.'s (2001) flow law. For the KSZ,

the highest values ($>1 \times 10^{-13}$ s^{-1}) correspond to samples located in the two mylonitic strands (LA30 and LA47). Intermediates values (around 1×10^{-14} s^{-1}) are obtained for samples located close (<200 m) to these mylonitic strands (LA26 and LA59). Finally, the lowest values ($\leq 1 \times 10^{-15}$ s^{-1}) correspond to samples located far from the mylonitic strands, either in the center of the shear zone (LA52) or outside of it (LA42/LA44). For sample LA52, a very low strain rate value is inferred from the nearly non deformed quartz, where no dynamic recrystallization could be identified (Fig. 4.4).

6.5.2.2 Quartz rheology vs. bulk rock rheology

We ensured that, at the scale of the quartz ribbon, quartz recrystallized freely by dislocation creep mechanism. As mentioned in Section 6.2.1, the purity of the quartz ribbon is important because foreign minerals can vary stress locally and partition strain (Stipp and Kunze 2008; Peternell et al. 2010). We evaluated the effect of strong

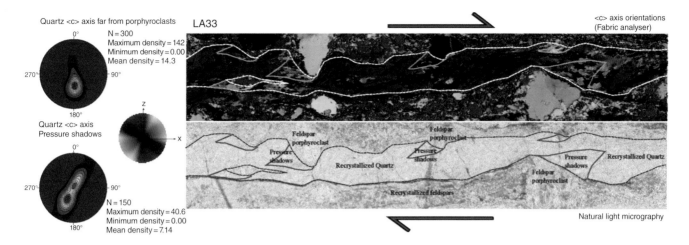

Fig. 6.12. Example of strain partitioning. Sample LA33 is located close to LA30 (see Fig. 6.3) but quartz recrystallization is disturbed by the presence of strong feldspars. Left: stereographic plots of crystallographic preferred orientation of quartz <c>-axis, far from the prophyroclasts (up) and in the pressure shadows (down). Density calculation: Cosine sums. Corsine exponent = 20. Contour intervals = 10. From minimum to maximum. Equal angle projection, lower hemisphere. Right: map of quartz <c>-axis CPOs (up) with corresponding color code, and corresponding natural light micrography of the quartz ribbon (down. with location of feldspars and pressure shadows during dextral shearing.

phases such as feldspar. We sampled a quartz ribbon (LA33) in which feldspar phenocrysts are present, within the mylonitic gneiss of the Tangtse strand of the KSZ located ~100 m from sample LA30. Figure 6.12 shows the <c>-axis orientations measured by Fabric Analyzer method. The feldspar prophyroclasts define pressure shadows, compatible with the dextral sense of shear. Inside the pressure shadows (Fig. 6.12), the recrystallized quartz grains are larger than elsewhere, and their CPOs rotated by an angle of ~30° compared to the CPOs outside of the pressure shadows. The CPOs outside of the pressure shadows resemble to those of the pure quartz ribbon LA30 located in the same area (Fig. 6.7). Thus, the local stress variations induced by the presence of strong prophyroclasts disturb quartz recrystallization and can induce errors in the QSR method.

It has been shown that micas aligned in continuous layers or foliations induce strain weakening (Park et al. 2006). Consequently, dynamic recrystallization of quartz modify at the contact of phyllosilicates. In most gneissic rocks where the quartz ribbons are sampled, the matrix consists of a mixture of quartz, feldspar, and micas, and one must avoid all measurements of grain size at the edges of the quartz ribbons. Finally, polymineralic assemblages (quartz + feldspar or micas) behave with a rheological law close to the diffusion creep one (Kilian et al. 2011a), whereas the QSR is based on dislocation creep laws.

For all these reasons, the QSR method should be applied in pure quartz ribbons. Few of our samples show small feldspars (YU29, YY33, YU61, and YY72, Fig. 6.4), but they are so elongated that we noticed no alteration of the shape and orientation of quartz around them.

At the scale of the sample, one has also to make sure that the quartz ribbon represents the bulk rheology. All our quartz ribbons from the ASRR and KFZ are formed by mineral segregation during deformation. If applied to a late quartz vein, even deformed, the QSR method would not constrain the rheology of the host rock. In this study, we selected quartz ribbons parallel to the main foliation, and we checked with the CPOs that the shear sense is compatible with that of the shear zone (Fig. 6.7). Handy (1990) described three end-member types of mechanical and microstructural behavior for polymineralic rocks: (1) strong minerals form a load-bearing framework that contains spaces filled with weaker minerals; (2) two or more minerals with low relative strengths control bulk rheology and form elongate boudins; (3) one very weak mineral governs bulk rheology, while the stronger minerals form clasts. Our gneissic samples (all ASRR samples and LA26 from KSZ) typically belong to group 2, with a phase of pure quartz, and a phase composed of a homogeneous mixing of micas, recrystallized quartz, and recrystallized feldspar. Both phases form elongated layers parallel to the foliation, indicating that they both control the bulk rheology. For the quartz ribbons surrounded by a schist matrix (all KSZ samples, except LA26), the phyllosilicate-rich layers can control the bulk rheology. However, when the difference of strength between the two phases is significant, the strong phase should form clasts and not ribbons (Mancktelow and Pennacchioni 2010). This indicates that even in the case of a schist matrix, the polymineralic rock deforms mostly in regime 2, and the measured strain rate represents the bulk rock strain rate.

6.5.3 Method limitations

Samples LA42 and LA44 (KSZ) illustrate well some limitations of the QSR method. First, is the minimum deformation rate needed to recrystallize quartz. The lowest value of strain rate measured by the QSR method in this study is 6.3×10^{-16} s^{-1} (LA42, KSZ). The strain rates value of this sample is to take with care, because neither the shape nor the size (Stipp et al. 2010) of the recrystallized grains allowed us to be sure that they recrystallized by SGR (Table 6.1): in this case, applying the QSR method calibrated for SGR is not adequate. For sample LA44, the strain rate is also not well constrained because no TitaniQ temperature was measured and we took the P–T values of the nearby sample LA42. A value of ~1×10^{-15} s^{-1} seems to be the lowest limit of strain rate that can be measured precisely by QSR-metry. Second, these two samples are taken outside of the KSZ, in the Pangong Metamorphic Complex, a range metamorphized and deformed during the late Cretaceous (Streule et al. 2009), and reheated by the KSZ during the Miocene (McCarthy and Weinberg 2010). Nevertheless, no clear dextral deformation can be inferred from the CPOs (Fig. 6.7), and the correlation between the temperature (T) and the deformation (σ) is not obvious. In that case, the calculated strain rate (ε) with the QSR method could be inaccurate. The only assertion we can make is that, during Miocene, the strain rate was lower than 1×10^{-15} s^{-1} at ~300 m northeast from the Mublig strand.

Considering the temperature and grain size uncertainties, as well as those of the piezometers and flow laws, yields relatively large error bars on the final result (Fig. 6.11, Table 6.1), the main error source being the uncertainty on the deformation temperature (Boutonnet et al. 2013). The error bars in strain rates are generally of the order of magnitude in log-scale, which has to be taken into account when absolute values are calculated. But the QSR method is robust when relative values are compared. In both our cases, train rates are obviously inhomogeneous across the shear zones.

6.5.4   Strain localization across shear zones

By using the QSR method calibrated by Boutonnet et al. (2013), we address the problem of localization of the deformation on two major shear zones, for which both fast and slow fault slip rates have been proposed (Fig. 6.11).

The Miocene slip rate of the 10 km-wide ASRR shear zone has been suggested to be rather fast, between 2.8 and 5.3 cm yr^{-1} using geological markers, plate tectonic reconstructions, and cooling histories (e.g. Leloup et al. 2001), or conversely to be slower than 1.4 cm yr^{-1} using different geological markers (e.g. Clift et al. (1997)). If deformation was homogeneous in space and time within a 10-km wide shear zone, this would correspond to shear rates between 8.9×10^{-14} s^{-1} and 1.7×10^{-13} s^{-1}, or below 4.4×10^{-14} s^{-1}, respectively. When plotted along a cross section of the shear zone, the measured strain rates show a progressive increase from 1.1×10^{-15} s^{-1} in the southwest to 2.0×10^{-12} s^{-1} in the north-east (Fig. 6.11), that can be approximated as a linear increase of log($\dot{\varepsilon}$). This suggests strong deformation localization along the northeast border of the shear zone, assuming that quartz represents the rock rheology (see Section 6.5.2.2).

According to the local P–T–t paths of the ASRR (Fig. 6.9), all the recrystallization used for QSR-metry occurred coevally, around 22 ± 1 Ma (Fig. 6.9). In the center of the shear zone (site C1), Sassier et al. (2009) showed that ductile deformation ceased just after the youngest 22.55 ± 0.25 Ma dyke emplaced. In the northeastern border of the shear zone, left-lateral, strike–slip ductile deformation probably ceased by about 20 Ma, followed by brittle deformation with a normal component along the Red River Fault (Harrison et al. 1992). Thus the considered recrystallization events correspond to the very last moments of ductile deformation. An integrated fault slip rate on the order of 4 cm yr^{-1}, valid at ~22 ± 1 Ma, is calculated across the shear zone. Such velocity is in the high range of the slip rates proposed for the ASRR.

Applying Equation 1, the differential stresses range is 28–47 MPa (Shimizu's (2008) piezometer), in the temperature range 350–550°C. The correlation between local temperatures and local strain rates is clear, with the lowest temperatures recorded in the south-western side and the highest ones in the north-eastern part (Table 6.1, Fig. 6.2). So, strain localization across the ASRR at ca. 22 Ma seems to be controlled mostly by the local temperatures.

The Neogene–Quaternary slip rate of the KFZ is disputed, with values deduced from geological and geodetic data range <0.5 cm yr^{-1} up to 1.1 cm yr^{-1} (e.g. Wright et al. 2004; Chevalier et al. 2005; Boutonnet et al. 2012). In the Tangtse area (India), deformation confined within the two narrow Tangtse and Muglib mylonitic strands (e.g. Boutonnet et al. 2012, Fig. 6.3). The six QSR measurements confirm this impression with values above 1.6×10^{-13} s^{-1} in the two mylonitic strands, and below 1.0×10^{-14} s^{-1} outside (Fig. 6.11). This suggests that strain localized in the two mylonitic strands, as suggested by the qualitative description (Boutonnet et al. 2012), assuming once more that quartz represents the rock rheology (see Section 6.5.2.2). The measured shear rates correspond to an integrated fault slip rate of 1.3 cm yr^{-1}, in the highest range of the previous geological and geophysical estimates.

According to the local P–T–t paths of the KSZ (Fig. 6.10), all the recrystallization events used for QSR-metry occurred at the same temperature, 380 ± 35°C, slightly above the ductile- brittle transition located at ~10 km depth (Boutonnet et al. 2012). As in the ASRR case, these recrystallization events correspond to the very last moments of ductile deformation. A diachronism of cooling ages is observed across the shear zone: ductile deformation ended at ~12 ± 1Ma in the Tangtse strand and 8.5 ± 0.5 Ma in the Muglib strand (Boutonnet et al. 2012). As no evidence of brittle deformation had been found in the Tangtse strand, we suggest that ductile deformation finally localized in the Muglib strand after 10 Ma, and became brittle when temperature decreased below ~250°C, as attested by the occurrence of brittle faults (Rutter et al. 2007).

According to Equation 1 and Shimizu (2008), the differential stresses range between 24 and 67 MPa, at a rather constant temperature (380 ± 35°C) across the shear zone. The highest stresses are recorded in the two mylonitic strands, especially in the Muglib strand, where brittle deformation finally localized. Thus, strain localization across the KSZ between 13 and 8 Ma seems to be associated to high stresses rather than temperature differences.

In the case of the ASRR and KSZ, deformation rates appear to be variable across strike, with narrow (a few kilometers wide) zones with strain rates of $\geq 10^{-13}$ s^{-1}, where most of the deformation localizes. This is in accordance with the qualitative field observations. For the two studied cases, the shear rates, when integrated across strike, are compatible with the fastest slip rates inferred from geologic and geodetic considerations. The strain rates in these kilometre-wide zones are more than 500 times higher than in the other parts of the exposed shear zones, and more than 1000 times higher than in the shear zone surroundings. This implies that a 1-km-wide zone of localized strain can accommodate as much deformation as a 1000-km-wide block. The temperature is often proposed to be the major cause of strain localization (e.g. Leloup et al. 1999), which seems to be the case in the ASRR shear zone at the end of its ductile history. Nevertheless, the strain localization in the Karakorum shear zone seems to be related to high stresses, which can themselves result from grain size reduction (e.g. Ricard and Bercovici 2009; Rozel et al. 2011).

6.6 CONCLUSIONS

The QSR method provides measurements of local strain rates, in terms of time and space, of ductilely deformed continental rocks. As quartz is one of the most ubiquitous mineral in continental crust, the measurements could be rather dense, provided that deformation is intense enough to recrystallize the quartz layers. Thus, a regular sampling of quartz ribbons should allow mapping the strain rate across or along a natural shear zone at a given moment of its history. Because of recent improvements of several methods that can be efficiently used for quartz measurements, such as the TitaniQ method (Wark and Watson 2006) or the Fabric Analyzer for the CPOs (Peternell et al. 2010), the QSR-metry method is fast and cheap. The quartz recrystallization regime and grain size are inferred from optical observations and crystallographic preferred orientations measurements. We combine the TitaniQ thermobarometer (Thomas et al. 2010) with the local P–T–t path of the shear zone in order to measure the thermodynamic conditions of the last recrystallization event. The accuracy can be improved by the use of the quartz fluid inclusions microthermometry (Boutonnet et al. 2013), but the time for analysis of each sample is much longer.

Nevertheless, several pitfalls are to be avoided: (1) in a context where pressure and temperature vary through time, such as during exhumation, a careful correlation between the different parameters (grain size, temperature,

pressure) has to be done; (2) quartz has to recrystallize freely, without interference with other minerals in pure quartz ribbons, and in a matrix of same strength, so that the quartz rheology represents the bulk rock rheology.

The obtained strain rates can be plotted on sections across strike and compared relatively. In the case of the ASRR shear zone, strain localization occurs in the northeastern side of the shear zone, and for the KSZ, two strands mostly localize the ductile deformation before the transition to brittle deformation. The quantitative results of strain localization agree with the field observations of mylonite localization. Moreover, despite the large error bars, the absolute shear rate, integrated across the entire shear zone, agree with longer term geological measurements. The two studied ductile shear zones thus localize large amounts of deformation.

Stress and strain rate profiles obtained through this method can be a reference for experimental studies, in order to test the main rheological laws and the piezometers, in real conditions. Indeed, most of the experimental studies are made at strain rates much higher than the natural ones (~eight orders of magnitude) and few experimental studies take into account the burial or exhumation processes. Finally, these profiles can be useful as a benchmark for numerical studies that deal with the strain localization inside shear zones (e.g. Leloup et al. 1999).

ACKNOWLEDGMENTS

We thank the French National Program 3F (INSU -CNRS) for funding. Emmanuelle Boutonnet thanks the ERC 258830. We are grateful to V. Gardien, C. Lefebvre, M. Montagnat, M. Peternell, Y. Ricard, C. Sassier, and K. Schulmann for discussions and advice. We thank Petr Jeřábek and Soumyajit Mukherjee for constructive review.

REFERENCES

Allen CR, Gillespie AR, Han Y, Sieh KE, Zhang B, Zhu C. 1984. Red river and associated faults, Yunnan province, china: Quaternary geology, slip rates and seismic hazard. Geological Society of America Bulletin 95, 686–700.

Anczkiewicz R, Viola G, Mutener O, Thirlwall MF, Villa IM, Quong NQ. 2007. Structure and shearing conditions in the day Nui Con Voi massif: Implications for the evolution of the red river shear zone in northern Vietnam. Tectonics 26(2), TC2002, doi: 10.1029/2006TC001972.

Banerjee P, Bürgmann R. 2002. Convergence across the northwest Himalaya from GPS measurements. Geophysical Research Letters 29(13), 30-1–30-4.

Beaumont C, Jamieson RA, Nguyen MH, Lee B. 2001. Himalayan tectonics explained by extrusion of a low-viscosity crustal channel coupled to focused surface denudation. Nature 414(6865), 738–742.

Behr WM, Platt J. P. 2011. A naturally constrained stress profile through the middle crust in an extensional terrane. Earth and Planetary Science Letters 303(3–4), 181–192.

BGMRY 1983. Geological map of Yunnan, 1:500000. Bureau of Geology and Mineral Resources of Yunnan.

Boutonnet E, Leloup PH, Arnaud JLPN, Davis WJ, Hattori K. 2012. Synkinematic magmatism, heterogeneous deformation, and progressive strain localization in a strike-slip shear zone: The case of the right-lateral karakorum fault. Tectonics 31, TC4012.

Boutonnet E, Leloup PH, Sassier C, GardienV, Ricard Y. 2013. Ductile strain rate measurements document long term strain localization in the continental crust. Geology 41, 819–822.

Briais A, Patriat P, Tapponnier P. 1993. Updated interpretation of magnetic anomalies and sea floor spreading stages in the South China Sea, implications for the tertiary tectonics of SE Asia. Journal of Geophysical Research 98(B4), 6299–6328.

Cherniak D, Watson E, Wark D. 2007. Ti diffusion in quartz. Chemical Geology 236(1–2) 65–74.

Chevalier M-L, Ryerson F, Tapponnier P, et al. 2005. Slip-rate measurements on the Karakorum fault may imply secular variations in fault motion. Science. 352(307), 411–414.

Christensen JN, Rosenfeld JL, DePaolo DJ. 1989. Rates of tectono-metamorphic processes from rubidium and strontium isotopes in garnet. Science 244, 1465–1469.

Clift PD, Long HV, Hinton R, Ellam RM, et al. 1997. Evolving East Asian river systems reconstructed by trace element and Pb and Nd isotope variations in modern and ancient red river-song hong sediments. Geochemistry Geophysics, and Geosystematics 9(4), doi:Q04039.

Derby B, Ashby M. 1987. On dynamic recrystallization. Scripta Metallurgica 21(6), 879–884.

Dunlap W, Weinberg R, Searle M. 1998. Karakorum fault zone rocks cool in two phases. Journal of the Geological Society 155, 903–912.

Ehiro M, Kojima S, Sato T, Ahmad T, Ohtani T. 2007. Discovery of Jurassic ammonoids from the Shyok suture zone to the northeast of Chang La Pass, Ladakh, northwest India and its tectonic significance. Island Arc 16, 124–132.

Fyhn MBW, Boldreel LO, Nielsen LH. 2009. Geological development of the central and South Vietnamese margin: Implications for the establishment of the South China Sea, Indochinese escape tectonics and cenozoic volcanism. Tectonophysics 478(3–4), 184–214.

Gapais D, Barbarin B. 1986. Quartz fabric transition in a cooling syntectonic granite (Hermitage Massif, France). Tectonophysics 125(4), 357–370.

Gaudemer Y, Tapponnier P, Turcotte D. 1989. River offsets across active strike-slip faults. Annales Tectonicae 3(2), 55–76.

Ghent E, Stout M. 1984. TiO_2 activity in metamorphosed pelitic and basic rocks: principles and applications to metamorphism in southeastern Canadian cordillera. Contribution to Mineral Petrology 86(3), 248–255.

Gilley L, Harrison T, Leloup P, Ryerson F, Lovera O, Wang J-H. 2003. Direct dating of left-lateral deformation along the Red River shear zone, China and Vietnam. Journal of Geophysical Research 108(B2), 1–21.

Gleason G, Tullis J. 1995. A flow law for dislocation creep of quartz aggregates determined with the molten salt cell. Tectonophysics 247(1–4), 1–23.

Grujic D, Stipp M, Wooden JL. 2011. Thermometry of quartz mylonites: importance of dynamic recrystallization on Ti-in-quartz reequilibration. Geochemistry Geophysics Geosystems 12(6), doi: 10.1029/2010GC003368.

Gueydan F, Mehl C, Parra T. 2005. Stress- strain rate history of a midcrustal shear zone and the onset of brittle deformation inferred from quartz recrystallized grain size. In Deformation Mechanisms, Rheology and Tectonics: from minerals to the lithosphere, edited by D. Gapais, J.P. Brun and P.R. Cobbold, Geological Society of London, Special Publications, 243, pp. 127–142.

Hammer O, Harper DAT, Ryan PD. 2001. Past: paleontological statistics software package for education and data analysis. Paleontologia Electronica 4(1), 9.

Handy MR. 1990. The solid-state flow of polymineralic rocks. Journal of Geophysical Research 95(B6), 8647–8661.

Harrison TM, Wenji C, Leloup PH, Ryerson FJ, Tapponnier P. 1992. An early miocene transition in deformation regime within the Red River fault zone, Yunnan, and its significance for Indo-Asian tectonics. Journal of Geophysical Research 97(B5), 7159–7182.

Harrison TM, Leloup PH, Ryerson FJ, et al. 1996. Diachronous initiation of transtension along the Ailao Shan-Red River shear zone, Yunnan and Vietnam. In The Tectonic Evolution of Asia, edited by A. Yin and T.M. Harrison, Cambridge University Press, Cambridge, pp. 208–226.

Helmcke D. 1985. The permo-triassic "paleotethys" in mainland southeast-Asia and adjacent parts of China. International Journal of Earth Sciences 74(2), 215–228.

Hirth G, Tullis J. 1991. Dislocation creep regimes in quartz aggregates. Journal of Structural Geology 14(2), 145–159.

Hirth G, Teyssier C, Dunlap W. 2001. An evaluation of quartzite flow laws based on comparisons between experimentally and naturally deformed rocks, International Journal of Earth Sciences 90(1), 77–87.

Holyoke CW, Kronenberg AK. 2010. Accurate differential stress measurement using the molten salt cell and solid salt assemblies in the Griggs apparatus with applications to strength, piezometers and rheology. Tectonophysics 494(1–2), 17–31.

Huang Z, Audedat A. 2012. The titanium-in-quartz (titaniQ) thermobarometer: A critical examination and recalibration. Geochimica et Cosmochimica Acta 84, 75–89.

Huang Z, Wang L, Xu M, Liu J, Mi N, Liu S. 2007. Shear wave splitting across the ailao shan-red river fault zone, SW China. Geophysical Research Letters 34(20), L20,301.

Jain A, Singh S. 2008. Tectonics of the southern Asian plate margin along the karakoram shear zone: Constraints from field observations and U/Pb shrimp ages. Tectonophysics 451(1–4), 186–205.

Jerabek P, Stunitz H, Heilbronner R, Lexa O, Schulmann K. 2007. Microstructural-deformation record of an orogen-parallel extension in the Vepor unit, west Carpathians. Journal of Structural Geology 29, 1722–1743.

Jolivet L, Beyssac O, Goffé B, et al. 2001. Oligo miocene midcrustal subhorizontal shear zone in Indochina. Tectonics 20(1), 46–57.

Kidder S, Avouac JP, Chan YC. 2013. Application of titanium-in-quartz thermobarometry to greenschist facies veins and recrystallized quartzites in the hsuehshan range, Taiwan. Solid Earth 4, 1–21.

Kilian R, Heilbronner R, Stunitz H. 2011a. Quartz grain size reduction in a granitoid rock and the transition from dislocation to diffusion creep. Journal of Structural Geology 33, 1265–1284.

Kilian R, Heilbronner R, Stunitz H. 2011b. Quartz microstructures and crystallographic preferred orientation: Which shear sense do they indicate? Journal of Structural Geology 33, 1446–1466.

Kohn M, Northrup C. 2009. Taking mylonites' temperatures. Geology 37, 47–50.

Lacassin R, Leloup PH, Tapponnier P. 1993. Bounds on strain in large tertiary shear zones of SE Asia from boudinage restoration. Journal of Structural Geology 15(6), 677–692.

Lacassin R, Valli F, Arnaud N, et al. 2004. Large-scale geometry, offset and kinematic evolution of the Karakorum Fault, Tibet. Earth and Planetary Science Letters 219(3–4), 255–269.

Leloup PH, Kienast JR. 1993. High-temperature metamorphism in a major strike-slip shear zone: the Ailao Shan-Red River, People's Republic of China. Earth and Planetary Science Letters 118(1–4), 213–234.

Leloup PH, Lacassin R, Tapponnier P, et al. 1995. The Ailao Shan-Red River shear zone (Yunnan, China). Tertiary transform boundary of Indochina. Tectonophysics 251(1–4), 3–84.

Leloup PH, Ricard Y, Battaglia J, Lacassin R. 1999. Shear heating in continental strike-slip shear zones: numerical modeling and case studies. Geophysical Journal International 136(1), 19–40.

Leloup PH, Arnaud N, Lacassin R, et al. 2001. New constraints on the structure, thermochronology and timing of the Ailao

Shan–Red River shear zone, SE Asia. Journal of Geophysical Research 106(B4), 6683–6732.

Leloup P, Tapponnier P, Lacassin R. 2007. Discussion on the role of the Red River shear zone, Yunnan and Vietnam, in the continental extrusion of SE Asia. Journal of the Geological Society 164, 1253–1260.

Leloup PH, Boutonnet E, Davis WJ, Hattori K. 2011. Long-lasting intracontinental strike-slip faulting: new evidence from the Karakorum shear zone in the Himalayas. Terra Nova 23(2), 92–99.

Luan FC, Paterson MS. 1992. Preparation and deformation of synthetic aggregates of quartz. Journal of Geophysical Research 97(B1), 301–320.

Mainprice D, Bouchez JL, Blumenfeld P, Tubia JM. 1986. Dominant c-slip in naturally deformed quartz: implications for dramatic plastic softening at high temperature. Geology 14, 819–822.

Mancktelow NS, Pennacchioni G. 2010. Why calcite can be stronger than quartz. Journal of Geophysical Research 115(B1), doi:10.1029/2009JB006526.

Matte P, Tapponnier P, Arnaud N, et al. 1996. Tectonics of Western Tibet between the Tarim and the Indus. Earth and Planetary Science Letters 142(3–4), 311–330.

McCarthy M, Weinberg R. 2010. Structural complexity resulting from pervasive ductile deformation in the Karakoram shear zone, Ladakh, NW India. Tectonics 29(3), doi:10.1029/2008TC002354.

Menegon L, Pennacchioni G, Heilbronner R, Pittarello L. 2008. Evolution of quartz microstructure and c-axis crystallographic preferred orientation within ductilely deformed granitoids (Arolla Unit, western Alps). Journal of Structural Geology 30, 1332–1347.

Mukherjee S. 2012. Simple shear is not so simple! Kinematics and shear senses in Newtonian viscous simple shear zones. Geological Magazine 149, 819–826.

Mukherjee S. 2013. Channel flow extrusion model to constrain dynamic viscosity and Prandtl number of the Higher Himalayan Shear Zone. International Journal of Earth Sciences 102, 1811–1835.

Mukherjee S, Mulchrone KF. 2013. Viscous dissipation pattern in incompressible Newtonian simple shear zones: an analytical model. International Journal of Earth Sciences 102, 1165–1170.

Müller W, Aerden D, Halliday AN. 2000. Isotopic dating of strain fringe increments, duration and rates of deformation in shear zones. Science 288, 2195–2198.

Nam TN, Toriumi M, Itaya T. 1998. P-T-t paths and post-metamorphic exhumation of the Day Nui Con Voi shear zone in Vietnam. Tectonophysics 290(3–4), 299–318.

Negrini M, Stünitz H, Berger A, Morales L, Menegon L. 2014. The effect of deformation on the titaniQ geothermobarometer an experimental study, submitted to Contributions to Mineralogy and Petrology 167, 982.

Park Y, Yoo SH, Ree JH. 2006. Weakening of deforming granitic rocks with layer development at middle crust. Journal of Structural Geology 28, 919–928.

Passchier CW, Trouw RAJ (eds). 1998. Microtectonics, 2nd edition. Springer, Berlin, Heidelberg, New York.

Paterson MS, Luan FC. 1990. Quartz rheology under geological conditions. In Deformation Mechanisms, Rheology and Tectonics, edited by R.J. Knipe, and E.H. Rutter, Geological Society Special Publications, 54, 299–307.

Pennacchioni G, Menegon L, Leiss B, Nestola F, Bromiley G. 2010. Development of crystallographic preferred orientation and microstructure during plastic deformation of natural coarse-grained quartz veins. Journal of Geophysical Research 115(B12), doi: 10.1029/2010JB007674.

Peternell M, Hasalova P, Wilson C, Piazolo S, Schulmann K. 2010. Evaluating quartz crystallographic preferred orientations and the role of deformation partitioning using EBSD and fabric analyser techniques. Journal of Structural Geology 32(6), 803–817.

Phillips R, Searle M. 2007. Macrostructural and microstructural architecture of the karakoram fault: relationship between magmatism and strike-slip faulting. Tectonics 26(TC3017).

Phillips R, Parrish R, Searle M. 2004. Age constraints on ductile deformation and long-term slip rates along the Karakoram fault zone, Ladakh. Earth and Planetary Science Letters 226(3–4), 305–319.

Ramsay JG. 1980. Shear zone geometry: a review. Journal of Structural Geology 2, 83–99.

Reichardt H, Weinberg R, Andersson U, Fanning M. 2010. Hybridization of granitic magmas in the source: The origin of the Karakoram batholith, Ladakh, NW India. Lithos 116(3–4), 249–272.

Replumaz A, Lacassin R, Tapponnier P, Leloup PH. 2001. Large river offsets and plio-quaternary dextral slip rate on the red river fault (Yunnan, China). Journal of Geophysical Research 106, 819–836.

Ricard Y, Bercovici D. 2009. A continuum theory of grain size evolution and damage, Journal of Geophysical Research 114(B01204).

Rolland Y, Mahéo G, Pêcher A, Villa I. 2009. Synkinematic emplacement of the pangong metamorphic and magmatic complex along the Karakorum fault (N Ladakh). Journal of Asian Earth Sciences 34(1), 10–25.

Roy P, Jain AK, Singh S. 2010. Microstructures of mylonites along the Karakoram shear zone, Tangste valley, Pangong mountains, Karakoram. Journal of the Geological Society of India. 75(5), 679–694.

Rozel A, Ricard Y, Bercovici D. 2011. A thermodynamically self-consistent damage equation for grain size evolution during dynamic recrystallization. Geophysical Journal International 184(2), 719–728.

Rutter EH, Brodie KH. 2004. Experimental intracrystalline plastic flow in hot-pressed synthetic quartzite prepared from Brazilian quartz crystals. Journal of Structural Geology 26(2), 259–270.

Rutter E, Faulkner D, Brodie K, Phillips R, Searle M. 2007. Rock deformation processes in the Karakoram fault zone, eastern Karakoram, Ladakh, NW India. Journal of Structural Geology 29, 1315–1326.

Sassier C, Leloup P, Rubatto D, Galland O, Yue Y, Lin D. 2009. Direct measurement of strain rates in ductile shear zones: a new method based on syntectonic dikes. Journal of Geophysical Research – Solid Earth 114(B1), B01406.

Schärer U, Zhang L-S, Tapponnier P. 1994. Duration of strike-slip movements in large shear zones: the Red River belt, China. Earth and Planetary Science Letters 126(4), 379–397

Searle M. 2006. Role of the Red River shear zone, Yunnan and Vietnam, in the continental extrusion of SE Asia. Journal of the Geological Society 163, 1025–1036.

Searle M, Phillips R. 2007. Relationships between right-lateral shear along the Karakoram fault and metamorphism, magmatism, exhumation and uplift: evidence from the K2–Gasherbrum–Pangong ranges, north Pakistan and Ladakh. Journal of the Geological Society 164, 439–450.

Searle M, Weinberg R, Dunlap W. 1998. Transpressional tectonics along the Karakoram fault zone, northern Ladakh: constraints on tibetan extrusion. In Continental Transpressional and Transtensional Tectonics, edited by R.E. Holdsworth, R.A. Strachan, and J.F. Dewey, Geological Society, London, Special Publications, 135, 307–326.

Shimizu I. 2008. Theories and applicability of grain size piezometers, the role of dynamic recrystallization mechanisms. Journal of Structural Geology 30, 899–917.

Slotemaker A, Bresser J. D. 2006. On the role of grain topology in dynamic grain growth – 2D microstructural modeling. Tectonophysics 427(1–4), 73–93.

Spear F, Wark D. 2009. Cathodoluminescence imaging and titanium thermometry in metamorphic quartz. Journal of Metamorphic Geology 27(3), 187–205.

Stipp M, Kunze K. 2008. Dynamic recrystallization near the brittle-plastic transition in naturally and experimentally deformed quartz aggregates. Tectonophysics 448(1–4), 77–97.

Stipp M, Tullis J. 2003. The recrystallized grain size piezometer for quartz. Geophysical Research Letters 30(21), 2088.

Stipp M, Stünitz H, Heilbronner R, Schmid SM. 2002a. Dynamic Recrystallization of Quartz: correlation between natural and experimental conditions, edited by D. De Meer, M.R. Drury, J.H.P. De Bresser, and G.M. Pennock, Geological Society, London, Special Publications 200, pp. 170–190.

Stipp M, Stünitz H, Heilbronner R, Schmid SM. 2002b. The eastern tonale fault zone: a 'natural laboratory' for crystal plastic deformation of quartz over a temperature range from 250 to 700°C. Journal of Structural Geology 24(12), 1861–1884.

Stipp M, Tullis J, Scherwath M, Behrmann J. 2010. A new perspective on paleopiezometry: dynamically recrystallized grain size distributions indicate mechanism changes. Geology 38(8), 759–762.

Streule M, Phillips R, Searle M, Waters D, Horstwood M. 2009. Evolution and chronology of the Pangong metamorphic complex adjacent to the Karakoram fault, Ladakh: constraints from thermobarometry, metamorphic modelling and U/Pb geochronology. Journal of the Geological Society 166, 919–932.

Tapponnier P, Molnar P. 1977. Active faulting and tectonics of China. Journal of Geophysical Research 82(20), 2905–2930.

Tapponnier P, Peltzer G, Armijo R, Coward M, Ries A. 1986. On the mechanics of the collision between India and Asia. In Collision Tectonics, edited by M. Coward and A.C. Ries, Geological Society Special Publications, 19, 115–157.

Tapponnier P, Leloup RLPH, Scharer U, et al. 1990. The Ailao Shan/Red River metamorphic belt: tertiary left-lateral shear between Indochina and south China. Nature 343, 431–437.

Tapponnier P, Zhiqin X, Roger F, et al. 2001. Oblique stepwise rise and growth of the Tibet plateau. Science 294, 1671–1677.

Thomas J, Watson EB, Spear F, Shemella P, Nayak S, Lanzirotti A. 2010. Titaniq under pressure, the effect of pressure and temperature on the solubility of Ti in quartz. Contributions to Mineralogy and Petrology 160, 743–759.

Tödheide K. 1972. Water at high temperatures and pressure. In Water, a Comprehensive Treatise, edited by F. Franks, Plenum Press, New York, pp. 463–514.

Twiss R. 1977. Theory and applicability of a recrystallized grain size paleopiezometer. Pure and Applied Geophysics 115(1–2), 227–244.

Valli F, Arnaud N. Leloup P.-H., et al. 2007. 20 million years of continuous deformation along the Karakorum fault, western Tibet: a thermochronological analysis. Tectonics 26(4) TC4004.

Wang PL, Lo CH, Lee TY, Chung SL, Lan CY, Yem NT. 1998. Thermochronological evidence for the movement of the Ailao Shan-Red river shear zone: a perspective from Vietnam. Geology 26, 887–890.

Wark D, Watson E. 2006. Titaniq: a titanium-in-quartz geothermometer. Contribution to Mineral Petrology 152, 743–754.

Weinberg R, Dunlap W, Whitehouse M, Khan M, Searle M, Jan M. 2000. New field, structural and geochronological data from the Shyok and Nubra valleys, northern Ladakh: Linking Kohistan to Tibet. In Tectonics of the Nanga Parbat Syntaxis and the Western Himalaya, edited by A. Khan, P.J. Treloar and M.P. Searle, Geological Society of London, Special Publications, 170, 253–275.

Wright T, Parsons B, England P, Fielding E. 2004. Insar observations of low slip rates on the major faults of western Tibet. Science 305, 236–239.

Zhang L-S, Schärer U. 1999. Age and origin of magmatism along the cenozoic Red River shear belt, China. Contributions to Mineralogy and Petrology 134(1), 67–85.

Chapter 7

Thermal structure of shear zones from Ti-in-quartz thermometry of mylonites: Methods and example from the basal shear zone, northern Scandinavian Caledonides

ANDREA M. WOLFOWICZ[1,2], MATTHEW J. KOHN[1], and CLYDE J. NORTHRUP[1]

[1] *Department of Geosciences, Boise State University, 1910 University Drive, Boise, ID 83725, USA*
[2] *Shell International Exploration and Production, Shell Woodcreek Complex, 200 N Dairy Ashford Rd, Houston, TX 77079, USA*

7.1 INTRODUCTION

Structural evolution of collisional orogens depends closely on how major shear zones develop and evolve. These shear zones, in turn, control aspects of the geometry and thermal structure of the orogenic wedge. Consequently, analysis of the thermal structure along major shear zones can discriminate among large-scale thermal–mechanical models. The Himalayan orogen provides one example where two competing orogenic models – channel flow vs. critical taper – predict vastly different thermal gradients along the underlying master decollement. Himalayan thrusts are not generally exposed for long distances across strike, however, so accurate determination of down-dip thermal gradients is difficult. Within this context, the Caledonian orogen in Norway and Sweden, which is often compared with the Himalaya, provides an unparalleled opportunity to investigate down-dip temperatures. Due to the unusual and consistent exposure of thrust surfaces for ~140 km across strike, thermometry along the deeply eroded remains of this late Silurian-early Devonian orogen (Gee and Sturt 1985) provides a unique opportunity to test competing models of the thermal and kinematic evolution of shear zones and orogens. Here, we report temperatures of deformation in ductilely sheared rocks (mylonites) from a northern transect across the Scandinavian Caledonides using the titanium-in-quartz thermobarometer (TitaniQ; Wark and Watson 2006; Thomas et al. 2010) to investigate the dynamics of quartz recrystallization during shear and to discriminate among competing thermal models for orogenic evolution.

TitaniQ offers several advantages over other thermobarometers. First, it can be applied over a wide range of rocks because quartz, the only phase that requires analysis, is stable over a large range of temperatures and pressures. Second, domains with different Ti-content can be readily identified and targeted for analysis because cathodoluminescence (CL) intensity correlates with trace element content (Rusk et al. 2008; Spear and Wark 2009; Kohn and Northrup 2009). Last, TitaniQ is unusually precise (±3°C at a specified pressure; Wark and Watson 2006).

We evaluated TitaniQ temperatures in mylonites (T_m) from the well-exposed basal thrust zone (BTZ) of the northern Scandinavian Caledonides (Fig. 7.1) and compared them with (1) TitaniQ thermometry in cross-cutting quartz veins (T_{qv}), (2) cation exchange thermometry, (3) metamorphic phase equilibria, (4) broad constraints from geochronology and conditions of the brittle-ductile transition, and (5) matrix quartz microstructures as calibrated experimentally and empirically. In principle, recrystallized quartz from granite might record temperatures attained prior to, during, or after the peak of metamorphism, or even relict temperatures inherited from the protolith. However, mylonites texturally postdate the peak of metamorphism, so quartz grains that recrystallized during mylonitization should record lower temperature than either the metamorphic peak or igneous cooling. Similarly, if temperatures were decreasing during deformation, then quartz veins that postdate deformation (Fig. 7.2) should yield $T_{qv} \leq T_m$. Because quartz veins precipitate directly from a fluid, they should not contain relict grains inherited from the protolith or reflective of earlier conditions, and instead provide a minimum bound on deformation temperatures.

Past work in Caledonian regional geology (Hodges 1982; Crowley 1985; Tilke 1986; Page 1990; Northrup 1996a, b) and quartz microstructures led to the following expectations. First, deformation was syn- to post-metamorphic, so we expected phase equilibrium and garnet-biotite Fe-Mg temperatures to exceed TitaniQ temperatures. Conversely, as temperatures evidently were waning during the latest stages of deformation, cross-cutting quartz veins should provide a minimum bound on deformation temperatures

Ductile Shear Zones: From Micro- to Macro-scales, First Edition. Edited by Soumyajit Mukherjee and Kieran F. Mulchrone.

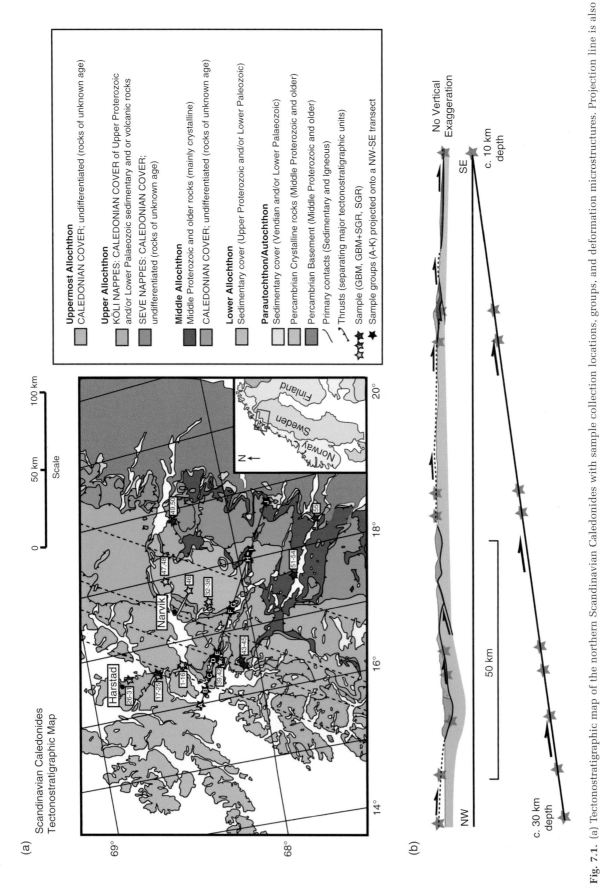

(a)

Scandinavian Caledonides
Tectonostratigraphic Map

Fig. 7.1. (a) Tectonostratigraphic map of the northern Scandinavian Caledonides with sample collection locations, groups, and deformation microstructures. Projection line is also line of cross-section in (c). Inset shows map of Scandinavia with field area in red rectangle. Adapted from Gee 1975. (c) Cross-section along projection line in (a) showing geological relationships and locations of samples along thrust surface. Schematic shows how samples were originally distributed with depth.

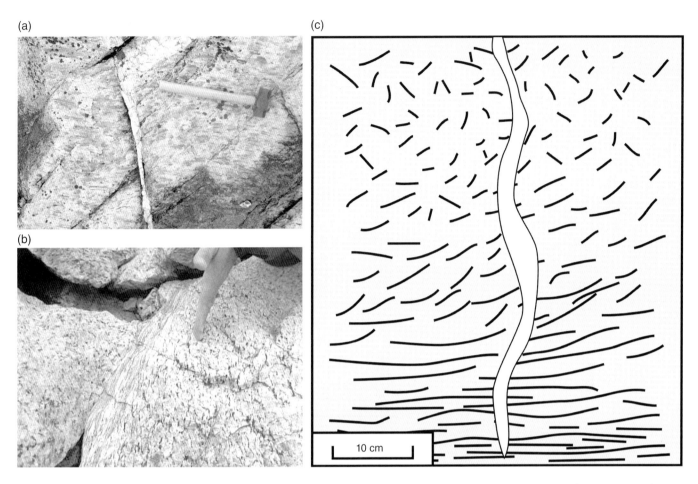

Fig. 7.2. Macroscopic textures. (a) Field photo of cross-cutting quartz vein. Hammer for scale. From SC10-6 locality. (b) Mylonitized basement granite. Hand for scale. From SC10-35 locality. (c) Summary sketch illustrating spatial association of undeformed granite (above) to mylonitized granite (below). Neoblastic quartz vein (white) cross-cuts the host rock and constrains the minimum temperature of deformation.

(i.e. should be at or below TitaniQ temperatures from mylonitic quartz). Second, the BTZ of the Scandinavian Caledonides formed at conditions ranging from the mid-amphibolite facies in the west to the brittle-ductile transition at the east, so we expected TitaniQ temperatures to decrease across the orogen, with maximum temperatures ranging from ~600°C in the west to ~350°C in the east. Last, quartz microstructures correspond with dynamic recrystallization regimes and associated temperatures both experimentally and empirically (Hirth and Tullis 1992; Stipp et al. 2002), so we expected some systematic correspondence between TitaniQ temperatures and microstructures.

7.2 BACKGROUND

7.2.1 The Scandinavian Caledonides

The following summary of Caledonian tectonics is taken from reviews of Gee (1975), Roberts and Gee (1985), Roberts (2003) and Gee et al. (2010), supplemented by a series of PhD studies in our specific research area (Hodges 1982; Crowley 1985; Tilke 1986; Page 1990; Northrup

1996b). In the early Paleozoic, closure of the Iapetus Ocean and the subduction of the western margin of Baltica beneath Laurentia formed a crustal scale composite allochthon that thrust eastward onto the Baltic craton. Erosion and extension removed or displaced much of the original allochthon and exposed deep crustal levels of the Caledonian Orogen. Today, the Scandinavian Caledonides consist of a relatively thin, but regionally extensive remnant of the original nappe stack that lies structurally above autochthonous or parautochthonous rocks of the Baltic craton and its pre-Caledonian sedimentary cover. The BTZ separates rocks of the composite Caledonian allochthon from parautochthonous structural basement. Of particular relevance to this study, modern exposures closely follow the base of the nappe stack from cold shallow levels in the SE to warm deep levels in the NW i.e. down-dip over distances exceeding 100 km (Fig. 7.1). Mylonites in the BTZ contain well-developed L-S fabrics and foliations parallel the structural contact at the base of the allochthon, with stretching lineations and inferred transport direction trending S60E. Extensional shear is restricted either to the overlying nappes or to minor shears with limited displacement within the basement (summing to less than a few km total). Proposed c. 30 km extension in the basement near Narvik (Fig. 7.1; Rykkelid and Andresen 1994) is refuted by detailed mapping along strike. Although several different orogenic events occurred in the Lower Paleozoic, the Scandian, at c. 425 Ma, was the only event to affect the craton and its immediate parautochthonous cover (Roberts and Gee 1985).

7.2.2   Channel flow, critical taper, and gravitational spreading of nappes

The channel flow extrusion model (Beaumont et al. 2001, 2004; Jamieson et al. 2004; Fig. 7.3a) links overall wedge geometry with a profoundly weak middle and lower crust (ca. 1×10^{19} Pa s). Through coupling to an erosional front, migmatitic rocks move upward and toward the foreland, advecting "extra heat" relative to other orogenic models

(Kohn 2008). Heat advection induces a high thermal gradient along the thrust surface (at least 40°C/km between 200 and 600°C), and migmatitic rocks propagate to within 50 km of the thrust front (Jamieson et al. 2004; Fig. 7.3a).

Davis et al. (1983) and Dahlen (1990) used the term "critical taper" to describe the geometry (taper or angle) that a wedge of deforming material develops in response to the frictional resistance along the wedge base and the internal compressive strength of the wedge material. They then considered cases of brittle deformation of the upper crust via Coulomb failure to explain features in shallow, distal portions of thrust wedges. Some researchers restrict the term critical taper to conditions satisfying critical Coulomb criteria, but others follow Dahlen's more general concept and apply the term to both brittle and viscous rheologies. Models with viscous rheologies, also referred to as critical taper, develop broadly analogous geometries as brittle rheologies (Willett 1999).

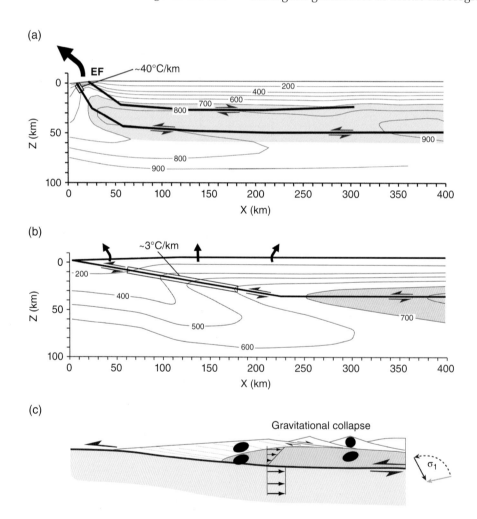

Fig. 7.3. Models of orogenic development based on Kohn (2008) and Northrup (1996a). (a) Channel Flow. Partial melt channel (gray zone, ≥700°C) couples with an erosional front (EF) and propagates forward until hot, partially molten rocks are brought close to the surface. Temperature gradient along thrust surface between 200 and 600°C averages ~40°C/km. (b) Critical wedge. As erosion uniformly removes material from on top of the section, underplating creates a series of in-sequence thrusts. A partial melt zone exists (gray zone), but remains far from the orogenic front. Temperature gradient along thrust surface between 200 and 600°C averages ~3°C/km. (c) Gravitational spreading of nappes (schematic). This model does not predict temperature distributions. Adapted from Kohn 2008 and Northrup 1996a.

What distinguishes critical taper from channel flow is deformation style and its impact on thermal structure. Critical taper follows classic models of orogenic wedge evolution (Royden 1993; Henry et al. 1997; Harrison et al. 1998; Huerta et al. 1998, 1999; Willett, 1999; Cattin and Avouac 2000; Bollinger et al. 2006; Herman et al. 2010; Fig. 7.3b), in which allochthonous sheets accumulate sequentially along discrete shear zones. Critical taper allows for weak portions of the crust and partial melts at depth, but partial melts do not flow in a pipe-flow sense and never approach the thrust front (Fig. 7.3b) so do not strongly influence the overall thermal and mechanical behavior of the orogen (Kohn 2008). In contrast to channel flow, critical taper models predict low thermal gradients along the thrust surface (ca. 3°C/km between 200 and 600°C; Henry et al. 1997; Cattin and Avouac 2000; Herman et al. 2010; Fig. 7.3b).

Some researchers argue qualitatively for a combination of channel flow and critical taper in the Himalaya to explain rock textures and overall styles of deformation (e.g. Larson et al. 2010; Mukherjee 2013). In the absence of quantitative estimates of each component's contribution to overall shortening, however, the impact on thermal structures and temperatures remains obscure. For example, a few tens of kilometers of channel flow might form characteristic textures without otherwise modifying thermal structures relative to critical taper – type models. Our work focuses on thermal structure, so addresses gross contributions of channel flow and critical taper to the Caledonian orogen. The basic question is: are down-dip thermal gradients high (e.g. 40°C/km, supporting channel flow), low (e.g. 3°C/km, supporting critical taper), or intermediate (e.g. 20°C/km, supporting combined channel flow and critical taper)?

In gravitational spreading of nappes (Ramberg 1977, 1981; Sanderson 1982) transport-parallel elongation during thrusting (Northrup 1996a) occurs because a strong upper crust compresses weakened middle and lower crust and induces flow in all directions (Fig. 7.3c). Unlike channel flow, general non-coaxial flow does not maintain a channel-like conduit and does not persist over geologic time. Flow can produce simultaneous foreland directed structural transport, penetrative thinning, and transport-parallel elongation at deep levels of the nappe stack (Northrup 1996a).

Different orogenic models have been proposed for the development of the Caledonian thrust wedge, including classic in-sequence thrusting akin to critical taper (Gee 1975), gravitational spreading (Northrup, 1996a), and channel flow (Gee et al. 2010). Similarities between the Caledonian and Himalayan orogens have been explicitly emphasized (Northrup 1996a; Gee et al. 2010), and a common view is that the deeper portions of the Caledonides may provide insight into processes within the Himalaya, and other major collisions, that are otherwise obscure. For example, advocates of channel flow in the Caledonides (Gee et al. 2010) emphasize the occurrence of high-grade migmatitic rocks (the Seve nappe) sandwiched between greenschist- to amphibolite-facies rocks of the basement below and the Köli nappe above.

7.2.3 Dynamic recrystallization

Dynamic recrystallization of quartz ranges from partially recrystallized microstructures with bimodal grain sizes to fully recrystallized microstructures with unimodal grain sizes. Partially recrystallized quartz exhibits aggregates of small grains with uniform extinction adjacent to larger grains with undulose extinction. The large grains can contain subgrains of approximately the same size as the small grains. Fully recrystallized quartz can be difficult to distinguish from a non-deformed, fine-grained rock, and is indicated by lattice preferred orientation (LPO), irregular grain boundaries due to pinning microstructures, window microstructures, dragging microstructures, and left-over grains (Fig. 7.4; Passchier and Trouw 1998, 2005).

Dynamic recrystallization mechanisms operative in quartz are subgrain rotation (SGR), grain boundary migration (GBM), and bulging (BLG; Fig. 7.4). Passchier and Trouw (1998, 2005) discuss textures and deformation mechanisms in detail, and the following summarizes concepts relevant to our study. In SGR, dislocation creep causes migration of dislocations to subgrain boundaries and progressive misorientation of subgrain crystal lattices. Distinctive, albeit not ubiquitous characteristics of SGR include core and mantle structures, in which small, recrystallized grains surround larger grains that contain subgrains of approximately the same size as the recrystallized grains. In GBM, differences in dislocation densities between neighboring grains drive growth of the less deformed grain at the expense of the more deformed grain. Grains that recrystallized by GBM are all approximately the same size and have straight but irregular grain boundaries. Like GBM, differences in dislocation densities between neighboring grains also drive grain boundary movement and BLG, but over a localized area, not throughout the entire grain boundary. Grains that recrystallized by BLG show irregular grain boundaries and bulges that sometimes pinch off and form smaller, separate, recrystallized grains.

Hirth and Tullis (1992) identified three different dislocation creep regimes in experimentally deformed quartz aggregates, operative over different ranges of temperature and strain rate. Likewise, Stipp et al. (2002) identified three zones with different dynamic recrystallization mechanisms in naturally deformed quartz veins in the Eastern Tonale strike–slip shear zone, Italian Alps. Most importantly, Stipp et al. (2002) constrained temperatures across the shear zone from metamorphic phase equilibria. The natural microstructures and temperatures, for a presumed strain rate, correlate with the dislocation creep regimes defined by Hirth and Tullis (1992): BLG dominated from 280 to 400°C, SGR dominated from 400 to 500°C, and both SGR and GBM dominated from 500 to the maximum temperature observed, 700°C. The temperatures reported

Fig. 7.4. Photomicrographs in crossed-polarized light showing different deformation textures and dominant dynamic recrystallization mechanisms. Tick-mark spacing on rulers is 100 microns. (a) Left-over grains, SC10-04. (b) Pressure shadow and inclusions indicate pre-deformational garnet. (c) Grain boundaries are highly irregular due to bulging, SC10-29. (d) Bulges and left-over grains indicate GBM, SC10-34. (e) Patchy extinction distinguishes subgrains, SC10-45. (f) Uniform extinction, uniform grain size and straight grain boundaries indicate GBM. (g) Ultramylonite containing K-spar porphyroclasts. (h) Classic core and mantle structure indicating SGR.

by Stipp et al. (2002) are commonly thought to constrain deformation temperatures in dynamically recrystallized quartz in other settings; for example, SGR is assumed to reflect deformation at 450 ± 50°C, etc.

For two reasons, large uncertainties (well above ±50°C) attend quartz microtexture-based thermometry. First, although each of these different dynamic recrystallization mechanisms is dominant in quartz at different temperatures, each mechanism, in fact, operates at all temperatures (Hirth and Tullis 1992; Stipp et al. 2002). Second, the mechanical behavior of quartz not only depends on temperature, but also on strain rate (Hirth and Tullis 1992). More precise temperature estimates would constrain strain rates and effective viscosities of crustal materials more accurately (Kohn and Northrup 2009).

7.2.4 The Ti-in-quartz (TitaniQ) thermobarometer

The tetravalent cation Ti^{4+} substitutes directly for Si^{4+} (Wark and Watson 2006) with an equilibrium concentration at a particular temperature and pressure that is governed by partitioning relative to rutile. For Ti-in-quartz (TitaniQ) thermometry, higher temperatures induce higher Ti contents. TitaniQ was calibrated at 600–1000°C and 0.5–2.0 GPa (Wark and Watson 2006; Thomas et al. 2010) to yield:

$$RT \ln X_{TiO2, Qtz} = -60952 + 1.520 \times T(K) \\ -1741 \times P(kbar) + RT \ln(a_{rt}) \quad (1)$$

where R is the ideal gas constant (8.3145 J/K), P is pressure, T is temperature, $X_{TiO2, Qtz}$ is the mole fraction of TiO_2 in quartz (Thomas et al. 2010), and a_{rt} is the activity of rutile. A more recent experimental attempt at calibrating TitaniQ (Huang and Audetat 2012) exploited rapid hydrothermal recrystallization of quartz in temperature gradients, but such kinetics likely bias results, which are scattered and spatially inhomogeneous. Application to metamorphic rocks yields implausibly high temperatures (e.g. Ashley et al. 2013). Although we prefer the Thomas et al. (2010) calibration, use of the Huang and

Audetat (2012) calibration would not substantially affect our interpretations. TitaniQ's main disadvantage is its moderate pressure dependence. Accurately assigning temperatures requires simultaneous application of another thermobarometer. We make a pressure correction by assuming a pressure at each location, in part based on conventional thermobarometry, but our conclusions are not significantly affected if a different pressure distribution is assumed.

7.3 METHODS

7.3.1 Sample collection and preparation

Fifty-five samples were collected across a ca. 140 km NW–SE transect from the Narvik/Harstad area in northwestern Norway towards the Kiruna region in northern Sweden (Fig. 7.1), emphasizing granite mylonites and quartz veins, but also including a few schists, marbles, and quartzites. Samples with rutile were preferred to minimize uncertainties regarding rutile activities (Equation 1). Polished oriented thin-sections were prepared perpendicular to foliation and parallel to lineation for petrographic analyses. Mineral assemblages, microstructures, and deformation textures were characterized for each sample (Tables 7.1 and 7.2).

The samples were organized into 10 groups (A through J) based on their collection location (Fig. 7.1; Table 7.3). These collection groups generally reflect single outcrops or associated outcrops (within 100 m structural position) and were projected onto a line that was drawn parallel to the S60E transport direction of the nappes (Fig. 7.1).

Table 7.1. Mineral assemblages from rocks of the Scandinavian Caledonides in northern Norway and Sweden

Sample	Group	Rock Type	Qtz	Kfs	Pl	Bt	Ms	Rt	Ttn	Chl	Grt	Cal	Amp	Ep	Cpx	Opaq
SC10–26	A	schist	X		X	X	X	X		X	X					X
SC10–29	A	quartz vein	X													
SC10–30	A	schist	X			X	X	X		X	X					X
SC10–31	A	schist	X	X	X	X	X	X		X		X	X	X		
SC10–18	B	schist	X		X	X	X	X		X	X					X
SC10–21	B	quartzite	X	X		X		X		X						X
SC10–25	B	quartz vein	X			X	X	X								X
SC10–03	C	mylonite	X	X	X	X	X	X								
SC10–04	C	quartz vein	X													
SC10–06	C	quartz vein	X	X	X	X	X	X	X							
SC10–12	C	mylonite	X	X		X	X	X	X	X						X
SC10–42	D	mylonite/ quartzite	X		X	X		X	X					X	X	X
SC10–45	E	quartz vein	X					X								
SC10–32	F	mylonite	X	X	X		X	X	X				X	X		
SC10–33	F	mylonite	X	X		X	X	X					X			
SC10–34	F	quartz vein	X	X												
SC10–35bii	F	mylonite	X	X		X	X	X		X			X			X
SC10–38	F	schist	X			X		X		X			X	X		
SC10–46	G	quartz vein	X													
SC10–48	H	mylonite	X	X	X	X	X	X		X						
SC10–51	I	ultramylonite	X			X	X									
SC10–49	J	quartzite ultramylonite	X			X	X	X		X		X				
SC10–55	K	quartzite ultramylonite	X					X				X				

Table 7.2. Thin section descriptions of rocks from the Scandinavian Caledonides in northern Norway and Sweden

Sample	Group	Rock Type	Thin Section Descriptions	GBM	SGR	BLG
				Recrystallization Mechanism		
SC10-26	A	schist	Grain boundaries are straight and grains are approximately the same size (~200 µm). Extinction is uniform.	X		
SC10-29	A	quartz vein	The vein is composed of mostly large quartz crystals (~3 mm) containing micro-cracks and fluid inclusion trails. The crystals have very irregular boundaries and there are many small bulges. Bulges and left-over microstructures are present. Some smaller crystals are ~500 µm. Patchy and sweeping undulose extinction.	X_D	X	
SC10-30	A	schist	Schistose texture and quartz ribbons. Some chlorite cross-cuts the fabric. There are a lot of quartz subgrains within larger grains. The larger grains There are a lot of quartz subgrains within larger grains. The larger grains are ~1 mm and the subgrains and smaller grains are ~100 µm). Undulose extinction.		X	
SC10-31	A	schist	Spaced foliation defined by ribbons of mica. This sample displays a partially recrystallized microtexture with a bimodal grain size distribution of quartz grains. Aggregates of small grains of approximately uniform size (100–300 µm) occur between large grains (5 mm). The larger grains contain subgrains of the same size as the small recrystallized grains. Undulose extinction	X_D	X	
SC10-18	B	schist	Spaced foliation that is defined by biotite ribbons. Grain boundaries are straight and grains are ~50–200 µm. Sweeping and patchy undulose extinction.		X	
SC10-21	B	quartzite	Quartz crystals are ~100–150 µm. Pinning microstructures, window microstructures and dragging microstructures contribute to highly irregular quartz grain boundaries. Some bulges along grain boundaries and small quartz nucleations within larger crystals. Mostly uniform extinction.	X_D	X	
SC10-25	B	quartz vein	Host rock displays spaced mylonitic foliation which is defined by ribbons of mica. This sample displays a partially recrystallized microtexture with a bimodal grain size distribution of quartz grains. Aggregates of small grains of approximately uniform size (~50 µm) occur between large grains (~1 mm). Large grains contain subgrains of the same size as the small grains. Undulose extinction.	X_D	X	
SC10-06	C	quartz vein	Host rock exhibits a mylonitic foliation. Quartz ribbons contain large quartz crystals (up to 10's of µm's). Pinning microstructures, window microstructures and dragging microstructures contribute to highly irregular quartz grain boundaries. Some bulges along grain boundaries and small quartz nucleations within larger crystals. Fluid inclusion trails are oriented normal to foliation. Mostly uniform extinction but some patchy and sweeping undulose extinction.	X		
SC10-12	C	mylonite	Spaced mylonitic foliation that is defined by quartz ribbons and mica ribbons. Pinning microstructures, window microstructures, and dragging microstructures contribute to highly irregular quartz grain boundaries. Some bulges along grain boundaries and small quartz nucleations within larger quartz grains. Quartz crystals are 200–400 µm. Mostly uniform extinction but some patchy and sweeping undulose extinction.		X_D	X
SC10-42	D	mylonite/ quartzite	Spaced foliation. Most of the grains have straight grain boundaries although dragging, pinning, and window microstructures are present. There are many small nucleations of quartz within larger grains. Quartz crystals are 100–300 µm. Mostly uniform extinction but few grains display patchy undulose extinction.	X_D	X	

Table 7.2. (*Continued*)

Sample	Group	Rock Type	Thin Section Descriptions	Recrystallization Mechanism		
				GBM	SGR	BLG
SC10-45	E	quartz vein	In host rock, many subgrains are present. Boundaries are highly irregular and there are many bulges. Rutile is almost completely concentrated in the grain boundaries or in the many microfractures. There are some fluid inclusion trails and left-over microstructures. The quartz crystals are 10's of mms. Undulose extinction.	X_D	X	
SC10-32	F	mylonite	Mylonitic foliation which is defined by a shape preferred orientation, ribbons of mica and amphibole, and quartz ribbons which contain large quartz crystals (1–2 mm's). The boundaries are mostly straight but there are some irregularities as well as pinning microstructures, window microstructures, and dragging microstructures. Smaller quartz crystals are ~200 μm. Mostly uniform extinction but some patchy and sweeping undulose extinction.	X_D	X	X_D
SC10-33	F	mylonite	Mylonitic foliation. Quartz ribbons contain large quartz crystals. Pinning microstructures, window microstructures and dragging microstructures contribute to highly irregular quartz grain boundaries. Some bulges along grain boundaries and small quartz nucleations (~100–200 μm within larger crystals (500 μm). Fluid inclusion trails are oriented normal to foliation. Mostly uniform buy some patchy and sweeping undulose extinction.	X_D	X	
SC10-34	F	quartz vein	The quartz grains have highly irregular grain boundaries. Many left-over grains and bulges, and nucleations of small quartz grains within the larger quartz grains (1 mm). There are long trails of fluid inclusions and micro-fractures. Mostly uniform but some patchy and sweeping undulose extinction.	X_D	X	
SC10-35bii	F	mylonite	Spaced mylonitic foliation. Dragging microstructures, bulges, and left-over grains contribute to the highly irregular grain boundaries. Some larger grains (1 mm) contain subgrains that are the same size as small recrystallized grains (~100 μm). Mostly uniform extinction.	X		
SC10-38	F	schist	Spaced foliation. The grain boundaries are regular and the quartz crystals are ~100–300 μm. Mostly uniform extinction but some sweeping undulose extinction.	X		
SC10-46	G	quartz vein	Large quartz crystals (~3 mm) with straight grain boundaries. There are a few smaller quartz crystals (~500 μm). Mostly uniform extinction.	X		
SC10-48	H	mylonite	Mylonitic foliation defined by quartz ribbons. The quartz in the ribbons is ~200 μm, and has straight grain boundaries. Large feldspar porphyroclasts. Uniform extinction.		X	
SC10-51	I	ultramylonite	Ultramylonite foliation: completely sheared, tiny grains (~10 μm). There are small quartz lenses that have slightly larger crystals (~200 μm). Undulose extinction.		X	
SC10-49	J	quartzite ultramylonite	Crenulation cleavage is defined by micas, chlorite, and quartz ribbons. Straight grain boundaries but there are window, dragging, and pinning microstructures. The quartz in the ribbons (~300 μm) is larger than the quartz in the matrix (~10–20 μm). Core and mantle texture in lenses. Uniform extinction.	X	X	
SC10-55	K	quartzite ultramylonite	Ultramylonite foliation: completely sheared, tiny grains (~10 μm). Classic Core and mantle structure in lenses with cores that are ~3mm mantled by crystals that are ~100–200 μm. There are many subgrains within the cores. Undulose extinction.	X	X	

Table 7.3. Titanium contents and calculated temperatures

Sample #:	Group	^{48}Ti/Si	±1σ	^{49}Ti/Si	±1σ	ppm Ti	P (kbar)	T (°C)	±2σ
SC10–03–01*	C	0.000180	27	0.0000114	70	4.12	7	439	16
SC10–03–02	C	0.000635	73	0.0000503	174	14.55	7	520	16
SC10–03–03*	C	0.000135	16	0.0000085	58	3.09	7	423	12
SC10–03–04*	C	0.000166	27	0.0000104	62	3.80	7	434	17
SC10–04–0^v	C	0.000226	34	0.0000150	66	5.17	7	452	17
SC10–04–02^v	C	0.000217	18	0.0000142	48	4.96	7	450	9
SC10–06–01^{v,}*	C	0.000103	23	0.0000070	52	2.36	7	408	21
SC10–06–02^{v,}*	C	0.000145	24	0.0000100	66	3.32	7	427	18
SC10–06–03^{v,}*	C	0.000122	18	0.0000086	51	2.79	7	417	15
SC10–06–04^{v,}*	C	0.000109	22	0.0000093	63	2.51	7	411	20
SC10–12–01	C	0.000370	15	0.0000297	66	6.85	7	469	5
SC10–12–02	C	0.000308	39	0.0000223	72	5.69	7	458	15
SC10–12–03*	C	0.000140	15	0.0000097	39	2.59	7	413	11
SC10–12–04	C	0.000406	32	0.0000288	93	7.51	7	475	10
SC10–18–01	B	0.000137	16	0.0000094	39	2.52	8	428	12
SC10–18–02*	B	0.000068	11	0.0000051	32	1.26	8	392	15
SC10–18–03*	B	0.000132	20	0.0000081	39	2.44	8	426	16
SC10–21–01*	B	0.000077	14	0.0000067	46	1.43	8	398	17
SC10–21–02*	B	0.000105	25	0.0000100	77	1.93	8	414	23
SC10–21–03*	B	0.000077	15	0.0000040	39	1.42	8	398	18
SC10–21–04	B	0.000160	21	0.0000131	59	2.96	8	437	14
SC10–25–01^{v,}*	B	0.000101	13	0.0000069	39	2.46	8	427	14
SC10–25–02^{v,}*	B	0.000095	15	0.0000063	40	2.31	8	423	16
SC10–25–03^v	B	0.000108	12	0.0000070	43	2.64	8	430	11
SC10–26–01	A	0.000216	18	0.0000146	50	5.26	8	470	10
SC10–26–02	A	0.000362	69	0.0000225	111	8.83	8	504	24
SC10–26–03	A	0.000276	27	0.0000195	68	6.74	8	486	12
SC10–29–01*	A	0.000141	13	0.0000112	51	3.43	8	445	10
SC10–29–02*	A	0.000138	13	0.0000105	37	3.38	8	444	10
SC10–29–03*	A	0.000143	18	0.0000113	58	3.50	8	446	14
SC10–29–04*	A	0.000148	17	0.0000100	46	3.61	8	448	12
SC10–30.1–01	A	0.000162	4	0.0000111	10	4.70	8	464	4
SC10–30.1–02	A	0.000165	13	0.0000111	27	4.80	8	465	9
SC10–30.1–03	A	0.000145	11	0.0000099	21	4.22	8	457	9
SC10–30.2–01	A	0.000141	8	0.0000102	27	4.11	8	456	6
SC10–30.2–02	A	0.000200	12	0.0000153	28	5.83	8	477	7
SC10–30.2–03	A	0.000179	6	0.0000124	29	5.21	8	470	4
SC10–30.3–01*	A	0.000130	8	0.0000077	18	3.77	8	451	7
SC10–30.3–02*	A	0.000101	6	0.0000071	21	2.94	8	436	6
SC10–30.3–03*	A	0.000111	8	0.0000071	6	3.22	8	442	8
SC10–31–01	A	0.000243	8	0.0000169	20	7.11	8	489	4
SC10–31–02	A	0.000235	12	0.0000161	22	6.89	8	487	6
SC10–31–03	A	0.000219	8	0.0000162	25	6.41	8	483	5
SC10–32–01	F	0.000064	16	0.0000044	14	1.86	6	380	23
SC10–32–02	F	0.000061	3	0.0000036	15	1.79	6	378	5
SC10–32–03	F	0.000069	6	0.0000050	17	2.03	6	384	8
SC10–32–04	F	0.000080	6	0.0000061	12	2.33	6	391	8
SC10–33–01	F	0.000043	5	0.0000030	8	1.26	6	361	10
SC10–33–02	F	0.000055	5	0.0000034	13	1.61	6	373	9
SC10–33–03	F	0.000048	7	0.0000030	9	1.40	6	366	14
SC10–33–04*	F	0.000030	4	0.0000024	8	0.89	6	345	11
SC10–34–01^v	F	0.000060	6	0.0000041	10	1.60	6	372	9
SC10–34–02^v	F	0.000068	8	0.0000039	11	1.84	6	379	11
SC10–34–03^v	F	0.000067	20	0.0000049	19	1.81	6	379	26
SC10–35bii-01	F	0.000077	11	0.0000051	21	2.06	6	385	13
SC10–35bii-02	F	0.000057	9	0.0000048	16	1.53	6	370	14
SC10–35bii-03	F	0.000050	10	0.0000031	15	1.34	6	364	18
SC10–38–01	F	0.000080	8	0.0000054	11	2.14	6	387	9
SC10–38–02	F	0.000072	8	0.0000042	17	1.95	6	382	11
SC10–38–03	F	0.000053	7	0.0000036	14	1.44	6	367	12
SC10–45–01	E	0.000340	15	0.0000226	30	8.24	6	463	6
SC10–45–02	E	0.000311	8	0.0000216	28	7.54	6	458	4

Table 7.3. (*Continued*)

Sample #:	Group	^{48}Ti/Si	±1σ	^{49}Ti/Si	±1σ	ppm Ti	*P* (kbar)	*T* (°C)	±2σ
SC10–45–03	E	0.000317	14	0.0000226	26	7.67	6	459	5
SC10–45–04	E	0.000340	12	0.0000226	27	8.24	6	463	4
SC10–46–01[v,*]	G	0.000045	5	0.0000031	13	1.08	5	339	9
SC10–46–02[v,*]	G	0.000057	7	0.0000040	11	1.37	5	349	11
SC10–46–03[v,*]	G	0.000036	4	0.0000025	12	0.86	5	329	10
SC10–46–04[v,*]	G	0.000043	6	0.0000030	12	1.04	5	337	11
SC10–48–01	G	0.000094	13	0.0000068	33	2.61	5	381	13
SC10–48–02*	G	0.000048	16	0.0000032	26	1.34	5	348	26
SC10–48–03*	G	0.000037	8	0.0000020	13	1.03	5	337	18
SC10–48–04*	G	0.000043	6	0.0000023	16	1.20	5	343	12
SC10–49–01*	I	0.000011	4	0.0000008	8	0.31	4	274	21
SC10–49–02	I	0.000064	35	0.0000047	28	1.77	4	346	42
SC10–49–03*	I	0.000017	4	0.0000011	7	0.47	4	289	15
SC10–49–04	I	0.000108	11	0.0000053	13	3.00	4	371	10
SC10–51–01	H	0.000576	29	0.0000432	44	20.40	4	486	7
SC10–51–02	H	0.000035	6	0.0000014	21	1.24	4	329	14
SC10–51–03*	H	0.000010	3	0.0000006	6	0.36	4	279	21
SC10–55–01	J	0.000075	18	0.0000051	23	2.66	3	349	21
SC10–55–02	J	0.000065	21	0.0000044	22	2.29	3	342	27
SC10–55–03	J	0.000071	10	0.0000053	15	2.52	3	346	13
SC10–55–04	J	0.000077	34	0.0000041	21	2.72	3	350	37

Note: ppm refers to concentrations on a weight basis, e.g. µg/g. Errors refer to uncertainty in the last decimal place reported. Superscript "v" indicates vein; "*" indicates analysis used in regression temperature vs. distance. GPS locations (WGS 84): SC10–3/4 = 33W0564807, 7593954; SC10–6 = 33W0564919, 7594189; SC10–12 = 33W0565311, 7594433; SC10–18 = 33W0563313, 7612971; SC10–21 = 33W0563295, 7612953; SC10–25 = 33W0563272, 7612932; SC10–26/29/30 = 33W0563792, 7629684; SC10–31 = 33W0563600, 7629798; SC10–32/33/34 = 33W0597314, 7577213; SC10–35 = 33W0597974, 7576471; SC10–38 = 33W0597226, 7577531; SC10–45 = 33W0564290, 7563234; SC10–46 = 33W0605921, 7587863; SC10–48 = 33W0613459, 7600603; SC10–49 = 34W0408393, 7584414; SC10–51 = 33W0609185, 7509864; SC10–55 = 34W0385748, 7488365;

Most samples in a group recrystallized with the same microstructure, and transitions between deformation regimes were drawn perpendicular to the main transport direction (Fig. 7.1). Twenty-three samples that contained rutile and that contained representative microstructures were petrographically characterized and selected for further analyses, including at least 1 sample from each group. Two samples also contained garnet, biotite, plagioclase, and muscovite. Samples were carbon coated and imaged with a JEOL T300 scanning electron microscope (SEM), using a Gatan Mini-cathodoluminescence (CL) detector, housed in the Department of Geosciences at Boise State University (Boise, ID). Locations for TitaniQ analyses were targeted using CL textures and intensities within and between quartz grains. Locations for major element compositions measurements were determined from X-ray maps.

7.3.2 Peak temperatures and pressures

X-ray maps and elemental compositions were collected using a Cameca SX100 electron probe housed in the Department of Earth and Environmental Sciences at Rensselaer Polytechnic Institute, Troy, NY. X-ray maps of Fe, Mg, Mn, Ca, and Al were collected on garnet using an accelerating voltage of 15 kV, a current of 200 nA, a pixel time of 30 ms, and a step size of 2–5 µm/pixel. These maps allowed us to target the location of fully quantitative line-scans and spot analyses. Natural and synthetic silicates and oxides were used for calibrations, and quantitative measurements were made using an accelerating voltage of 15 kV and a current of 20 nA. A minimum beam size was used on garnet at 10 µm intervals along the line. A 10 µm beam size was used on plagioclase and micas to minimize beam damage. Peak count times were 10 s (Na, Ca, Fe, Mn, Si, Al), and 20 s (Mg, Ti, K) (Table 7.4).

Pressure–temperature conditions of final equilibration in groups A and B were calculated using the garnet-biotite thermometer (Ferry and Spear 1978; Berman 1990) and garnet-plagioclase-muscovite-biotite barometer (Hoisch 1990). The lowest Mn and Fe/(Fe+Mg) values, associated with a trough in garnet compositions from core to rim, were selected for temperature and pressure estimates. Other estimates of peak temperature are provided by phase equilibria (the prograde stabilization of kyanite + biotite assemblages at ca. 600°C; Spear and Cheney 1989), other thermobarometric studies (Hodges and Royden 1984; Crowley 1985; Tilke 1986; Steltenpohl and Bartley 1987), regional muscovite ^{40}Ar/^{39}Ar geochronology near the eastern edge of the transect (Page 1992) coupled with estimates of the closure temperature of muscovite with respect to Ar diffusion (Harrison et al. 2009), and estimates of the temperature of the brittle-ductile transition (350 ± 100°C; Chen and Molnar 1983).

We estimated depths at the western end of the transect from thermobarometry (ca. 8 kbar; this study; Hodges and Royden 1984; Steltenpohl and Bartley 1987) and

Table 7.4. Representative electron probe analyses of garnet, plagioclase, and micas

	Garnet							
Sample	SC10–18	SC10–18	SC10–18	SC10–18	SC10–26	SC10–26	SC10–26	SC10–26
Si	2.961	2.972	2.958	2.972	2.988	2.981	2.998	2.995
Al	2.001	1.980	1.988	1.981	1.980	2.016	1.993	2.006
Ti	0.002	0.004	0.003	0.000	0.003	0.003	0.004	0.003
Mg	0.301	0.326	0.314	0.324	0.403	0.380	0.366	0.354
Fe	2.282	2.357	2.335	2.331	2.010	2.002	1.992	1.965
Mn	0.094	0.079	0.079	0.094	0.064	0.066	0.072	0.067
Ca	0.396	0.315	0.368	0.335	0.571	0.561	0.577	0.609
Wt% Total	100.20	99.77	99.92	99.84	100.91	101.79	101.71	101.47
Alm	0.742	0.766	0.754	0.756	0.659	0.665	0.663	0.656
Grs	0.129	0.102	0.119	0.109	0.187	0.186	0.192	0.203
Prp	0.098	0.106	0.102	0.105	0.132	0.126	0.122	0.118
Sps	0.031	0.026	0.025	0.03	0.021	0.022	0.024	0.022
Fe/(Fe+Mg)	0.883	0.878	0.881	0.878	0.833	0.84	0.845	0.847

Cations normalized to 12 oxygens.

	Plagioclase					
Sample	SC10–18	SC10–18	SC10–18	SC10–18	SC10–26	SC10–26
Xan(%)	0.161	0.164	0.164	0.162	0.391	0.393
Wt% Total	100.78	99.05	99.60	99.03	99.14	100.01

	Muscovite and biotite								
Sample	SC10–18	SC10–18	SC10–18	SC10–18	SC10–26	SC10–26	SC10–26	SC10–26	SC10–26
Mineral	Bt	Bt	Bt	Ms	Bt	Bt	Bt	Bt	Ms
Si	2.799	2.792	2.767	3.143	2.784	2.790	2.795	2.786	3.149
Al	1.552	1.550	1.520	2.591	1.580	1.587	1.587	1.562	2.623
Ti	0.102	0.104	0.105	0.031	0.093	0.093	0.093	0.108	0.031
Mg	1.283	1.292	1.268	0.147	1.529	1.516	1.513	1.521	0.152
Fe	1.103	1.110	1.227	0.087	0.876	0.868	0.864	0.880	0.069
Mn	0.013	0.012	0.011	0.001	0.007	0.007	0.008	0.008	0.000
Ca	0.003	0.001	0.002	0.003	0.000	0.001	0.000	0.000	0.000
Na	0.017	0.019	0.024	0.091	0.032	0.033	0.024	0.030	0.087
Ba	0.001	0.000	0.001	0.004	0.004	0.004	0.003	0.004	0.007
K	0.917	0.919	0.909	0.955	0.890	0.883	0.889	0.884	0.867
Wt% Total	94.70	95.23	94.78	95.50	95.29	95.31	96.09	95.80	95.10
Fe/(Fe+Mg)	0.463	0.462	0.464	0.372	0.364	0.364	0.363	0.366	0.31

Cations normalized to 11 oxygens.
* Biotite composition corrected for retrograde net-transfer reaction (Kohn and Spear, 2000).

elsewhere by assuming that the BTZ dipped ~8° towards N60W during the main Caledonian orogeny, similar to the current dip magnitude of the basal thrust in the Himalaya (ca. 9°; Hauck et al. 1998). Assuming a relatively flat surface, this dip angle corresponds to a pressure of 3 kbar (10 km) at the present easternmost exposure of the thrust. Rocks at this location are greenschist facies and exhibit brittle structures, whereas ductile structures occur immediately west (i.e. lie close to the brittle–ductile transition; Tilke 1986; Page 1990), so must lie west of the original thrust front. The BTZ could not have dipped more steeply than 12° on average because it would otherwise project to the surface west of the current exposures of the thrust. Conversely, a dip of ~5° is the minimum observed for thicker portions of modern accretionary wedges (e.g. Dahlen 1990). Calculated

TitaniQ temperatures and temperature gradients are insensitive to changes in the dip angle between 5° and 12° NW (ca. ±40°C maximum difference at the easternmost exposure of the thrust).

7.3.3 Secondary ion mass spectrometer (ion microprobe) analysis

Carbon coats were removed by polishing with 0.3 μm alumina, and selected areas were drilled from the polished thin sections using a drill press and a diamond studded drill bit, producing 5 mm diameter disks. The polished disks were then remounted in one inch epoxy mounts along with a Herkimer "diamond" [a low temperature quartz that formed at 150–200°C (Smith 2006), so should contain Ti concentrations on a weight basis of

3 ± 2.5 ppb; Kohn and Northrup 2009], and a natural quartz crystal from a Himalayan migmatite (LT01-15), whose Ti content was determined by electron and ion microprobes (39 ± 6 ppm Ti on a weight basis; Kohn and Northrup 2009; Fig. 7.5). The Herkimer "diamond" and LT01-15 quartz acted as a "blank" and a standard, respectively, and were mounted and polished before remounting with the samples. A total of nine mounts each contained two to three sample disks and one to two grains of each standard (Fig. 7.5). The disks and standards were positioned within the inner 1.1 cm of the mount. The mounts were then photographed in reflected light and specific locations on each sample were selected for ion probe spot analyses. Prior selection of specific locations for analyses permitted greater microstructural control and analytical efficiency.

The nine mounts were gold coated, and Ti concentrations in quartz were measured with a Cameca 6F ion microprobe at the Arizona State University Secondary Ion Mass Spectrometry Laboratory. Analyses followed Behr et al. (2011), with the first field aperture set to 1800 μm and the second contrast aperture set to 150 μm. Masses of ^{27}Al, ^{30}Si, ^{40}Ca, ^{48}Ti, and ^{49}Ti were collected using an energy window of 250 eV and a mass resolving power of 2800. Ratios of ^{49}Ti/^{30}Si and ^{48}Ti/^{30}Si were consistent with relative abundance of ^{48}Ti and ^{49}Ti, indicating no mass interferences on Ti isotopes. ^{40}Ca and ^{27}Al were collected to monitor for contamination and microinclusions. Mass cycles that showed excess ^{40}Ca and ^{27}Al were removed during data processing.

At least one spot analysis of the Herkimer "diamond" and three to seven spot analyses of the natural quartz standard were obtained per mount. In initial tests, the Ti/Si ratio of the Herkimer quartz drifted steadily downward throughout analysis, indicating surface contamination. To eliminate surface contamination, we first rastered

an area of 50 × 50 μm, at 17–20 nA, and then reanalyzed in the center of the rastered area with a focused, 25 × 25 μm spot at 4–8 nA. Peak count times were 15 minutes during the initial raster and 20 minutes per spot analysis. All data reported here for standards and unknowns followed this analytical protocol.

Three to four spots on single or adjacent grains were analyzed per sample. The average ^{48}Ti/^{30}Si ratios in LT01-15 standard for each block were calculated (range = 0.0011–0.0017) and normalized to its known Ti concentration (39 ppm). The resulting conversion factors (range = 22900–35400) were then multiplied by the ^{48}Ti/^{30}Si concentrations measured in the sample to obtain Ti concentrations (Table 7.3). Reproducibility on LT01–15 quartz was ~±10% (2σ), which propagates to temperature uncertainties less than ±10°C at a specified pressure. Uncertainties of ±1 kbar propagate to temperature uncertainties of ~±15°C.

7.4   RESULTS

Two samples analyzed for garnet-biotite Fe–Mg exchange thermometry [SC10-26 (group A) and SC10-18 (group B)] show similar or lower TitaniQ temperatures for the same rock (Table 7.3; Fig. 7.7): SC10-26$_{Grt-Bt}$ = 525 ± 35°C, and SC10-26$_{TitaniQ}$ = 470–500°C at P = 8 kbar; SC10-18$_{Grt-Bt}$ = 525 ± 20°C and SC10-18$_{TitaniQ}$ = 390–430°C at P = 8 kbar. The occurrence of the assemblage garnet + kyanite + staurolite + biotite + muscovite at the highest metamorphic grades at the west end of the transect imply minimum temperatures of ~600 ± 25°C (Spear and Cheney 1989), so our measured Grt-Bt and TitaniQ temperatures must be reset. Because isopleths of Ti in quartz and Fe–Mg exchange between garnet and biotite are similar, a different assumed pressure would change absolute temperatures, but not relative differences.

(a)   (b)

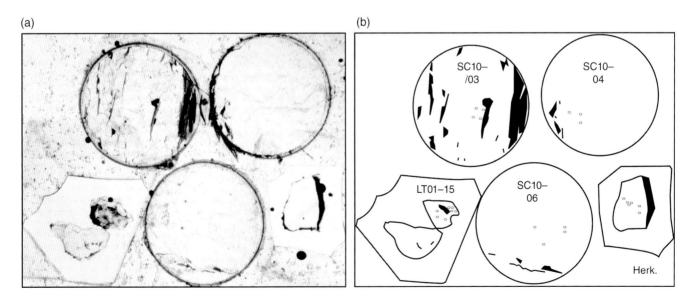

Fig. 7.5. Example of a mount containing three samples, and two standards. (a) Photomicrograph. (b) Sketch. "Herk" = Herkimer quartz standard blank. Small elliptical dots show locations of ion probe rasters. Each sample disk is 5 mm in diameter.

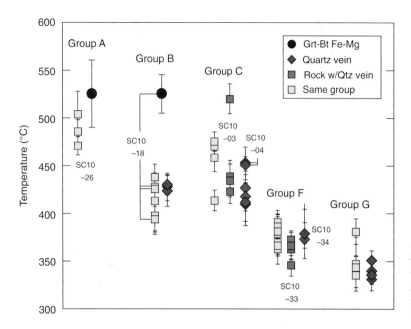

Fig. 7.6. Comparisons of thermometric results. Garnet biotite thermometry gives higher temperatures than TitaniQ. Temperatures obtained from two cross-cutting quartz veins (SC10–04 and SC10–34) mimic temperatures of their respective host rocks (SC10–33 and SC10–34, respectively).

In all instances, cross-cutting quartz vein temperatures mimic TitaniQ temperatures of the host rocks and host rock groups (Fig. 7.6): Group B quartz vein = 420–430°C, mylonite = 390–440°C; Group C quartz vein = 410–450°C, mylonite = 410–520°C; Group F quartz vein = 370–380°C, mylonite = 340–390°C; Group G quartz vein = 330–350°C, mylonite = 340–380°C. Directly comparing cross-cutting quartz veins to their host rocks shows quartz vein SC10-04 = ~450°C, and host rock SC10-03 = 420–520°C; and quartz vein SC10-34 = 370–380°C, and host rock SC10-33 = 340–370°C.

Metamorphic mineral assemblages constrain peak temperatures at the west end of the transect to ca. 600°C, whereas ductile and brittle thrust-sense structures found near the thrust front (Tilke 1986; Page 1990) indicate conditions approaching the brittle–ductile transition, possibly 10–15 km depth and 350 ± 100°C (e.g. Chen and Molnar 1983). Minimum TitaniQ temperatures, which may represent the latest movement on the BTZ, range from <300°C towards the thrust front to >400°C at the deepest crustal levels (Fig. 7.7). Regression of minimum temperatures at each location (Table 7.3; Fig. 7.7) implies a thermal gradient of 1.5 ± 0.1°C/km along the shear zone (Fig. 7.7), with extrapolated temperatures of 233 ± 14°C at the easternmost exposure of the thrust and 442 ± 23°C at the deepest crustal levels. Higher TitaniQ temperatures are preserved in some grains and approach peak metamorphic conditions deduced from thermometric estimates (Fig. 7.7). The maximum possible gradient is 3.2°C/km (600°C peak metamorphic temperatures at the western end of the transect and 280°C mylonite temperature for a Group I sample), but would imply extremely low temperatures at the easternmost exposure of the thrust (~150°C). Peak metamorphic conditions imply a thermal gradient of ~1.8°C/km (Fig. 7.7).

Most TitaniQ temperatures differ considerably with temperatures commonly reported for quartz microstructures (Fig. 7.7). TitaniQ temperatures for GBM and SGR are as low as 340°C and 280°C respectively, far below previously published temperatures for GBM (500 to at least 700°C) and SGR (400 to at least 700°C; Hirth and Tullis 1992; Stipp et al. 2002). Many samples exhibiting SGR, however, do record temperatures above 400°C.

7.5   DISCUSSION

7.5.1   Temperatures of deformation

For reasons explained later, we take the lowest statistically consistent group of temperatures recorded by recrystallized quartz as the temperature of final shearing, a cluster of temperatures near this minimum as the range of temperatures over which the shear zone moved (at a specific location) and anomalously high temperatures as possibly relict protolith or peak metamorphic grains or domains whose Ti contents were incompletely erased by later deformation (Fig. 7.7). The following observations explain these choices. First, most TitaniQ deformation temperatures fall well below the metamorphic peak, so must either predate or postdate it. Yet they cannot reflect relict prograde quartz because they are texturally linked to post-metamorphic shearing. Second, TitaniQ temperatures in cross-cutting quartz veins must also post-date metamorphism and deformation, and these mimic host rock minimum temperatures. The simplest explanation is that shearing resets Ti contents of quartz (Kohn and Northrup 2009), at least over most temperatures that the BTZ experienced, and that cross-cutting quartz veins formed at effectively the same temperature soon after thrusting stopped. The higher temperatures that are preserved in each location approach peak metamorphic

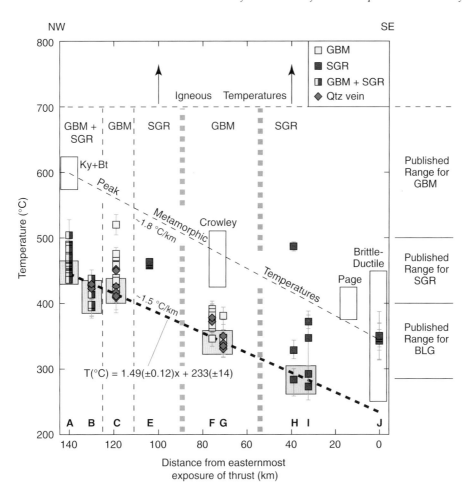

Fig. 7.7. TitaniQ T vs. distance. Thin, dashed vertical lines indicate precise boundaries between GBM and SGR, and thick, dashed vertical lines indicate less well-constrained boundaries between SGR and GBM. Lowest T's in gray boxes are interpreted as the minimum temperatures of deformation. These project to ~230 °C at the thrust front, and ~440 °C for deepest structural levels, with a slope (down-dip temperature gradient) of ~1.5°C/km. Temperatures of peak metamorphism are based on phase equilibria (Spear and Cheney 1989) thermobarometry (Crowley 1985; Tilke 1986), regional cooling patterns (Page 1992) linked to the closure temperature of muscovite (Harrison et al. 2009), and general crustal seismicity delimiting the brittle-ductile transition (Chen and Molnar 1983). TitaniQ does not appear to reequilibrate at temperatures below ~275°C, and SGR does not reset TitaniQ as effectively as GBM.

estimates (Fig. 7.7) so probably represent relict, near-peak grains. Relatively slow diffusion rates of Ti in quartz (Cherniak et al. 2006) rule out diffusive resetting of TitaniQ *T*'s over the range of temperatures recorded in this study. If minimum *T*'s reflect the best estimate of the final temperatures of deformation (Fig. 7.7), final shear zone movement occurred at an extrapolated *T* of 233 ± 14°C at the easternmost exposure of the thrust front (brittle–ductile transition) to 442 ± 23°C at the deepest exposed crustal levels.

Anomalously high *T*'s are recorded in many rocks, especially those that were deformed at temperatures of ~275°C and below. Resetting of Ti contents in quartz does not appear effective below that temperature, where SGR dominates. For example, group J does not plot on the lower bound of temperatures and groups E, H and I, which recrystallized by SGR, record numerous anomalously high temperatures (Fig. 7.7). CL images of sample SC10-45

(group E) reveal quartz grains with bright cores and dark rims, even on a subgrain scale, which suggests a high concentration of Ti in the cores (i.e., relict compositions; anomalously high *T*'s in Fig. 7.7 and Fig. 7.8). Kohn and Northrup (2009) and Ashley et al. (2014) report similar CL textures and compositional behavior. Alternatively, in fine-grained rocks (Fig. 7.9), high *T*'s could reflect analytical contamination from grain boundaries because the ion probe spot invariably overlaps multiple grains. That is, recrystallized grains have low Ti contents but Ti simply migrated to the nearest local grain boundary and precipitated submicroscopic rutile, which was included in the region sputtered by the ion beam (Fig. 7.9). It is also conceivable that we misinterpreted some fine-grained quartzites as ultramylonites so that, despite evidence for deformation (e.g. lineations), original relics of high-*T* metamorphic or detrital grains resisted post-metamorphic shear-induced recrystallization.

(a)

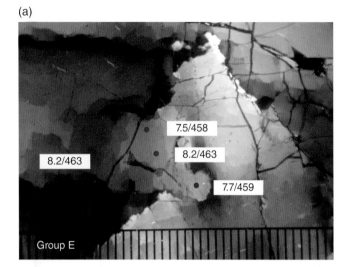

(b)

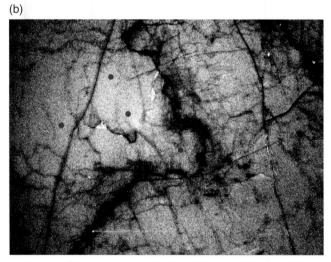

Fig. 7.8. (a) Photomicrograph and (b) CL image of sample SC10-45, group E. Red dots are analytical spots, with concentrations in ppm by weight, and T in °C. Tick marks on scale bar are 100 μm. The dominant dynamic recrystallization mechanism is SGR. CL image reveals bright cores and dark rims, even on a subgrain scale, which suggests a higher concentration of Ti in the cores (i.e. relict compositions).

7.5.2 Temperatures of deformation microstructures

Commonly, microstructures associated with different recrystallization mechanisms are presumed to correspond with specific temperature intervals (for a presumed strain rate): BLG = 280–400°C, SGR = 400–500°C, and GBM or combined SGR and GBM = 500 to at least 700°C (Hirth and Tullis 1992; Stipp et al. 2002). Application of TitaniQ to mylonites from the Eastern Tonale shear zone corroborates these temperature intervals (Grujic et al. 2011). Thus, Tonale zone results contrast markedly from ours, where BLG, as a dominant recrystallization mechanism, is observed in only one sample, SGR is observed at temperatures well below 400°C, and GBM is observed at temperatures as low as 340°C (Fig. 7.7). Strain rate and temperature both contribute to textures observed in quartz, and therefore

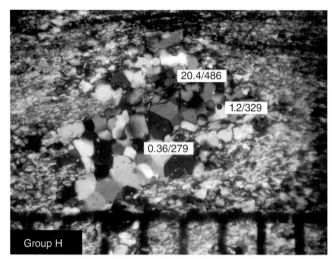

Fig. 7.9. Photomicrograph of ultramylonite SC10–51, group H. Red dots are analytical spots, with concentrations in ppm by weight, and T in °C. Tic marks on scale bar are 100 μm. High T's could reflect analytical contamination from grain boundaries or relict grains.

temperatures of mylonite formation should not be based solely on quartz microstructures. In the Caledonides, a much lower strain rate during final thrust movement may have stabilized SGR and GBM at lower temperatures than anticipated elsewhere.

7.5.3 Comparison to orogenic models

Numerous thermal, metamorphic and chronologic criteria have been offered to distinguish the geodynamics of critical taper from channel flow (e.g. see Kohn 2008). Here, we focus on quantitative observables of our study – temperature distributions along the thrust plane and locations of partially molten rocks relative to the original thrust front. Overall, our data are better explained by critical taper, rather than channel flow. Channel flow predicts high thermal gradients along the thrust surface (ca. 40°C/km between 200 and 600°C; Jamieson et al. 2004), whereas critical taper predicts low thermal gradients (ca. 4°C/km between 200 and 600°C; Cattin and Avouac 2000). The observed gradient of only 1.5°C/km (strict limit of <3.2°C/km) clearly favors critical taper. Later deformation is unlikely to have modified this along-thrust thermal gradient substantially. Extension at the level of the BTZ is far subordinate to thrust-sense shearing, so probably did not decrease the gradient. The basal thrust itself does exhibit some degree of imbrication (Fig. 7.1), but correction for this effect would decrease the estimated thermal gradient.

Both channel flow and critical taper are consistent with the presence of partial melts (migmatites), but channel flow predicts their occurrence within 50 km of the thrust front, whereas critical taper models predict distances from 150 to ≥300 km (Henry et al. 1997; Cattin and Avouac 2000; Herman et al. 2010). The distance to the original thrust front is unknown, but extrapolating our inferred temperature gradient to 25°C (Fig. 7.7) implies it lay ~140 km farther east than the easternmost exposure today. The highest grade rocks in the hinterland, 140 km west of the easternmost exposures, did not even reach migmatite-grade conditions yet were 250–300 km from the inferred front. Thus, migmatitic rocks, if they ever existed, must be still farther west. Such a gross discrepancy between channel flow predictions (migmatites ≤50 km from the thrust front) and observations (migmatites ≥250 km from the thrust front) effectively rules out channel flow at this location in the Caledonides and supports critical taper. Overall, channel flow does not explain metamorphic distributions along the thrust surface well, and instead a colder orogenic model (e.g., critical taper) is preferred.

Critical wedge dynamics linked with general non-coaxial flow in the mid-crust readily accounts for foreland-directed transport, subvertical thinning, and transport-parallel elongation of the rock mass at depth (Northrup 1996a), and could explain coeval upper crustal extension and contractional thrusting at deeper crustal levels. In the Caledonides, general non-coaxial flow occurred prior to and during the main Caledonian orogeny (Northrup 1996a) and is evidenced by spreading lineations in all directions. Therefore, we suggest that the Caledonides dominantly formed by critical wedge kinematics accompanied by gravitational spreading of the nappes, but without significant thermal contributions from channel flow.

7.5.4   Future work

Our results provide a new approach for understanding the dynamics of thrust movement. Most studies of temperatures and temperature gradients focus on (oblique) cross-sections upwards or downwards through a section. Not all such gradients help discriminate among thermal models, for example the temperature gradient across the Main Central (basal) thrust of the Himalaya can be explained by nearly any model, including channel flow or critical taper (Kohn 2008). TitaniQ in deformed quartz now allows exploration of thermal gradients *along* the transport direction of a thrust to be determined. In the case of critical taper vs. channel flow, the transport-parallel gradient is far more diagnostic than cross-structure gradients. Future work could explore this novel approach in other orogens to discriminate mechanisms of heat transport and models of crustal deformation and evolution.

ACKNOWLEDGMENTS

We thank R. Hervig and L. Williams for ion probe assistance and training, P.H. Leloup, D. Grujic, and S. Mukherjee for their helpful reviews, H. Stunitz for suggesting we analyze quartz veins, S.L. Corrie for collecting electron probe data, and K. Matthews, D. Sandford, and F. John of John Wiley & Sons for support. Funded by NSF grants EAR0810242, EAR1048124 and EAR1321897 to M.J.K., NSF Instrumentation and Facilities grant EAR0622775 for support of the Arizona State University SIMS facility, and Boise State University.

REFERENCES

Ashley KT, Webb LE, Spear FS, Thomas JB. 2013. P-T-D histories from quartz: A case study of the application of the TitaniQ thermobarometer to progressive fabric development in metapelites. Geochemistry Geophysics Geosystems 5, doi: 10.1002/ggge.20237.

Ashley KT, Carlson WD, Law RD, Tracy RJ. 2014. Ti resetting in quartz during dynamic recrystallization: Mechanisms and significance. American Mineralogist 99, 2025–2030.

Beaumont C, Jamieson RA, Nguyen MH, Lee B. 2001. Himalayan tectonics explained by extrusion of a low-viscosity crustal channel coupled to focused surface denudations. Nature 414, 738–742.

Beaumont C, Jamieson RA, Nguyen MH, Medvedev S. 2004. Crustal channel flows: 1. Numerical models with applications to the tectonics of the Himalayan Tibetan Orogen. Journal of Geophysical Research 109, B06406.

Behr WM, Thomas JB, Hervig RL. 2011. Calibrating Ti concentrations in quartz for SIMS determinations using NIST silicate glasses and application to the TitaniQ geothermobarometer. American Mineralogist 96(7), 11001106.

Berman RG. 1990. Mixing properties of Ca-Mg-Fe-Mn garnets. American Mineralogist 75, 328–344.

Bollinger L, Henry P, Avouac JP. 2006. Mountain building in the Nepal Himalaya: thermal and kinematic model. Earth and Planetary Science Letters 244, 58–71.

Cattin R, Avouac JP. 2000. Modeling mountain building and the seismic cycle in the Himalaya of Nepal. Journal of Geophysical Research 105, 13389–13407.

Chen W-P, Molnar P. 1983. The depth distribution of intracontinental and intraplate earthquakes and its implications for the thermal and mechanical properties of the lithosphere. Journal of Geophysical Research 88, 4183–4214.

Cherniak DJ, Watson EB, Wark DA. 2006. Ti diffusion in quartz. Chemical Geology 236, 65–74.

Crowley PD. 1985. The structural and metamorphic evolution of the Sitas Area, northern Norway and Sweden. PhD thesis, Massachusetts Institute of Technology.

Dahlen FA. 1990. Critical taper model of fold-and-thrust belts and accretionary wedges. Annual Review of Earth and Planetary Sciences 18, 55–99.

Davis D, Suppe J, Dahlen FA. 1983. Mechanics of fold-and-thrust belts and accretionary wedges. Journal of Geophysical Research 88, 1153–1172.

Ferry JM, Spear FS, 1978. Experimental calibration of the partitioning of Fe and Mg between biotite and garnet. Contributions to Mineralogy and Petrology 66, 113–117.

Gee DG. 1975. A tectonic model for the central part of the Scandinavian Caledonides. American Journal of Science 275-A, 468–515.

Gee DG, Sturt BA. 1985. The Caledonide Orogen – Scandinavia and Related Areas. John Wiley & Sons, Chichester.

Gee DG, Juhlin C, Pascal C, Robinson P. 2010. Collisional orogeny in the Scandinavian Caledonides (COSC). Geologiska Föreningen i Stockholm Förhandlingar 132, 29–44.

Grujic, D., Stipp M., and Wooden, J.L., 2011, Thermometry of quartz mylonites: Importance of dynamic recrystallization on Ti-in-quartz reequilibration: Geochemistry Geophysics Geosystems, v. 12, no. 6, doi: 10.1029/2010GC003368.

Harrison TM, Grove M, Lovera OM, Catlos EJ. 1998. A model for the origin of Himalayan anatexis and inverted metamorphism. Journal of Geophysical Research 103, 27017–27032.

Harrison TM, Célérier J, Aikman AB, Hermann J, Heizler MT. 2009. Diffusion of ^{40}Ar in muscovite. Geochimica et Cosmochimica Acta 73, 1039–1051.

Hauck ML, Nelson KD, Brown LD, Zhao WJ, Ross AR. 1998. Crustal structure of the Himalayan orogen at ~ 90° east longitude from Project INDEPTH deep reflection profiles. Tectonics 17(4), 481–500.

Henry P, Le Pichon X, Goffe B. 1997. Kinematic, thermal and petrological model of the Himalayas: Constraints related to metamorphism within the underthrust Indian crust and topographic evolution. Tectonophysics 273, 31–56.

Herman F, Copeland P, Avouac J-P, et al. 2010. Exhumation, crustal deformation, and thermal structure of the Nepal Himalaya derived from inversion of thermochronological and thermobarometric data and modeling of the topography. Journal of Geophysical Research 115, doi:10.1029/2008/JB006126.

Hirth G, Tullis J. 1992. Dislocation creep regimes in quartz aggregates. Journal of Structural Geology 14(2), 145–159.

Hodges KV. 1982. The tectonic evolution of the Aefjord-Sitas area, Norway-Sweden. PhD thesis, Massachusetts Institute of Technology.

Hodges KV, Royden L. 1984. Geologic thermobarometry of retrograde metamorphic rocks: an indication of the uplift trajectory of a portion of the northern Scandinavian Caledonides. Journal of Geophysical Research 89, 7077–7090.

Hoisch TD. 1990. Empirical calibration of six geobarometers for the mineral assemblage quartz+muscovite+biotite+plagioclase+garnet. Contributions to Mineralogy and Petrology 104(2), 225–234.

Huang R, Audetat A. 2012. The titanium-in-quartz (Ti-taniQ) thermobarometer: a critical examination and re-calibration. Geochimica et Cosmochimica Acta 84, 75–89.

Huerta AD, Royden LH, Hodges KV. 1998. The thermal structure of collisional orogens as a response to accretion, erosion, and radiogenic heating. Journal of Geophysical Research 103, 15287–15302.

Huerta AD, Royden LH, Hodges KV. 1999. The effects of accretion, erosion, and radiogenic heat on the metamorphic evolution of collisional orogens. Journal of Metamorphic Geology 17, 349–366.

Jamieson RA, Beaumont C, Medvedev S, Nguyen MH. 2004. Crustal channel flows: 2. Numerical models with implications for metamorphism in the Himalayan-Tibetan orogen. Journal of Geophysical Research 109, B06406.

Kohn MJ. 2008. P-T-t data from central Nepal support critical taper and repudiate large-scale channel flow of the Greater Himalayan Sequence. Geological Society of America Bulletin 120(3/4), 259–273.

Kohn MJ, Northrup CJ. 2009. Taking mylonites' temperature. Geology 37(1), 47–50.

Larson KP, Godin L, Price RA. 2010. Relationships between displacement and distortion in orogens: Linking the Himalayan foreland and hinterland in central Nepal. Geological Society of America Bulletin 122, 1116–1134.

Mukherjee S. 2013. Higher Himalaya in the Bhagirathi section (NW Himalaya, India): its structures, backthrusts and extrusion mechanism by both channel flow and critical taper mechanisms. International Journal of Earth Sciences 102, 1851–1870.

Northrup CJ. 1996a. Structural expressions and tectonic implications of general noncoaxial flow in the midcrust of a collisional orogeny: the northern Scandinavian Caledonides. Tectonics 15, 490–505.

Northrup CJ. 1996b. Tectonic evolution of the Caledonian collisional system, Ofoten-Efjorden, North Norway. PhD thesis, Massachusetts Institute of Technology.

Page LM. 1990. Structure, metamorphism, and geochronology of the Singis-Nikkaluokta region, arctic Scandinavian Caledonides. PhD thesis, Massachusetts Institute of Technology.

Page LM. 1992. [40]Ar/[39]Ar geochronologic constraints on timing of deformation and metamorphism of the central Norrbotten Caledonides, Sweden. Geological Journal 27, 127–150.

Passchier CW. Trouw RAJ. 1998. Microtectonics. Springer, New York.

Passchier CW, Trouw RAJ. 2005. Microtectonics, 2nd edition. Springer, New York.

Ramberg H. 1977. Some remarks on the mechanism of nappe movement. Geologiska Föreningen i Stockholm Förhandlingar 99, 110–117.

Ramberg H. 1981. The role of gravity in orogenic belts, in thrust and Nappe Tectonics. Special Publication Geological Society, London 9, 125–140.

Roberts D. 2003. The Scandinavian Caledonides: event chronology, palaeogeographic settings and likely modern analogues. Tectonophysics 365, 283–299.

Roberts D, Gee DG. 1985. An introduction to the structure of the Scandinavian Caledonides. In The Caledonide Orogen – Scandinavia and related areas, edited by D.G. Gee and B.A. Sturt, John Wiley & Sons, Chichester, pp. 55 68.

Royden, L.H., 1993, The steady-state thermal structure of eroding orogenic belts and accretionary prisms: Journal of Geophysical Research, B, Solid Earth and Planets, v. 98, p. 4487–4507.

Rusk BG, Lowers HA, Reed MH. 2008. Trace elements in hydrothermal quartz: relationships to cathodoluminescent textures and insights into vein formation. Geology 36(7), 547–550.

Rykkelid E, Andresen A. 1994. Late Caledonian extension in the Ofoten area, northern Norway. Tectonophysics 231, 157–169.

Sanderson DJ. 1982. Models of strain variation in nappes and thrust sheets; a review. Tectonophysics 88(3–4), 201–233.

Smith LB, Jr. 2006. Origin and reservoir characteristics of Upper Ordovician Trenton–Black River hydrothermal dolomite reservoirs in New York. American Association of Petroleum Geologists (AAPG) Bulletin, 90, 1691–1718.

Spear FS, Cheney JT. 1989. A petrogenic grid for pelitic schists in the system SiO_2-Al_2-FeO-MgO-K_2O-H_2O. Contributions to Mineralogy and Petrology 101(2), 149–164.

Spear FS, Wark DA. 2009. Cathodoluminescence imaging and titanium thermometry in metamorphic quartz. Journal of Metamorphic Geology 27, 187–205.

Steltenpohl M, Bartley JM. 1987. Thermobarometric profile through the Caledonian nappe stack of Western Ofoten, North Norway. Contributions to Mineralogy and Petrology 96, 93–103

Stipp M, Stunitz H, Heilbronner R, Schmid SM. 2002. The eastern Tonale fault zone: a 'natural laboratory' for crystal plastic deformation of quartz over a temperature range from 250–700°C. Journal of Structural Geology 24(12), 1861–1884.

Thomas JB, Watson EB, Spear FS, Shemella PT, Nayak SK, Lanzirotti A. 2010. TitaniQ under pressure: the effect of pressure and temperature on the solubility of Ti in quartz. Contributions to Mineralogy and Petrology 160, 743–749.

Tilke PG. 1986. Caledonian structure, metamorphism, geochronology, and tectonics of the Sitas-Singis area, Sweden. Earth, Atmospheric and Planetary Sciences. PhD thesis, Massachusetts Institute of Technology, Cambridge.

Tilke PG. 1986. Caledonian structure, metamorphism, geochronology, and tectonics of the Sitas-Singis area, Sweden. PhD thesis, Massachusetts Institute of Technology.

Wark DA, Watson EB. 2006. TitaniQ: a titanium-in-quartz geothermometer. Contributions to Mineralogy and Petrology, 152, 743–754.

Willett SD. 1999. Orogeny and orography: the effects of erosion on the structure of mountain belts. Journal of Geophysical Research 104, 28957–28981.

Chapter 8

Brittle-ductile shear zones along inversion-related frontal and oblique thrust ramps: Insights from the Central–Northern Apennines curved thrust system (Italy)

PAOLO PACE[1], FERNANDO CALAMITA[1], and ENRICO TAVARNELLI[2]

[1] *Dipartimento di Ingegneria e Geologia, Università degli Studi "G. D'Annunzio" di Chieti-Pescara, Via dei Vestini 31, 66013, Chieti Scalo (CH), Italy*

[2] *Dipartimento di Scienze Fisiche, della Terra e dell'Ambiente, Università degli Studi di Siena, Via Laterina 8, 53100, Siena, Italy*

8.1 INTRODUCTION

High-strain deformation within the Earth's crust often occurs in localized, narrow, and sub-parallel wall-sided zones known as shear-zones, which accommodate differential movement during the deformation of the lithosphere. They may be related to any tectonic regime (compression, extension, or strike–slip), varying in width from microns/millimeters (grain-scale) to kilometers (mega-shears). The heterogeneous character of natural deformation in shear zones produces characteristic fault rocks as mylonites and cataclasites, developed under deep-seated (10–25 km deep) ductile (viscous) or shallow-crustal (0–15 km deep) brittle–ductile (frictional–viscous) deformation regimes, respectively (e.g. Ramsay and Graham 1970; Sibson 1977, 1983; Ramsay 1980; Alsop and Holdsworth 2004).

The analysis of brittle–ductile and ductile shear zones exhumed and/or extruded and exposed at the surface through a variety of approaches and across a range of scales is essential for unraveling deformation histories. Deciphering the kinematic significance of deformation fabrics within fault rocks and reconstructing the regional tectonics contribute profoundly to understand how localized crustal deformation occurs (e.g. Casas and Sàbat 1987; Alsop et al. 2004; Carosi et al. 2004; Iacopini et al. 2008; Mukherjee 2007, 2010a,b, 2011, 2013a, b, c, 2014a, b; Mukherjee and Koyi 2010a,b; Calamita et al. 2012a; Tesei et al. 2013).

In this chapter the geometric and kinematic characteristics of shear deformation fabrics associated with frontal and oblique ramps belonging to curve-shaped thrusts are described. A detailed mesoscale structural and kinematic analysis is presented by examining some remarkable examples of brittle–ductile thrust shear zones related to regional-scale frontal and oblique thrust ramps in the Central–Northern Apennines of Italy.

8.1.1 Thrust shear zone fabrics

Brittle–ductile and ductile shear zones related to thrust faults generally develop under dominant simple shear deformation (Mukherjee 2012a,b, 2014c; Mukherjee and Mulchrone 2013; Mukherjee and Biswas 2014; and many others). At shallow crustal levels, pressure-solution cleavage codevelops with shear fracturing, due to simple shear commonly related to compression/thrusting, developing foliation-dominated cataclastic fault rocks (i.e. foliated cataclasites, Chester et al. 1985; Jefferies et al. 2006). The rock volume involved in such a deformation is characterized by typical planar fabrics, the so-called S-C fabric (ten Grotenhuis et al. 2003; Mukherjee 2011; Fig. 8.1), comprising two main sets of strain-sensitive planar anisotropies, which are strictly related to the shear sense and are reliable kinematic indicators (Ramsay and Graham 1970; Berthé et al. 1979a; Lister and Snoke 1984; Blenkinsop and Treloar 1995). One set of planar anisotropy is represented by cleavage planes that are related to the accumulation of finite strain in the S-surfaces (S: "schistosité"). The second set is formed by discrete displacement shear planes occurring in localized high shear strains into the C-surfaces (C: "cisaillement") (Berthé et al. 1979a; Jegouzo 1980; Ponce de Leon and Choukroune 1980). The S planes represent a penetrative developed strain-sensitive flattening, likely corresponding to the XY plane of the finite strain ellipsoid (Ramsay and Graham 1970; Hanmer and Passchier 1991). The C planes are discrete narrow shear zones or slip surfaces with the same shear sense as that of the overall shear zone (Simpson and Schmid 1983; Hanmer and Passchier 1991; Passchier and Trouw 2005). At various scales, in sections perpendicular to the shear planes and parallel to the shear directions (the *XZ*-section), the S foliation is systematically comprised in between the spaced C shear planes. The trace of the S surface tends to curve into the bounding pair of C surfaces, identifying

Ductile Shear Zones: From Micro- to Macro-scales, First Edition. Edited by Soumyajit Mukherjee and Kieran F. Mulchrone.
© 2016 John Wiley & Sons, Ltd. Published 2016 by John Wiley & Sons, Ltd.

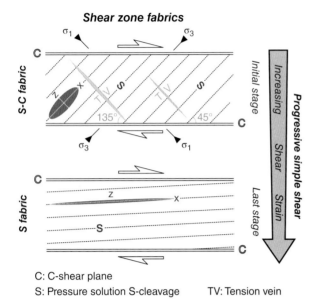

C: C-shear plane
S: Pressure solution S-cleavage   TV: Tension vein

Fig. 8.1. Schematized representation of the two types of deformation fabrics (S-C and S) characterizing thrust shear zones. Data from McClay 1987 and Logan et al. 1992.

typical sigmoid-shaped lithons, as the intensity of shear progressively increases from the middle toward the edges of the shear zone (e.g. Berthé et al. 1979a; Lister and Snoke 1984; Hanmer and Passchier 1991). The C surfaces initiate and remain parallel to the main shear zone boundary (e.g. the thrust fault), whereas the S-surfaces start to develop at 45° to the C planes at the initial deformation stage, developing synchronously. This angular relationship between the two fabric surfaces deduces shear sense (Simpson and Schmid 1983). As the rotation increases progressively, this angle falls. The C-surfaces intensify, become more closely-spaced and both synthetic (R) and antithetic (R′) Riedel shears occur cutting obliquely across the S cleavage at an angle of ~15° and ~75° to the C displacement shear plane, respectively, mostly in brittle and brittle–ductile shear zones (Riedel 1929; Tchalenko 1968; Misra et al. 2009). These Riedel shears are known as C′ and C″ planes in purely ductile shear zones (Passchier and Trouw 2005). During the last stages of progressive simple shear, the angle between the S and C surfaces decreases significantly, and the S cleavages tend to sub-parallel the shear zone boundary slightly, matching the C shear planes and defining a planar foliated cataclasite (Fig. 8.1) known as S-tectonites (Ramsay and Graham 1970; Logan et al. 1992; Ragan 2009).

Nevertheless, in some cases, the relatively simple geometry of these shear fabrics (S-C and S tectonites) characterizing sheared fault rocks typically developed within thrust shear zones may be more complex due to the occurrence of flanking structures (Mukherjee and Koyi 2009; Passchier 2001; Mukherjee 2014a) or folded fabrics, thus producing composite fabrics. Both synthetic

and antithetic discrete internal shear surfaces, developing oblique to the shear zone boundaries, are often observed within thrust shear zones, suggesting that a component of coaxiality acted within the flow (i.e. sub-simple shear, reviewed in Mukherjee 2007). They can be restricted within the shear fabrics associated with the thrust or sometimes they may also affect the entire shear zone by displacing the thrust sheet. When these subsidiary extensional shear planes rest within the thrust sheet or the thrust shear zone, they have been interpreted variously as either developed during: (i) advanced stages of deformation when extension cannot occur along earlier foliation planes (Watts and Williams 1979, White et al. 1980); (ii) a later stage flattening with a sub-coaxial deformation developing the so-called extensional crenulation cleavage (Platt and Vissers 1980); or (iii) when the flow departs from progressive simple shear and the bulk flow field partitions into slip along discrete planes producing coaxial stretching along the foliation (Platt 1984). Alternatively, these conjugate sets of shear planes may develop within simple shear strain when slip in favored along pre-existing foliation planes (Harris and Cobbold 1984). In other cases, synthetic meso-scale extensional planes have been explained by Riedel-type shear planes synchronous with thrusting (Yin and Kelty 1991; Butler 1992). Moreover, extensional fault zones that post-date the compressive structures displacing thrust faults have been also documented and mostly explained as due to syn-thrusting gravity collapse or spreading as in the Moine Thrust belt of NW Scotland (Coward 1982, 1983; Holdsworth et al. 2006).

A further complexity in the geometry of thrust shear zones may be the occurrence of meso-scale folds affecting the pressure solution S-cleavage (Casas and Sàbat 1987; Holdsworth 1989; Powell and Glendinning 1990; Alsop 1991; Calamita et al. 1991; Tavarnelli 1999) and intrafolial folds (Mukherjee et al. 2016, Chapter 12), which seems to be not directly related to the simple shear deformation due to thrusting. This may occur when the folded pressure–solution cleavage is restricted between a pair of shear planes with opposite movement, as recognized for the Alpine thrusts in the Balearic Isles (Casas and Sàbat 1987) or associated with a shear-sense reversal (Mukherjee 2013c; Mukherjee and Koyi 2010b; Chattopadhyay, 2016; Pamplona et al, 2016; Sengupta and Chatterjee, 2016, Chapters 9, 10, and 13) causing down-dip extensional reactivation of previously formed thrust surfaces (Holdsworth 1989; Powell and Glendinning 1990; Alsop 1991; Calamita et al. 1991; Tavarnelli 1999).

8.1.2   Kinematic vorticity analysis

Kinematic vorticity analyses are valuable for studying and seeking to quantify the kinematics of flow in shear zones. This quantitative approach applies in ductile and brittle–ductile naturally sheared rocks by several methods (review by Xypolias 2010). The kinematic vorticity number (W_k), a dimensionless number, describes the kind of flow. Pure shear is described by $W_k = 0$, and simple shear by $W_k = 1$ (Means et al. 1980; Passchier 1986; Simpson and De Paor 1993; Xypolias 2010). General shear/sub-simple shear represents intermediate flows between the pure and simple shear end members with $0 < W_k < 1$ (De Paor 1983). Since general shear or sub-simple shear seems to be the rule rather than the exception in natural deformation flows (Coward 1976; Platt and Vissers 1980; Kurz and Northrup 2008; Sullivan 2009), vorticity studies have been principally devoted to deciphering and trying to quantify the degree of non-coaxiality of deformation through relationships between the vortical and the stretching components of flow (Means et al. 1980; Ghosh 1987; Xypolias 2010, and references therein). However, due to the complexity of natural deformation and too many assumptions, methods of vorticity estimates still rely on a certain degree of systematic error/uncertainty (e.g. Tikoff and Fossen 1995; Iacopini et al. 2011).

Synthetic shear bands or flanking structures cross-cutting obliquely the shear zone boundary and dipping in the direction of the shear sense are one of the most observed structures within shear zones (e.g. Platt and Vissers 1980; Passchier 2001; Grasemann et al. 2003). Such structures are used frequently as kinematic indicators to deduce the shear sense and may constrain kinematic vorticity of flow as well. Simpson and De Paor (1993) proposed that synthetic and antithetic shear bands nucleate approximately parallel with the acute and obtuse bisectors of the stable flow eigenvectors (apophyses), respectively during general shear. As per Platt and Visser's (1980) non-coaxial deformation model, both conjugate shear bands rotate toward the flow/shear plane; consequently, as deformation proceeds, the low-angle synthetic shears rotate very slowly and remain active longer than the high-angle antithetic ones. Recently, Kurz and Northrup (2008) verified this hypothesis by analyzing naturally deformed rocks. For two-dimensional (2D) deformation with no net external rotation, W_k can be defined simply as the cosine of the acute angle (α) between the flow eigenvectors (Xypolias 2010, and references therein). Given that the synthetic shear planes bisect the acute angle (α) between the flow apophyses, in other words, they nucleate approximately at an angle $\beta = 0.5\alpha$ (Kurz and Northrup 2008), W_k is estimated as:

$$W_k = \cos(2\beta) = \cos(\alpha) \qquad (1)$$

Here β is the largest angle between the synthetic shear bands and the flow plane of the simple shear deformation. Considering the boundary conditions, we sought to estimate the kinematic vorticity analysis by applying this criterion, since in the studied examples gently dipping synthetic extensional shear planes were always identified along 2D planes of observation, parallel to the XZ plane of the finite strain.

8.2 REGIONAL GEOLOGY

The Apennines of Italy are a foreland-directed fold-and-thrust belt that have developed since Oligocene, following the closure of the Mesozoic Tethys Ocean and the consequent collision between Africa and Europe (Boccaletti et al. 2005). The orogenesis involved thick Triassic–Miocene sedimentary piles deposited in different paleogeographic domains (e.g. Latium–Abruzzi platform and Umbria–Marche basin) of the Mesozoic Adria continental margin (Ciarapica and Passeri 2002). This process has been accompanied by a general hinterland-to-foreland "in-sequence" propagation of the thrust-fold system with associated foredeep and thrust-top basins as documented by the age of the syn-orogenic siliciclastic sediments, which become progressively younger eastward (Ricci Lucchi 1986; Boccaletti et al. 1990). The structural architecture of the Apennines thrust system is composed of two major arcs, the Northern Apennine Arc and the Southern Apennines Arc, linking up, respectively toward S and N, in the Central Apennines through two regionally important cross-strike discontinuities (Olevano-Antrodoco-Sibillini and Sangro-Volturno; Rusciadelli et al. 2005; Calamita et al. 2011; Satolli et al. 2014; Fig. 8.2).

The study area is located within the outer zone of the thrust belt at the Central–Northern Apennines boundary, known as the Umbria–Marche Apennines. The outcropping sedimentary sequence constitutes Mesozoic–Cenozoic well-bedded pelagic carbonates with interbedded cherts and marls that deposited in the deep Umbria-Marche basin (Speranza et al. 2005). The pelagic succession lies over thick massive shallow-water Lower Jurassic platform carbonates and is overlain unconformably by Miocene foredeep siliciclastic sediments (Fig. 8.3). The Jurassic–Eocene pre-orogenic sequence displays abrupt lateral thickening and facies variations controlled by syn-sedimentary Jurassic and Cretaceous–Paleogene normal faults (Alvarez 1990; Santantonio 1993; Bosellini 2004). Moreover, Miocene syn-flexural sets of normal faults have also been documented within syn-orogenic basins (Calamita et al. 1998; Scisciani et al. 2001). From Middle–Late Miocene onward, deformation switched from extension to compression (i.e. positive inversion tectonics; Tavarnelli et al. 2004; Butler et al. 2006). Pre-thrusting normal faults were reactivated, truncated, rotated and/or modified by the Messinian–Lower Pliocene thrusting and related folding. The resulting structural geometries and interference patterns of positive inversion were described in a wide range of scales (e.g. Tavarnelli 1996c; Tavarnelli and Peacock 1999; Scisciani et al. 2002; Tavarnelli and Alvarez 2002; Calamita et al. 2009; Pace et al. 2011, 2014). Within this context of inversion, favorably oriented N–S-trending pre-existing extensional discontinuities reactivated in transpression leading to regional NNW–SSE-trending oblique thrust ramps (e.g. Olevano-Antrodoco-Sibillini oblique thrust ramp; Di Domenica et al. 2012), whereas the misoriented NW-SE-trending inherited normal faults

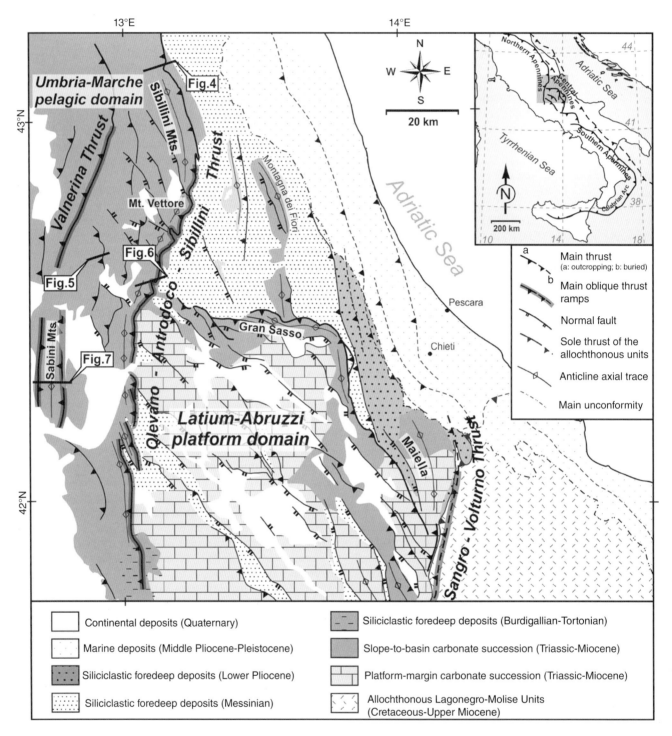

Fig. 8.2. Simplified geological and structural map showing the location of the studied examples along the frontal and the oblique thrust ramps within the Central–Northern Apennines thrust belt.

truncated and passively translated in the hanging-wall of shortcut propagating thrusts along frontal thrust ramps (Calamita et al. 2011, 2012b; Di Domenica et al. 2014). The thrust belt geometries have been therefore strongly influenced by the inherited extensional fault template resulting in characteristic curved map traces of the thrust system. Folds and thrusts interacted with a mechanically heterogeneous multilayer characterized by strong rheo-

logical contrasts and detachment levels within the Umbria-Marche sedimentary pile (Fig. 8.3). Contractional deformation was accommodated by high-amplitude thrust-related folds related to fault-bend folding, fault-propagation folding, and push-up structures (Calamita 1990; Tavarnelli 1993a; Calamita et al. 2012b; Pace et al. 2012; Pace and Calamita 2014) also accompanied by diffusive pressure-solution at mesoscale (Alvarez et al. 1976).

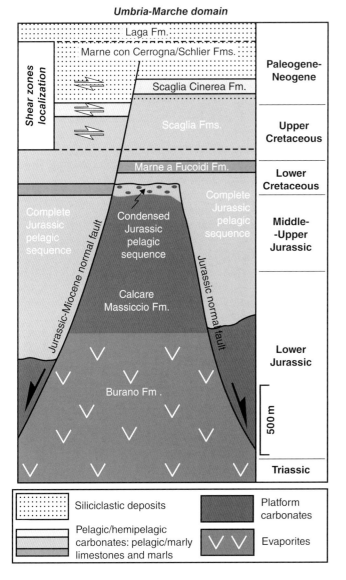

Umbria-Marche domain

Shear zones localization

Laga Fm.	Paleogene-Neogene
Marne con Cerrogna/Schlier Fms.	
Scaglia Cinerea Fm.	
Scaglia Fms.	Upper Cretaceous
Marne a Fucoidi Fm.	Lower Cretaceous
Condensed Jurassic pelagic sequence / Complete Jurassic pelagic sequence	Middle--Upper Jurassic
Calcare Massiccio Fm.	Lower Jurassic
Burano Fm.	
	Triassic

Jurassic-Miocene normal fault

Jurassic normal fault

Complete Jurassic pelagic sequence

500 m

Legend:
- Siliciclastic deposits
- Pelagic/hemipelagic carbonates: pelagic/marly limestones and marls
- Platform carbonates
- Evaporites

Fig. 8.3. Schematic stratigraphy of the Umbria–Marche sedimentary sequence cropping out in the study area. Adapted from Pierantoni 1998.

During the Quaternary, post-orogenic extension, characterized by hinterland-dipping NW–SE-trending normal faults with associated intermontane basins and seismicity, affected the axial zone of the Central–Northern Apennines carbonate mountain belt (Calamita et al. 2000).

8.3 BRITTLE–DUCTILE THRUST SHEAR ZONES IN THE APENNINES: A BRIEF REVIEW

Foliated cataclasites and shear deformation fabrics within well-foliated shear zones related to thrust faults have been widely described in the Umbria–Marche sector of the Northern Apennines of Italy (Koopman 1983;

Lavecchia 1985; Calamita et al. 1987, 2012a; Calamita 1991; Alberti et al. 1996; Tavarnelli 1997).

Shear zone fabrics are mostly well developed within the upper section (Eocene–Miocene) of the Umbria–Marche sedimentary sequence (Fig. 8.3) composed of mixed hemipelagic carbonate-terrigenous lithologies, made up of fine-grained pelagic limestones, marly limestones, and marls (Scaglia Rossa-, Scaglia Cinerea-, Schlier-, and Marne con Cerrogna Formations).

Both S-C and S fabrics have been recognized within the well-foliated shear zones related to the imbricate thrust system of the Umbria–Marche Apennines (Koopman 1983; Tavarnelli 1997; Calamita et al. 2012a; Tesei et al. 2013). In some cases, the relatively simple geometries of the S and S-C fabrics might be more complex, due to the occurrence of superimposed extensional mesoscale structures (extensional crenulation cleavage; Koopman 1983; Calamita 1991) or folded pressure-solution S cleavage surfaces. Such folded pressure-solution S fabric seems common along some oblique thrust ramps in the Central–Northern Apennines. The occurrence of this folding phenomenon has been mostly attributed to a shear-sense reversal, from a foreland-ward NE-directed thrusting, to a hinterland-ward top-to-the-SW movement, occurring on the pre-existing C-shear planes (Calamita et al., 1991; Tavarnelli 1999; Bigi 2006). This interpretation led to important implications for thrusting–normal fault relationships that considered a post-orogenic negative inversion tectonics model, in which the Quaternary hinterland-dipping normal faults reactivate the former Neogene thrust planes (Calamita et al., 1991; Tavarnelli 1999; Bigi 2006).

Recently, efforts to understand the deformation mechanisms and seismic behavior of carbonate-bearing thrust faults have been carried out by analyzing analogue exhumed shear zones exposed along some regional-scale thrusts in the Umbria–Marche Apennines (Collettini et al. 2013; Tesei et al. 2013).

8.4 DEFORMATION SHEAR FABRICS WITHIN CURVED THRUSTS

8.4.1 Shear zone fabrics along frontal thrust ramps

8.4.1.1 *The Sibillini Mountains thrust shear zone*

The NW–SE-trending frontal thrust ramp of the Sibillini Mountains links toward the south the NNE-SSW-trending oblique thrust ramp. These two structures form the Olevano-Antrodoco-Sibillini (OAS) thrust (Fig. 8.2), a curved regional thrust representing the outer front of the Northern Apennines (Calamita and Deiana 1988). The frontal thrust ramp, with associated hanging-wall fold structures and shear zones, has been studied widely (Lavecchia 1985; Calamita et al. 1987, 2011, 2012a, b; Pierantoni et al. 2005, 2013; Mazzoli et al. 2005). To the N, the NW–SE frontal thrust (Sibillini thrust) emplaces the Jurassic–Cretaceous Umbria–Marche carbonate platform and pelagic sequence (Calcare Massiccio-Scaglia Formations; Fig. 8.3) onto the Oligocene–Miocene hemipelagic marly–calcareous pelagic succession and siliciclastic deposits (Scaglia Cinerea, Marne con Cerrogna, and Laga Formations). To the S, along the oblique thrust ramp, the Latium–Abruzzi carbonate platform domain represents the footwall (Fig. 8.2). In the analyzed area, an overturned box-fold anticline characterizes the hanging-wall thrust-related fold structure (Lavecchia 1985). The high-amplitude fold shows a flat crest joining the overturned forelimb through an angular hinge (Fig. 8.4a,b; Calamita et al. 2012b). Both SW-dipping Mesozoic–Tertiary pre-thrusting normal faults and NE-dipping Jurassic normal faults truncated and translated passively in the hanging-wall of the Sibillini Mountains thrust within the development of a "shortcut anticline" (Butler 1989; Fig. 8.4a,b) characterizing the overall frontal thrust ramp (Calamita et al. 2011, 2012b).

Superb outcrops along the naturally exposed section in the Fiastrone Valley (Fig. 8.4a) allow to investigate the thrust-related fold structures and shear zone characterizing the Sibillini Mountains thrust. The main thrust is associated with a brittle–ductile shear zone, which is decametric-scale thick, well developed in the footwall block, and is in tectonic lenses of rock bodies sandwiched in a pair of splay thrust surfaces. Both outcrop and sample-scale kinematic and geometric analyses of the sheared rocks mostly involving the micritic pelagic limestones of the Scaglia Rossa Formation revealed that the shear zone texture is dominantly of S-C fabric. Centimeter- to decimeter-spaced striated discrete displacement shear planes (C) developed sub-parallel to the main thrust shear zone boundary and localized the formation of a syn-kinematic centimeter-spaced stylolitic pressure solution S-cleavage (Fig. 8.4c,d). The well-developed stylolitic S-cleavage surfaces identify spaced and elongated sigmoidal-shaped lithons of calcareous rock bodies, whose asymmetry makes an average angle of ~42° with respect to the C-shears (Fig. 8.4d,e) allowing deduction of the sense of movement that is confirmed by kinematic data on striated shear surfaces to be N60° (Fig. 8.4d). Millimeter- to centimeter-scale syn-tectonic tension veins with sparry calcite infill opened and grew perpendicular to the S-foliation whereas calcite-bearing shear veins lay on the C-planes (Fig. 8.4c,d). Low-angle synthetic shear bands, interpreted as R-Riedel planes, are observable and display an average angle ~16° with the C-shears (Fig. 8.4d,e). Taking into account the maximum angular relationships (17°) between the flanking synthetic R-Riedel shears and the C-shear planes, the resulting kinematic vorticity number is $W_k = 0.83$ (Fig. 8.4e).

8.4.1.2   The Mount Coscerno thrust shear zone

The Mount Coscerno structure is a NNW–SSE-trending shortcut anticline belonging to the frontal thrust ramp of a regionally curved thrust known as the Mount-Coscerno–Rivodutri thrust (Tavarnelli 1993b, 1994), located 40 km

SW to the previously described Sibillini Mountains thrust (Fig. 8.2). The wide anticline displays a box-shaped profile with a vertical-to-overturned forelimb and hinterland-verging folds and backthrusts in the backlimb involving the Jurassic–Lower Cretaceous section of the Umbria–Marche sequence (Barchi and Lemmi 1996; Fig. 8.5a). Thrust ramps localized due to Tethyan-derived pre-existing extensional discontinuities producing typical shortcut geometries (Tavarnelli 1996a,b,c). In the studied section, the Mount Coscerno thrust emplaces the Upper Cretaceous–Eocene well-bedded pelagic limestones of the Scaglia Formation onto the Eocene–Oligocene marly limestones and marls of the Scaglia Variegata and Scaglia Cinerea Formations.

The shear zone is developed mostly within the less competent marly limestones and marls cropping out in the thrust footwall. The thrust shear zone fabric exhibits a foliated cataclasite organized in a very clear S-C fabric (Fig. 8.5b,c), as also documented by other workers (Barchi and Lemmi 1996; Tesei et al. 2013). Gently SW-dipping centimeter-spaced C-planes localize narrow zones dominated by a diffuse and pervasive millimeter- to centimeter-spaced pressure-solution S-cleavage. The well-developed S-foliation identifies high-angle SW-dipping sigmoidal lithons of elongated marly rock bodies. The angular relationship between the C- and S- surfaces is ~43° on average and the thrust kinematics measured on striated C-shear planes directs NE, striking ~N62° (Fig. 8.5d). Frequent low-angle shear planes dipping towards shear sense were interpreted as synthetic R-shears. These extensional flanking structures have an average angle of ~15° with the C-planes (Fig. 8.5d). The kinematic vorticity estimate obtained by using the highest R-C angle (16°) measured from the collected data is $W_k = 0.85$ (Fig. 8.5d).

8.4.2   Shear zone fabrics along oblique thrust ramps

8.4.2.1   The Mount Boragine thrust shear zone

The Mount Boragine structure lies along the NNE–SSW-trending oblique ramp of the regionally important Olevano-Antrodoco-Sibillini (OAS) thrust (Fig. 8.2). This high-angle oblique thrust ramp emplaces pelagic carbonates of the so called Umbria–Marche–Sabina pelagic domain over the platform carbonates belonging to the Latium–Abruzzi domain, resulting from the transpressional reactivation of an inherited W-dipping Mesozoic normal fault (Di Domenica et al. 2012, with references therein). Hanging-wall thrust-related folding accompanied the NE-verging oblique thrusting and transpressional reactivation processes occurred along the pre-existing extensional ramp by developing fault-bend folds all over the NNE–SSW segment (Calamita et al. 2012b; Pace and Calamita 2014). The shortening achieved by the oblique ramp (4–5 km) is coherent with the previously described Sibillini structure along the frontal ramp (Calamita et al. 2012b). NNW–SSE-trending

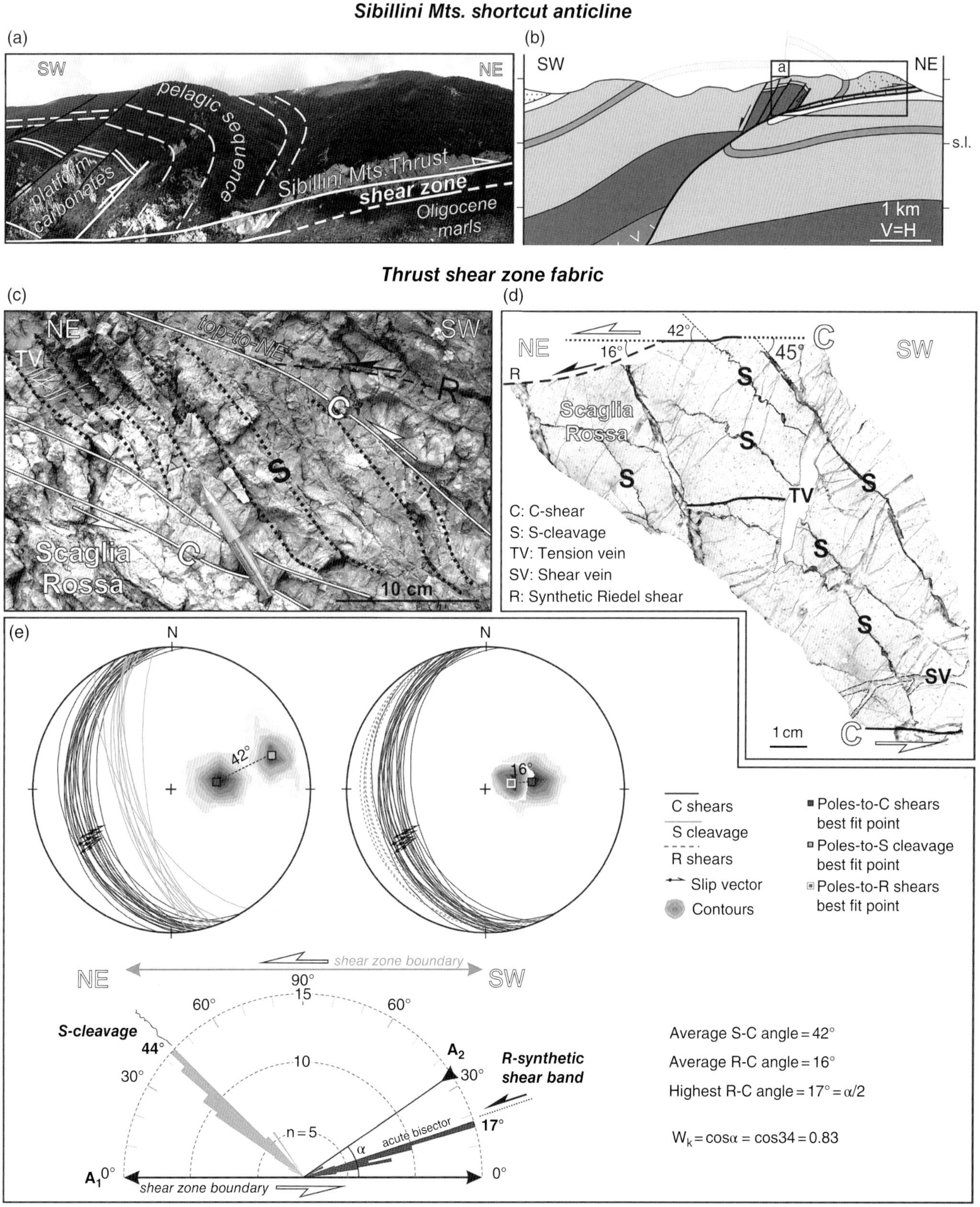

Sibillini Mts. shortcut anticline

Thrust shear zone fabric

C: C-shear
S: S-cleavage
TV: Tension vein
SV: Shear vein
R: Synthetic Riedel shear

C shears
S cleavage
R shears
Slip vector
Contours

Poles-to-C shears best fit point
Poles-to-S cleavage best fit point
Poles-to-R shears best fit point

Average S-C angle = 42°

Average R-C angle = 16°

Highest R-C angle = 17° = α/2

$W_k = \cos\alpha = \cos 34 = 0.83$

Fig. 8.4. Thrust shear zone fabric along the NW–SE-trending Sibillini Mountains shortcut-related frontal thrust ramp. (a) Panoramic view along the Fiastone valley and (b) geological cross-section (key to colors in Fig. 8.3) showing the Sibillini Mountains shortcut anticline structure. Adapted from Calamita et al. 2012b. Outcrop (c) and scan (d) of the polished hand sample cut parallel to the XZ plane of the finite strain ellipsoid displaying the remarkable S-C fabric of the brittle-ductile thrust shear zone within the Scaglia Rossa Formation (e) Equal-area lower-hemisphere stereographic projections of the data collected along the Sibillini Mountains thrust shear zone showing the average angle relationships between the C-shear planes and the S-cleavage and between C- and synthetic R-shear planes. Rose diagram (inspired by Xypolias 2010) showing the orientations of S-cleavage surfaces and synthetic R-type shear bands; n: number of data.

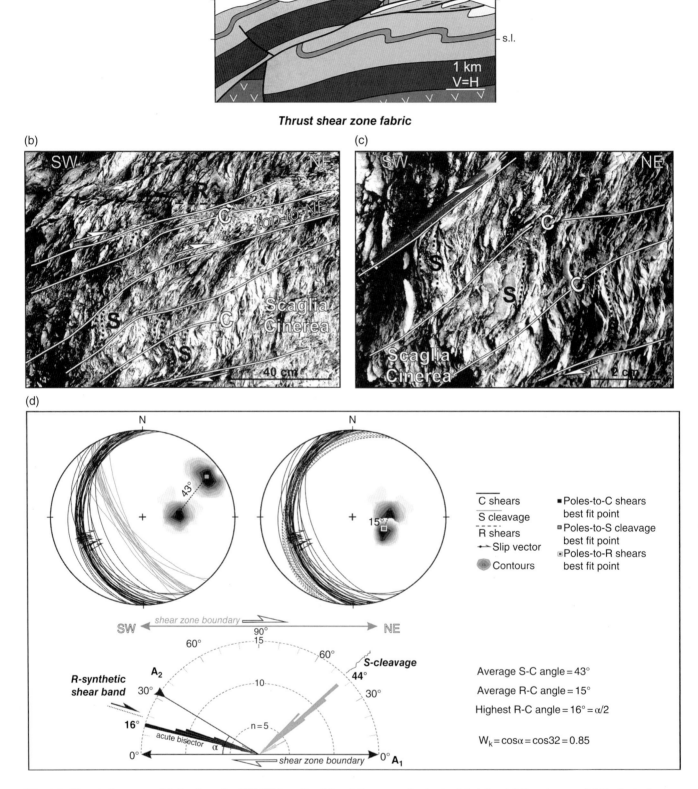

Fig. 8.5. Thrust shear zone fabric along the SW–NE-trending Mount Coscerno shortcut-related frontal thrust ramp. (a) Geological cross-section (key to colors in Fig. 8.3) across the Mount Coscerno shortcut anticline structure. Adapted from Tavarnelli 1993b. (b and c) Outcrops of the S-C thrust shear fabric developed within the Scaglia Cinerea Formation (d) Equal-area lower-hemisphere stereographic projections of the data collected along the Mount Coscerno thrust shear zone showing the average angle relationships between the C-shear planes and the S-cleavage and between C- and synthetic R-shear planes. Rose diagram with orientations of S-cleavage surfaces and synthetic R-type shear bands.

thrust-related anticlines developed in the footwall of the previously emplaced NNE–SSW-striking OAS thrust and deformed the shallow oblique thrust sheet within an in-sequence thrusting (Calamita et al. 2012a; Fig. 8.6a,b).

Superb exposures in the footwall of the Mount Boragine structure allow investigation of the deformation fabric associated with thrust shear zone of the oblique ramp. In the studied outcrop, the thrust surface emplaces the pelagic carbonates of Scaglia Rossa Formation onto the marls and shales of the Marne con Cerrogna Formation (Fig. 8.6a). Outcrop-scale and sample-scale observations revealed that the thrust shear zone fabric is characterized by a penetrative S-cleavage that sub-parallels (~10°) the main thrust plane (Fig. 8.6c,d). This geometric relationship suggests that an S fabric dominates the thrust shear zone and is consistent all over the shear zones belonging to the oblique thrust ramp (Calamita et al. 2012a). Kinematic data collected along the striated thrust shear surfaces reveal a top-to-the NE thrusting trending N70° (Fig. 8.6e). S-cleavages are defined by stylolitic surfaces related to pressure-solution identifying sub-horizontal marly–calcareous lens-shaped lithons and/or by syn-tectonics calcite shear veins (Fig. 8.6d). The S-foliation appears decimeter-spaced within the Scaglia Rossa Formation in the hanging-wall whereas it is more pervasive and close-spaced (cm to mm) in the marly footwall rocks, developing a well-foliated cataclasite dominated by an S fabric. Both synthetic and antithetic extensional flanking structures associated with calcite shear veins developed within the S-fabric shear zone (Fig. 8.6c,d) characterizing the oblique thrust ramp, as previously pointed out also by Koopman (1983), Calamita (1991) and Calamita et al. (2012a). The extensional flanking structures are mostly developed in the footwall appearing pervasive and closely-spaced and are confined by major S-cleavage surfaces within the shear zone. The highest angle relationship with the main thrust plane is of 37° and the resulting calculated kinematic vorticity number is $W_k = 0.27$ (Fig. 8.6e). Moreover, meter-spaced foreland-dipping high-angle extensional planes affect and displace the main thrust surface and the associated shear zone (Fig. 8.6a). Kinematic vorticity estimates using this other set of extensional shear planes have been previously performed by Calamita et al. (2012a) revealing $W_k = 0$–0.17.

8.4.2.2 *The Sabini Mountains thrust shear zone*

The Sabini Mountains are located immediately W of the OAS oblique thrust ramp within the Mesozoic Umbria-Marche pelagic domain (Fig. 8.2). This structure is constituted by a N–S-trending box-shaped anticline developed in the hanging-wall of two (W- and E-dipping) oblique thrust ramps, which reactivated the pre-existing normal faults bounding a symmetric Jurassic basin (Scisciani 2009; Calamita et al. 2011). The overall geometry of the inversion structure is defined by a broad anticline related to upward-diverging high-angle transpressive thrust faults describing a push-up structure

(Pace and Calamita 2014). The interaction between the pre-existing Mesozoic N–S-trending extensional discontinuities and the subsequent NE–SW-oriented Neogene compression promoted transpression, with strong strain partitioning along the inversion-related oblique thrust ramps (Pierantoni 1997; Calamita et al. 2011). Specifically, along the E-dipping high-angle transpressive Sabina fault (Alfonsi 1995) the compressive deformation has been accommodated by a complex combination of different kinematics ranging from reverse oblique/ transpressive to right-lateral strike-slip (e.g. Pierantoni 1997). For this reason, shear fabrics along this structure have not been clearly recognized.

However, although the Sabini Mountains structure developed along two oblique thrust ramps, geological and structural analyses revealed a shear zone exhibiting remarkable S-C fabrics in the footwall of the foreland-dipping Sabina fault. Detailed mapping allowed identification of the fact that this S-C fabric is related to a subordinate and localized low-angle back-thrust fault (i.e. Cottanello back-thrust) characterized by a top-to-the-SW movement branching off as a splay fault from the more regionally important Sabina fault (Fig. 8.7a). The tectonic significance of this splay back-thrust fault is consistent with the overall geometry of the Sabini Mountains inversion push-up, which is characterized by an upward-branching arrangement of several thrust and back-thrust faults (e.g. Pace and Calamita 2014). The shear zone mostly involves the Scaglia Rossa Formation and it is exposed spectacularly in a quarry of the Roman period, in which the strong fault rock, improperly known as Red Cottanello Marble, has been excavated and used to decorate (Fig. 8.7b). Both outcrop and sample-scale analyses revealed that the shear fabric is defined by discrete centimeter-spaced C-shear surfaces among which a millimeter- to centimeter-spaced pressure-solution S-cleavage developed in between (Fig. 8.7c,d). The S-foliation is at ~42° to C-shear planes. Frequent low-angle synthetic shear planes representing R-type Riedel structures cross-cut obliquely the shear zone at ~20° to the C-surfaces (Fig. 8.7e). The kinematic vorticity value derived from the highest R-C angle (22°) measured among the collected data is $W_k = 0.72$ (Fig. 8.7d).

8.5 DISCUSSIONS AND CONCLUSIONS

Brittle–ductile thrust shear zones, associated with frontal and oblique ramps belonging to curved thrust systems from the Central–Northern Apennines, were analyzed in this study. The curved thrusts characterizing the central area of the Apennines thrust belt have been influenced strongly in their geometry by an extensional fault template heritage (e.g. Tavarnelli et al. 2004; Butler et al. 2006; Calamita et al. 2009). Specifically, the NNE–SSW-trending oblique ramps developed by reactivating in transpression the favorably oriented pre-existing normal faults; whereas the NW–SE-trending frontal thrust ramp

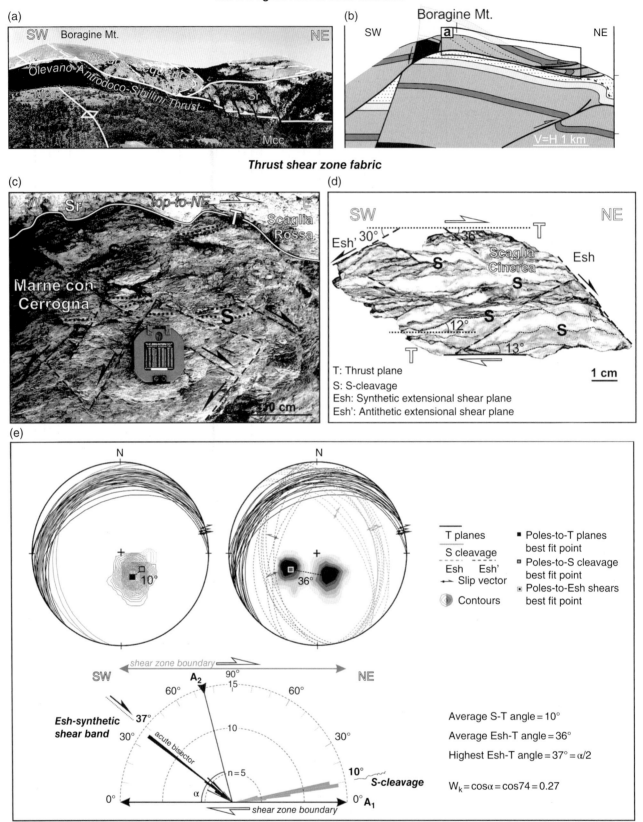

Fig. 8.6. Thrust shear zone fabric along the NNE–SSW-trending Mount Boragine inversion-related oblique thrust ramp. (a) Panoramic view and (b) geological cross-section (key to colours in Fig. 8.3) showing the Mount Boragine reactivated anticline structure. Adapted from Calamita et al. 2012b. Outcrop (c) and scan (d) of the polished hand sample (collected along the oblique thrust ramp within the Scaglia Cinerea Formation northward to Mount Boragine) cut parallel to the XZ plane of the finite strain ellipsoid displaying the remarkable S fabric of the brittle–ductile thrust shear zone with associated conjugate high-angle extensional shear planes. The thrust emplaces the Scaglia Rossa Formation (Sr) onto the Marne con Cerrogna Formation (Mcc). (e) Equal-area lower-hemisphere stereographic projections of the data collected along the Mount Boragine thrust shear zone showing the average angle relationships between the C-shear planes and the S-cleavage and between C- and synthetic extensional shear planes (Esh). Rose diagram with orientations of S-cleavage surfaces and synthetic extensional shear bands (Esh).

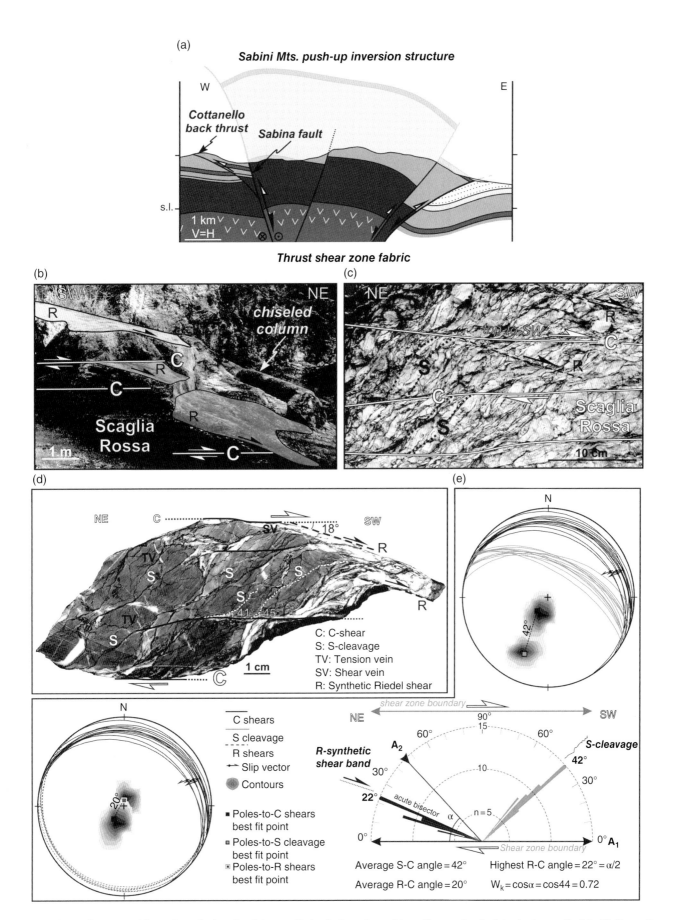

Fig. 8.7. Thrust shear zone fabric along the localized Cottanello back-thrust branching off as a splay fault in the footwall of the N–S-trending Sabina inversion-related oblique thrust ramp. (a) Geological cross-section (key to colors in Fig. 8.3) across the Sabini Mountains push-up inversion structure. Adapted from Pace and Calamita 2014. Outcrops (b and c) and scan of the polished hand sample cut parallel to the XZ plane of the finite strain ellipsoid (d) displaying the stunning S-C fabric of the brittle-ductile thrust shear zone within the Scaglia Rossa Formation (e) Equal-area lower-hemisphere stereographic projections of the data collected along the Cottanello thrust shear zone showing the average angle relationships between the C-shear planes and the S-cleavage and between C- and synthetic R-type shear planes. Rose diagram with orientations of S-cleavage surfaces and synthetic R-type shear bands.

propagated, following shortcut trajectories, through the inherited extensional discontinuities (Calamita et al. 2011, 2012b).

The geological and structural studies of some remarkable examples allowed identification of two main types of shear fabrics along the differently oriented frontal and oblique thrust ramps (Fig. 8.8). Both outcrop and sample-scale structural analyses of the well-foliated cataclasites, developed within marly–calcareous rocks, revealed that the shear zones along the shortcut-related frontal thrust ramps are dominated by S-C fabrics (e.g. Sibillini and Coscerno thrusts; Figs 8.4 and 8.5). These fault rocks are characterized by close-spaced gently dipping C-planes, parallel to the main thrust, and by high-angle pressure solution stylolitic S-cleavage surfaces localizing sigmoid rock lenses. Conversely, the shear zones along the reactivation-related oblique thrust ramps exhibit predominantly S-fabrics, which are characterized by a pervasive stylolitic S-foliation sub-paralleling the main

thrust surface (e.g. Mount Boragine thrust; Fig. 8.6). The S-fabric development along the Olevano-Antrodoco-Sibillini oblique thrust ramp does not deal with a progressive simple shear history from a S-C- to a S-fabric (Ragan 2009) because the shortening realized along the oblique ramp is coherent with the one achieved by the frontal ramp (i.e. Sibillini Mountains thrust). A plausible explanation is instead that such S-fabric developed during the transpressive reactivation of the pre-existing normal fault characterizing the oblique ramp. In addition, in the footwall of retro-verging N–S-trending high-angle transpressive/oblique thrust ramps, S-C fabrics can be developed when related to localized gently-propagating splay back-thrust faults characterizing the back-limb of push-up inversion structures (e.g. Sabini Mountains; Fig. 8.7).

In all analyzed cases, synthetic extensional shear bands cross-cutting the shear zone and dipping towards the shear sense were always identified in 2D along the

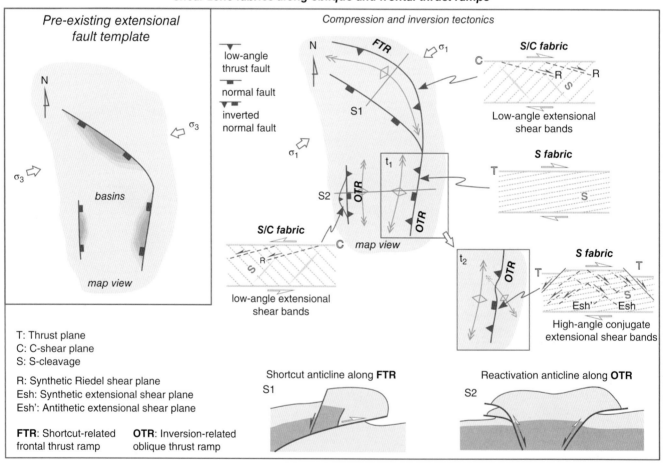

Fig. 8.8. Summary diagram of the two different types of shear deformation fabric characterizing shear zones related to frontal and oblique thrust ramps based on the examples presented in this study. S-C fabrics dominate along the shortcut-related frontal thrust ramps (schematic cross-section S1), whereas along the inversion-related oblique thrust ramps (schematic cross-section S2) a S fabric is mostly developed (t1). Conjugate high-angle extensional shear planes may affect the S fabric dominated shear zone along the OTR due to their interaction with subsequent footwall anticlines within in-sequence deformation (t2). Subordinate S-C fabrics may be associated to localized gently dipping splay thrusts developing in the footwall of inversion-related OTR within push-up inversion structures.

XZ plane of finite strain. Adopting the shear bands/flanking structures method (Kurz and Northrup 2008; Xypolias 2010), we estimated the bulk of the kinematic vorticity. Kinematic vorticity estimates along the shortcut-related frontal thrust ramps (Sibillini and Coscerno thrusts), with 0.83–0.85 W_k, values suggest a predominant simple shear deformation. The gently dipping (usually 15–20°) synthetic shear planes within S-C-dominated shear zones were interpreted as R-type Riedel shears (Fig. 8.4 and 8.5) and the overall structural assemblage is consistent with a simple shear-dominated deformation. These results match the structures commonly observed in simple shear-dominated natural shear zones (e.g. Berthé et al. 1979a, b) and deals with the vorticity values (0.6–0.8) characterizing flows in which gently inclined (<30°) synthetic shear bands develop (Grasemann et al. 2003).

Low kinematic vorticity values ($W_k = 0.27$) calculated along the reactivation-related oblique thrust ramps (e.g. Mount Boragine thrust) suggest that simple shear has been accompanied by a pure shear component (i.e. together, a sub-simple shear). In this case the synthetic extensional shear bands dip >30° and are associated systematically with antithetic structures (Fig. 8.6). The conjugate extensional shear planes testify that the S-fabric dominated shear zone experienced a significant component of pure shear coaxial deformation that accommodated foliation-normal contraction and transport-parallel elongation. This matches numerical studies revealing that conjugate sets of shear bands develop in flows dominated by a pure shear component with kinematic vorticity values <0.6 (Grasemann et al. 2003). These conjugate sets affecting the brittle-ductile shear zone of the Mount Boragine thrust along with the more spaced and inclined ones involving the forelimb of the fold (Fig. 8.5a) have been interpreted in terms of outer arc extension structures related to the growth of an anticline in the footwall of the previously emplaced one within in-sequence thrusting (Calamita et al. 2012a). The results of the kinematic vorticity analysis performed in this study along the Olevano-Antrodoco-Sibillini thrust are consistent with the estimates evaluated in previous works by using the tension veins and tectonic stylolites method (Calamita et al. 2012a). Furthermore, our study revealed that localized sub-simple shear ($W_k = 0.72$) may occur along back-thrusts as gently propagating splay faults associated with reactivation-related oblique thrust ramps (e.g. Sabini Mountains push-up inversion structure, Fig. 8.7).

Given the many uncertainties and assumptions in obtaining kinematic vorticity estimates (e.g. Iacopini et al. 2011; Xypolias 2010), the values presented here do not probably exactly represent accurate estimates of the mean vorticity of flow. However, these estimates combined with macro- and mesoscale geological and structural analyses would help to understand the degree of non-coaxiality within thrust shear zones deserving efforts in deciphering their tectonic significance.

Collectively, this study revealed that two types of shear fabric may be encountered along frontal and oblique thrust ramps within curved thrust systems that are related to different positive inversion tectonics (Fig. 8.8). S-C fabrics with associated synthetic R-type Riedel shear bands characterize the brittle–ductile shear zones of shortcut-related frontal thrust ramps and are dominated by simple shear. Whereas along the oblique thrust ramps resulting from the transpressive reactivation of pre-existing normal faults, a S-fabric characterized by a pure shear component is developed within the associated brittle–ductile shear zones. High-angle conjugate extensional shear planes producing a significant component of pure shear within the deformation flow may develop along such oblique thrust ramps because of the interaction of differently oriented thrust-related structures developing within in-sequence thrusting. In addition, S-C fabrics associated with sub-simple shear deformation may develop in the footwall of reactivation-related oblique thrust ramps along gently-propagating splay thrusts within push-up inversion structures.

This study provides insights and potentially represents analogs when examining shear fabrics of brittle–ductile shear zones associated to frontal and oblique ramps within curved thrust systems belonging to thrust belts that enjoyed structural inheritance of extensional faults, as in the Apennines.

ACKNOWLEDGMENTS

The authors thank Soumyajit Mukherjee and Kieran Mulchrone for inviting us to write this contribution. Insightful reviews by Soumyajit Mukherjee and Giovanni Toscani are acknowledged. Stimulating and fruitful discussions with David Iacopini during the period that P.P. spent at University of Aberdeen were gratefully appreciated and are acknowledged. Stereographic projections were performed with Stereonet 9.0 software (http://www.geo.cornell.edu/geology/faculty/RWA/programs/stereonet.html) and Rose diagrams with GeoRose 0.3.0 software (http://www.yongtechnology.com/georose-geological-rose-diagram-program/). This research was supported by ex-60% University 'G. D'Annunzio' grant awarded to F.C.

REFERENCES

Alberti M, Decandia FA, Tavarnelli E. 1996. Modes of propagation of the compressional deformation in the Umbria-Marche Apennines. Memorie della Società Geologica Italiana 51, 71–82.

Alfonsi L. 1995. Wrench tectonics in Central Italy, a segment of the Sabina Fault. Bollettino della Società Geologica Italiana 114, 411–421.

Alsop GI. 1991. Gravitational collapse and extension along a mid-crustal detachment: the Lough Derg Slide, northwest Ireland. Geological Magazine 128, 345–354.

Alsop GI, Holdsworth RE. 2004. Shear zones – an introduction and overview. In Flow Processes in Faults and Shear Zones, edited by G.I. Alsop, R.E. Holdsworth, K.J.W. McCaffrey, and M. Hand, Geological Society of London Special Publications, London.

Alsop GI, Holdsworth RE, McCaffrey KJW, Hand M. 2004. Flow Processes in Faults and Shear Zones. Geological Society of London Special Publications, London, pp. 1–10.

Alvarez W. 1990. Pattern of extensional faulting in pelagic carbonates of the Umbria–Marche Apennines of central Italy. Geology 18, 407–410.

Alvarez W, Engelder T, Lowrie W. 1976. Formation of spaced cleavage and folds in brittle limestones by dissolution. Geology 4, 698–701.

Barchi MR, Lemmi M. 1996. Geologia dell'area del M.Coscerno-M. di Civitella (Umbria sud-orientale). Bollettino della Società Geologica Italiana 115, 601–624.

Berthé D, Choukroune P, Jegouzo P. 1979a. Orthogneiss, mylonite and non-coaxial deformation of granites: the example of the South Armorican shear zone. Journal of Structural Geology 1, 31–42.

Berthé D, Choukroune P, Gapais D. 1979b. Orientations préférentielles du quartz et orthogneissification progressive en régime cisaillant: l'exemple du cisaillement sudarmoricain. Bulletin de Minéralogie 102, 265–272.

Blenkinsop TG, Treloar PJ. 1995. Geometry, classification and kinematics of S-C and S-C' fabrics in the Mushandike area, Zimbabwe. Journal of Structural Geology 17, 397–408.

Bigi S. 2006. An example of inversion in a brittle shear zone. Journal of Structural Geology 28, 431–443.

Boccaletti M, Calamita F, Deiana G, et al. 1990. Migrating foredeep thrust belt systems in the northern Apennines and southern Alps. Palaeogeography, Palaeoclimatology, Palaeoecology 77, 41–50.

Boccaletti M, Calamita F, Viandante MG. 2005. La Neo-Catena litosferica appenninica nata a partire dal Pliocene inferiore come espressione della convergenza Africa-Europa. Bollettino della Società Geologica Italiana 124, 87–105.

Bosellini A. 2004. The western passive margin of Adria and its carbonate platforms. In The Geology of Italy, edited by U. Crescenti, S. S. D'Offizi, Merlini, and R. Sacchi, Società Geologica Italiana, Italy, Special Volume 1, pp. 79–92.

Butler RWH. 1989. The influence of pre-existing basins structure on thrust system evolution in the western Alps. In Inversion Tectonics, edited by M. Cooper and G.D. Williams, Geological Society of London Special Publications, vol. 44, Geological Society of London, London, pp. 105–122.

Butler RWH. 1992. Thrust zone kinematics in a basement-cover imbricate stack: Eastern Pelvoux massif, French Alps. Journal of Structural Geology 14, 29–40.

Butler RWH, Tavarnelli E, Grasso M. 2006. Structural inheritance in mountain belts: An Alpine-Apennine perspective. Journal of Structural Geology 28, 1893–1908.

Calamita F. 1990. Thrusts and fold-related structures in the Umbria-Marche Apennines (Central Italy). Annales Tectonicae 4, 83–117.

Calamita F. 1991. Extensional mesostructures in thrust shear zones: examples from the Umbro-Marchean Apennines. Bollettino della Società Geologica Italiana 110, 649–660.

Calamita F, Deiana G. 1988. The arcuate shape of the Umbria-Marche-Sabina Apennines (Central Italy). Tectonophysics 146, 139–147.

Calamita F, Deiana G, Invernizzi C, Mastrovincenzo S. 1987. Analisi strutturale della linea "Ancona-Anzio Auctorum" tra Cittareale e Micigliano (Rieti). Bollettino della Società Geologica Italiana 106, 365–375.

Calamita F, Decandia FA, Deiana G, Fiori AP. 1991. Deformation of S-C tectonites in the Scaglia Cinerea Formation in the Spoleto area (South-East Umbria): Bollettino della Società Geologica Italiana 110, 661–665.

Calamita F, Pizzi A, Ridolfi M, Rusciadelli G, Scisciani V. 1998. Il buttressing delle faglie sinsedimentarie pre-thrusting sulla strutturazione neogenica della catena appenninica: l'esempio della M.gna dei Fiori (Appennino Centrale esterno). Bollettino della Società Geologica Italiana 117, 725–745.

Calamita F, Coltorti M, Piccinini D, et al. 2000. Quaternary faults and seismicity in the Umbro-Marchean Apennines (central Italy). Journal of Geodynamics 29, 245–264.

Calamita F, Esestime P, Paltrinieri W, Scisciani V, Tavarnelli E. 2009. Structural inheritance of pre- and synorogenic normal faults on the arcuate geometry of Pliocene–Quaternary thrusts: Examples from the Central and Southern Apennine Chain. Italian Journal of Geosciences 128, 381–394.

Calamita F, Satolli S, Scisciani V, Esestime P, Pace P. 2011. Contrasting styles of fault reactivation in curved orogenic belts: examples from the Central Apennines (Italy). Geological Society of America Bulletin 123, 1097–1111.

Calamita F, Satolli S, Turtù A. 2012a. Analysis of thrust shear zones in curve-shaped belts: Deformation mode and timing of the Olevano-Antrodoco-Sibillini thrust (Central/Northern Apennines of Italy). Journal of Structural Geology 44, 179–187.

Calamita F, Pace P, Satolli S. 2012b. Coexistence of fault-propagation and fault-bend folding in curve-shaped foreland fold-and-thrust belts: examples from the Northern Apennines (Italy). Terra Nova 24, 396–406.

Carosi R, Di Pisa A, Iacopini D, Montomoli C, Oggiano G. 2004. The structural evolution of the Asinara Island (NW Sardinia, Italy). Geodinamica Acta, 17, 309–329.

Casas JM, Sàbat F. 1987. An example of three-dimensional analysis of thrust-related tectonites. Journal of Structural Geology 9, 647–657.

Chattopadhyay N, Ray S, Sanyal S, Sengupta P. 2016. Mineralogical, textural and chemical reconstitution of granitic rock in ductile shear zone: A study from a part of the South Purulia Shear Zone, West Bengal, India. In Ductile Shear Zones: From Micro- to Macro-scales, edited by S. Mukherjee and K.F. Mulchrone, John Wiley & Sons, Chichester.

Chester F, Friedman M, Logan J. 1985. Foliated cataclasites. Tectonophysics 111, 139–146.

Ciarapica G, Passeri L. 2002. The palaeogeographic duplicity of the Apennines. Bollettino della Società Geologica Italiana Special Volume 1, 67–75.

Collettini C, Viti C, Tesei T, Mollo S. 2013. Thermal decomposition along natural carbonate faults during earthquakes: Geology 41, 927–930.

Coward MP. 1976. Strain within ductile shear zones. Tectonophysics 34, 181–197.

Coward MP. 1982. Surge zones in the Moine thrust zone of NW Scotland. Journal of Structural Geology 4, 247–256.

Coward MP. 1983. The thrust and shear zones of the Moine thrust zone and the NW Scottish Caledonides. Journal of the Geological Society 140, 795–811.

De Paor DG. 1983. Orthographic analysis of geological structures – I. Deformation theory. Journal of Structural Geology 5, 255–277.

Di Domenica A, Turtù A, Satolli S, Calamita F. 2012. Relationships between thrusts and normal faults in curved belts: new insight in the inversion tectonics of the Central-Northern Apennines (Italy). Journal of Structural Geology 42, 104–117.

Di Domenica A, Bonini L, Calamita F, Toscani G, Galuppo C, Seno S. 2014. Analogue modeling of positive inversion tectonics along differently oriented pre-thrusting normal faults: An application to the Central–Northern Apennines of Italy. Geological Society of America Bulletin. First published online March 20, 2014, . doi:10.1130/B31001.1.

Ghosh SK. 1987. Measure of non-coaxiality. Journal of Structural Geology 9, 111–113.

Grasemann B, Stüwe K, Vannay JC. 2003. Sense and non-sense of shear in flanking structures. Journal of Structural Geology 25, 19–34.

Hanmer S, Passchier CW. 1991. Shear sense indicators: a review. Geological Survey of Canada 90–17, 1–72.

Harris LB, Cobbold PR. 1984. Development of conjugate shear bands by bulk simple shearing. Journal of Structural Geology 7, 37–44.

Holdsworth RE. 1989. Late brittle deformation in a caledonian ductile thrust wedge: new evidence for gravitational collapse in the Moine Thrust sheet, Sutherland, Scotland. Tectonophysics 170, 17–28.

Holdsworth RE, Strachan RA, Alsop GI, Grant CJ, Wilson RW. 2006. Thrust sequences and the significance of low-angle, out-of-sequence faults in the northernmost Moine Nappe and Moine Thrust Zone, NW Scotland. Journal of the Geological Society, London 163, 801–814.

Iacopini D, Carosi R, Montomoli C, Passchier CW. 2008. Strain analysis and vorticity of flow in the Northern Sardinian Variscan Belt: recognition of a partitioned oblique deformation event. Tectonophysics 446, 77–96.

Iacopini D, Frassi C, Carosi R, Montomoli C. 2011. Biases in three-dimensional vorticity analysis using porphyroclast system: limits and application to natural examples. In Deformation Mechanisms, Rheology and Tectonics: Microstructures, Mechanics and Anisotropy, edited by D.J. Prior, E.H. Rutter, and D.J. Tatham, Geological Society of London, Special Publications, London, 360, pp. 301–318.

Jefferies SP, Holdsworth RE, Shimamoto T, Takagi H, Lloyd GE, Spiers CJ. 2006. Origin and mechanical significance of foliated cataclastic rocks in the cores of crustal-scale faults: Examples from the Median Tectonic Line, Japan. Journal of Geophysical Research 111, B12303.

Jegouzo P. 1980. The South Armorican shear zone. Journal of Structural Geology 7, 37–44.

Koopman, A. (1983) Detachment tectonics in the Central Apennines, Italy. Geologica Ultraiectina 30, 1–55.

Kurz GA, Northrup CJ. 2008. Structural analysis of mylonitic rocks in the Cougar Creek Complex, Oregon–Idaho using the porphyroclast hyperbolic distribution method, and potential use of SC'-type extensional shear bands as quantitative vorticity indicators. Journal of Structural Geology 30, 1005–1012.

Lavecchia G. 1985. Il sovrascorrimento dei Monti Sibillini: analisi cinematica e strutturale. Bollettino della Società Geologica Italiana 104, 161–194.

Lister GS, Snoke AW. 1984. S-C mylonites. Journal of Structural Geology 6, 617–638.

Logan JM, Dengo CA, Higgs NG, Wang ZZ. 1992. Fabrics of experimental fault zones: their development and relationship to mechanical behavior. In Fault Mechanics and Transport Properties of Rocks, edited by B. Evans and B. Wong, Academic press, London, pp. 33–67.

Mazzoli S, Pierantoni PP, Borraccini F, Paltrinieri W, Deiana G. 2005. Geometry, segmentation pattern and displacement variations along a major Apennine thrust zone, central Italy. Journal of Structural Geology 27, 1940–1953.

McClay KR. 1987. The Mapping of Geological Structures. John Wiley & Sons, Chichester, UK.

Means WD, Hobbs BE, Lister GS, Williams PF. 1980. Vorticity and non-coaxiality in progressive deformations. Journal of Structural Geology 2, 371–378.

Misra S, Mandal N, Chakraborty C. 2009. Formation of Riedel shear fractures in granular materials: Findings from analogue shear experiments and theoretical analyses. Tectonophysics 471, 253–259.

Mukherjee S. 2007. Geodynamics, deformation and mathematical analysis of metamorphic belts of the NW Himalaya. PhD thesis. Indian Institute of Technology, Roorkee.

Mukherjee S. 2010a. Microstructures of the Zanskar Shear Zone. Earth Science India 3, 9–27.

Mukherjee S. 2010b. Structures in meso- and micro-scales in the Sutlej section of the Higher Himalayan shear zone, Indian Himalaya. e-Terra 7, 1–27.

Mukherjee S. 2011. Mineral fish: their morphological classification, usefulness as shear sense indicators and genesis. International Journal of Earth Sciences 100, 1303–1314.

Mukherjee S. 2012a. Simple shear is not so simple! Kinematics and shear senses in Newtonian viscous simple shear zones. Geological Magazine 149, 819–826.

Mukherjee S. 2012b. Tectonic implications and morphology of trapezoidal mica grains from the Sutlej section of the Higher Himalayan Shear Zone, Indian Himalaya. The Journal of Geology 120, 575–590.

Mukherjee S. 2013a. Channel flow extrusion model to constrain dynamic viscosity and Prandtl number of the Higher Himalayan Shear Zone. International Journal of Earth Sciences 102, 1811–1835.

Mukherjee S. 2013b. Higher Himalaya in the Bhagirathi section (NW Himalaya, India): its structures, backthrusts and extrusion mechanism by both channel flow and critical taper mechanisms. International Journal of Earth Sciences 102, 1851–1870.

Mukherjee S. 2013c. Deformation Microstructures in Rocks. Springer, Heidelberg, pp. 1–111.

Mukherjee S. 2014a. Atlas of Shear Zone Structures in Meso-scale. Springer, Cham, pp. 1–124.

Mukherjee S. 2014b. Review of flanking structures in meso- and micro-scales. Geological Magazine 151, 957–974.

Mukherjee S. 2014c. Kinematics of "top-to-down" simple shear in a Newtonian rheology. The Journal of Indian Geophysical Union 18, 273–276.

Mukherjee S, Biswas R. 2014. Kinematics of horizontal simple shear zones of concentric arcs (Taylor–Couette flow) with incompressible Newtonian rheology. International Journal of Earth Sciences 103, 597–602.

Mukherjee, S., Punekar JN, Mahadani T, Mukherjee R. 2016. Intrafolial folds: Review and examples from the western Indian Higher Himalaya. In Ductile Shear Zones: From Micro- to Macro-scales, edited by S. Mukherjee and K.F. Mulchrone, John Wiley & Sons, Chichester.

Mukherjee S, Koyi HA. 2009. Flanking microstructures. Geological Magazine 146, 517–526.

Mukherjee S, Koyi HA. 2010a. Higher Himalayan Shear Zone, Sutlej section: structural geology and extrusion mechanism by various combinations of simple shear, pure shear and channel flow in shifting modes. International Journal of Earth Sciences 99, 1267–1303.

Mukherjee S, Koyi HA. 2010b. Higher Himalayan Shear Zone, Zanskar Indian Himalaya: microstructural studies and extrusion mechanism by a combination of simple shear and channel flow. International Journal of Earth Sciences 99, 1083–1110.

Mukherjee S, Mulchrone KF. 2013. Viscous dissipation pattern in incompressible Newtonian simple shear zones: an analytical model. International Journal of Earth Sciences 102, 1165–1170.

Pace P, Calamita F. 2014. Push-up inversion structures v. fault-bend reactivation anticlines along oblique thrust ramps: examples from the Apennines fold-and-thrust belt (Italy). Journal of the Geological Society 171, 227–238.

Pace P, Scisciani V, Calamita F. 2011. Styles of Plio-Quaternary Positive Inversion Tectonics in the Central-Southern Apennines and in the Adriatic Foreland. Rendiconti Online Società Geologica Italiana 15, 88–91.

Pace P, Satolli S, Calamita F. 2012. The control of mechanical stratigraphy and inversion tectonics on thrust-related folding along the curved Northern Apennines thrust front. Rendiconti Online Società Geologica Italiana 22, 162–165.

Pace P, Di Domenica A, Calamita F. 2014. Summit low-angle faults in the Central Apennines of Italy: younger-on-older thrusts or rotated normal faults? Constraints for defining the tectonic style of thrust belts. Tectonics 33, 756–785.

Pamplona J, Rodrigues BC, Llana-Fúnez S, et al. 2016. Structure and Variscan evolution of Malpica-Lamego ductile shear zone (NW of Iberian Peninsula). In Ductile Shear Zones: From Micro- to Macro-scales, edited by S. Mukherjee and K.F. Mulchrone, John Wiley & Sons, Chichester.

Passchier CW. 1986. Flow in natural shear zones – the consequences of spinning flow regimes. Earth and Planetary Science Letters 77, 70–80.

Passchier CW. 2001. Flanking structures. Journal of Structural Geology 23, 951–962.

Passchier CW, Trouw RAJ. 2005. Microtectonics, second edition. Springer, Berlin, pp. 1–366.

Pierantoni PP. 1997. Faglie trascorrenti sin-thrusting come ripartizione della deformazione: L'esempio della Faglia Sabina (Appennino Centrale). Studi Geologici Camerti 14, 279–289.

Pierantoni PP. 1998. Il modello deformativo di sovrascorrimento cieco per lo sviluppo di anticlinali coricate e rovesciate nell'Appennino umbro-marchigiano. Studi Geologici Camerti 14, 291–303.

Pierantoni PP, Deiana G, Romano A, Paltrinieri W, Borraccini F, Mazzoli S. 2005. Geometrie strutturali lungo la thrust zone del fronte montuoso umbro-marchigiano-sabino. Bollettino della Società Geologica Italiana 124, 395–411.

Pierantoni PP, Deiana G, Galdenzi, S. 2013. Stratigraphic and structural features of the Sibillini Mountains (Umbria-Marche Apennines, Italy). Italian Journal of Geosciences 132, 497–520.

Platt JP. 1984. Secondary cleavages in ductile shear zones. Journal of Structural Geology 6, 439–442.

Platt JP, Vissers RLM. 1980. Extensional structures in anisotropic rocks. Journal of Structural Geology 2, 397–410.

Ponce de Leon MJ, Choukroune P. 1980. Shear zones in the Iberian arc. Journal of Structural Geology 2, 63–68.

Powell D, Glendinning RW. 1990. Late Caledonian extensional reactivation of a ductile thrust in NW Scotland: Journal of the Geological Society 147, 979–987.

Ragan DM. 2009. Structural Geology, An Introduction to Geometrical Techniques. Cambridge University Press, Cambridge.

Ramsay JG. 1980. Shear zone geometry: a review. Journal of Structural Geology 2, 83–101.

Ramsay JG, Graham RH. 1970. Strain variation in shear belts. Canadian Journal of Earth Sciences 7, 786–813.

Ricci Lucchi F. 1986. The Oligocene to recent foreland basins of the northern Apennines. Special Publication of the International Association for Sediment 8, 105–139.

Riedel W. 1929. Zur mechanik geologischer Brucherscheinungen. Zentralblatt für Mineralogie. Geologie und Paläontologie B, 354–368.

Rusciadelli G, Viandante MG, Calamita F, Cook AC. 2005. Burial-exhumation history of the central Apennines (Italy), from the foreland to the chain building: thermochronological and geological data. Terra Nova 17: 560–572.

Santantonio M. 1993. Facies associations and evolution of pelagic carbonate platform/basin systems: examples from the Italian Jurassic. Sedimentology 40, 1039–1067.

Satolli S, Pace P, Viandante MG, Calamita F. 2014. Lateral variations in tectonic style across cross-strike discontinuities: an example from the Central Apennines belt (Italy). International Journal of Earth Sciences 103, 2301–2313.

Scisciani V. 2009. Styles of positive inversion tectonics in the Central Apennines and in the Adriatic foreland: Implications for the evolution of the Apennine chain (Italy). Journal of Structural Geology 31, 1276–1294.

Scisciani V, Tavarnelli E, Calamita F. 2001. Styles of tectonic inversion within syn-orogenic basins: Examples from the Central Apennines, Italy. Terra Nova 13, 321–326.

Scisciani V, Tavarnelli E, Calamita F. 2002. The interaction of extensional and contractional deformations in the outer zones of the Central Apennines, Italy. Journal of Structural Geology 24, 1647–1658.

Sengupta S, Chatterjee SM. 2016. Microstructural variations in quartzofeldspathic mylonites and the problem of vorticity analysis using rotating porphyroclasts in the Phulad Shear Zone, Rajasthan, India. In Ductile Shear Zones: From Micro- to Macroscales, edited by S. Mukherjee and K.F. Mulchrone, John Wiley & Sons, Chichester.

Sibson, R.H. (1977) Fault rocks and fault mechanisms. Journal of the Geological Society 133, 191–213.

Sibson RH. 1983. Continental fault structure and the shallow earthquake source. Journal of the Geological Society 140, 741–767.

Simpson C, De Paor DG. 1993. Strain and kinematic analysis in general shear zones. Journal of Structural Geology 15, 1–20.

Simpson C, Schmid SM. 1983. An evaluation of criteria to deduce the sense of movement in sheared rocks. Geological Society of America Bulletin 94, 1281–1288.

Speranza F, Satolli S, Mattioli E, Calamita F. 2005. Magnetic stratigraphy of Kimmeridgian-Aptian sections from Umbria-Marche (Italy): New details on the M polarity sequence. Journal of Geophysical Research: Solid Earth 110, B12109.

Sullivan WA. 2009. Kinematic significance of L tectonites in the footwall of a major terrane-bounding thrust fault, Klamath Mountains, California, USA. Journal of Structural Geology 31, 1197–1211.

Tavarnelli E. 1993a. Evidence for fault propagation folding in the Umbria-Marche-Sabina Apennines (Central Italy). Annales Tectonicae 7, 87–99.

Tavarnelli E. 1993b. Carta geologica dell'area compresa fra la Valnerina e la Conca di Rieti (Umbria sud-orientale ed alto Lazio). Dipartimento di Scienze della Terra, Università di Siena, scale 1:100 000, 1 sheet.

Tavarnelli E. 1994. Analisi geometrica e cinematica dei sovrascorrimenti compresi fra la Valnerina e la conca di Rieti (Appennino Umbro-Marchigiano-Sabino). Bollettino della Società Geologica Italiana 113, 249–259.

Tavarnelli E. 1996a. Tethyan heritage in the development of the Neogene Umbria-Marche fold-and-thrust belt, Italy: A 3D approach. Terra Nova 8, 470–478.

Tavarnelli E. 1996b. Ancient synsedimentary structural control on thrust ramp development: an example from the Northern Apennines, Italy. Terra Nova 8, 65–74.

Tavarnelli E. 1996c. The effects of preexisting normal faults on thrust ramp development: An example from the Northern Apennines, Italy. Geologische Rundschau 85, 363–371,

Tavarnelli E. 1997. Structural evolution of a foreland fold-and-thrust belt: the Umbria-Marche Apennines, Italy. Journal of Structural Geology 19, 523–534.

Tavarnelli E. 1999. Normal faults in thrust sheets: pre-orogenic extension, post-orogenic extension, or both? Journal of Structural Geology 21, 1011–1018.

Tavarnelli E, Alvarez W. 2002. The mesoscopic response to positive tectonic inversion processes: an example from the Umbria-Marche Apennines, Italy. Bollettino della Società Geologica Italiana 1, 715–727.

Tavarnelli E, Peacock DCP. 1999. From extension to contraction in syn-orogenic foredeep basins: the Contessa section, Umbria-Marche Apennines, Italy. Terra Nova 11, 55–60.

Tavarnelli E, Butler RWH, Decandia FA, et al. 2004. Implication of fault reactivation and structural inheritance in the Cenozoic tectonic evolution of Italy. In The Geology of Italy, edited by U. Crescenti, S. D'Offizi, S. Merlini, and R. Sacchi, Società Geologica Italiana, Italy, Special Volume 1, pp. 209–222.

Tchalenko JS. 1968. The evolution of kink-bands and the development of compression textures in sheared clays. Tectonophysics 6, 159–174.

ten Grotenhuis SM, Trouw RAJ, Passchier CW. 2003. Evolution of mica fish in mylonitic rocks. Tectonophysics 372, 1–21.

Tesei T, Collettini C, Viti C, Barchi MR. 2013. Fault architecture and deformation mechanisms in exhumed analogues of seismogenic carbonate-bearing thrusts. Journal of Structural Geology 55, 1–15.

Tikoff B, Fossen H. 1995. The limitations of three-dimensional kinematic vorticity analysis. Journal of Structural Geology 17, 1771-1784.

Watts MJ, Williams GD. 1979. Fault rocks as indicators of progressive shear deformation in the Guingamp region, Brittany. Journal of Structural Geology 1, 323–332.

White SH, Burrows SE, Carreras J, Shaw MD, Humphreys FJ. 1980. On mylonites in ductile shear zones. Journal of Structural Geology 2, 175–187.

Xypolias P. 2010. Vorticity analysis in shear zones: a review of methods and applications. Journal of Structural Geology 32, 2072–2092.

Yin A, Kelty TK. 1991. Development of normal faults during emplacement of a thrust sheet: an example from the Lewis allochthon, Glacier National Park, Montana (USA). Journal of Structural Geology 13, 37–47.

Chapter 9

Microstructural variations in quartzofeldspathic mylonites and the problem of vorticity analysis using rotating porphyroclasts in the Phulad Shear Zone, Rajasthan, India

SUDIPTA SENGUPTA and SADHANA M. CHATTERJEE

Department of Geological Sciences, Jadavpur University, Kolkata 700032, India

9.1 INTRODUCTION

The Phulad Shear Zone (PSZ) is situated within the Delhi mobile belt in the state of Rajasthan in western India (Fig. 9.1). The shear zone is marked by the occurrence of mylonite, containing layers of metamorphosed siliciclastic rocks within a calcareous matrix (Ghosh et al. 1999, 2003; Golani et al. 1998; Roy and Jakhar 2002; Sengupta and Ghosh 2004, 2007). The mylonitic foliation has a general attitude of 035°/70°E. There is strong down-dip stretching and striping lineation parallel to the transport direction. Our detailed study of mesoscopic structures in this area shows the presence of at least three generations of reclined folds and a later generation of subhorizontal folds (Ghosh et al. 1999, 2003; Sengupta and Ghosh 2004, 2007). These folds developed in the mylonitic foliation during progressive ductile shearing with their axial surface parallel to the foliation. Asymmetry of mesoscopic structures indicates a thrusting sense of movement towards NW and a subhorizontal vorticity vector (Ghosh et al. 1999; Sengupta and Ghosh 2004).

Mylonitic lineation is parallel to the hinges of the earliest reclined folds. This lineation has been deformed by later folds. The lineation is more or less parallel on the limbs of the later reclined folds but makes a high angle over the hinges. On the unrolled surface of the folds, the lineation shows a U-pattern (Ghosh et al 1999; Sengupta and Ghosh 2004). The U-pattern may be symmetric or asymmetric (fig. 5 of Sengupta and Ghosh 2004). On the unrolled surface of the fold, in case the apex of the U-pattern of deformed lineation lies on the hinge line: the pattern is symmetric. If the apex lies on one side of the deformed lineation, the pattern is asymmetric. Theoretical analysis by Ghosh et al. (1999) showed that a symmetric U-pattern of deformed lineation on the unrolled surface of a fold cannot develop in a simple shear dominated deformation. In PSZ, symmetric patterns are more common than asymmetric U-patterns of lineation. Dominance of symmetrical U-patterns indicates

that the deformation has a strong component of pure shear. Our earlier studies (Ghosh et al 1999; Sengupta and Ghosh 2004) also reported the presence of at least three generations of sheath folds. These sheath folds have their apices parallel to the direction of down-dip stretching lineation.

Quartzofeldspathic mylonites are intensely deformed, and are mostly ortho- or ultramylonites. However, an idea of the progressive deformation can be obtained from successive stages of deformation of pegmatites that are emplaced syntectonically at different stages of shearing. The bodies range in size from a few cm to a few m and are flattened parallel to the foliation. These pegmatite lenses are generally elongated with their longer direction parallel to the direction of stretching (Sengupta and Ghosh 2004). The newly initiated shear zone foliation in these pegmatite lenses occurs at a very low angle with the mylonitic foliation. This feature indicates that the bulk of the deformation is not dominated by simple shear but there is a simultaneous component of pure shear (Sengupta and Ghosh 2004). Comparing results of theoretical analysis with natural mesoscopic structures, they concluded that the deformation in PSZ was transpressional with stretching along the transport direction being much greater than the direction of the vorticity vector.

The present study deals with a detailed analysis of the microstructures of quartzofeldspathic mylonites to complement our mesoscopic observations. Microstructures of mylonite provide several distinctive signatures of different types of deformation. During progressive deformation in a ductile shear zone, such microstructures usually show later stages of deformation as older structures are gradually replaced by later ones. However, some of the older structures are still preserved at places and we may get a range of features, which would indicate successive stages of development of mylonitic structures from the initial to the final stage. In the next section, a variety of microstructures of quartzofeldspathic mylonites of PSZ is analyzed.

Ductile Shear Zones: From Micro- to Macro-scales, First Edition. Edited by Soumyajit Mukherjee and Kieran F. Mulchrone.

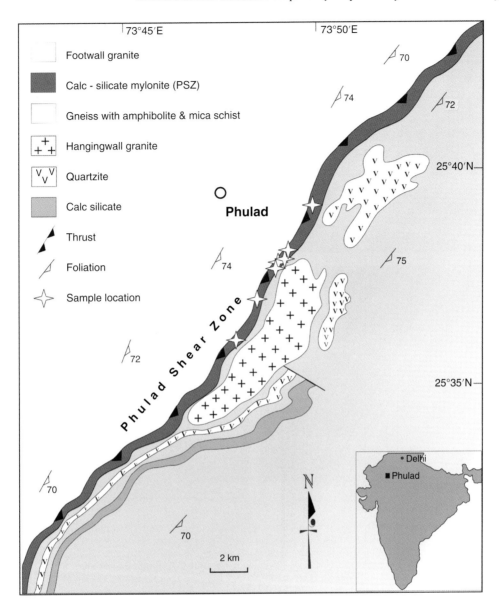

Fig. 9.1. Location of Phulad Shear Zone. Adapted from Ghosh et al. (1999).

An attempt is also made to carry out a vorticity analysis of the deformation with the help of rotating porphyroclasts (Passchier 1987; Simpson and De Paor 1993, 1997) to quantify the nature of transpressional deformation in PSZ. Noncoaxiality of a deformation may be expressed in terms of kinematic vorticity number (W_k), which is essentially a ratio between simple shear and pure shear component of flow (Passchier and Urai 1988). In recent years, certain structural elements especially rigid porphyroclasts are used to determine the kinematic vorticity number (W_k) of natural shear zones (e.g. Ghosh 1987; Passchier 1987; Simpson and De Paor 1993, 1997; Jessup et al. 2007). However, such an analysis is based on many assumptions, which may not be met in natural situations (Passchier and Trouw 2005; Xypolias 2010; Li and Jiang

2011). In Section 9.3, the problem of using rigid porphyroclasts for vorticity analysis in PSZ is discussed.

9.2 MICROSTRUCTURAL VARIATION IN QUARTZOFELDSPATHIC MYLONITES

Thin sections were prepared perpendicular to mylonitic foliation and parallel to the stretching lineation (*XZ* section of strain). This section is also perpendicular to the vorticity vector. A few sections were prepared with orientation parallel to the vorticity vector and perpendicular to both mylonitic foliation and lineation (*YZ* section of strain). When otherwise not mentioned, the sections that are parallel to the lineation are referred.

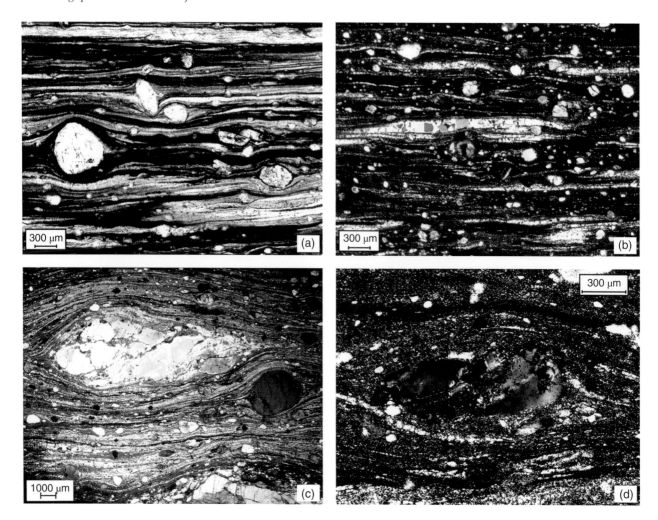

Fig. 9.2. Photomicrograph showing quartzofeldspathic mylonites.All photos show dextral shear sense (direction of thrusting in PSZ). (a) Orthomylonite with well developed banding. Feldspar porphyroclasts occur in a matrix of fine grained quartz, feldspar and calcite. PPL. (b) Polycrystalline quartz ribbons in an ultramylonite with a few small porphyroclasts of feldspar. (c) A large porphyroclast of feldspar (aspect ratio = 3) elongated parallel to foliation is divided into several smaller units by internal microshears. The internal shear zones make an acute angle with the direction of shear and contain very small grained recrystallized feldspars. (d) Highly strained elongated feldspar porphyroclast with microshears. Asymmetric pattern around the clast shows sinistral rotation in a dextral shearing. b–d in CPL.

Quartzofeldspathic mylonites are very well foliated and usually show very fine bands of constituent minerals. Most of the sheared rocks described below are orthomylonites or ultramylonites (Fig. 9.2a, b). The banding is generally defined by layers of recrystallized quartz alternating with layers of recrystallized feldspar. Bands of extremely fine grained calcite occur parallel to the mylonitic foliation. The feldspar is mostly a plagioclase but microcline is also present. Both the varieties show similar deformational features (Passchier and Trouw 2005, p. 58). In the following description we shall refer to them as feldspar.

9.2.1 Deformation of quartz

In the proper shear zone of the Phulad area, the rock is mostly orthomylonite or ultramylonite. Quartz occurs as fine polygonal equant grains or as thin polycrystalline ribbons (Fig. 9.2b). Quartz megacrysts are absent in these

mylonites. However, earlier stages of mylonitization can be observed in pegmatites, which were syntectonically emplaced during ductile shearing. All stages from the protomylonitic to ultramylonitic through to the orthomylonitic stage can be observed in different parts of these deformed pegmatite lenses. In the protomylonitic stage, the quartz is deformed by crystal plastic processes showing subgrain formation and recrystallization. The deformed quartz shows irregular, sinuous, sutured grain boundaries indicating dynamic recrystallization. The temperature of deformation is estimated to be ~500°C (Stipp et al. 2002). In the orthomylonitic stage, quartz shows a very fine grained polygonal granoblastic fabric (Fig. 9.2a). However, it also occurs as polycrystalline ribbons (Fig. 9.2b). Individual grains within these ribbons are elongated parallel to the mylonitic foliation and show subgrain formation indicating dynamic recrystallization.

9.2.2 Deformation of feldspar

The only megacrysts observed in mylonites in PSZ are large deformed grains of feldspar. The feldspar layers have been clearly derived from progressive deformation and show a range from large megacrysts about 1 cm size to small remnant clasts of about 0.01 mm (Fig. 9.2a–d). Smaller residual clasts can still be distinguished from the recrystallized feldspar grains by their larger grain size. The recrystallized feldspar grains are much smaller in size.

Feldspar behaved in a more competent manner than quartz during deformation. It was deformed essentially by crystal plastic processes, however brittle fractures can also be observed in some grains. Feldspar mostly behaved in a ductile manner showing undulose extinction, bulging boundaries, subgrain formation and recrystallization. Kinking, tapered lamellae and brittle intracrystralline fractures are also common. Porphyroclasts of feldspar may be equant or ellipsoidal with average aspect ratios (long axis/ short axis of the clasts) ranging between 1.5 and 2 (Fig. 9.3a). However, feldspar grains with larger aspect ratios also occur (Fig. 9.2c, d). The majority of the feldspar clasts are generally parallel to or make a low-angle with the mylonitic foliation. However, there is a wide range of orientation of these clasts with the direction of shear (Fig. 9.3b).

The larger feldspar clasts usually have a large aspect ratio and the smaller clasts are more equant (Fig. 9.2c, d), which indicates that the aspect ratio decreases with progressive deformation. The larger grains show successive stages of deformation where they are subdivided by microscopic shear zones and divided into smaller grains. Even during the initial stages, the microshear formed a low-angle with the mylonitic foliation indicating that the deformation was not on account of simple shear alone and that an additional component of compression perpendicular to the shear zone also contributed to the deformation (Sengupta and Ghosh 2004). Figure 9.2c shows such a large feldspar grain, with an aspect ratio of 4.5, subdivided into smaller lenticular units by thin microshears. The acute angle between the microshears and the foliation is sympathetic to the overall sense of shear. The large grains, showing undulose extinction are flattened parallel to the mylonitic foliation and are elongated parallel to the lineation. Grains within the microshears are much smaller but are larger than the matrix. The matrix consists of very fine-grained quartz, feldspar, and calcite. The contrast in grain size between the porphyroclast and the matrix is large. The tails of the porphyroclast show a gradual grain size reduction and finally merge with the matrix.

Figure 9.2d shows another elongated megacryst of feldspar with an aspect ratio of nearly 3, oriented parallel to the mylonitic foliation and lineation. The grain is dissected by internal shear oriented at an acute angle to the foliation. The sense of shear deduced from the microshear is dextral and sympathetic to the general direction

(a)

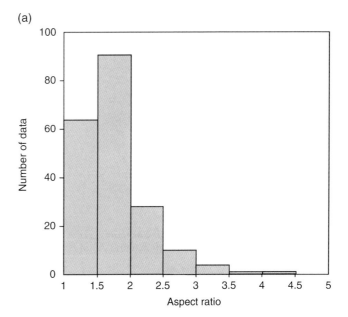

(b)

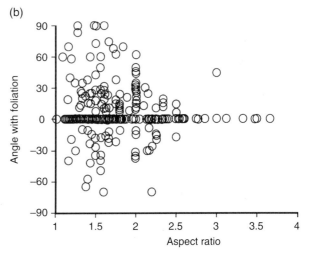

Fig. 9.3. (a) Histogram showing the distribution of aspect ratios of feldspar porphyroclasts in quartzofeldspathic mylonite of Phulad shear zone. (b) Data plot showing orientation of feldspar clasts with respect to mylonitic foliation.

of shear (thrusting towards NW). However, the asymmetry of the foliation indicates a sinistral rotation. Perhaps the clast was oriented antithetic to the shear direction and back rotated to the present position. Such a structure is expected when a component of pure shear is added to the simple shear. A back-rotated elliptical clast can also be seen in Fig. 9.2a (central part).

Feldspar of the quartzofeldspathic mylonites in PSZ generally shows a core-and-mantle structure with recrystallized tails (White 1976). The grain size within the recrystallized mantle is very small compared to the core. Figure 9.4a shows a feldspar clast with a mantle of recrystallized feldspar grains around a deformed core. There is no transitional zone with subgrain formation between the core and the recrystallized mantle. The core

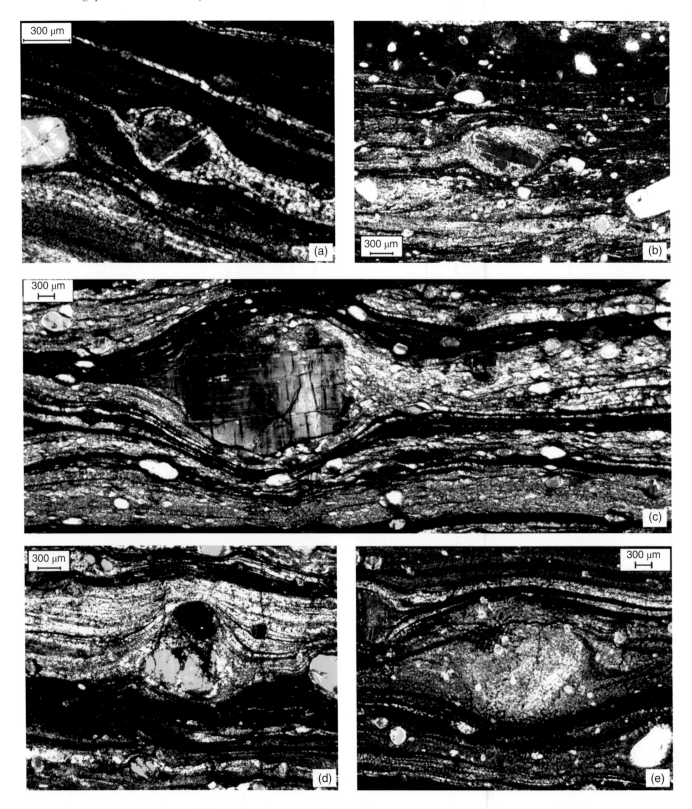

Fig. 9.4. Progressive destruction of feldspar clasts by recrystallization. (a) Oval shaped feldspar with a strained core within a mantle of neocrystallized grains. The recrystallized mantle material is drawn out parallel to the foliation as a distinct band of feldspar grains. (b) A residual feldspar porphyroclast within a mantle of recrystallized neoblasts oriented antithetically to the dextral shear direction. (c) A feldspar porphyroclast showing undulose extinction subgrain formation occurs within a mantle of recrystallized grains. The tail in the right side merges with a long band of recrystallized feldspar grains. The mylonitic foliation marked by quartz ribbons swerves around the clast. (d) A feldspar porphyroclast containing two remnants of a deformed core within a mantle of recrystallized feldspar grains. (e) A feldspar porphyroclast completely replaced by small recrystallized neoblasts. The presence of the old grain can be recognized by the oppositely bent foliation. Recrystallized tails of the old grain indicate dextral shear.

is transacted by a thin zone of microshear. The tails of the feldspar clasts are elongated parallel to the mylonitic foliation and gradually merge with the bands of the recrystallized feldspar. The grain size of the newly formed feldspar bands is larger than the grain size of the matrix. Other feldspar layers with similar grain size contain a few small residual feldspar clasts.

Figure 9.4b shows another clast with a fine grained recrystallized mantle around the core. The remnant clast along with the mantle is parallel to the mylonitic foliation. The asymmetric tails show an σ-pattern indicating a dextral sense of shear. However, the residual rectangular clast is antithetically inclined against the sense of shear. This antithetic orientation of the remnant porphyroclast is not on account of rotation of the grain in opposite sense but due to uneven recrystallization of the original grain. The size of the recrystallized grains in the tails is slightly larger than the matrix. The tails gradually form a banding parallel to the mylonitic foliation. In the lower right corner, another rectangular porphyroclast is oriented synthetically with the shear direction. The asymmetric tails around the grain indicate a synthetic rotation.

Tails of feldspar porphyroclasts may vary in size and shape on either side of the clast. The feldspar porphyroclast in Fig. 9.4c shows a long tapering mantle on the right side with partial recrystallization at the edges and around the upper right corner. The recrystallized grains within the tail on the right side are also elongated parallel to the foliation indicating dynamic recrystallization. The tail continues to form a regular band parallel to the foliation. On the left side, the tail is much shorter. The remnant core shows wavy extinction and subgrain formation. The presence of a fine grained mantle with or without a transitional zone between them indicates that the shearing occurred under medium grade conditions at around 450 to 600°C (Passchier and Trouw 2005, p. 58).

With progressive grain refinement, the size of the mantle increases gradually where islands of residual clasts occur within an apron of recrystallized grains. Figure 9.4d shows such residual clasts of a feldspar megacryst within a mantle of feldspar neoblasts. The contrast in grain size between the remnant feldspar and the recrystallized grains in the mantle is large. With continued deformation, the size of the residual clasts decreases and the whole porphyroclast is replaced by recrystallized neoblasts of feldspar (Fig. 9.4e). At the advanced stage of mylonitization, with a reduction in the percentage of porphyroclasts, the rock is transformed to an ultramylonite (Fig. 9.3b). The dominant deformation mechanism at this stage is probably by grain boundary sliding and the rock becomes superplastic (Boullier and Guéguen 1975; White et al 1980).

Porphyroclasts, without an apron of recrystallized grains around them, also occur. Figure 9.5(a,b) shows feldspar grains with wavy extinction and subgrain development, but without any prominent mantle of recrystallized grains. The small tails on either side of the

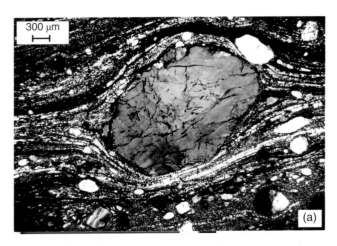

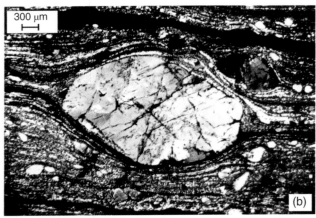

Fig. 9.5. (a) Oval shaped feldspar porphyroclast at low angle to the foliation with a thin mantle around the core. The asymmetry of foliation in both cases indicates a dextral shear. (b) Feldspar clast without any noticeable mantle occurs with long axis subparallel to mylonitic foliation.

grains contain feldspar neoblasts of different grain size. The fine-grained matrix shows a prominent banding with a few larger residual grains lying parallel to the banding. Polycrystalline quartz ribbons swerve around the clasts. Both clasts are from the same thin section. However, the clast in Fig. 9.5a is synthetically oriented with the dextral shear direction while the clast in Fig. 9.5b is antithetically oriented and possibly back rotated to the present position.

Internal kinking and deformation twin lamellae are present in some deformed feldspar megacrysts (Fig. 9.6). Figure 9.6a shows feldspar grain flattened parallel to the foliation with internal kink bands. The kink plane is subparallel to the foliation. The presence of internal kink bands indicates that the deformation of feldspar was on account of crystal plastic processes. Tapered deformation twin lamellae can be seen in two feldspar grains in Fig. 9.6b. These grains occur within a fine grained recrystallized mantle of feldspar neoblasts. The foliation marked by polycrystalline quartz ribbons swerves around the feldspar grains. The quartz grains within the ribbons are elongated parallel to foliation indicating dynamic

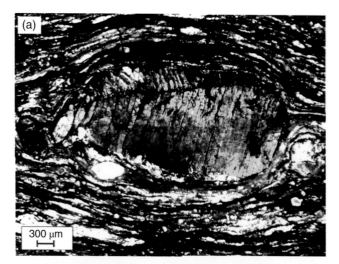

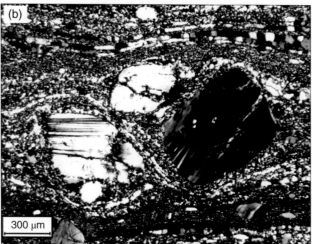

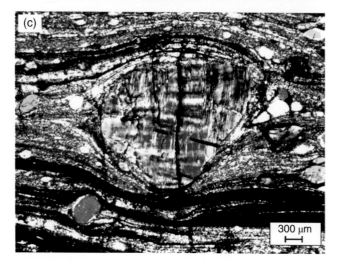

Fig. 9.6. (a) A flattened feldspar grain oriented parallel to the mylonitic foliation shows internal kinking. The kink plane is parallel to the foliation. (b) Deformation twin lamellae in two feldspar clasts within a mantle of recrystallized quartz. The foliation marked by polycrystalline quartz ribbons swerves around the grains. (c) Deformed cross-hatched twinning within a K-feldspar clast.

recrystallization. Figure 9.6c shows a K-feldspar with deformed cross-hatched twinning and a thin apron of fine-grained mantle. The porphyroclast shows undulose extinction with some recrystallization parallel to the twin bands.

A few feldspar porphyroclasts are deformed by brittle fractures. The fractures usually occur in smaller clasts indicating that brittle deformation occurred during later stages of deformation. These micro-fractures occur at various angles. The fractures are probably guided by the orientation of the cleavage within the feldspar. When such micro-faults form a low angle with the mylonitic foliation (Figs 9.7a and 9.3a), the displacement along the fault is sympathetic or synthetic with the shear direction (Ghosh 1993; Ramsay and Huber 1987; Passchier and Trouw 2005; Trouw et al. 2010). However, when the fracture occurs at high angles to the foliation, the displacement can be either synthetic or antithetic (Ghosh 1993, p. 520). Figure 9.7b shows a feldspar grain with microfaults oriented at an angle of 50° with the mylonitic foliation. The slip along the microfault is synthetic to the dextral sense of shear but the grain shows antithetic rotation. This is due to the transpressional nature of deformation in PSZ (compare with fig. 21.13b of Ghosh 1993, p. 521). In another instance, Fig. 9.7c shows a feldspar clast with a microfault at an angle of about 45° to the foliation with an antithetic slip along the fault.

Intracrystalline microfaults within feldspar clasts are often guided by the orientation of the cleavage planes of the grain. Figure 9.7d shows a clast with two sets of fractures at high angles to one another possibly following the cleavage planes. The small broken fragments on one side of the clast form a mosaic (Trouw et al. 2010) and are drawn out parallel to the foliation as a tail. The shapes of these fragments are mostly square or rectangular. The presence of kink bands, tapered twin lamellae, and intracrystalline faults within feldspar grains, indicates that temperature of deformation at this stage was between 400 and 500°C (Pryer 1993).

9.2.3 Sense of shear

There is a wide range of variation in the orientation of the clasts with respect to the mylonitic foliation. The long axes of majority of clasts are oriented either parallel or at a low angle to the foliation (Fig. 9.3b). There are two types of asymmetry or rolling structures (Van den Driessche and Brun 1987) around the clasts. In the first type the tails of the porphyroclast show the asymmetry around the clasts (Fig. 9.8) and in the second type there is no recognizable tail but the asymmetry of the pre-existing foliation indicates rotation of the porphyroclast (Fig. 9.5a,b). In the first type, the grain size of the recrystallized tails of the porphyroclast is larger than the grain size of the matrix. As a result, the tails can be distinguished from the matrix. In the second type, the grain size of the recrystallized mantle is so small that it cannot be distinguished from the matrix. However, in both cases, the sense of

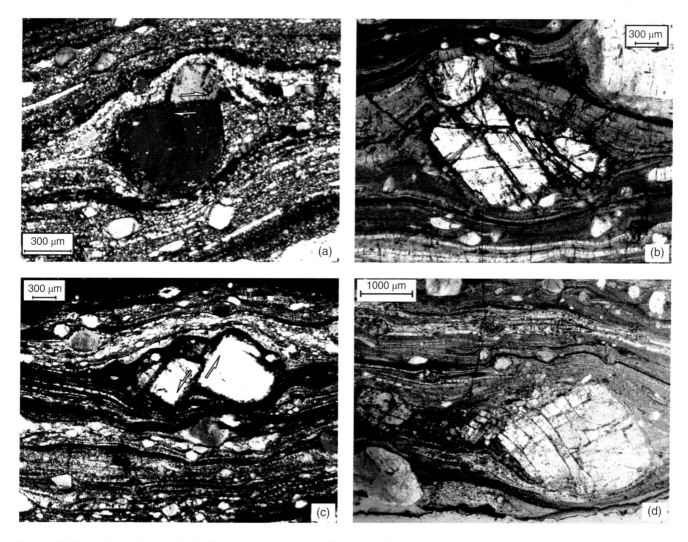

Fig. 9.7. Feldspar clasts showing brittle fractures. Shear sense is dextral in all photos. (a) The sense of movement along the fault within a small feldspar clast is synthetic to the sense of shear. (b) Domino type fragmented porphyroclast. The movement along the fault is synthetic sense of shear. The rotation of the faulted units is antithetic to the shear direction. (c) Fragmented porphyroclast of feldspar with internal faults at high angle to the foliation. The sense of movement along the faults is antithetic. (d) Feldspar porphyroclast with two sets of internal faults. The small fragments in the right side forms a mosaic and extended as a tail of the clast.

shear can be deduced from the asymmetry of the tail or the foliation. The sense of shear from the asymmetric pattern always indicates a thrusting sense of movement towards NW. In all the photomicrographs, dextral sense of shear points to the up-dip direction of the lineation.

Within these mylonites, a large number of porphyroclasts occur with symmetric tails (Figs 9.3c and 9.4c). However, σ or δ tails (Passchier and Simpson 1986) are also present; σ tails being more common than δ. Figure 9.8(a, b) shows an oval feldspar clast oriented at an angle to the mylonitic foliation. The recrystallized tails show an σ shaped tail indicating a dextral sense of shear. Figure 9.8(c and d) shows another clast with low aspect ratio oriented parallel to the mylonitic foliation and the δ shaped tail indicates larger rotation of the grain in a dextral shear. In the same thin section, there is another grain (Fig. 9.9a) with its long axis oriented antithetically at a high angle to the foliation. The tails

indicate a complex rotational behavior. The tails of the clast indicate both forward and backward rotation of the clast in a dextral shearing. Such a situation may arise in when there is a change in the ratio of rates of simple shear and pure shear (Ghosh and Ramberg 1976) in a transpressional deformation.

9.2.4 Folds in mylonite

Mylonitic foliation in PSZ has been repeatedly folded in progressive deformation. These folds are also present in microscopic scale. In sections perpendicular to the mylonitic foliation and parallel to the mylonitic lineation, we mostly see the profiles of subhorizontal and inclined folds. Most of these folds are very tight or isoclinal (Fig. 9.9b). The folds may be symmetrical or asymmetrical. The asymmetry of the folds always indicates an upward thrusting sense. Some of these have very sharp

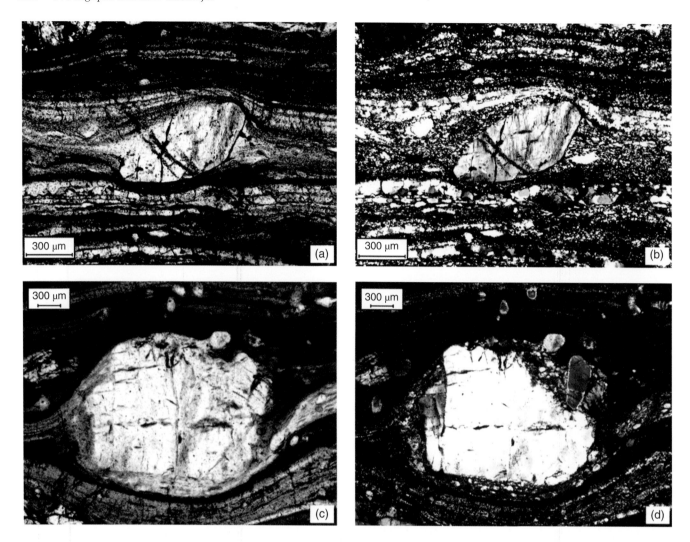

Fig. 9.8. (a) Oval shaped porphyroclast of feldspar oriented at low angle to the foliation. The tails show an σ pattern indicating dextral shear. (b) Same grain under crossed nicols. (c) Feldspar porphyroclast with δ type of tails indicating dextral shear sense. (d) Same grain under crossed nicols.

hinges and long limbs. With progressive deformation, the hinges may get obliterated and it may be difficult to identify the adjacent bands as the two limbs of a single fold. On the other hand, sometimes an isolated hinge can be present within the mylonitic matrix as a rootless intrafolial fold.

In sections parallel to the down-dip stretching lineation, the profiles of reclined folds cannot be seen. However, these sections contain the parallel limbs of isoclinal reclined folds. The sense of rotation of the porphyroclasts in the adjacent limbs of an isoclinal fold will be opposite in the initial stages of fold formation. However, once the fold becomes isoclinal, the two limbs will rotate to become parallel to each other. The sense of rotation within the limbs will then be controlled by the overall sense of shear. A porphyroclast in one limb will keep rotating in the same sense (say dextral as in the upper limb) whereas in the adjacent limb, the sense of rotation of the porphyroclast will be reversed (say sinistral to dextral). If the hinge is obliterated, two

adjacent bands will contain porphyroclasts with different degrees of rotation. Reversal in ductile shear sense has been reported from different tectonic contexts (Chattopadhyay et al., 2016; Pace et al., 2016; Pamplona et al., 2016).

The profiles of reclined folds can be seen in sections perpendicular to both the foliation and lineation (YZ section of strain). These sections are also parallel to the vorticity vector in PSZ. Figure 9.9c shows such a section showing the profile of an isoclinal fold. The limbs of the isoclinal fold contain a number of feldspar megacrysts. The large clast at the upper limb (left) shows an apparent dextral rotation. The other clasts show either symmetrical or apparent dextral or sinistral rotation (cf. Sengupta and Ghosh 2004).

Sheath folds are also common in the YZ sections (Fig. 9.10). Figure 9.10 shows a series of sheath folds in microscopic scale showing closed outcrops. Closed outcrops of at least four microscopic sheaths can be seen in the top central part of the photomicrograph. On either

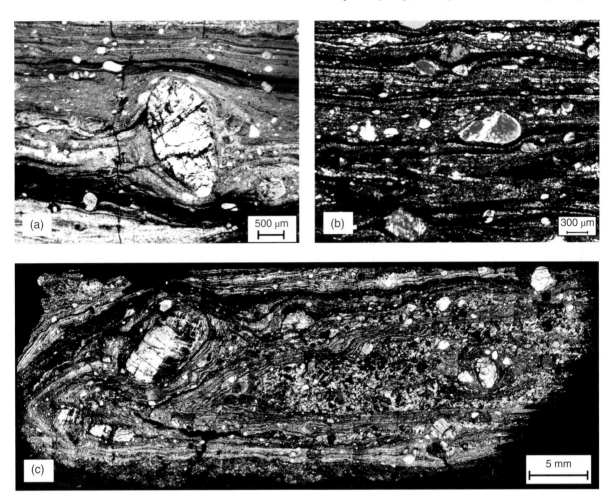

Fig. 9.9. (a) An oval porphyroclast oriented antithetically at a high angle to the foliation. The tails of the clast shows a complex pattern. This feature indicates that the deformation is a combination of simple and pure shear. (b) An ultramylonite with small feldspar clasts. The mylonitic foliation is folded into tight isoclinals folds. Section is parallel to *XZ*. (c) Section perpendicular to mylonitic lineation (*YZ*) shows an isoclinals reclined fold. The feldspar clast within the limb shows varying senses of rotation.

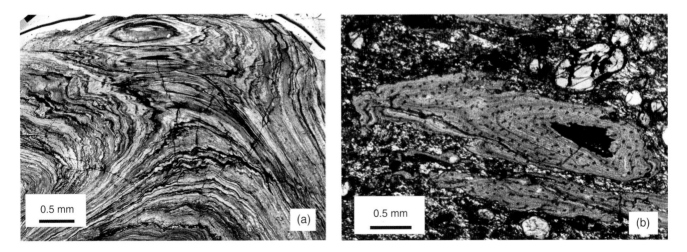

Fig. 9.10. (a) Microscopic sheath folds in a section perpendicular to mylonitic lineation. At least four sheaths can be identified in the central part. The minor folds on either side of the sheaths show opposite senses of asymmetry. The folds in the lower central part are symmetrical. (b) Microscopic sheaths within an ultramylonite.

side of these sheaths, minor folds indicate opposite senses of asymmetry. Similarly, there may an opposite sense of rotation of clasts on either sides of a sheath fold. Indeed, in sections parallel to the vorticity vector, in addition to symmetric tails, porphyroclasts with both dextral and sinistral senses of rotation have been observed (Sengupta and Ghosh 2004).

9.3 PROBLEM OF USING ROTATING PORPHYROCLASTS FOR VORTICITY ANALYSIS IN THE PHULAD SHEAR ZONE

The study of noncoaxiality in deformation is commonly referred to as vorticity analysis. Noncoaxiality of flow can be expressed by the kinematic vorticity number W_k (Truesdell 1954, Means et al. 1980). It is a dimensionless constant that measures the degree of rotationality of deformation with respect to the internal reference frame. In transpressional deformation, it gives the ratio between rotation and stretch. For simple shear, W_k is 1, for coaxial deformation W_k is 0 and in a combination of pure shear and simple shear deformation, W_k lies between 0 and 1.

In recent years, detailed studies in natural shear zones by several workers indicate that progressive deformation in natural shear zones is not by simple shear alone but there is an additional component of pure shear. To quantify the noncoaxiality of flow, it is necessary to carry out the vorticity analysis of the deformation. A convenient method of carrying out vorticity analysis in shear zones is by observing rotating rigid inclusions in mylonites.

Rigid elliptical objects undergoing penetrative ductile shear will rotate with respect to the kinematic frame of the bulk flow. Ghosh and Ramberg (1976), in their pioneering work, have shown that even when the flow is complex and noncoaxial, with simultaneous simple and pure shear, the rotation of the rigid objects can be predicted. Depending on the ratio of intensity of pure and simple shear (S_r), the axial ratio R (semimajor/semiminor axes of the elliptical inclusion, Ghosh and Ramberg 1976) of the inclusion and its initial orientation, the inclusion may rotate synthetically forward or antithetically backward or even remain stationary. A spherical inclusion will always rotate forward but an elongate inclusion may rotate forward or backward. Again, depending on critical values of S_r and R (Ghosh and Ramberg 1976), the total rotation can be very large or restricted. The inclusion may reach a stable position with the longer axis of the inclusion nearly parallel to the direction of shear. If such a stable position can be recognised in naturally deformed rocks, the shape and orientation of the inclusion can provide valuable information regarding the character of the deformation.

Since then, several studies have been carried out to measure and analyze the noncoaxiality of complex deformation in shear zones. Theoretical studies were complemented by studies to measuring W_k from naturally deformed rocks (e.g. Ghosh 1987; Passchier 1987,

1997; Passchier and Urai 1988; Fossen and Tikoff 1993; Tikoff and Fossen 1993, 1995; Simpson and De Paor 1993, 1997; Jessup et al 2007). A detailed review of vorticity analysis can be found in Xypolias (2010).

For vorticity analyses using rigid clasts in natural shear zones, the porphyroclast aspect ratio (PAR) method of Passchier (1987) and the porphyroclast hyperbolic distribution (PHD) method of Simpson and De Paor (1993, 1997) are widely used. Both these methods are based on the theory of Ghosh and Ramberg (1976) using the stable position of elongate clasts. However, there are certain assumptions for using these methods. The methods require that the deformation should be homogeneous and steady state. The other prerequisites are as follows: (1) the shape of the clasts should be ellipsoidal and the aspect ratio of the clasts should remain constant during deformation; (2) there should be no slip between the clasts and the matrix; (3) clasts should be embedded in a homogeneous and fine-grained matrix; (4) there should be a large number of clasts with variable aspect ratios; (5) clasts should be free to rotate; and (6) strain must be large for clasts to reach a stable position (cf. Xypolias 2010).

Although very few of these prerequisites are met in natural deformations, the PAR and PHD methods have been used extensively by a large number of researchers to determine the kinematic vorticity number of noncoaxial deformations (Xypolias 2010, and references therein) . However, the methods are still not very accurate and need further modifications (Passchier and Trouw 2005, p. 252; Xypolias 2010; Li and Jiang 2011). The deformation in natural shear zones is usually heterogeneous in time and space (Li and Jiang 2011; Vitale and Mazzoli, 2016). In an exhaustive analysis, Li and Jiang (2011) have shown that these methods are inadequate to quantify the noncoaxiality of the deformation as the prerequisites are never met in most cases. With the help of numerical analysis, they have showed that when there is a dominant pure shear component, the kinematic vorticity number may not represent the true noncoaxiality of the deformation even when the assumptions are met. Piazzolo et al. (2002) considered the kinematic vorticity number was locality dependant and could be significantly different from the kinematic vorticity number of the bulk flow.

In order to carry out the vorticity analysis, we assessed the suitability of the samples of quartzofeldspathic mylonites of PSZ. We have found that the prerequisites are difficult to apply in PSZ. Like most natural shear zones, the deformation was heterogeneous and non steady (Sengupta and Ghosh 2004; Jiang 1994; Jiang and Williams 1999). The number of suitable back rotated clasts with synthetic tails are limited. It is unrealistic to assume that the shape of the clasts remained constant during progressive deformation. The microstructures of the quartzofeldspathic mylonites of PSZ show that the aspect ratio and shape of the clasts continually changed during successive stages of shearing. The larger clasts have been dissected by internal shear zones or brittle fractures (Figs. 9.2c and 9.7). The aspect ratios of these newly formed smaller clasts vary from their parent clasts (Figs. 9.2 and 9.4). With progressive deformation, these clasts behaved as separate inclusions and showed a different history of deformation and rotation. Deformation needs to have progressed for a considerable length of time till the clasts attain a suitable ellipsoidal shape. Progressive recrystallization also continuously changed the shapes of the clasts and the aspect ratios (Fig. 9.4). In certain cases, an ellipsoidal clast may become rectangular and the orientation of the longer axis of the clast was not caused by rotation but by partial recrystallization with progressive deformation (Fig. 9.4b). Moreover, a large number of clasts were found flattened parallel to the foliation, indicating thereby that they were deformed in a plastic manner and not by rigid rotation. It should also be noted that rotation of a rigid inclusion is a three-dimensional problem (Ghosh et al. 2003; Sengupta and Ghosh 2004; Passchier and Trouw 2005, p.252; Forte and Bailey 2007; Li and Jiang 2011). Tikoff and Fossen (1994) showed that W_k may be useful in two-dimensional strain analysis, but it is less precise and hence less useful in three-dimensional deformation.

Another factor that may interfere with vorticity analysis is the prevalence of isoclinal folds and sheath folds (Figs. 9.9b,c and 9.10). Passchier (1987) has mentioned that samples from limbs of large scale folds are unsuitable for vorticity analysis. He also suggested that it is better to use samples from a straight regular shear zone. Large-scale folds do not occur in PSZ. The shear zone runs for a few kilometres with a more or less constant strike. However, the quartzofeldspathic bands show profuse development of mesoscopic folds. The bands show at least three generations of isoclinal reclined folds and sheath folds (Ghosh et al. 1999; Sengupta and Ghosh 2004). These folds are ubiquitous even in microscopic scales. We have selected samples from straight unfolded bands for the analysis. However, even in sections with no apparent mesoscopic folds, microscopic scale folds are present (Fig. 9.9b). In addition, a plane (XZ section) perpendicular to the vorticity axis also contains parallel limbs of reclined folds. So, adjacent parallel bands may actually represent two limbs of an isoclinal reclined fold. In that case, such bands may have suffered repeated folding and hinge line rotation and the clasts within the adjacent limbs possibly had very different histories of deformation and rotation. It would therefore be erroneous to carry out vorticity analysis using rotation of rigid clasts from such samples. Considering all the factors above, quartzofeldspathic mylonites from PSZ are unsuitable for vorticity analysis using rotating clasts.

9.4   CONCLUSIONS

Most microstructures of sheared quartzofeldspathic rock of the PSZ show orthomylonitic or ultramylonitic stages of deformation. The microstructures of the mylonites

indicate that shearing occurred under middle to lower amphibolites facies conditions. Both quartz and feldspar have been deformed by the crystal plastic process. Quartz shows intensive grain refinement and polygonization and usually occurs as thin polycrystalline ribbons or as small equant recrystallized grains. Feldspar has behaved in a more competent manner and shows pronounced recrystallization by subgrain formation and grain boundary bulging. A few feldspar grains are flattened plastically and become parallel to the mylonitic foliation. A core-and-mantle structure is common in feldspar porphyroclasts. With progressive deformation, the size of the mantle increases at the cost of the deformed core and the mantle is drawn out to form bands of recrystallized feldspar grains. Individual neoblasts in such bands are elongated parallel to foliation, indicating dynamic recrystallization. Semi-brittle and brittle deformational features have been observed in some feldspar porphyroclasts during later stages of shearing. The size and shape of the feldspar grains change with progressive deformation. The aspect ratio of the feldspar megacryst decreases with progressive shearing. The grains became more ellipsoidal or globular with continued deformation. However, an elliptical grain may also become rectangular at a later stage due to uneven recrystallization. Gradually the rock becomes an ultramylonite with a pronounced banded structure. The mylonites frequently show isoclinal as well as sheath folds in the microscopic scale. Presence of antithetically rotated feldspar clasts with synthetic sense of shear indicates that the deformation was not by simple shear alone but there was an additional component of pure shear. This observation supports our earlier findings from mesoscopic structures in the PSZ (Ghosh et al. 1999; Sengupta and Ghosh 2004).

PAR and PHD methods using rotating porphyroclasts have been widely applied for vorticity analysis in natural shear zones to determine the noncoaxiality of the deformation. These methods are found to be inadequate for quantification of noncoaxial strain in natural shear zones in PSZ. For the quartzofeldspathic mylonites in PSZ, these methods may give erroneous results due to the presence of several generations of folds, heterogeneity of deformation, and also on account of the continuously changing aspect ratios of porphyroclasts. To avoid oversimplification, it would be better to use several criteria, both in mesoscopic and in microscopic scale, to get a clear picture of the noncoaxiality of the shear zone. Existing methods of vorticity analysis need modification and improvement to include the complexities of natural deformation.

ACKNOWLEDGMENTS

We thank Soumyajit Mukherjee, Saibal Gupta, and an anonymous reviewer for suggestions to improve the manuscript. We are grateful to Nibir Mandal for his constructive comments and Meenakshi Sarkar for her help during preparation of the manuscript. S.S. and S.M.C. acknowledge financial support from Indian National Science Academy and Department of Science and Technology (Scheme No: SR/FTP/ES-21/2008), respectively.

REFERENCES

Boullier AM, Guéguen Y. 1975. SP-mylonites: origin of some mylonites by superplastic flow. Contribution to Mineralogy and Petrology 50, 93–104.

Chattopadhyay N, Ray S, Sanyal S, Sengupta P. 2016. Mineralogical, textural and chemical reconstitution of granitic rock in ductile shear zone: A study from a part of the South Purulia Shear Zone, West Bengal, India. In Ductile Shear Zones: From Micro- to Macro-scales, edited by S. Mukherjee and K.F. Mulchrone, John Wiley & Sons, Chichester.

Forte AM, Bailey CM. 2007. Testing the utility of porphyroclast hyperbolic distribution method of kinematic vorticity analysis. Journal of Structural Geology 29, 983–1001.

Fossen H, Tikoff B. 1993. The deformation matrix for simultaneous simple shearing, pure shearing and volume change, and its application to transpression- transtension tectonics. Journal of Structural Geology 15, 413–422.

Ghosh SK. 1987. Measure of non-coaxiality. Journal of Structural Geology 9, 111–113.

Ghosh SK. 1993. Structural Geology: Fundamentals and Modern Developments. Pergamon Press, Oxford.

Ghosh SK, Ramberg H. 1976. Reorientation of inclusions by combination of pure shear and simple shear. Tectonophysics 34, 1–70.

Ghosh SK, Hazra S, Sengupta S. 1999. Planar, non-planar and refolded sheath folds in Phulad shear zone, Rajasthan, India. Journal of Structural Geology 21, 1715–1729.

Ghosh SK, Sen G, Sengupta S. 2003. Rotation of long tectonic clasts in transpressional shear zones. Journal of Structural Geology 25, 1083–1096.

Golani PR, Reddy AB, Bhattacharjee J. 1998. The Phulad Shear Zone in Central Rajasthan and its tectonostratigraphic implications. In The Indian Precambrian, edited by B.S. Paliwal, Scientific Publishers (India), Jodhpur, pp. 272–278.

Jessup MJ, Law RD, Frassi C. 2007. The rigid grain net (RGN): an alternative method for estimating mean kinematic vorticity number (Wm). Journal of Structural Geology 29, 411–421.

Jiang D. 1994. Flow variation in layered rocks subjected to bulk flow of various kinematic vorticities: theory and geological implications. Journal of Structural Geology 16, 1159–1172.

Jiang D, Williams PF. 1999. A fundamental problems with kinematic interpretation of geological structures. Journal of Structural Geology 21, 933–937.

Li C, Jiang D. 2011. A critique of vorticity analysis using rigid clasts. Journal of Structural Geology 33, 203–219.

Means WD, Hobbs BE, Lister GS, Williams PF. 1980. Vorticity and non-coaxiality in progressive deformation. Journal of Structural Geology 2. 371–378.

Pace P, Calamita F, Tarvanelli E. 2016. Brittle-ductile shear zones along inversion-related frontal and oblique thrust ramps: Insights from the Central-Northern Apennines curved thrust system (Italy). In Ductile Shear Zones: From Micro- to Macro-scales, edited by S. Mukherjee and K.F. Mulchrone, John Wiley & Sons, Chichester.

Pamplona J, Rodrigues BC, Llana-Fúnez S, et al. 2016. Structure and Variscan evolution of Malpica-Lamego ductile shear zone (NW of Iberian Peninsula). In Ductile Shear Zones: From Micro- to Macro-scales, edited by S. Mukherjee and K.F. Mulchrone, John Wiley & Sons, Chichester.

Passchier CW. 1987. Stable positions of rigid objects in non-coaxial flow – a study in vorticity analysis. Journal of Structural Geology 9, 679–690.

Passchier CW. 1997. The fabric attractor. Journal of Structural Geology 19, 113–127.

Passchier CW. Simpson C. 1986. Porphyroclast systems as kinematic indicators. Journal of Structural Geology 8, 831–844.

Passchier CW, Trouw RAJ. 2005. Microtectonics, 2nd edition. Springer, Berlin.

Passchier CW, Urai JL. 1988. Vorticity and strain analysis using Mohr diagrams. Journal of Structural Geology 10, 755–763.

Piazzolo S, Bons PD, Passchier CW. 2002. Influence of matrix rheology and vorticity on fabric development of populations of rigid objects during plane strain deformation. Tectonophysics 351, 315, 329.

Pryer LL. 1993. Microstructures in feldspars from a major crustal thrust zone: the Grenville Front, Ontario, Canada. Journal of Structural Geology 15, 21–36

Ramsay JG, Huber MI. 1987. The techniques of modern structural geology, 2: Folds and fractures. Academic Press, London.

Roy AB, Jakhar SR. 2002. Geology of Rajasthan (Northwestern India), Precambrian to Recent, Scientific Publication. 421.

Sengupta S, Ghosh SK. 2004. Analysis of transpressional deformation from geometrical evolution of mesoscopic structures from Phulad shear zone, Rajasthan, India. Journal of Structural Geology 26, 1961–1976.

Sengupta S, Ghosh SK. 2007. Origin of striping lineation and transposition of linear structures in shear zones. Journal of Structural Geology 29, 273–287.

Simpson C, De Paor DG. 1993. Strain and kinematic analysis in general shear zones. Journal of Structural Geology 15, 1–20.

Simpson C, De Paor DG. 1997. Practical analysis of general shear zones using the porphyroclast hyperbolic distribution method: an example from the Scandinavian Caledonides. In Evolution of Geological Structures in Micro- to Macro-Scales, edited by S. Sengupta, Chapman & Hall, London, pp. 169–184.

Stipp M, Stuniz H, Heilbronner R, Schmid S. 2002. The eastern Tonale fault zone: a natural laboratory for crystal plastic deformation of quartz over a temperature range from 250° to 700°C. Journal of Structural Geology 24, 1861–1884.

Tikoff B, Fossen H. 1993. Simultaneous pure shear and simple shear: the unifying deformation matrix. Tectonophysics 217, 267–283.

Tikoff B, Fossen H. 1995. The limitations of three-dimensional kinematic vorticity analysis. Journal of Structural Geology 17, 1771–1784.

Truesdell CA. 1954. The kinematics of vorticity. Indiana University Press, Bloomington.

Trouw RAJ, Passchier CW, Wiersma DJ. 2010. Atlas of mylonites and related microstructures. Springer, Berlin.

Van den Driessche J, Brun JP. 1987. Rolling structures at large shear strain. Journal of Structural Geology 9, 691–704.

Vitale S, Mazzoli S. 2016. From finite to incremental strain: insights into heterogeneous shear zone evolution. In Ductile Shear Zones: From Micro- to Macro-scales, edited by S. Mukherjee and K.F. Mulchrone, John Wiley & Sons, Chichester.

White SH. 1976. The effects of strain on microstructures, fabrics and deformation mechanisms in quartzites. Philosophical Transactions of the Royal Society of London, A 283, 69–86.

White SH, Burrows SE, Carreraras J, Shaw ND, Humphreys FJ. 1980. On mylonites in ductile shear zone. Journal of Structural Geology 2, 175–187.

Xypolias P. 2010. Vorticity analysis in shear zones: A review of methods and applications Journal of Structural Geology 32, 2072–2092.

Chapter 10

Mineralogical, textural, and chemical reconstitution of granitic rock in ductile shear zones: A study from a part of the South Purulia Shear Zone, West Bengal, India

NANDINI CHATTOPADHYAY, SAYAN RAY, SANJOY SANYAL, and PULAK SENGUPTA

Department of Geological Sciences, Jadavpur University, Kolkata 700032, India

10.1 INTRODUCTION

Ductile shear zones are storehouse of deformation fabrics that provide critical information on the mechanism of deformation operated at meso- to microscale, the rheology of rocks at different physicochemical conditions, and the processes that help exhumation of deep-seated rocks at shallower crustal levels (Vernon 2004; Passchier and Trouw 2005; Mukherjee 2012, 2013). Ductile shear zones also act as conduits for extraneous fluids and melts (Vernon 2004; reviewed in Harlov and Austrheim 2013). These fluids are not in equilibrium with the rocks they infiltrate and hence induce significant changes in chemistry (major, trace, and isotope), mineralogy, and rheology, and seismic properties of the host rocks (cf. Harlov and Austrheim 2013 and Vernon 2004). Fluid flow along crustal-scale ductile shear zones, therefore, has important consequences for thermal and chemical evolution of the continental crust (cf. Harlov and Austrheim 2013). Granitic rocks in the hydrated ductile shear zone display a plethora of microstructures and reaction textures that provide valuable insight about fluid flow, deformation kinematics, and mass transport in shear zones (cf. Vernon 2004; Passchier and Trouw 2005; Harlov and Austrheim 2013; Mukherjee 2013). The South Purulia Shear Zone (SPSZ) of the East Indian shield exposes complexly folded, sheared, and dismembered rock suites including metapelites, metabasites, granitoids, alkaline rocks, and carbonatite. No age data exist to fix the timing of deformation and metamorphism. One deformed and metamorphosed alkaline rock suite at Sushina has been dated to be 0.93 Ga (Reddy et al. 2009; Chatterjee et al. 2013).

In this chapter we present microstructures and reaction textures of a porphyritic granite that emplaced and subsequently deformed and metasomatized within the SPSZ. Deformation and petrologic attributes of the porphyritic granite are thus the subject of the present communication. Integrating the results of microstructural analyses, textural modeling, mass balance calculations,

and numerical modeling of representative bulk rock composition, we tried to trace the physicochemical changes of porphyritic granite during the evolution of the SPSZ.

10.2 GEOLOGY OF THE AREA

The SPSZ, a part of the ~E W to ESE–WNW trending Tamar-Porapahar lineament of the east Indian shield occurs at the interface between two crustal blocks with contrasting geological and geochronological characteristics (reviewed in Mahadevan 1992; Acharyya et al. 2006; Mahato et al. 2008; Sanyal and Sengupta 2012). The southern block, referred as the North Singhbhum Fold Belt (NSFB), consisting of an ensemble of sedimentary and volcanic protoliths, have been metamorphosed in greenschist to amphibolites facies by ca. 1.5–1.8 Ga tectonothermal events (Mahato et al. 2008; Sengupta et al. 2011; reviewed in Sanyal and Sengupta 2012). This NSFB girdles the Meso- to Neo Archaean Singhbhum cratonic nucleus of the East Indian shield (reviewed in Mahadevan 1992 and Acharyya et al. 2006). The Meso- to Neoproterozoic Chotanagpur Granite Gneiss Complex (CGGC) north of the SPSZ consists of amphibolite to granulite facies granite gneisses with enclaves of metapelite, metabasic and metacalcareous rocks (reviewed in Sanyal and Sengupta 2012). The arcuate SPSZ that runs >150 km along the strike, represents highly deformed, dismembered, and hydrothermally altered rocks, akin to both NSFB as well as CGGC (Acharyya et al. 2006). Additionally, a variety of tuff-like rocks, tourmalinite, carbonatite, nepheline-bearing syenite, apatite deposits, and granitoids, constitute an important part of the litho-ensemble that are also strongly deformed and metamorphosed along with the other rock types of the SPSZ (Acharyya et al. 2006). Abundance of tuffs and carbonatite-alkaline-apatite-rich rocks were produced by rift-related extensional tectonics during the early history of the SPSZ (reviewed in Acharyya et al. 2006; Chakrabarty and Sen 2010; Ray et al. 2012). Around 0.93 Ga, a U-Pb zircon age of one nepheline-bearing syenite from the western part of the SPSZ provides the timing of the rift activity (Reddy et al., 2009; Chatterjee et al. 2013). Intense deformation, shear, and metamorphism of this early rock ensemble mark the closure of the rift zone (Acharyya et al., 2006; Ray et al. 2012; Talukdar et al. 2012). The SPSZ, which affects the rocks of the NSFB more than the rocks of the CGGC, shows ductile deformation with late rejuvenation in brittle regime (Acharyya et al. 2006; Talukdar et al. 2012). Hydrothermal alteration during brittle deformation produced ferruginous and siliceous breccias (Talukdar et al. 2012).

10.3   STUDY AREA

The terrain is located to the eastern extremity of the SPSZ near Porapahar region (Fig. 10.1a) and exposes three rock suites (Fig. 10.1b). The litho-package of NSFB constitutes metapelite (commonly phyllite) with inter-layered quartzite. Muscovite, biotite, chlorite, and quartz are the dominant minerals in metapelite. Thin layers of carbonaceous materials are found in the metapelite. The schistosity (S_1) of the metapelite trends 055–095° and dips 50–70° towards SE to S. At places, S_1 is first tightly folded with E–W axial trace (F_2) and then by a set of open folds with ~N–S axial traces (F_3). Bands of ferruginous quartz breccias occur locally along the axial planes of the F_3 open folds. Tropical weathering

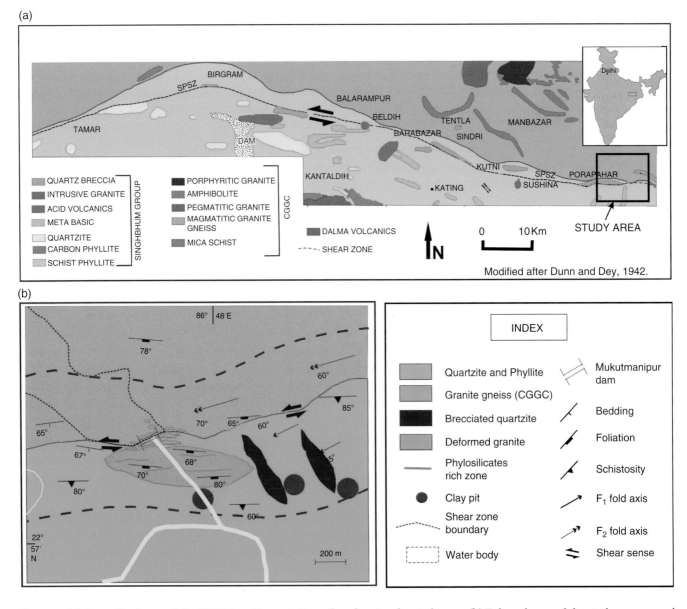

Fig. 10.1. (a) Generalized map of the SPSZ from Tamar to Porapahar showing the study area. (b) Enlarged map of the study area around Porapahar, near the Mukutmanipur Dam.

produced many clay pockets in the metapelites. The CGGC is represented by a medium-grained homophanous granitoid with rare amphibole and/or biotite porphyroblasts. An indistinct E–W trending foliation is defined by elongated quartz and feldspar grains. A suite of porphyritic granite intruded the NSFB as well as the homophanous granitoid of the CGGC. The reservoir of Kangsabati river dam covers a substantial portion of the CGGC. The contact between the NSFB and the CGGC is a ~1 km wide zone (Fig. 10.1b), where rocks of both the units including the porphyritic granite are sheared intensely. The mylonitic foliation in the sheared rocks is sub-parallel to the S$_1$ foliation of the metapelite lying outside the intensely deformed zone. Laterally discontinuous bands and lenses of tourmalinite are observed in the shear zone.

10.4   FIELD FEATURES OF THE STUDIED GRANITE

Magmatic structures of the augen gneiss got obliterated by superposed deformation and shearing. Nevertheless, in domains of feeble deformation, a relict porphyritic structure with tabular and simple twinned K-feldspar phenocrysts is discernible (Fig. 10.2a). As strain intensifies, feldspar phenocrysts flattened parallel to each other with a distinct rind of recrystallized grains of same mineral (Fig. 10.2b). In high-strain domains, the augen gneiss converted to granite mylonite with a few K-feldspar porphyroclasts. A prominent down-dip stretching lineation, defined by feldspar aggregates, developed on the shear plane (Fig. 10.2c). General strike of the mylonitic foliation ranges 55–95° and dip

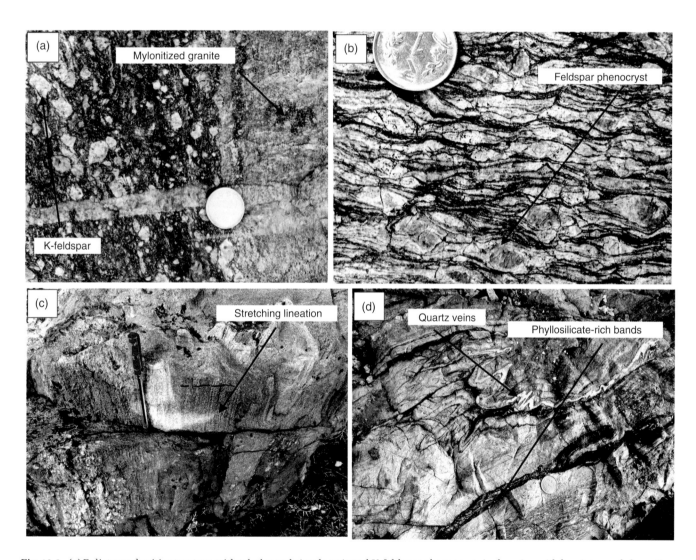

Fig. 10.2. (a) Relict porphyritic structure with tabular and simple twinned K-feldspar phenocrysts in domains with less intense deformation. Note the sharp contact between the two types of granite in a single outcrop. (b) Feldspar phenocrysts, flattened parallel to each other with a distinct rind and tail of recrystallized grains of same mineral in the highly strained part of the granite. (c) A prominent down dip stretching lineation defined by feldspar aggregates on the shear plane. (d) Centimeter- to decimeter-thick dark phyllosilicate-rich bands parallel to the mylonitic foliation of the enclosing granite mylonite that are dissected by different generations of quartz veins of variable thickness which are co-folded with the mylonitic foliation.

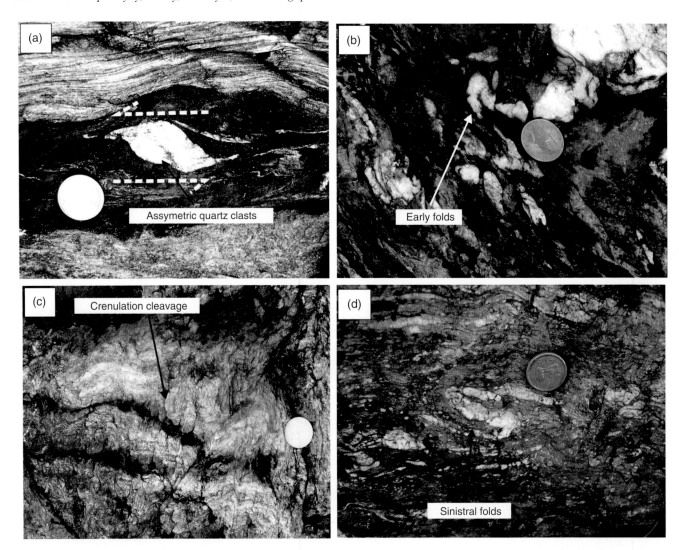

Fig. 10.3. (a) Asymmetricity shown by detached quartz clasts or "Quartz fish" (Mukherjee 2011a) with shear sense top-to-left, in phyllosilicate-rich bands within the mylonitic granite. (b) Quartz veins define early folds (F_1) and schistosity concordant with mylonitic foliation in the phyllosilicate bands develop paralled to the axial plane of the folds. (c) The granite mylonite and the interlayered phyllosilicate bands produce a crenulation cleavage, which are made prominent by thin layers of phyllosilicate –rich materials (S_2). (d) Vergence of asymmetric rootless folds in the weathered portions of the granite indicates a sinistral sense of shear.

50–70° towards S to SE (S_1). The mylonitic foliation sub-parallels the regional schistosity of the pelitic rocks of the adjoining NSFB. Commonly, the less strained domains grade abruptly into high-strain domains within a few centimeters (Fig. 10.2a). Centimeter- to decimeter-thick dark phyllosilicate-rich bands develop parallel to the mylonitic foliation of the enclosing granite mylonite (Fig. 10.2d and Fig. 10.3a). The phyllosilicates develop a distinct schistosity that parallels the mylonitic foliation. The granite mylonite and the phyllosilicate-rich bands are dissected by different generations of quartz veins of varied thickness (Fig. 10.2d and Fig. 3b,c). Mylonitic foliation and schistosity in phyllosilicate bands develop parallel to the axial plane of F_1 folds, defined by early quartz veins (Fig. 10.3b,d).

The granite mylonite and the interlayered phyllosilicate bands are folded (F_2, Fig. 10.2d and Fig. 10.3c). Locally, a crenulation cleavage developed, which are made prominent by thin layers of phyllosilicate–rich materials (S_2). Rarely, a N–S trending broad warp developed on the limbs of the F_2 folds (Fig. 10.2d). F_1 and F_2 are coaxial folds, whereas F_3 occur as cross folds. Although no pervasive planar structures develop with F_3, discrete brittle faults locally filled with ferruginous materials are developed locally along the axial plane of the F_3 fold. Acharyya et al. (2006) described a similar sequence of folding from the eastern part of the SPSZ. Our study corroborates the view of these workers that the main phase of mylonitization in the SPSZ occurred during the F_1 folding.

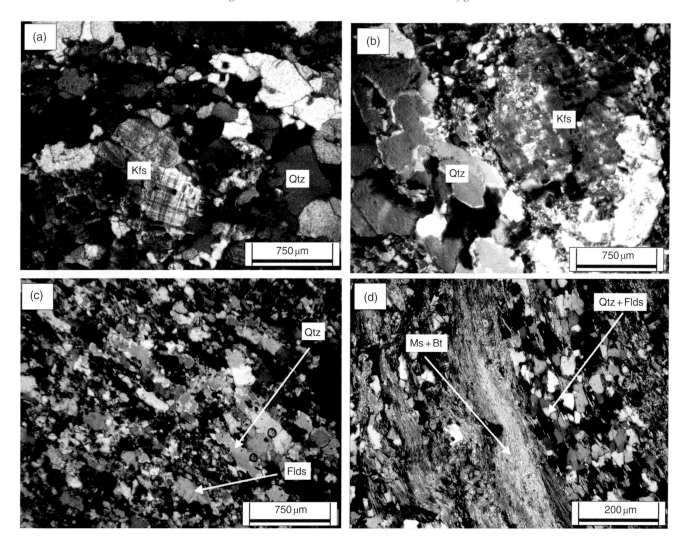

Fig. 10.4. (a) Tabular to mildly elliptical grains of feldspar in the less deformed portions of the granite. (b) Tabular K feldspar and anhedral quartz grains in the less deformed portion of the granite. (c) Magmatic feldspar grains and quartz are flattened to define a prominent planar fabric. (d) Commonly millimeter- to centimeter-thick layers with abundant phyllosilicate minerals alternate with phyllosilicate-poor bands in the deformed rock.

10.5 MICROSTRUCTURES AND MINERAL DEFORMATION BEHAVIOR

10.5.1 Evolution of microfabric

The studied granite shows several textures and structures, encompassing magmatic crystallization to post-magmatic deformation. Though the superimposed deformation obliterated most of the magmatic features, some of these primary features are present in weakly deformed lenses within the strongly sheared rocks. This may be related to strain accumulation in softer regions (Vernon 2004; Mukherjee 2014a). Commonly, in these low-strain domains, tabular to mildly elliptical K-feldspar, quartz, and rare plagioclase form the framework where quartz and biotite are present in the interstitial space (Fig. 10.4a,b). Progressive changes of

the magmatic mineralogy, induced by superimposed deformation and fluid-rock interaction developed several microfabrics and abundant phyllosilicates. Commonly, flattened feldspar and quartz define a prominent planar fabric parallel to the regional foliation (S_1, Fig. 10.4c). Millimeter- to centimeter-thick phyllosilicate-rich (>70 vol%) layers with alternate phyllosilicate-poor bands is a common feature of the granite mylonite. (Fig. 10.4d). Thin folia, made up of muscovite, biotite, and chlorite, swerve around flattened feldspar porphyroclasts with pressure shadow of quartz aggregates at the two ends (Fig. 10.5a). Sigma type of porphyroclast tails (Fig. 10.5b) and development of an oblique foliation defined by elongated quartz grains (Fig. 10.5c) suggest that the planar structure is a mylonitic foliation (S_1). Locally, this mylonitic foliation is axial planar to a set of folds defined by quartz veins (Fig. 10.6a). In many places, the mylonitic

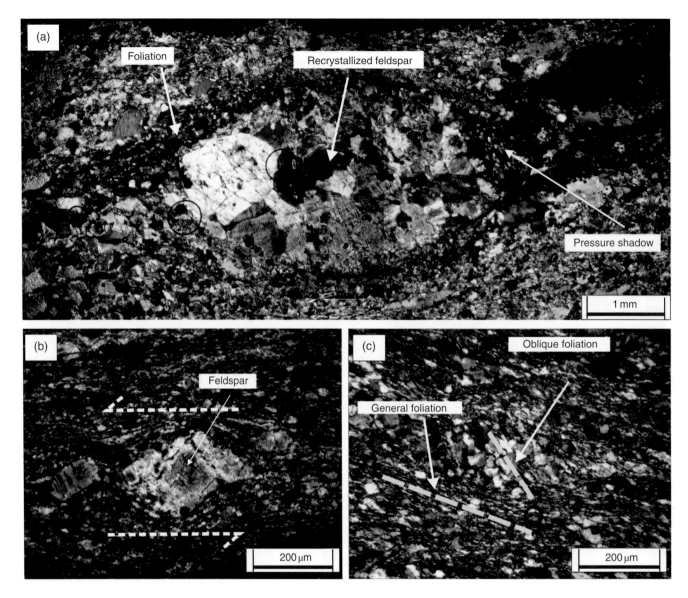

Fig. 10.5. (a) Thin folia of phyllosilicates (chlorite and mica) swerve around the flattened K-feldspar porphyroclasts, which form elliptical recrystallized aggregates at places, with pressure shadow at the two ends. (b) Asymmetric feldspar porphyroclasts often develop tails of dynamically recrystallized grains of same feldspar and produce σ-type porphyroclasts. (c) Quartz lens showing a foliation defined by flattened quartz and feldspars oblique to the main mylonitic foliation. The oblique fabric is denoting a top-to-left (up) shear.

foliation is crenulated (Fig. 10.6b). Commonly, a crenulation cleavage defined by stretched quartz, biotite, and muscovite, when present, is developed parallel to the axial planes of the fold (S_2, Fig. 10.6c,d and Fig. 10.7a). The following sections presents the response of different minerals to the superimposed deformation.

10.5.2 Deformation of minerals

10.5.2.1 *Feldspar*

A number of features suggest that both plagioclase and K-feldspar porphyroclasts were deformed under ductile as well as brittle regimes. These features are:

1 K-feldspar porphyroclasts are girdled by smaller grains of the same mineral thereby producing core-and-mantle structure (Fig. 10.7b). The average thickness of the rind is much smaller than the porphyroclast core. The latter shows a serrated outline and records internal deformation in the form of patchy/undulatory extinction and development of pericline twin (in K-feldspar). In places, augen-shaped aggregates of recrystallized K-feldspar porphyroclasts are swerved by a foliation (S_1) defined by oriented grains of mica and chlorite (Fig. 10.5a). Commonly, the feldspar porphyroclasts showing core-and-mantle structure, resulted presumably from recyrstallization and grain boundary rotation (Fig. 10.7b, Vernon 2004). Deformed

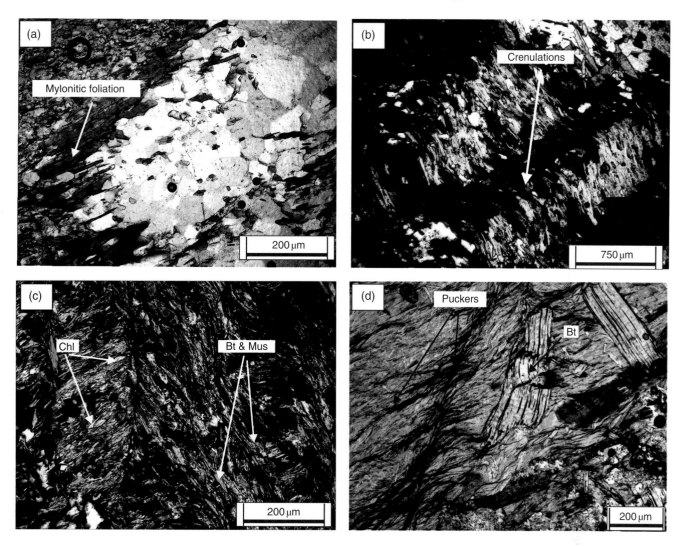

Fig. 10.6. (a) Mylonitic foliation is axial, planar to some folds defined by quartz micro veins. (b) F_2 crenulations on the schistosity defined by mostly biotite and some muscovite. They are formed parallel to the axial plane of the folds. (c) Biotite and muscovite grains are rotated and aligned parallel to the F_2 folds to producing a prominent crenulation cleavage. New biotite and muscovite grains also developed along the crenulation cleavage. (d) Two sets of intersecting crenulations on the schistosity surface. The "pucker" can be redefined as flanking structure (Mukherjee and Koyi 2009; Mukherjee 2011b, 2014b). Near a biotite cross-cutting element, biotites of a different generation in the matrix are drag folded.

feldspar porphyroclasts are swerved by quartz ribbon and produce an eye-like (augen) structure (Fig. 10.7c).

2 K-feldspar porphyroclasts that align along the lineation on the mylonitic foliation are pinched and swelled (Fig. 10.7d). Rarely, feldspar porphyroclasts are boudinaged with boudin neck filled up with recrystallized quartz and K-feldspar (Fig. 10.7d). This resembles fig 4.7 of Mukherjee (2014a). The detached feldspar clasts in the boudin neck show a highly serrated boundary against the finer-grained quartzofeldspathic aggregates (Fig. 10.7d). When in contact with mica, feldspar porphyroclasts show planar margin (Fig. 10.8a).

3 In places, plagioclase porphyroclasts are intersected by a set of parallel microfaults, (Fig. 10.8b) to produce imbricate structure (Fig. 10.8c). Recrystallization of

the displaced feldspar fragments are noted at places along the fault trace (Fig. 10.8c). The feldspar porphyroclasts also develop curvilinear fractures without any relative displacement (Fig. 10.8d).

4 Commonly, feldspar porphyroclasts show undulatory extinction with prolific subgrain formation owing to misfits in crystallographic properties within a grain (Fig. 10.9a, Vernon 2004). The undulatory extinction can be sweeping as well as patchy. Bending and kinking of lamellar twin planes of plagioclase is seen at a few places (Fig. 10.9b). Primary twin lamellae in plagioclase are locally offset by microfaults (Fig. 10.8b).

5 Brittle deformation of feldspar porphyroclasts produces fractures that are at high-angle to the mylonitic foliation (Fig. 10.9c). Serrated outline of the broken fragments and the feature showing the aggregate of

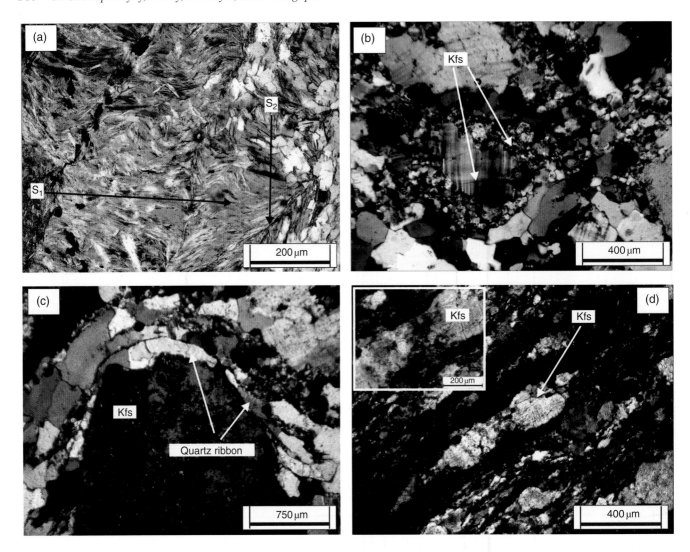

Fig. 10.7. (a) The crenulation cleavage (S$_2$) folded the (S$_1$) mylonitic foliation in the phylosilicate rich bands. (b) Large K-feldspar porphyro-clasts are girdled by smaller grains of the same mineral thereby producing typical core-and-mantle structure. (c) Feldspar porphyroclasts are swerved by quartz ribbons. (d) Plagioclase feldspar porphyroclasts flattened parallel to mylonitic foliation, produce pinch-and-swell struc-ture. Boudin neck filled up with recrystallized quartz and feldspar. The detached feldspar clasts show highly serrated boundary against the quartzofeldpathic mass.

recrystallized feldspar grains protruding into the por-phyroclast suggest that the broken fragments too underwent crystal–plastic deformation.

10.5.2.2   Quartz

Unlike feldspar, microfabrics of quartz grains are con-sistent with deformation under ductile regime. Both static and dynamic recrystalization of this mineral are observed. Some important features in support of this view are:

1 Quartz grains are flattened parallel to the mylonitic foliation to produce ribbon structure (Fig. 10.9d; as in fig. 4d of Mukherjee and Koyi 2010). The quartz ribbons show strain free segments with linear contacts (Fig. 10.9d).

2 Quartz shows an oblique grain shape fabric (Vernon 2004), which is characterized by lenses of elongated quartz aggregates at an angle to the mylonitic foliation defined by elongated quartz with or without oriented phyllosilicate minerals (Fig. 10.5c). Elongated quartz grains, products of dynamic recrystallization, show undulatory extinction (Fig. 10.10a). A similar feature is reported from many ductile shear zones (Vernon 2004).

3 Like the feldspar porphyroclasts, quartz-quartz grain contacts are also serrated (Fig. 10.10b). Bulging of quartz grains is common and the feature attest to grain boundary migration (Fig. 10.10b). Static recrys-tallization of quartz is manifested by strain-free polygonal quartz aggregates with interfacial angles ~120° (Fig. 10.10c), which denotes more stable conditions.

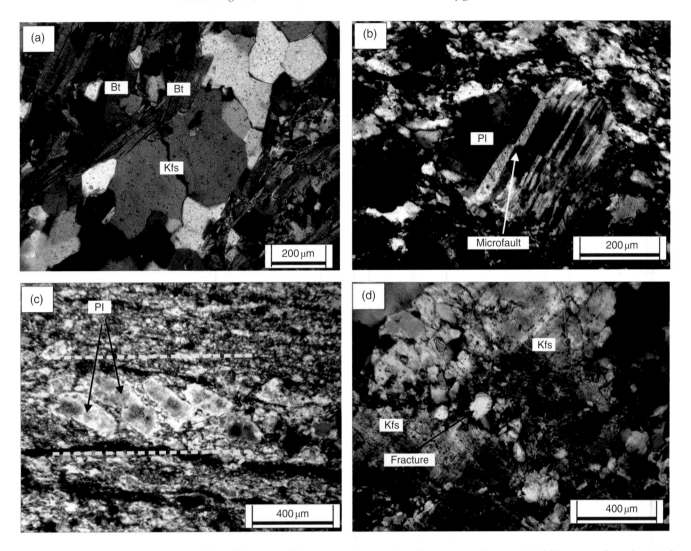

Fig. 10.8. (a) A prominent foliation is defined by oriented biotite grains. Note that planar contact between the feldspar porphyroclasts and biotite. See text for details. (b) Plagioclase porphyroclasts are intersected by microfaults along which the different fragments show relative displacement. (c) Relative displacement of the plagioclase feldspar fragments separated by microfaults produce imbricate structure. Recrystallization of the displaced feldspar fragments are noted along the fault plane. (d) The feldspar porphyroclasts also develop fractures without any displacement along the fracture. Note the crystallisation of quartzofeldspathic minerals along the fractures.

All these microstructures indicate that dislocation creep was the important process of quartz deformation along with grain boundary area reduction (GBAR; Passchier and Trouw 2005). At places the mylonitic foliations along with an oblique foliation represent S-C structures (Fig. 10.10d) within the dynamically recrystallized quartzofeldspathic mass.

10.5.2.3 Mica

Biotite shows two modes of growth. Commonly, biotite grains define the mylonitic foliation (Fig. 10.6b). In the second mode, stumpy biotites either define the crenulation cleavage or orient haphazardly (decussate texture) on the mylonitic foliation (Fig. 10.11a). Biotite grains that define the mylonitic foliation and the crenulation cleavage shows kinks (Fig. 10.6d). Commonly, muscovite grains define crenulations cleavage along with biotite (Fig. 10.6c).

The aforesaid microstructures constrain the mechanism and thermal regime during deformation. Formation of core–mantle structure indicates progressive misorientation of subgrains from core to recrystallized rind/ mantle. Owing to subgrain rotation, recrystallization concentrates along the former grain boundary region of the aggregates (reviewed in Vernon 2004). Dislocation creep and recovery of the strained porphyroclasts plausibly results in core-to-mantle structure (Vernon 2004). For a geologically reasonable strain rate, $\geq500^{\circ}$C is required to form core-and-mantle structure in feldspar porphyroclasts (Rosenberg and Stuenitz 2003). Temperature of $\geq500^{\circ}$C also explains recrystallization of K-feldspar porphyroclasts into coarse aggregates (Fig. 10.5a, Vernon, 2004). Serrated quartz-quartz and feldspar–feldspar grain contacts suggest dynamic recrystallization through grain-boundary migration (cf. Vernon 2004). The process of grain boundary migration is dependent upon the in

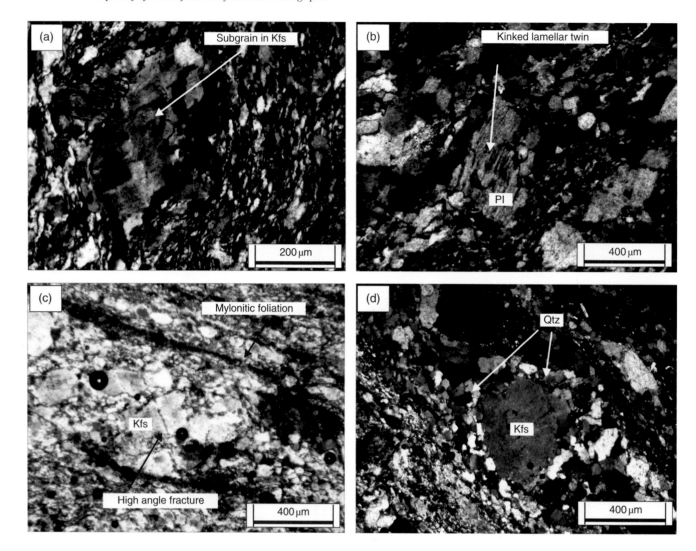

Fig. 10.9. (a) Feldspar porphyroclasts commonly show undulatory extinction with pronounced subgrain formation. (b) Bending and kinking of lamellar twin planes of plagioclase is seen in a few places. (c) Feldspar porphyroclasts, dissected by microcracks that develop at high angle with the mylonitic foliation are found in the granite, underwent brittle deformation. Feldspar porphyroclasts thus show grain size reduction through fracturing and thereafter separation of those broken fragments. (d) Recrystallized strain free quartz grains showing planar contact are flattened to produce ribbon structure, swerves around feldspar porphyroclast.

situ activity of a_{H2O} (cf. Vernon 2004). In absence of suitable mineral assemblage, computing a_{H2O} of the deformed domains is difficult. However, presence of abundant hydrous phases (and lack of carbonate minerals) suggests high water activity that might augment grain boundary migration (cf. Vernon 2004). Several studies demonstrated that at geologically reasonable strain rates, quartz and feldspar undergo grain boundary migration recrystallization at temperatures >400 °C and ≥500 °C, respectively, in low-grade hydrous rocks (cf. Vernon 2004). Planar contact between feldspar and mica indicate fast grain boundary migration due to localization of strain (Tullis and Yund 1980). Brittle deformation in these minerals formed later, presumably at a lower temperature. This may result from exhumation of the ductile rocks to brittle regime (cf. Vernon 2004).

10.6 REACTION TEXTURES

The most conspicuous reaction texture is manifested by feldspars replaced extensively by chlorite (Fig. 10.11b). In phyllosilicate-rich domains (>70 vol% phyllosilicates), K-feldspar is converted completely to chlorite. Rafts of plagioclase porphyroclasts occurs as corroded relics amidst cluster of chlorite (Fig. 10.11c). Measured chemical compositions (Table 10.1 and Table 10.2) of this mineral shows that albite occupy 80 vol% of these grains whereas volumetrically minor oligoclase occurs as irregular patches in the core of albite (Fig. 10.11d). Stumpy biotite grains replace chlorite and albite (Fig. 10.12a). Coarse biotite grains contain patches of chlorite (Fig. 10.12b), and less commonly, albite. Successive replacement of plagioclase to chlorite to biotite is seen in few places (Fig. 10.12a). Muscovite occurs in restricted domains

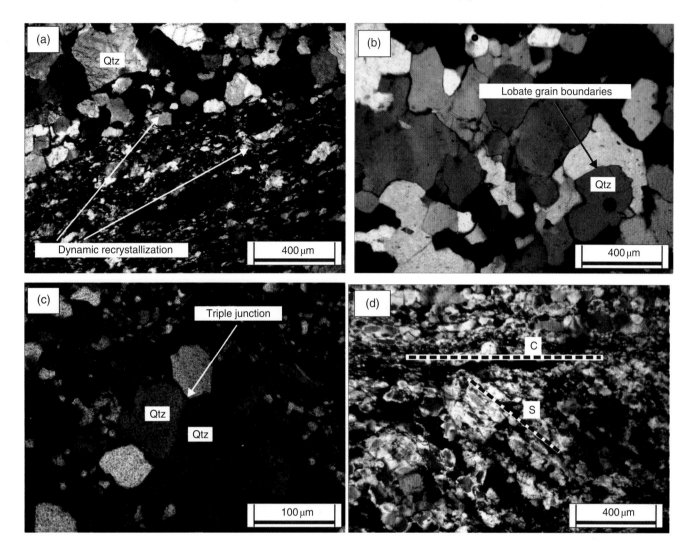

Fig. 10.10. (a) Large quartz porphyroclast, dynamically recrystallized into aggregates of elongated quartz showing undulose extinction are aligned parallel to the mylonitic foliation. (b) Highly irregular and lobate grain boundaries of quartz: one grain engulfs partially the other. (c) Static recrystallization of quartz is manifested by recrystallization of a quartz porphyroclast into polygonal strain free grains with interfacial angle of ~120°. (d) Mylonitic foliation along with an oblique foliation represent S-C structures within the dynamically recrystallised quartzo feldspathic minerals.

where chlorite is volumetrically minor or absent. Along with biotite, muscovite replaces both K-feldspar and plagioclase porphyroclasts (Fig. 10.12c).

Muscovite, biotite, and chlorite coexist rarely in the crenulation cleavage (Fig. 10.6c). Clusters of randomly oriented euhedral ilmenite crystals that are replaced variably by rutile are common (Fig. 10.12d). Allanite developed close to the Fe-Ti oxide phases. Allanite and calcite are common accessory minerals in some phyllosilicate bands.

10.7 MINERAL CHEMISTRY

Compositions of minerals were analyzed with a CAMECA SX100 microprobe at the Geological Survey of India, Kolkata. Natural and synthetic standards were used and the raw analyses were corrected with ZAF. Beam diameters were varied with the 1–3 µm with accelerating voltage 15 kV and 12 nA beam current. Analytical variation for primary standards was kept at or below ±1.5% for primary standards and ±3% for secondary standards. Representative compositions of biotite, muscovite, chlorite, plagioclase, K- feldspar, ilmenite, and rutile are presented in Tables 10.1 and 10.2.

- Biotite: The biotite shows variable X_{Mg}, Mg/(Mg + Fe$_{total}$) ~ 0.34 to 0.66) depending upon the sample location. TiO$_2$ content of biotite varies from 1.36 to 2.64 wt% with Ti-rich biotite compositions developing proximal to rutile and ilmenite.
- Chlorite: This mineral is distinctly more magnesian (X_{Mg} ~ 0.62–0.64) compared to co-existing biotite. Chlorite compositions of this study fall in the field of ripidolite (Hey 1954).

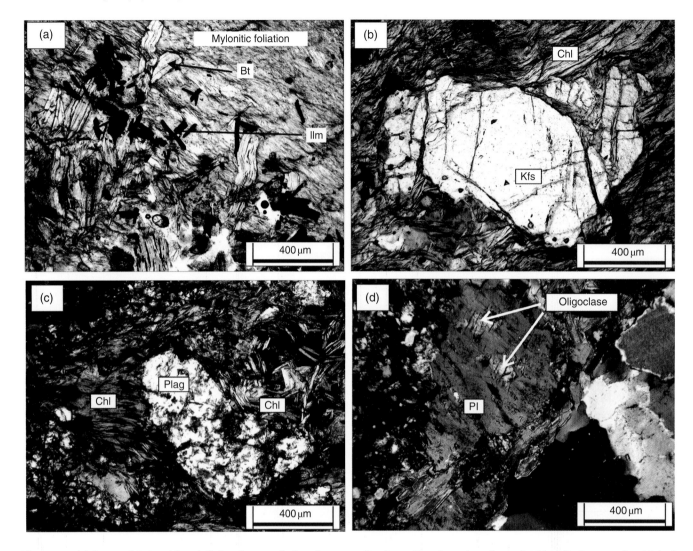

Fig. 10.11. (a) Stumpy biotite either defining the crenulation cleavage or haphazardly oriented on the mylonitic foliation are often kinked. (b) Extensive replacement of K - feldspars by chlorite in the phyllosilicate rich zone. (c) Islands of plagioclase porphyroclasts occurs as corroded relicts amidst cluster of chlorite flakes. (d) Altered plagioclase crystals are mostly albite whereas volumetrically minor oligoclase occurs as irregular patches in the core of albite.

- Plagioclase: Compositions of plagioclase fall in two distinct clusters. Commonly plagioclase grains show albitic composition (An_{2-4}). A few scattered grains and cores of some plagioclase are compositionally oligoclase (An_{71-82}).
- Ilmenite: It is almost $FeTiO_3$ with small Fe_2O_3 (0–0.48 wt%) and MnO (1.2–2 wt%) contents.
- Rutile, quartz, and K-feldspar have virtually end-member compositions.

10.8   CHEMICAL REACTIONS AND MASS TRANSPORT DURING MYLONITIZATION

Textures and microstructures observed in the deformed granite attest to intense fluid–rock interaction. These processes stabilized abundant phyllosilicate minerals in the mylonitized granite. In this section we first constrain the chemical reactions that triggered the growth of the phyllosilicate minerals. This follows estimation of loss or gain of different chemical species during mylonitization.

10.8.1   Chemical reactions

Textural features that attest to following sequence of mineralogical changes during and subsequent to shearing:

1 K – feldspar + oligoclase + quartz → chlorite + albite → biotite

2 K – feldspar + oligoclase + quartz → muscovite + biotite + albite

To obtain stoichiometrically balanced reactions among the participating phases, we performed textural

Table 10.1. Composition of feldspar and chlorite

	Feldspar											Chlorite	
Sample No.	**PP1/2**							**PP18**			**PP1/2**	**PP1/2**	
DataSet/Point	57/1.*	58/1.	60/1.	61/1.*	62/1.	63/1.	40/1.*	2/1.	16/1.	17/1.	19/1.	47/1.	56/1.*
Species	Ab	Olig	Ab	Olig	Ab	Olig	K-feldspar	Olig	Olig	Olig	Olig		
SiO_2	68.66	61.94	67.79	60.91	68.33	60.84	64.19	61.42	64.10	61.43	62.34	25.72	25.68
TiO_2	0.00	0.04	0.00	0.00	0.04	0.00	0.03	0.00	0.00	0.00	0.03	0.12	0.00
Al_2O_3	19.93	23.92	20.42	23.73	19.24	24.45	18.65	23.68	22.80	24.23	24.26	21.37	22.83
Cr_2O_3	0.00	0.13	0.11	0.08	0.10	0.09	0.01	0.07	0.00	0.02	0.05	0.04	0.03
MgO	0.00	0.03	0.01	0.00	0.04	0.01	0.00	0.00	0.02	0.00	0.00	19.10	19.22
FeO	0.14	0.14	0.18	0.12	0.11	0.12	0.00	0.09	0.00	0.09	0.00	20.61	19.67
CaO	0.70	5.47	0.89	5.99	0.31	6.08	0.00	5.02	3.81	6.21	5.56	0.07	0.04
MnO	0.00	0.00	0.00	0.06	0.00	0.00	0.00	0.03	0.00	0.00	0.06	0.16	0.15
Na_2O	11.04	8.71	11.31	8.79	11.43	8.68	0.49	8.86	9.58	8.42	8.87	0.02	0.00
K_2O	0.09	0.04	0.06	0.05	0.05	0.09	16.92	0.02	0.10	0.12	0.08	0.05	0.02
Total	**100.56**	**100.42**	**100.77**	**99.73**	**99.65**	**100.36**	**100.29**	**99.19**	**100.41**	**100.52**	**101.26**	**87.26**	**87.64**
Oxygen basis	8	8	8	8	8	8	8	8	8	8	8	28	28
# Si	2.982	2.739	2.948	2.722	2.996	2.702	2.974	2.747	2.818	2.720	2.735	5.318	5.247
# Ti	0.000	0.001	0.000	0.000	0.001	0.000	0.001	0.000	0.000	0.000	0.001	0.019	0.000
# Al	1.021	1.247	1.047	1.250	0.995	1.280	1.019	1.249	1.182	1.265	1.255	5.209	5.499
# Cr	0.000	0.005	0.004	0.003	0.003	0.003	0.000	0.002	0.000	0.001	0.002	0.007	0.005
# Mg	0.000	0.002	0.001	0.000	0.004	0.001	0.000	0.000	0.001	0.000	0.000	5.885	5.852
# Fe	0.005	0.005	0.007	0.004	0.004	0.004	0.000	0.003	0.000	0.003	0.000	3.564	3.361
# Ca	0.033	0.259	0.041	0.287	0.015	0.289	0.000	0.241	0.179	0.295	0.261	0.016	0.009
# Mn	0.000	0.000	0.000	0.002	0.000	0.000	0.000	0.001	0.000	0.000	0.002	0.028	0.026
# Na	0.930	0.747	0.954	0.762	0.972	0.747	0.044	0.768	0.817	0.723	0.755	0.008	0.000
# K	0.005	0.002	0.003	0.003	0.003	0.005	1.000	0.001	0.006	0.007	0.004	0.013	0.005
Cation	**4.975**	**5.008**	**5.005**	**5.034**	**4.991**	**5.033**	**5.038**	**5.012**	**5.002**	**5.012**	**5.015**	**20.066**	**20.004**
$X_{An}=Ca/(Ca+Na+K)$	**0.034**	**0.257**	**0.042**	**0.273**	**0.015**	**0.278**	**0.000**	**0.238**	**0.179**	**0.288**	**0.256**		
$X_{Ab}=Na/(Ca+Na+K)$	**0.961**	**0.741**	**0.955**	**0.724**	**0.982**	**0.717**	**0.042**	**0.761**	**0.815**	**0.706**	**0.739**		
$X_{Or}=K/(Ca+Na+K)$	**0.005**	**0.002**	**0.003**	**0.003**	**0.003**	**0.005**	**0.958**	**0.001**	**0.006**	**0.007**	**0.004**		
XMg												0.623	0.635
Xfe												0.377	0.365
Al/(Al+Si)												0.495	0.512

*Composition used in C-space.

Table 10.2. Representative composition of biotite, muscovite, rutile, and ilmenite

	Biotite							Muscovite				Rutile	Ilmenite	
Sample #	PP1/2			PP18				PP18				PP1/2	PP18	
Point #	48/1*	51/1	52/1	5/1	10/1	15/1	21/1	33/1	44/1	49/1*	7/1	49/1*	43/1*	51/1
SiO_2	38.07	37.67	36.95	34.30	33.09	34.03	35.50	45.96	45.51	46.13	44.92	0.02	0.03	0.02
TiO_2	1.57	1.36	1.65	2.59	2.29	2.64	2.55	0.80	0.44	0.79	0.22	98.8	52.1	53.34
Al_2O_3	15.96	16.13	16.37	15.96	15.96	15.91	16.46	29.94	30.71	30.36	31.26	0.02	0	0.01
Cr_2O_3	0.00	0.10	0.10	0.05	0.06	0.00	0.00	0.03	0.00	0.00	0.02	0.22	0.04	0.04
MgO	15.26	15.19	15.15	7.31	7.32	7.65	7.41	1.22	1.15	1.08	1.40	0.00	0.05	0.11
FeO	14.09	14.18	14.78	24.93	24.67	24.37	24.77	4.91	5.10	4.73	4.59	0.65	45.63	46.18
CaO	0.02	0.28	0.03	0.00	0.01	0.00	0.00	0.00	0.00	0.00	0.00	0.02	0.32	0.04
MnO	0.04	0.10	0.00	0.13	0.10	0.09	0.04	0.03	0.00	0.09	0.00	0.00	1.96	1.17
NiO				0.00	0.00	0.00	0.02	0.00	0.00	0.00	0.00	0.00	0.00	0.00
Na_2O	0.13	0.14	0.09	0.06	0.11	0.08	0.06	0.27	0.24	0.27	0.24	0.02	0.01	0.04
K_2O	10.22	7.61	9.86	10.19	9.99	10.16	10.12	11.53	11.66	11.41	11.42	0.00	0.03	0.00
Total	**95.36**	**92.76**	**94.98**	**95.52**	**93.6**	**94.93**	**96.93**	**94.69**	**94.81**	**94.86**	**94.07**	**99.76**	**100.17**	**100.94**
Oxygen basis	11	11	11	11	11	11	11	11	11	11	11	2	6	6
# Si	2.830	2.840	2.769	2.711	2.676	2.701	2.745	3.166	3.136	3.165	3.109	0.000	0.002	0.001
# Ti	0.088	0.077	0.093	0.154	0.139	0.158	0.148	0.041	0.023	0.041	0.011	0.994	1.980	2.003
# Al	1.399	1.434	1.446	1.487	1.521	1.489	1.500	2.432	2.495	2.456	2.551	0.000	0.000	0.001
# Cr	0.000	0.006	0.006	0.003	0.004	0.000	0.000	0.002	0.000	0.000	0.001	0.002	0.002	0.002
# Mg	1.691	1.707	1.692	0.861	0.882	0.905	0.854	0.125	0.118	0.110	0.144	0.000	0.004	0.008
#Fe	0.876	0.894	0.926	1.648	1.668	1.618	1.602	0.283	0.294	0.271	0.266	0.007	1.929	1.928
# Ca	0.002	0.023	0.002	0.000	0.001	0.000	0.000	0.000	0.000	0.000	0.000	0.000	0.017	0.002
# Mn	0.003	0.006	0.000	0.009	0.007	0.006	0.003	0.002	0.000	0.005	0.000	0.000	0.084	0.049
# Ni	0.000	0.000	0.000	0.000	0.000	0.000	0.001	0.000	0.000	0.000	0.000	0.000	0.000	0.000
# Na	0.019	0.020	0.013	0.009	0.017	0.012	0.009	0.036	0.032	0.036	0.032	0.001	0.000	0.002
#K	0.969	0.732	0.943	1.027	1.031	1.029	0.998	1.013	1.025	0.999	1.008	0.000	0.001	0.000
Cation	7.876	7.739	7.89	7.909	7.946	7.917	7.860	7.100	7.122	7.084	7.124	1.005	4.018	3.996
X Mg	0.659	0.656	0.646	0.343	0.346	0.359	0.348							
Fe^{3+}													0.020	0.000
Fe^{2+}													1.909	1.928
Pyrophanite													0.042	0.025
Hematite													0.010	0.000
Ti- muscovite								0.021	0.011	0.020	0.006			
Phlogopite-annite								0.000	0.000	0.000	0.000			
Celadonite								0.183	0.195	0.171	0.199			
Pyrophyllite								0.238	0.236	0.241	0.240			
Paragonite								0.018	0.016	0.018	0.016			
Margarite								0.000	0.000	0.000	0.000			
Muscovite								0.540	0.542	0.550	0.539			
K/(K+Ca+Na)								0.966	0.970	0.965	0.969			

*Composition used in C-space.

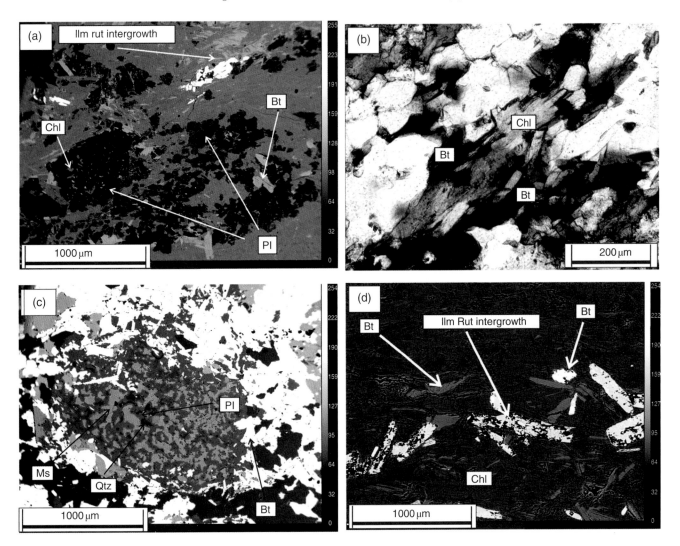

Fig. 10.12. (a) Successive replacement of plagioclase to chlorite to biotite is seen in a few places. Stumpy biotite grains often replace both chlorite and albite. (b) Coarse biotite grains contain patches of chlorite, which has replaced chlorite to varying degrees. (c) Along with biotite, muscovite replaces both K-feldspar and plagioclase porphyroclasts in the phyllosilicate rich bands. (d) Clusters of randomly oriented euhedral ilmenite crystals, variably replaced by rutile. Allanite, a common accessory mineral, developed near the Fe-Ti oxide phases.

modeling studies integrating the observed textures and the compositions of the phases (as in Torres-Roldan et al. 2000; Lang et al. 2004; Sengupta and Dasgupta 2009). The computer program C-space (Torres-Roldan et al. 2000) was used to solve the compositional matrix. Sengupta and Dasgupta (2009) and Chowdhury et al. (2013) described the procedure in detail.

Modelling studies return the following stoichiometrically balanced reactions:

1 a) Granite (K-feldspar-oligoclase-quartz) to chlorite schist (chlorite + albite)

$195.7122 \, \text{Kfs} + 41.4760 \, \text{Olig} + 28.1328 \, \text{SiO}_2 + 8.0000 \, \text{H}_2\text{O} + 183.6549 \, \text{Na} + 4.3987 \, \text{Fe} + 5.8520 \, \text{Mg} = 240.7214 \, \text{Ab} + 1.0000 \, \text{Chl} + 194.6281 \, \text{K} + 3.9508 \, \text{Ca}$

$\Delta Vs = -166 \, \text{J bar}^{-1} \, \text{mol}^{-1}$

ΔVs is change of volume of rock mass.

b) Chlorite to biotite in chlorite-biotite schist

$147.1501 \, \text{Chl} + 45.1289 \, \text{Ilm} + 805.8119 \, \text{SiO}_2 + 10.4263 \, \text{Na} + 530.9592 \, \text{K} + 64.1830 \, \text{Mg} = 547.1930 \, \text{Bt} + 627.0597 \, \text{H}_2\text{O} + 102.5758 \, \text{Fe} + 1.0000 \, \text{Ca}$

$\Delta Vs = 333 \, \text{J bar}^{-1} \, \text{mol}^{-1}$

2 Granite (K-feldspar-oligoclase) to biotite-muscovite schist (biotite + muscovite + albite)

$40.8580 \, \text{Kfs} + 53.9388 \, \text{Olig} + 1.0000 \, \text{Ilm} + 41.1986 \, \text{H}_2\text{O} + 13.3651 \, \text{Fe} + 15.6861 \, \text{Mg} = 15.0186 \, \text{Ab} + 7.0577 \, \text{Bt} + 34.1409 \, \text{Ms} + 95.5203 \, \text{SiO}_2 + 27.5686 \, \text{Na} + 14.9877 \, \text{Ca}$

$\Delta Vs = -137 \, \text{J bar}^{-1} \, \text{mol}^{-1}$

10.9 MASS TRANSFER

10.9.1 Theory

Gresens (1967) composition-volume relation forms the basis of quantification of mass change during fluid-rock interaction. The theory of Gresens (1967) was subsequently modified by a number of workers (Grant 1986; Ague 1994; Ague and Van Haren 1996; Philpots and Ague 2009; Bucholz and Ague 2010 and references therein). The equation used here to quantify mass changes follow Ague (1994), Ague and Van Haren (1996), and Philpotts and Ague (2009), and references therein. The gain and loss of elements during the metamorphism/metasomatism of granite to deformed granite will change the rock mass. Some elements may have been gained but these gains were outweighed by the losses of other elements. (Bucholz and Ague 2010).

Ague (1994) defined a term, T_i to calculate the fraction of rock mass change during metasomatic alteration. Total rock mass change was determined using a reference immobile species i by the equation:

$$T_i = \left[\frac{\text{Final mass} - \text{Initial mass}}{\text{Initial mass}} \right] = \left[\frac{C°i}{C'i} - 1 \right] \quad (1)$$

where i is a reference species, $C°i$ and $C'i$ are the initial (protolith) and final (altered state) concentration of i (Ague, 1994). Negative and positive T_i values denote mass loss and mass gain respectively. Changes in the masses of any mobile species j can be quantified by using the transport function τ of Brimhall et al. (1988). The mass changes of j computed using the reference species i, denoted as $^j\tau_i$ can be written as (Ague 1994; Philpotts and Ague 2009):

$$^j\tau_i = \left[\frac{\text{Final mass j} - \text{Initial mass j}}{\text{Initial mass j}} \right] = \left[\left(\frac{C°i}{C'i} \right) \left(\frac{C'j}{C°j} \right) - 1 \right] \quad (2)$$

Volume strain (ε_i) during chemical alteration of a rock can be computed as a function of the concentration of a reference immobile species i, by the equation (cf. Philpotts and Ague 2009):

$$\varepsilon_i = \left[\frac{\text{Final volume} - \text{Initial volume}}{\text{Initial volume}} \right] = \left[\left(\frac{C°i}{C'i} \right) \left(\frac{\rho°}{\rho'} \right) - 1 \right] \quad (3)$$

where $\rho°$ and ρ' are bulk densities of initial rock and final rock respectively at the time of metasomatism.

The graphical technique of Ague (2003) was followed to depict loss or gain of different chemical species during fluid–rock interaction. In this scheme, ratio (CR) of concentration of each element in altered (C') and pristine rocks ($C°$), has been plotted along the Y-axis; and different species, elements, and oxides are plotted along the X-axis (Figs 10.13 and 10.14). Two pair of rocks, a least deformed granite and phyllosilicate-rich bands, are chosen to test the chemical behavior of different species during fluid–rock alteration (Table 10.3). According to Equation 1, the magnitude of CR of any

immobile species will indicate loss (for CR > 1) or gain (CR < 1) of rock mass (T_i). To calculate the loss or gain of mobile species, at least one immobile species should be identified (Equation 2). There is no prior criterion that can identify an immobile species, as every element has finite solubility in metasomatic fluids under certain physicochemical condition (Manning 2007). However, several studies have demonstrated that Zr and Al are nominally insoluble in a wide range of crustal and mantle fluids and hence, these species can be used as reference for geochemical modeling (cf. Philpotts and Ague 2009). If this is the case, then CR of Zr and Al_2O_3 will have similar values (see Equation 1). Within the analytical uncertainties, the CR of Zr and Al_2O_3 in the two pairs of altered and unaltered rocks, chosen for the modeling, show very similar values (~0.9 in both the sets, Figs 10.13 and 10.14). Observing this, we considered Zr and Al_2O_3 to be immobile species and a horizontal line that passes through the two CR values is taken as reference frame (Figs 10.13 and 10.14). CR of elements that are plotted above the reference line will indicate gain in concentrations during the genesis of phyllosilicate-rich layers. To calculate the volume strain (ε_i), we have computed the densities of granite and phyllosilicate band with the computer program PERPLEX 07 (Connolly 2005) run at the inferred pressure–temperature (*P–T*) conditions of metasomatism.

10.9.2   Results

A CR value of ~0.9 for immobile Zr and Al_2O_3 translates to a gain of rock mass by ~11% (Equation 1). This follows that formation of phyllosilicate-rich bands did not significantly alter the rock mass. Combining the density data of granite and phyllosilicate bands with the CR of Zr, volume strain is calculated using Equation 3. The result shows small volume strain ~10% (#PP1a and b) and ~4% (for #PP2a and b). The X–Y diagram was contoured for different percentage change, using CR of Zr and Al_2O_3 as the reference line. Figures 10.13 and 10.14 show that, in both sets, MgO, P_2O_5, TiO_2, Fe_2O_3 (total Fe calculated in trivalent state), LOI, Ni, Cu, Ba, V, and Sr show gains whereas K_2O, Pb, Th, U, Y, and Nb show losses (Figs 10.13 and 10.14). CaO and Eu record gain and loss in #PP1 and #PP2, respectively (Figs 10.13 and 10.14). Significant gain in Ce is noted in #PP2 (Fig. 10.14). Marked gains in MgO, Fe_2O_3, TiO_2, and LOI and losses in K_2O are consistent with extensive replacement of K-feldspar and plagioclase by chlorite, biotite, and Fe-Ti oxides. A gain in Ce and total rare earth elements (REE) in #PP2 can be explained by development of allanite, a sink of Ca and REE. The behaviors of CaO and Eu are not properly understood. However, the presence of calcite in variable proportion in the two phyllosilicate bands may explain the anomaly, at least partially.

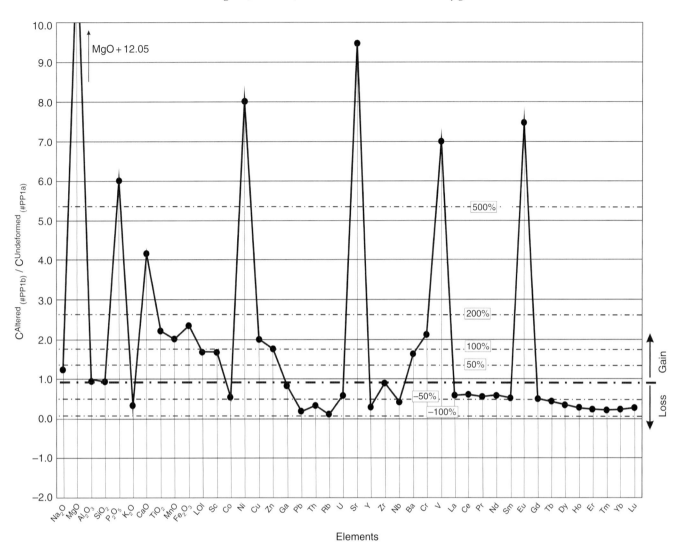

Fig. 10.13. Concentration ratio (CR) diagrams for phyllosilicate bands along the profile relative to least deformed granite. Geochemical reference frame denoted by thick dashed line. MgO, P_2O_5, TiO_2, Fe_2O_3, LOI, Ni, Cu, Ba, V, and Sr show gains whereas K_2O, Pb, Th, U, Y, and Nb show loss. CaO records gain.

10.10   PHYSICAL CONDITIONS OF SHEARING

Mineral assemblages in the deformed granite are not suitable for precise quantitative geothermobarometry. Nevertheless, coexistence of muscovite, biotite, and chlorite and absence of K-feldspar suggest that the metamorphic condition was in the biotite grade and that the reaction K-feldspar + chlorite → biotite + muscovite + H_2O were driven to the right. Precise temperature at which this reaction operates in the KFMASH (K_2O-FeO-MgO-Al_2O_3-SiO_2-H_2O) system is not well understood and studies showed that kinetics play an important role in destabilizing K-feldspar + chlorite and formation of biotite + muscovite (reviewed in Spear 1993). However, many studies have demonstrated that the reaction occurs within the stability of pyrophyllite and hence, the

metamorphic assemblage of the studied granite is likely to exceed 400°C. Non KFMASH components (e.g. CaO, Na_2O, MnO, TiO_2, etc.) is expected to change the stability field of K-feldspar + chlorite. To constrain the optimum *P–T* stability field of the assemblage muscovite + biotite + chlorite + albite + oligoclase + quartz, common in the mylonitic granite, a phase diagram is computed numerically using a representative bulk composition (Fig. 10.15). The phase diagram was computed using the computer program PERPLEX (Connolly 2005, updated in January 2012) along with thermodynamic data of Holland and Powell (1998, updated in 2002). Solid solution models in different minerals are taken from Holland and Powell (1998, updated in 2002). The computational detail is presented in Sengupta and Dasgupta 2009, Sengupta et al. (2009). The computed phase diagram is

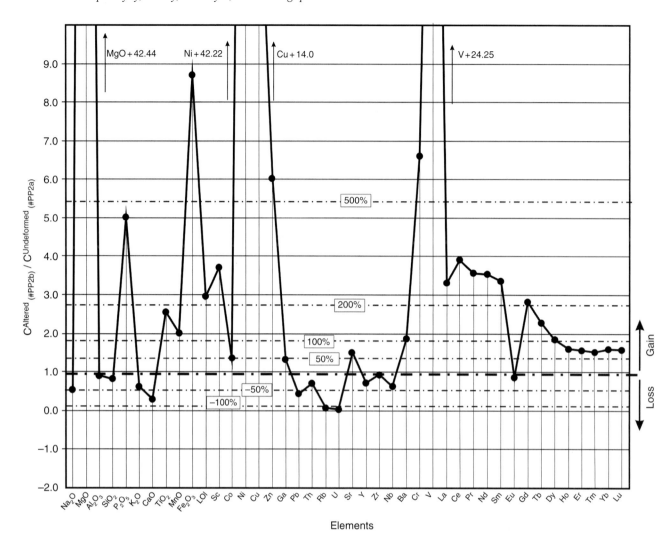

Fig. 10.14. Concentration ratio (CR) diagrams for phyllosilicate bands along the profile relative to least deformed granite. Geochemical reference frame denoted by thick dashed line. In both Figs. 13 and 14 MgO, P_2O_5, TiO_2, Fe_2O_3, LOI, Ni, Cu, Ba, V, and Sr show gains whereas K_2O, Pb, Th, U, Y and Nb show loss. Eu records loss.

shown in Fig. 10.15. Figure 10.15 suggests that stability of the assemblage muscovite + biotite + chlorite + albite + oligoclase + quartz is strongly temperature dependent. The constructed pseudosection fixes the temperature for this assemblage in the range of 470 ± 50 C at ~ 4 kbar. The estimated temperature range changes a little for a large variation of pressure (±2 kbar). The estimated temperature agrees well with the temperature at which feldspar deforms crystal–plastically by dislocation creep (Vernon 2004; Passchier and Trouw 2005 and references). This also follows that shear heating (Mukherjee and Mulchrone 2013) was not significant to influence the stability of minerals. Unfortunately, the aforesaid mineral assemblage does not provide any clue about the pressure of metamorphism. Considering a geothermal gradient of 25–30°C km⁻¹, common in fold-thrust belts (Winter 2001), 18–20 km depth of formation is suggested for the exposed section of mylonite. Mica schist of the study area is devoid of garnet and staurolite.

However, kyanite-quartz segregations were reported from the neighboring Haripaldih area. At the inferred temperature of 470 ± 50°C, pressure should exceed 4 kbar (~15 km lithostatic pressure). It therefore seems reasonable that the SPSZ exposes 18–20 km deep middle continental crust.

10.11   DISCUSSION

The porphyritic granite of the studied area underwent intense textural, mineralogical, and chemical changes, in response to ductile shear along the SPSZ. Transformation of weakly porphyritic granite (euhedral simple twinned K-feldspar) to augen gneiss to fine-grained granite mylonite is attributed to progressive increase in the strain intensity. Origin of augen gneiss is a debated issue. Opinions vary from syn-deformation growth of feldspar (e.g. Dickson 1996) to residual magmatic phenocryst

Table 10.3. Bulk Composition of major oxides (wt%) and trace elements and REE (ppm) used for mass balance calculations

Sample Description

		Na$_2$O	MgO	Al$_2$O$_3$	SiO$_2$	P$_2$O$_5$	K$_2$O	CaO	TiO$_2$	MnO	Fe$_2$O$_3$	LOI
PP1a	Least deformed granite	2.95	0.17	14.06	76.96	0.04	5.25	0.27	0.10	0.01	1.43	0.88
PP1b	Phyllosilicate bands	3.62	2.05	13.01	71.59	0.24	1.67	1.12	0.22	0.02	3.34	1.47
PP2a	Least deformed granite	2.62	0.23	12.31	77.80	0.01	5.59	0.91	0.11	0.02	0.93	1.12
PP2b	Phyllosilicate bands	1.42	9.76	11.05	64.19	0.05	3.47	0.27	0.28	0.04	8.08	3.30

		Sc	Co	Ni	Cu	Zn	Ga	Pb	Th	U	Rb	Sr	Y	Zr	Nb	Ba	Cr	V
PP1a	Least deformed granite	4.9	81.6	1	1	8	16	11	104	20.7	394	1.9	78	180	29	264	9	3
PP1b	Phyllosilicate bands	8.2	42.5	8	2	14	13	2	33	11.9	39	18	22	160	12	429	19	21
PP2a	Least deformed granite	4.6	77.5	0.9	3	5	15	12	90	19.8	282	8	68	167	25	217	10	4
PP2b	Phyllosilicate bands	17	105	38	42	30	19.79	5.2	62.86	0.59	20	12	48	153	15.6	403	66	97

		La	Ce	Pr	Nd	Sm	Eu	Gd	Tb	Dy	Ho	Er	Tm	Yb	Lu
PP1a	Least deformed granite	97	185	20	69	13.6	0.15	14.96	2.25	12.03	2.59	6.81	1.1	6.05	0.93
PP1b	Phyllosilicate bands	56	111	11	40	6.96	1.12	7.51	0.97	4.16	0.69	1.59	0.23	1.36	0.25
PP2a	Least deformed granite	33	59	7	26	5.09	2.09	5.78	0.92	5.09	1.14	2.93	0.45	2.36	0.38
PP2b	Phylosilicate bands	109	230	25	92	17.1	1.78	16.2	2.09	9.4	1.81	4.55	0.68	3.74	0.6

* Bulk compositions are calculated with an XRF ( ) for major and trace and ICP-MS (PerkinElmer SCIEX ELANDRC-e) for REE from Wadia Institute of Himalayan Geology. Dehradun, India. For details of methodology please refer to Khanna et.al. 2009. Himalayan Geology 30(1), 95–99.

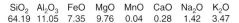

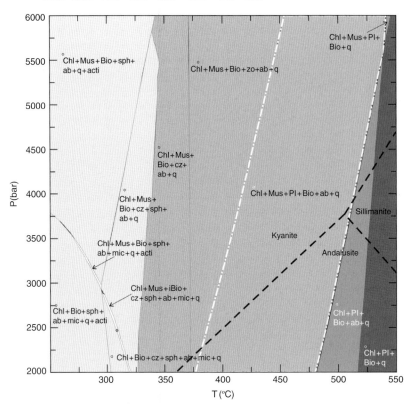

Fig. 10.15. The numerically computed phase diagram using a representative bulk composition of the deformed granite suggests a *P–T* window of 470 ± 50°C at ~4 kbar from the stability of this assemblage.

(cf. Vernon 2004). Progressive transformation of porphyritic granite to augen gneiss supports, as seen in this study, the second option. This is further corroborated by the formation of a mantle of recrystallized grains around relict clast and its internal deformation. Ductile shearing in the studied area is manifest by extensive grain size reduction of quartz and feldspar porphyroclasts and various shear sense indicators such as, imbrications of the segmented porphyroclasts (Fig. 10.8c), σ-type tails of recrystallized feldspar porphyroclasts (Fig. 10.5b), oblique foliation of quartz lens (Fig. 10.5c), vergence of asymmetric folds (Fig. 10.3d), asymmetry of detached quartz clasts in phyllosilicate-rich bands (Fig. 10.3a), and development of S-C fabric (Fig. 10.10d)

The shear markers connote a monotonous sinistral shear. Consistency of shear sense together with extensive recrystallization of porphyroclasts is considered to be the hallmark of regional shear zones (reviewed in Vernon 2004). The sense of shear movement is consistent with a south-over-north reverse movement along the south dipping planes of the SPSZ. Presence of down-dip mineral lineation suggests a contraction-dominated deformation regime with overriding of rocks of NSFB over rocks of the CGGC. Vergence of thrusting along the SPSZ has been debated. However, most workers support that the rocks of CGGC thrust over the rocks of NSFB along N dipping thrust planes (Dun and Dey 1942; Bhattacharya 1992; Acharya et al. 2006, and references therein). In contrast,

Mahato et al. (2008) reported S dipping thrusting with NSFB overriding the CGGC. These workers postulate that initially the NSFB subducted beneath the CGGC (along N-directed thrust). Subsequently, during exhumation, both N and S dipping thrust developed at the interface of the two blocks with S dipping thrust plane at the southern extremity of the CGGC (fig.18 of Mahato et al. 2008). In many areas where reverse shear sense was observed, reactivated hook fabrics are found (Wennberg 1996; Mukherjee 2007, 2010a,b, 2013; Mukherjee and Koyi 2010). This is not found in the present study area. A critical evaluation of the existing opinions regarding the direction of thrusting – north-over-south and vice versa – is beyond the scope of this study. Such reverse shear senses have been noted from different terrains (e.g. Pace et al. 2016 (Chapter 8); Pamplona et al. 2016 (Chapter 13); Sengupta and Chatterjee 2016 (Chapter 9)).

Infiltration-driven metamorphism that accompanied ductile shearing in the SPSZ converted the mylonitic granite to mica-chlorite schist. Syntectoinc quartz veining also supports infiltration of aqueous fluid in the shear zone (Yardley et al. 2000; Philpotts and Ague 2009). Restriction of phyllosilicate-rich bands in centimeter- to decimeter-thick layers suggests that fluid ingress was extremely channelized. Similar channelized flow of aqueous fluids and concomitant metasomatic alteration has been reported from many ductile shear zones (reviewed in Vernon 2004). Stoichiometrically balanced reactions from textural modeling studies suggest large negative solid volume changes for reaction 1a (chlorite + albite forming) and reaction 2 (muscovite-biotite forming). Operation of these reactions, therefore, enhances porosity and helps extraneous fluids to infiltrate and metasomatize the host rocks (reaction-enhanced permeability). Reaction 1b (chlorite+albite → biotite), on the other hand, has positive solid volume change. Operation of this reaction, therefore, diminishes the existing porosity and hence hinders the biotite growth. Preponderance of biotite in phyllosilicate-rich bands requires that porosity of the rock to be created continuously by some other means that facilitate continuous fluid ingress (Yardley et al. 2000 and Vernon 2004). Reviewing the existing literature of Vernon (2004) demonstrates that in hydrated shear zones, grain size reduction, together with transformation of feldspar to phyllosilicate minerals, causes "reaction softening" due to changes in the rheology of the rock. Infiltrated fluids promote micro cracking if fluid pressure exceeds the minimum normal compressive stress and tensile strength of the rock. These microcracks (Fig. 10.8d) and preferred orientation of the grains of phyllosilicate minerals (Fig. 10.4d) provide access to the extraneous fluids and drive hydration reactions (reviewed in Vernon 2004). Deformation enhances exchange of chemical components and hence expedites the chemical and isotopic exchange in rocks (Yund and Tullis 1991). These cross-feedback mechanisms, as has happened in the SPSZ, play a critical role in evolution of many ductile shear zones. Textural modeling of the phyllosilicate-rich bands and mass-balance calculations suggest that chemical potential of a large number of elements were controlled by infiltrating fluids (open system behavior). Losses and gains of different chemical species, as predicted from the mass-balance calculation, nicely correspond to the minerals present in the phyllosilicate bands. Very similar CRs of Al_2O_3 and Zr support that the theory that these two species behaved as immobile components. Mass-balance calculations also support that the metasomatic process that led to phyllosilicate-rich bands induced ~11% change in rock mass. Volume strain during metasomatic alteration as calculated from chemical parameters is low (≤10%). Quantitative geothermometry, together with a numerically computed phase diagram (pseudosection), and ductile behavior of plagioclase porphyroclast, constrain that shearing and concomitant metasomatism in the studied part of the SPSZ occurred at 18–20 km depth where the attendant temperature was ~470 ± 50°C. The estimated pressure and temperature indicates that the SPSZ is a crustal-scale shear zone. A geothermal gradient of 25–30°C km^{-1}, reported many from fold belts (cf. Philpotts and Ague 2009) explains the estimated pressure and temperature. In absence of geochronologic data, timing of shearing in the SPSZ is indeterminate. Nevertheless, in an adjoining area, 0.93 Ga old suites of alkaline rocks are sheared (Reddy et al. 2009 and Chatterjee et al. 2013).

Thus, a Neoproterozoic tectono-thermal event along the SPSZ is understood.

ACKNOWLEDGMENTS

N.C. and S.R. acknowledge financial assistance from Department of Science and Technology, (New Delhi) in terms of a Women Scientist Program: SR/WOS-A/ES-12/2009, SR/WOS-A/ES-32/2013 and the Council of Scientific and Industrial Research (New Delhi). Research grants from UPE II, Jadavpur University supported the work of S.S. and P.S. We thank Sayantan Sarkar, Maitrayee Chakraborty, and Sayan Biswas for help in the field. We thank the anonymous reviewer for critical constructive suggestions that enhanced the clarity. Thanks to Soumyajit Mukherjee for inviting us to contribute in this volume and for reviewing this work. Support by Ian Francis, Delia Sandford, and Kelvin Matthews (Wiley Blackwell) is appreciated.

ABBREVIATIONS

Abbreviation used in the chapter for mineral names and list of abbreviations for mineral names mainly follows the list published by Kretz (1983) with recommendations by the IUGS Sub-commission on the systematics of metamorphic rocks (SCMR). The abbreviations are as follows:

Alanite	Aln
Albite	Ab
Biotite	Bt
Chlorite	Chl
Feldspar	Flds
Ilmenite	Ilm
K-feldspar	Kfs
Muscovite	Ms
Oligoclase	Olig (Not in Kretz 1983)
Plagioclase	Pl
Quartz	Qtz
Rutile	Rt

REFERENCES

Acharyya A, Roy S, Chaudhuri BK, Basu SK, Bhaduri SK, Sanyal SK. 2006. Proterozoic rock suites along south Purulia shear zone, eastern India: Evidence for rift related setting. Journal of Geological Society of India 68, 1069–1086.

Ague JJ, van Haren JLM. 1996. Assessing metasomatic mass and volume changes using the bootstrap, with application to deep-crustal hydrothermal alteration of marble. Economic Geology 91, 1169–1182.

Ague JJ. 1994. Mass transfer during Barrovian metamorphism of pelites, south-central Connecticut. I: evidence for changes in composition and volume. American Journal of Science 294, 989–1057.

Bhattacharyya DS. 1992. Early proterozoic metallogeny, tectonics and geochronology of the singhbhum Cu-U belt, eastern India. Precambrian Research 58(1), 71–83.

Brimhall GH, Lewis CJ, Ague JJ, et al. 1988. Metal enrichment in bauxites by deposition of chemically mature Aeolian dust. Nature 333, 819–824.

Bucholz CE, Ague JJ. 2010. Fluid flow and Al transport during quartz-kyanite vein formation,Unst, Shetland Islands, Scotland. Journal of Metamorphic Geology 28, 19–39.

Chakrabarty A, Sen AK. 2010. Enigmatic association of the carbonatite and alkalipyroxenite along the Northern Shear Zone, Purulia, West Bengal: a saga of primary magmatic carbonatite. Journal of the Geological Society of India 76, 403–413.

Chatterjee P, De S, Ranaivoson M, Mazumder R, Arima M. 2013. A review of the w1600 Ma sedimentation, volcanism, and tectono-thermal events in the Singhbhum craton, Eastern India. Geoscience Frontiers 4, 277–287.

Chowdhury P, Talukdar M, Sengupta P, Sanyal S, Mukhopadhyay D. 2013. Control of P-T path elements mobility on the formation of corundum pseudomorphs in palaeoproterozoic high-pressure anorthosite from Sittampundi, Tamilnadu, India. American Mineralogist 98, 1725–1737.

Connolly J. 2005. Computation of phase equilibria by linear programming: A tool for geodynamic modeling and its application to subduction zone decarbonation. Earth and Planetary Science Letters 236, 524–541.

Dickson FW. 1996. Porphyroblasts of barium-zoned K-feldspars and quartz, Papoose Flat, Inyo Mountains, California, genesis and exploration implications. In Geology and ore deposits of the American Cordillera, edited by A.R. Cooper and P.L. Fahey, Geological Society of Nevada, Reno, pp. 909–924.

Dunn JA, Dey AK. 1942. The geology and petrology of eastern Singhbhum and surrounding areas. Memoir. Geological Survey of India 69, 281–450.

Grant JA. 1986. Theisocon diagram – a simple solution to Gresens_ equation for metasomatic alteration. Economic Geology 81, 1976–1982.

Gresens RL. 1967. Composition–volume relations of metasomatism. Chemical Geology 2, 47–65.

Harlov DE, Austrheim H. 2013. Metasomatism and the Chemical Transformation of Rock. Lecture notes in Earth System Sciences, Springer-Verlag, Berlin, Heidelberg.

Hey MH. 1954. A new review of the chlorites. Mineralogical Magazine 30, 277–292.

Holland TJB, Powell R. 1998. An internally consistent thermodynamic data set for phases of petrological interest. Journal of Metamorphic Geology 16, 309–343.

Kretz R. 1983. Symbols for rock-forming minerals. American Mineralogist 68(1–2), 277–279

Lang HM, Wachter AJ, Peterson VL, Ryan JG. 2004. Coexisting clinopyroxene/spinel and amphibole/spinel symplectites in metatroctolites from the Buck Creek ultramafic body, North Carolina Blue Ridge. American Mineralogist 89, 20–30.

Mahadevan TM. 1992. Geological evolution of the Chotonagpur gneissic complex in part of Purulia district, West Bengal. Indian Journal of Geology 64, 1–22.

Mahato S, Goon S, Bhattacharya A, Misra B, Bernhardt HJ. 2008. Thermotectonic evolution of the north Singhbhum mobile belt: a view from the western part of the belt. Precambrian Research 162, 102–107.

Manning CE. 2007. Solubilty of corundum+kyanite in H_2O at 700°C and 10 kbar: evidence for Al-Si complexing at high pressure temperature. Geofluids 7, 258–269.

Mukherjee S. 2007. Geodynamics, deformation and mathematical analysis of metamorphic belts of the NW Himalaya. PhD Thesis, Indian Institute of Technology, Roorkee, pp. 1–267.

Mukherjee S. 2010a. Structures in meso-and micro-scales in the Sutlej section of the Higher Himalayan Shear Zone, Indian Himalaya. e-Terra 7, 1–27.

Mukherjee S. 2010b. Microstructures of the Zanskar shear zone. Earth Science India 3, 9–27.

Mukherjee S. 2011a. Mineral fish: their morphological classification, usefulness as shear sense indicators and genesis. International Journal of Earth Sciences 100, 1303–1314.

Mukherjee S. 2011b. Flanking microstructures from the Zanskar Shear Zone, NW Indian Himalaya. YES Bulletin 1, 21–29.

Mukherjee S. 2012. Simple shear is not so simple! Kinematics and shear senses in Newtonian viscous simple shear zones. Geological Magazine 149, 819–826.

Mukherjee S. 2013. Deformation Microstructures in Rocks, Springer, Berlin.

Mukherjee S. 2014a. Atlas of Shear Zone Structures in Meso-Scale. Springer, Berlin.

Mukherjee S. 2014b. Review of flanking structures in meso-and micro-scales. Geological Magazine 151, 957–974.

Mukherjee S, Koyi HA. 2009. Flanking microstructures. Geological Magazine 146, 517–526.

Mukherjee S, Koyi HA. 2010. Higher Himalayan Shear Zone, Zanskar Indian Himalaya: microstructural studies and extrusion mechanism by a combination of simple shear and channel flow. International Journal of Earth Sciences 99, 1083–1110.

Mukherjee S, Mulchrone K. 2013. Viscous dissipation pattern in incompressible Newtonian simple shear zones – an analytical model. International Journal of Earth Sciences 102, 1165–1170.

Pace P, Calamita F, Tarvanelli E. 2016. Brittle-ductile shear zones along inversion-related frontal and oblique thrust ramps: Insights from the Central-Northern Apennines curved thrust System (Italy). In Ductile Shear Zones: From Micro- to Macro-scales, edited by S. Mukherjee and K.F. Mulchrone, John Wiley & Sons, Chichester, Chapter 8.

Pamplona J, Rodrigues BC, Llana-Fúnez S, Simões PP, Ferreira N, Coke C, Pereira E, Castro P, Rodrigues J. 2016. Structure and Variscan evolution of Malpica-Lamego Ductile Shear Zone (NW of Iberian Peninsula). In Ductile Shear Zones: From Micro- to Macro-scales, edited by S. Mukherjee and K.F. Mulchrone, John Wiley & Sons, Chichester, Chapter 8.

Passchier CW, Trouw RAJ. 2005. Microtectonics, 2nd edition. Springer, Berlin.

Philpotts A, Ague JJ. 2009. Principles of Igneous and Metamorphic Petrology, 2nd edition, Cambridge University Press, Cambridge, UK.

Ray S, Biswas S, Chakraborty M, Sanyal S, Sengupta P. 2012. Mass transport during fennitization of granite at the contact of carbonatite at Beldihi, Purulia, West Bengal, In: N.V. ChalapathiRao and K. Surya PrakashRao, National Seminar on recent advances and future challenges on Geochemistry and Geophysics:the Indian scenario, BHU, Abstract Volume, pp. 114.

Reddy SM, Clarke C, Mazumder R. 2009. Temporal constraints on the evolution of the Singhbhum Crustal Province from U-Pb SHRIMP data. In: D. Saha and R. Mazumder, Paleoproterozoic Supercontinents and Global Evolution, International Association for Gondwana Research Conference Series, Abstract volume, 9, pp. 17–18.

Rosenberg CL, Stuenitz H. 2003. Deformation and recrystallization of plagioclase along a temperature gradient; an example from the Bergell tonalite. Journal of Structural Geology 25, 389–408.

Sanyal S, Sengupta P. 2012. Metamorphic evolution of the Chotanagpur Granite Gneissic Complex (CGGC) of Eastern Indian Shield: current status. Geological Society, London, Special Publication on Palaeoproterozoic of India, 365, pp. 117–145.

Sengupta P, Dasgupta S. 2009. Modelling of metamorphic textures with C-space: evidence of pan-african high-grade reworking in the eastern ghat belt, India. Indian National Science Academy 1, 29–39.

Sengupta S, Chatterjee SM. 2016. Microstructural variations in quartzofeldspathic mylonites and the problem of vorticity analysis using rotating porphyroclasts in the Phulad Shear Zone, Rajasthan, India. In Ductile Shear Zones: From Micro- to Macro-scales, edited by S. Mukherjee and K.F. Mulchrone, John Wiley & Sons, Chichester, Chapter 8.

Spear FS. 1993. Metamorphic phase equilibria and pressure tem-perature time paths. Mineralogical Society of America, Washington, D.C.

Talukdar M, Chattopadhyay N, Sanyal S. 2012. Shear controlled fe-mineralization from parts of South Purulia Shear Zone. Journal of Applied Geochemistry 14, 496–508.

Torres-Roldan RL, Garcia-Casco A, Garcia-Sanchez P. 2000. CSpace: An integrated workplacefor the graphical and algebraic analysis of phase assemblages on 32-bit wintel platforms. Computers and Geosciences 26, 779–793.

Tullis J, Yund RA. 1980. Hydrolytic weakening of experimentally deformed Westerly granite and Hale albite rock: Journal of Structural Geology 2, 439–451.

Vernon HR. 2004. A Practical Guide to Rock Microstructure. Cambridge University Press, Cambridge.

Wennberg OP. 1996. Superimposed fabrics due to reversal of shear sense: an example from the Bergen Arc Shear Zone, western Norway. Journal of Structural Geology 18(7), 871–889

Winter JD. 2001. An introduction to Igneous and Metamorphic Petrology. Prentice-Hall Inc., New Jersey.

Yardley BWD, Gleeson S, Bruce S, Banks D. 2000. Origin of retro-grade fluids in metamorphic rocks. Journal of Geochemical Exploration 69–70, 281–285.

Chapter 11

Reworking of a basement–cover interface during Terrane Boundary shearing: An example from the Khariar basin, Bastar craton, India

SUBHADIP BHADRA[1] and SAIBAL GUPTA[2]

[1] *Department of Earth Sciences, Pondicherry University, R.V. Nagar, Kalapet, Puducherry 605014, India*
[2] *Department of Geology and Geophysics, Indian Institute of Technology Kharagpur, Kharagpur 721302, West Midnapore, West Bengal, India*

11.1 INTRODUCTION

Zones of tectonic convergence or divergence may both develop sedimentary basins (Watts 1992). In convergent zones, crustal loading by thrusting commonly flexes the lithosphere and forms foreland basins (Fowler 1990; Naylor and Sinclair 2008). On the other hand, rifting is an alternative important mechanism that may form basins of extensional origin (McKenzie 1978; Roberts and Bally 2012). Since thrusting and rifting are commonly associated with processes of continental collision and break-up, large sedimentary basins may sometimes be correlated with global-scale tectonic events. In such cases, the time–space evolution of an amalgamated assembly within a Precambrian shield, or of previously adjacent, now disintegrated landmasses currently located in geographically separated continents, can be inferred from the spatial and temporal evolution of these sedimentary basins.

Important in this respect are the Proterozoic sedimentary successions of Peninsular India, preserved in the Vindhyanchal-, Cuddapah-, Chhattisgarh-, Khariar-, Indravati-, Pranhita-Godavari, Bhima-, and Kaladgi basins (Fig. 11.1), that are collectively referred to as "Purana basins" (Holland 1906; Ramakrishnan 1987; Kale 1991; Chaudhuri et al. 1999; Ramakrishnan and Vaidyanadhan 2008). The association of these basins with the underlying basement rocks of the Bastar, Dharwar, and Bundelkhand craton make them particularly interesting, since the cratonic rocks are in juxtaposition with the lithologic ensemble of the Proterozoic Eastern Ghats Mobile Belt (EGMB) in an intensely tectonized set up. The evolutionary history of the polychronous, multiply deformed EGMB is marked by a peak Grenvillian age, granulite facies metamorphism (ca. 1000 Ma, Kelly et al. 2002; Mezger and Cosca 1999; see Gupta 2012 for review) that affected almost the entire lithologic ensemble of the northern segment of the granulite belt: this segment is referred to as the Eastern Ghats Province (Dobmeier and

Raith 2003). This was followed by upper-amphibolite to granulite facies reworking (Bhadra et al. 2003) related to the juxtaposition of the Eastern Ghats Province (EGP) with the cratonic nucleus of peninsular India during the Pan-African orogeny (ca. 550 Ma). The western boundary of the EGP has been described as a thrust (Gupta et al. 2000; Bhadra et al. 2004), although the northern boundary with the Singhbhum craton is more complex, the most recent interpretation being that of a thrust modified by Cambro-Ordovician dextral strike-slip shear (Lisker and Fachmann 2001; Misra and Gupta 2014). In a global tectonic perspective, the first granulite facies tectonothermal event has been correlated with assembly of the supercontinent Rodinia (Simmat and Raith 2008; Bose et al. 2011; Das et al. 2011a). The later tectonic reworking event in the EGP, on the other hand, is considered to mark the final assembly of a later supercontinent, the Gondwanaland (Bhadra et al. 2004; Biswal et. al. 2000; Nasipuri and Bhadra 2013). This later reworking was an outcome of intracontinental orogenesis (Nanda and Gupta 2012) followed by decompression to shallow crustal levels at relatively high temperatures (Nanda et al. 2014).

Interestingly, the evolutionary span of the Purana basins, within a time frame of 1700 to 700 Ma (Vinogradov et al. 1964; Crawford 1969; Crawford and Compston 1970, 1973; Kreuzer et al. 1977; Chaudhuri and Howard 1985; Murti 1987; Chaudhuri and Chanda 1991; Kale 1991), corresponds well with that of the polychronous, multiply deformed Proterozoic mobile belts that comprise the EGMB. Rapid variation in lithofacies, presence of stacked cyclothems (Patranabis-Deb and Chaudhuri 2002) and localized intense deformation along the margins of these basins (Kale 1991; Patranabis-Deb and Chaudhuri 2002; Chaudhuri and Chanda 1991; Murti 1987, Valdiya 1982; Meijerink et al. 1984; Jayaprakash et al. 1987) clearly reflect tectonic activity in their evolutionary history. However, whether this tectonism is discrete in its nature affecting the basins alone, or is an outcome of far-field stresses induced concomitant to the

Ductile Shear Zones: From Micro- to Macro-scales, First Edition. Edited by Soumyajit Mukherjee and Kieran F. Mulchrone.

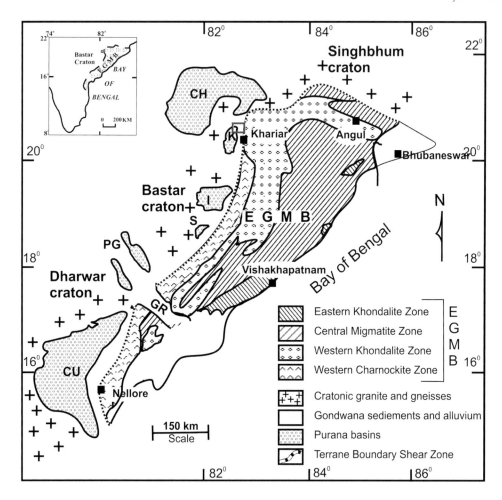

Fig. 11.1. Generalized geological map of India with disposition of Eastern Ghats Mobile Belt (EGMB) and adjoining cratons, and distribution of Purana basins of India; in EGMB. The location of the EGMB is demarcated in the inset map of India for reference. Red rectangle: the study area. Abbreviations: CH, Chhattisgarh basin; CU, Cuddapah basin; EGMB, Eastern Ghats Mobile Belt; GR, Godavari rift; I, Indravati basin; K, Khariar basin; PG, Pranhita-Godavari basin; S, Sukma basin. Adapted from Ratre et al. 2010.

juxtaposition of the cratonic basement with the mobile belt is yet to be deciphered.

Existing stratigraphic and sedimentological works on these basins supports their intracratonic origin (Patranabis-Deb and Chaudhuri 2002; Chaudhuri and Chanda 1991) related to basement rifting, and extensional features in terms of widespread faulting. However, additional complexities may arise due to the operation of two or more tectonic processes at different stages of evolution of the basin. Therefore, an integrated study of correlation of structures of the sedimentary formations within the basin as well as that of the provenance is important for understanding its tectonic evolution. Such an approach seems to be pertinent in the present context since it has been unanimously accepted, through palaeocurrent analyses (Murti 1987; Kale 1991) and lithofacies assemblages, that the basement gneisses were the provenance during deposition.

In this study, we attempt to correlate the deformation pattern across the cratonic basement and the metasedimentary cover of the Khariar basin (Fig. 11.1). Earlier data

on the Bastar craton-EGMB assembly (Bhadra et al. 2003, 2004; Nasipuri and Bhadra 2013) has also been integrated to investigate the effects of the far-field stress induced by the final stage of westward thrusting of the EGMB (more specifically, the EGP), if any, on the metasediments. The study is expected to reveal the evolutionary history of the Khariar Purana basins and the nature of basement–cover reworking during Terrane Boundary shearing.

11.2   GEOLOGICAL SETTING

A ~40 km long corridor, including the Eastern Ghats Mobile Belt (easternmost part of the study area), Bastar craton (central portion) and the Khariar metasedimentary basin (westernmost part) has been chosen in this study (Fig. 11.2). In the westernmost part of the study area, the metasediments of the Khariar basin comprising metaquartzites and slate (corresponding respectively with upper and lower standstone and shale units in Fig. 11.3)

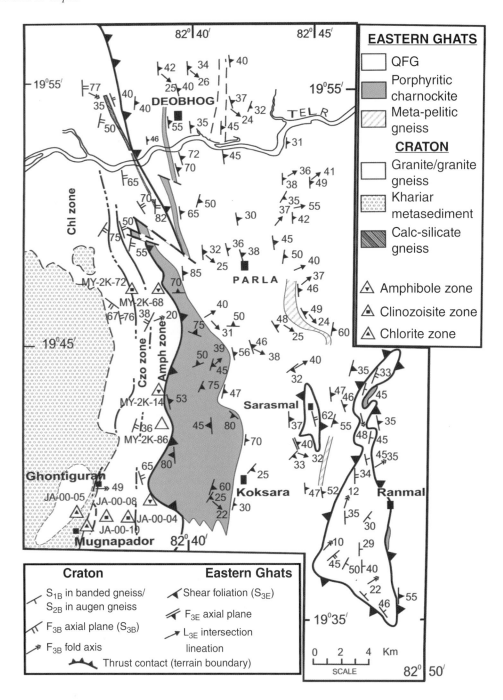

Fig. 11.2. Lithological and structural map of the study area demarcated in Fig. 11.1. Sample locations used for *P–T* estimates are shown for reference. Note waning temperature gradient in the foreland granite as manifested by the stabilization of different index minerals along the thrust-induced fabric. Adapted from Bhadra, 2003 and from Bhadra, et al. 2003.

occur as a N-S trending linear belt with high topographic relief. The metasediments also occur as isolated lenses (outliers) within the basement granites of the Bastar craton to the east (Fig. 11.2) forming plateaus. Further eastward, variably deformed and metamorphosed cratonic granites/granite gneisses (3500–3000 Ma, Sarkar et al. 1993), hitherto referred to as foreland granites, are in contact with the lithologic ensemble of the Eastern Ghats Mobile Belt. A thin, km-wide N-S trending band of

porphyritic charnockite separates the foreland granite from the EGMB lithologies that are dominated by orthopyroxene-bearing migmatitic quartzofeldspathic gneisses (QFG). The foreland granites in contact with the porphyritic charnockite are mylonitic. The only penetrative fabric within the foreland granite is characterized by the stabilization of different index metamorphic minerals, from which the metamorphic zonation across the foreland domain is inferred (Fig. 11.2). Cratonic rocks

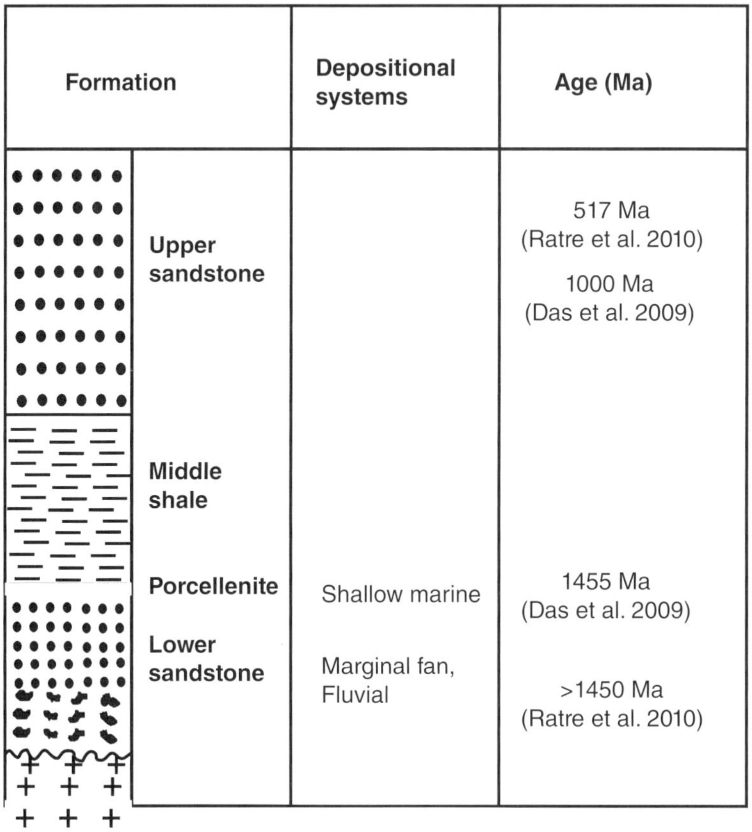

Formation	Depositional systems	Age (Ma)
Upper sandstone		517 Ma (Ratre et al. 2010) 1000 Ma (Das et al. 2009)
Middle shale		
Porcellenite	Shallow marine	1455 Ma (Das et al. 2009)
Lower sandstone	Marginal fan, Fluvial	>1450 Ma (Ratre et al. 2010)

Fig. 11.3. Generalized stratigraphic sequence of the Khariar metasediments. Reported ages of the sedimentary formations are given for reference. Adapted from Chakraborty et al. 2012.

also occur in two "windows" (Bhadra et al. 2004) within the main mass of the EGMB lithologies (Fig. 11.2), around the villages of Ranmal (Ranmal window) and Sarasmal (Sarasmal window). The QFG of the EGMB hosts distended bands of mafic granulites and metapelitic gneisses (including khondalites *sensu-stricto*, high Mg-Al cordierite-biotite-bearing metapelites, and sapphirine granulites), calc-silicate gneisses and several intrusives like metagabbros and porphyritic charnockites.

11.3   DEFORMATION PATTERN AND FABRIC RELATIONSHIP

11.3.1   Khariar metasediments

The metasediments have been studied in detail in two places, around the villages of Mugnapador and Ghontigurah (Fig. 11.2). In both places, the metasediments preserve evidence of intense deformation in contact with the cratonic granites. A detailed outcrop-scale map (Fig. 11.4a) from the Mugnapador area reveals repeated folding, (isoclinal followed by asymmetric) of the bedding plane that contrasts sharply with the barely perceptible foliation in the underlying granites. The asymmetric folds are open in nature with westward vergence and top-to-west shear sense (Fig. 11.4b), associated with an easterly dipping axial planar slaty cleavage (010°/60°E). In Ghontigurah, long limbs of the asymmetric folds in the metasediments

(shale) are strongly attenuated and almost indistinguishable from the axial planar slaty cleavage. Also, compared to the barely perceptible fabric of the basement granite in Mugnapador, the axial planar slaty cleavage in the metasediments at Ghontigurah parallels the penetrative mylonite foliation in the underlying basement. In microscopic scale, the slaty cleavage defines S-C structures with a top-to-west shear (Fig. 11.4c), which is conformable with macroscopic (Fig. 11.5) and microscopic (discussed below) shear sense indicators in the foreland granites.

Poles to the bedding planes of the metaquartzite at Mugnapador describe a well-defined girdle (Fig. 11.6a). This is fairly consistent with the observed open nature (Fig. 11.4b) of the hinges of the asymmetric folds. The three clusters in the stereoplot (Fig. 11.6a) correspond to data from the two limbs and the hinge. Though the attitude of the axial plane could not be measured directly due to the open nature of the late asymmetric folds, the attitudes of the fold limbs (Fig. 11.6a) suggest a NE–SW trend of the axial plane. The latter corresponds well with the distribution of bedding plane *vis-à-vis* axial planar cleavage of the shale unit at Ghontigurah, which also describes a well-defined girdle (Fig. 11.6b) with a northeasterly plunging β-axis (37°→057°). At Ghontigurah, the attitude of the slaty cleavage (Fig. 11.6b) is conformable with that of the penetrative fabric in the basement granites (Fig. 11.6c). Contrasting deformation styles, e.g. easily identifiable mesoscopic folds in one place versus absence of such

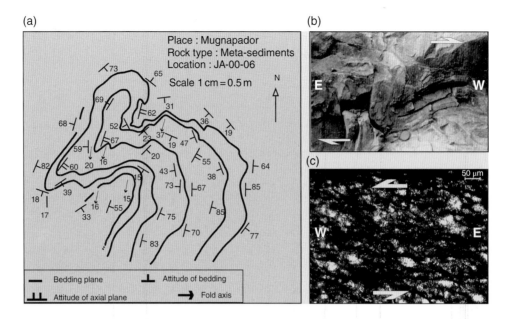

Fig. 11.4. (a) Outcrop-scale map of the mesoscopic fold on the bedding plane of the metaquartzite at Mugnapador. (b) West-verging asymmetric fold on the bedding plane of metaquartzite at Mugnapador. Photograph facing north. Exposure length: 1 m, width: 0.8 m. (c) S-C structure with top-to-the west shear sense in the slaty cleavage of the Middle shale unit (Fig. 11.3) from Ghontigurah (Fig. 3b of Bhadra et al. 2004, Reproduced with permission of Elsevier).

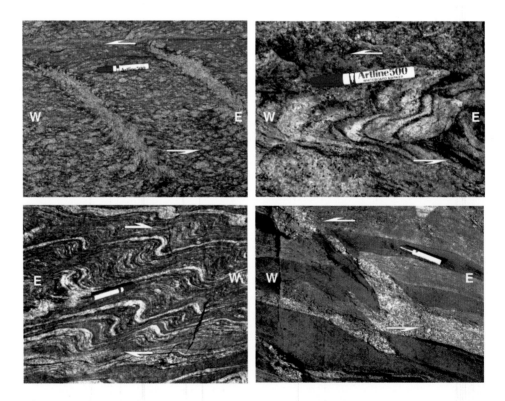

Fig. 11.5. Field photograph showing consistent top-to-the west shear sense in the deformed cratonic granites from: (top left) interior foreland, (top right) intermediate, and (bottom) frontal foreland zone. Note: cratonic granites are gneissic only in the frontal zone (near Terrain Boundary Shear Zone). Foliation spacing and tightness of the west-verging asymmetric folds decreases away from the thrust contact. Marker length: 15 cm.

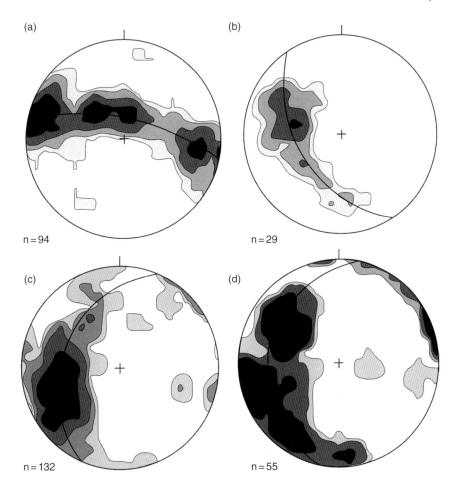

Fig. 11.6. Stereographic projection of pole to the (a) bedding plane in metaquartzite at Mugnapador. Contour interval: 1%, 2%, 4%, 8%, 16%. Calculated girdle 102°/73°N. (b) bedding plane/axial planar cleavage in shale at Ghontigurah. Contour interval: 3%, 6%, 12%, 24%. Calculated girdle, b-axis 147°/53°W and 37°→057°. (c) Foliation in the foreland granite. Contour interval: 1%, 2%, 4%, 8%. Calculated girdle and b-axis 023°/41W and 49°→113° respectively (Fig. 2d of Bhadra et al. 2004). (d) Foliation in the porphyritic charnockite. Contour interval: 1%, 2%, 4%, 8%. Calculated girdle and b-axis 192°/31°W and 59°→102° respectively (Fig. 2c of Bhadra et al. 2004). Reproduced with permission of Elsevier.

folds in the other, between the Mugnapador and the Ghontigurah outcrops, possibly point towards spatial variation in strain intensity or deformation partitioning within the metasedimentary unit.

Also important is the progressive decrease in the intensity of the deformation towards west. Mesoscopically, this can be documented in terms of sub-horizontal (<10°) dip of beds in the interior of the sedimentary hills as opposed to the near vertical (~80°) bedding plane attitude near the contact with the granitic basement. This is evidenced microscopically by profound grain size reduction in the meta-quartzite from about 400 μm in the west to 20 μm near the contact.

11.3.2 Foreland granites

Based on the mesoscopic structural development and intensity of structural fabric development, the entire foreland domain is divided into three zones. From east (near the thrust contact, Fig. 11.2) to west (in contact with

metasedimentary rocks, Fig. 11.2) these are hitherto referred as the frontal zone, intermediate foreland zone and interior foreland zone respectively. In the frontal zone, west of the easterly dipping thrust contact, a single penetrative fabric is defined by K-feldspars augens wrapped by foliation-defining mafic phases, dominantly amphibole. Mesoscopic scale folding on the augen-defined foliation is less conspicuous. Near the thrust contact, that is, along the western margin of the porphyritic charnockite unit the granite is proto- to ultramylonite. The foliation spacing increases substantially as the distance from the thrust contact increases, and the fabric varies from anastomosing through zonal in the intermediate foreland zone, finally grading into an undeformed zone (interior foreland zone) where the fabric is sparse. This apparently undeformed zone is characterized by infrequent, narrow, discontinuous, N–S trending ductile shear zones within which the rock is mylonitized with distinct foliation. Poles to all measured foliations (Fig. 11.6c) within the foreland granites west of the thrust contact with the EGMB show a girdle distribution that closely compares with the foliation pole girdle for the porphyritic charnockite (Fig. 11.6d). The waning phase of foliation development across the foreland granite corresponds with the systematic increase in grain size. These observations can be attributed to a strain gradient that was possibly impressed in the footwall concomitant to westward transport of the EGMB thrust sheet on to the craton. Stabilization of varying mineral assemblages in the thrust- induced fabric also indicates a temperature gradient. While the mylonitic zone near the contact records amphibolite facies conditions, the interior foreland zone near the contact with the metasediments preserves a chlorite zone assemblage (Bhadra et al. 2003). Bhadra et al. (2004) attributed sequential fabric development in the cratonic footwall to the migration of the deformation front consequent to hot over cold thrusting of the EGMB onto the Bastar craton.

In this study, syn-deformational temperatures across the cratonic footwall, especially in the foreland zone, were constrained additionally based on the crystallographic preferred orientation (CPO) of quartz and mineral thermobarometry. With regard to the CPO of quartz, the basal <a> slip is generally accepted to be predominant at lower grades, with rhomb <a>, prism <a>, and finally, prism <c> slip becoming active at progressively higher temperatures (Bouchez 1977; Tullis 1977; Lister and Dornsiepen 1982; Mainprice et al. 1986; Culshaw 1987; Law 1987; Wenk et al. 1989; Jessel and Lister 1990; Joy and Saha 2000; Hippertt et al. 2001).

Samples chosen for quartz c-axis determinations are all X–Z sections, i.e. parallel to the lineation and perpendicular to the foliation plane. The X-direction coincides with the lineation. The c-axes orientations were determined with a 3-axis Leica Universal Stage fitted to a Leica Orthoplan polarizing microscope.

11.4   MICROSTRUCTURE AND QUARTZ C-AXIS PATTERNS

11.4.1   Frontal zone

The cratonic granite in the frontal zone is characterized by a well-developed segregation banding comprising predominantly amphibole (occasionally 3-amphibole: cummingtonite-anthophyllite-tremolite) + biotite – bearing mafic layers and discrete quartz and feldspar-bearing layers (Fig. 11.7a). Biotites commonly describe S-C structures, with the C-plane concordant with the segregation banding. K-feldspars show sub-grain rotation recrystallization in the form of 'core–mantle' structures, suggesting temperatures >500°C (Paaschier and Trouw 2005). The c-axis preferred orientations of quartz grains (Fig. 11.7b) in this location (MY-2K-68, Fig. 11.2) show clustering parallel to the Y-axis, and in intermediate zones between Y and Z, suggesting the operation of

prism <a> (Wilson 1975) and rhomb <a> (Bouchez and Pecher 1981) slip mechanisms, respectively. These imply temperatures >700°C, characteristic of the amphibolite facies (Bunge and Wenk 1977; Schmid and Casey 1986; Blumenfeld et al. 1986). The Y-maxima in the c-axis patterns have also been reported in greenschist facies rocks deformed under high fluid activity conditions (e.g. Lister and Dornsiepen 1982; Joy and Saha 2000). However, microstructural evidence such as the lack of unaltered feldspars (no saussuritization), and pressure solutions do not support any pervasive fluid flux in this zone.

Occasionally, in parts of the frontal zone (MY-2K-14, Fig. 11.2) quartz has highly serrated boundaries characteristic of grain boundary migration recrystallization (Fig. 11.7c). Large, relict microcline grains show sweeping extinction, while most feldspar porphyroclasts show core–mantle microstructure with a mosaic of medium-sized equidimensional grains along the periphery of relict clasts (Fig. 11.7c). This suggests dominance of sub-grain rotation recrystallization. c-axes of quartz grains in this location show two maxima – around Y, and at a small angle to X (Fig. 11.7d). These suggest activity along the prism <a> and <c> slip systems respectively during simple shear deformation, compatible with high-T deformation microstructures, that is, GBM recrystallization in quartz (Stipp et al. 2002a, b).

11.4.2   Intermediate foreland zone

Further west, lepidoblastic biotites define the foliation. Quartz grains flattened to discontinuous ribbons (defining a stretching lineation), within which deformation bands are near orthogonal, akin to twist-wall subgrain boundary (Passchier and Trow 2005), to the ribbon boundaries (Fig. 11.7e). Feldspars (microcline) survive as large augens (= lenticular fish of Mukherjee 2011), which deformed mainly cataclastically (Fig. 11.7e). Weak undulose extinction in microcline, at places, represents the only evidence of plastic deformation, consistent with recovery at low temperatures (Vernon 2004). Plagioclase grains are highly saussuritized, and in places almost completely replaced, connoting enhanced activity of hydrous fluids. Relict grains indicate cataclasis. Quartz ribbons wrap around feldspar clasts (e.g. Mukherjee 2013). c-axis of quartz grains is characterized by weak maxima between Y and Z, and a major concentration displaced 30° clockwise from Z (Fig. 11.7f). The former can be correlated with the operation of rhomb <a> slip, while the latter suggests the operation of basal <a> slip during simple shear.

In this zone, ductile flow of quartz and predominantly brittle behavior of feldspar suggests that deformation temperature did not exceed 550°C: the approximate limit for brittle-ductile transition in feldspar (Tullis 1983; Zadins and Mitra 1986; Pryer 1993; Srivastava and Mitra 1996). The inference of low deformation temperatures is supported by the operation of basal <a> slip in quartz (Bouchez 1977; Blumenfeld et al. 1986).

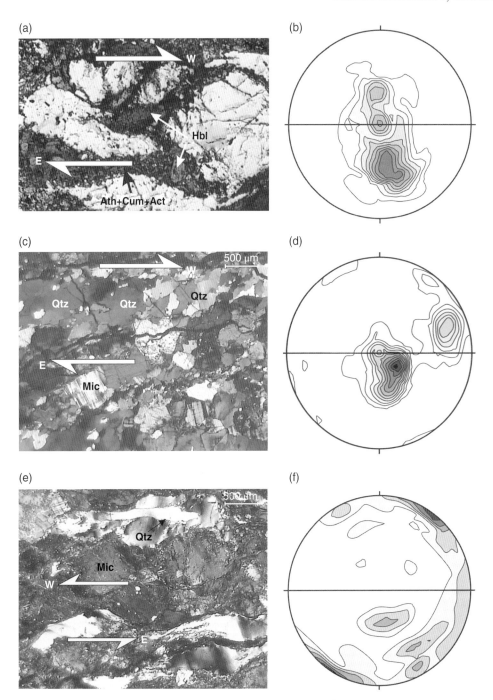

Fig. 11.7. Photomicrographs and plots of c-axis preferred orientation of quartz grains in mylonite zones within cratonic rocks. (a) Frontal Zone (mylonite zone): Quartz and feldspar segregated into distinct layers. Quartz recrystallized to a strain-free equigranular mosaic, while K-feldspars preserve core-and-mantle structures (Sample My-2k-68a). (b) Quartz c-axis maxima in Sample My-2k-68a cluster around Y and in intermediate positions between Y and Z. Contour intervals 1, 2, 3, 4, 5, 6, 7, 8 and 9%, $n = 203$. (c) Frontal Zone (mylonite zone): Quartz in segregated domains shows serrated grain boundaries, while feldspars recrystallized to equidimensional mosaic. Relict microcline porphyroclasts show sweeping extinction (Sample My-2k-14b). (d) Quartz c-axis maxima in Sample My-2k-14b cluster around Y and X. Contour intervals 1, 2, 3, 4, 5, 6, 7, 8, 9, 10, and 11%, $n = 199$. (e) Intermediate foreland zone: Discontinuous ribbons of quartz with deformation bands at high angles to ribbon boundaries. Note peripheral recrystallization along margins of quartz ribbons. Large microcline porphyroclasts are deformed by cataclasis and show bookshelf structures (Sample My-2k-33). (f) Quartz c-axis maxima in Sample My-2k-33 cluster at around 30° from Z, with a sub-maxima intermediate between Y and Z. Contour intervals 1, 2, 3, 4, and 5 %, $n = 257$. (g) Interior foreland zone: Quartz ribbons with deformation bands at high angles to ribbon boundaries, but with no peripheral recrystallization. Feldspars are fractured with ferruginous matter in fractures indicating enhanced fluid activity. Note chlorite and biotite alignment parallel to foliation (Sample My-2k-102). (h) Quartz c-axis maxima in Sample My-2k-102 cluster at around Z. Contour intervals 1, 2, 3, and 4 %, $n = 234$. (i) Interior foreland one: feebly deformed granite with large microclines and unrecrystallized quartz grains with limited strain (Sample My-2k-50). (j) Orientation of foliation and lineation on the stereographic projection. S = foliation, L = lineation.

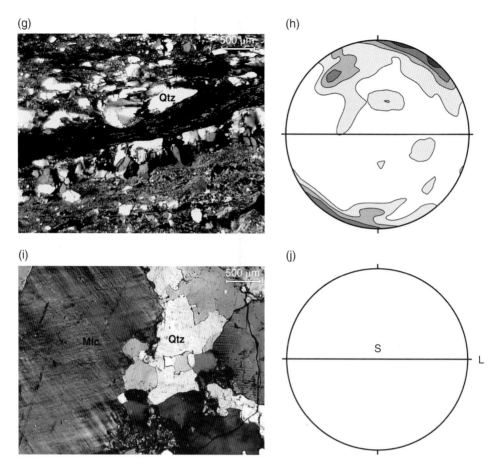

Fig. 11.7. (*Continued*)

11.4.3 Interior foreland zone

This zone is characterized by stabilization of chlorite. In high strain domains within this zone, quartz recrystallized dynamically and deformation bands are common. Here, too, deformation bands occur at high angles to stretched quartz boundaries that lack any peripheral recrystallization (Fig. 11.7g). All feldspars deform by brittle fracturing, with V-pull-apart (Hippertt 1993; Mukherjee 2010, 2013; Mukherjee and Koyi 2010) microstructures being particularly common suggesting low-grade condition ($T \ll 550°C$). Chlorite and biotite parallel the fabric and in places define S-C fabrics (Mukherjee 2011). c-axis preferred orientation of quartz grains (Fig. 11.7h) shows maxima displaced 10° clockwise from Z. This can be interpreted as basal <a> slip during simple shear. Increase in shear strain may also rotate the c-axes progressively towards Z from a position intermediate between Z and X.

In adjacent low-strain domains in the interior foreland zone, porphyritic to seriate igneous textures preserved in the basement granite and the foliation is weak, that is, spaced disjunctive in nature (Fig. 11.7i). Occasional shears dissecting large amphiboles form locales for stabilization of biotite, chlorite, and epidote. Quartz shows undulose extinction but no evidence of recrystallization.

Feldspars are undeformed, but show considerable hydrous alteration.

11.5 *P–T* CONDITION ACROSS THE FORELAND

Pressure (*P*) and temperature (*T*) were estimated from close pair rim–rim amphibole-plagioclase analyses (Table 11.1) using the thermometric formulation of Holland and Blundy (1994) and the barometric formulation of Bhadra and Bhattacharya (2007). Mineral chemical analyses were carried out at IIT Roorkee using a 4 channel WDS Cameca Electron Microprobe. Gupta et al. (2000) presented detailed analytical procedures. In the frontal zone, amphibole and plagioclase define the mylonitic foliation. However, in the interior foreland zone, granites are characterized by relict or partially modified amphiboles. Consequently, rim-rim amphibole-plagioclase pairs ensure that the retrieved *P*, *T* conditions closely represent the deformation condition, though shear heating could contribute partially to the temperature rise (Mukherjee and Mulchrone 2013). Figure 11.8 shows the retrieved *P*, *T* conditions. The estimated temperature of 740°C near the thrust contact matches the observed c-axis distribution pattern

Table 11.1. Electron probe data of plagioclase (a) and amphibole (b) used for *P–T* estimation

(a)	Chl Zone		Ep zone			Amph zone		
	JA05	JA10E	JA04	JA08	MY72	MY68A	MY68B	MY86
SiO_2	59.31	63.73	68.55	70.24	63.03	62.21	63.26	58.88
Al_2O_3	25.15	22.72	22.68	18.97	23.52	24.04	23.47	24.42
Fe_2O_3	0.00	0.00	0.36	0.00	0.09	0.09	0.00	0.24
CaO	5.07	4.41	2.06	0.12	5.40	5.30	4.06	6.78
Na_2O	9.49	9.30	9.26	11.02	8.81	8.93	8.87	7.82
K_2O	0.43	0.15	0.12	0.05	0.07	0.30	0.27	0.04
Total	**99.46**	**100.31**	**103.02**	**100.4**	**100.92**	**100.87**	**99.94**	**98.18**
Si	2.67	2.81	2.90	3.04	2.77	2.74	2.79	2.68
Al	1.33	1.18	1.13	0.97	1.22	1.25	1.22	1.31
Ca	0.24	0.21	0.09	0.01	0.25	0.25	0.19	0.33
Na	0.83	0.80	0.76	0.93	0.75	0.76	0.76	0.69
K	0.03	0.01	0.01	0.00	0.00	0.02	0.02	0.00
Σ	**5**	**5**	**5**	**5**	**5**	**5**	**5**	**5**
Ab	0.75	0.79	0.87	0.99	0.74	0.74	0.79	0.68
An	0.22	0.21	0.12	0.01	0.25	0.24	0.20	0.32
Or	0.02	0.01	0.01	0.00	0.01	0.02	0.01	0.00

(b)	Chl Zone		Ep zone			Amph zone		
	JA05	10E	JA04	JA08	MY72	MY68A	MY68B	MY86
SiO_2	46.59	42.66	41.05	38.14	42.42	40.58	43.67	41.98
TiO_2	0.22	0.76	0.15	0.38	0.96	1.09	0.51	0.65
Al_2O_3	4.00	10.04	10.13	12.35	10.3	11.23	9.25	10.67
Cr_2O_3	0.00	0.28	0.00	0.32	0.27	0.00	0.24	0.01
Fe_2O_3	3.06	3.95	9.59	6.64	4.12	3.50	6.17	9.21
FeO	21.47	20.65	16.35	23.17	19.88	20.45	16.45	9.84
MnO	0.2	0.42	0.64	0.3	0.22	0.33	0.33	0.29
MgO	8.27	5.78	6.72	2.38	6.47	5.85	8.26	10.9
CaO	11.36	11.99	11.36	11.27	11.88	11.28	11.63	11.86
Na_2O	1.02	0.94	1.13	0.94	1.04	1.4	0.99	1.22
K_2O	0.22	1.02	1.07	1.82	1.08	1.57	0.82	0.45
Total	**96.39**	**98.49**	**98.19**	**97.72**	**98.64**	**97.28**	**98.32**	**97.08**
Si	7.25	6.56	6.34	6.1	6.49	6.34	6.61	6.31
Al	0.73	1.44	1.66	1.9	1.51	1.66	1.39	1.69
Ti	0.02	0.00	0.00	0.00	0.00	0.00	0.00	0.00
Σ	**8**	**8**	**8**	**8**	**8**	**8**	**8**	**8**
Al	0	0.38	0.17	0.42	0.35	0.41	0.26	0.2
Ti	0.01	0.09	0.02	0.05	0.11	0.13	0.06	0.07
Cr	0	0.03	0	0.04	0.03	0	0.03	0
Fe^{3+}	0.36	0.46	1.14	0.8	0.47	0.41	0.7	1.04
Mg	1.92	1.32	1.53	0.57	1.48	1.36	1.86	2.44
Fe^{2+}	2.71	2.65	2.09	3.1	2.54	2.67	2.08	1.24
Mn	0	0.06	0.06	0.02	0.02	0.01	0.01	0.01
Σ	**5**	**4.99**	**5**	**5**	**5**	**5**	**5**	**5**
Fe^{2+}	0.08	0.00	0.00	0.00	0.00	0.00	0.00	0.00
Mn	0.03	0	0.02	0.02	0.01	0.03	0.03	0.03
Ca	1.89	1.97	1.86	1.93	1.95	1.89	1.89	1.91
Na	0	0.03	0.12	0.05	0.04	0.08	0.08	0.06
Σ	**2**	**2**	**2**	**2**	**2**	**2**	**2**	**2**
Na	0.31	0.25	0.22	0.25	0.27	0.35	0.21	0.29
K	0.04	0.2	0.21	0.37	0.21	0.31	0.16	0.09
ΣA	**0.35**	**0.45**	**0.43**	**0.62**	**0.48**	**0.66**	**0.37**	**0.38**

After Holland and Powell (1998): (a) based on 8 oxygens; (b) based on 23 oxygens.

Fe_2O_3 is calculated based on stoichiometry.

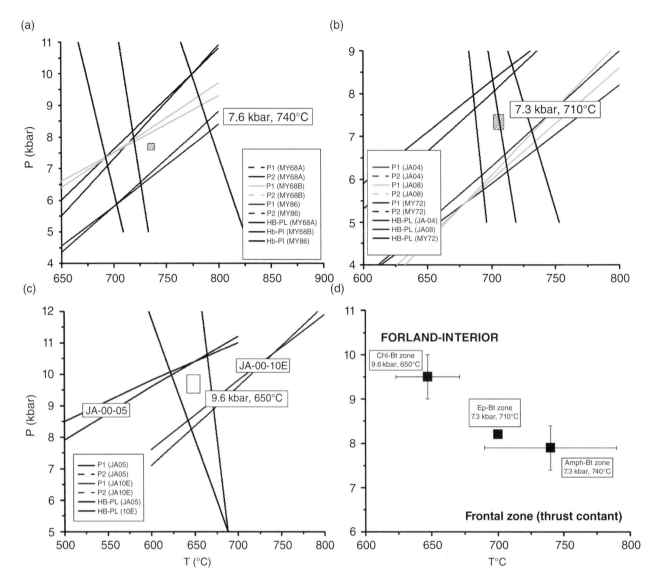

Fig. 11.8. Thermobarometric results for (a) frontal zone, (b) intermediate foreland zone, and (c) interior foreland zone. (d) Summary of *P–T* results across the foreland. *P, T* values shown in the diagram is the mean of the four points (dots) generated from the intersection of geothermometers and geobarometers.

(Fig. 11.7b), and high-*T* deformation microstructures in quartz, as discussed earlier. Nevertheless, the estimated *T* values clearly indicate a waning temperature gradient towards the interior foreland, where deformation temperatures were slightly above the brittle-ductile transition temperature for feldspar and low-*T* basal <a> slip mechanism in quartz. Such discrepancy in the estimated *T* and observed microstructures possibly reflect incomplete compositional re-equilibration of igneous amphiboles and/or plagioclase in the interior foreland zone due to low prevailing *T* and waning deformation. The deduced *P, T* conditions from selected locations along the E–W traverse suggest that an inverted metamorphic field gradient was impressed concomitant to westward thrusting of the EGMB over the craton.

11.6   DISCUSSION

11.6.1   Nature of reactivation of the basement: consequence of Terrane Boundary shearing

Quartz c-axis patterns of granites from high strain zones across the foreland indicate that development of penetrative fabric in the foreland domain is coeval and characteristic of non-coaxial deformation (Passchier and Trouw 2005). The magnitude of non-coaxiality (vorticity, W_m) was determined for samples from the frontal and intermediate foreland zones using the δ/β method (δ: angle between oblique grain shape fabric and main foliation, β: angle between perpendicular to c-axis girdle and main fabric and $W_m = \sin 2(\delta+\beta)$, Fig. 11.9) of Xypolias (2009). W_m for the frontal and intermediate

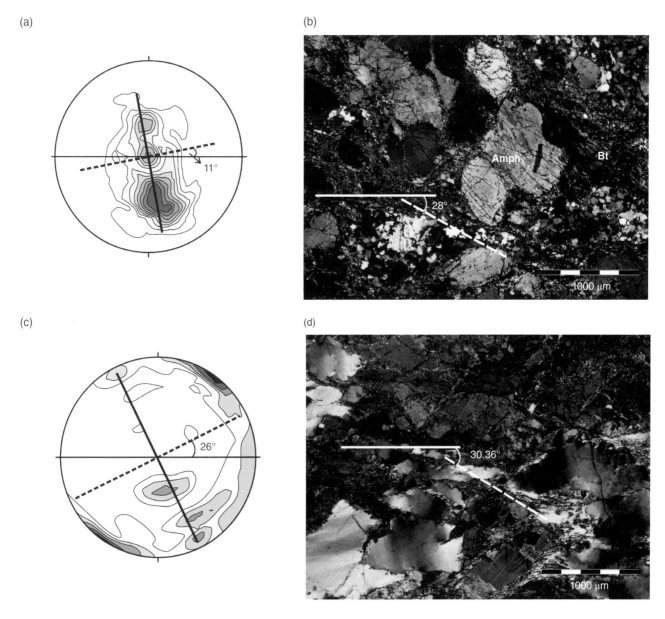

Fig. 11.9. Qualitative vorticity (W_m) analyses for two representative samples from frontal (a, b) and intermediate (c, d) foreland zone using the δ/β method of Xypolias (2009). In (a) and (c) the angles refer to the values of β subtended by perpendicular (black dotted line) to c-axis girdle (black solid line) and main fabric (E–W black line through center of the sterogram. In (b) and (d) the angles refer to the value of δ subtended by oblique grain shape fabric (dotted white line) and main foliation (solid white line). Vorticity (W_m) = sin 2(δ+β),

foreland zone are estimated to be 0.98 and 0.92 respectively, implying dominant simple shear along the thrust contact between the foreland granite and EGMB, and simple to general shear dominated deformation in the intermediate foreland zone. Such partitioning of deformation regimes, from simple to pure shear dominated, through an intermediate general shear domain, characterizes oblique continental collision zones/transpression zones (Fossen and Tikoff 1993; Dewey et al. 1998). Since the contact between the foreland granite and EGMB represents a Terrane Boundary Shear Zone (Biswal et al. 2000; Bhadra et al. 2004), the observed c-axis patterns together with vorticity analyses suggest that reactivation of the cratonic footwall (i.e., the basement to the Khariar

metasediments) was synchronous with Terrane Boundary shear. Microstructural evolution and mineral thermobarometric results place additional constraints on the nature of basement reactivation and suggest that the deformation temperature progressively decreased away from the thrust contact, that is, towards the interior of the foreland. The deduced pressure values, on the other hand, show a steady increase from near the thrust contact (frontal zone) into the interior foreland. The obtained *P, T* values confirm the earlier proposition that an inverted metamorphic gradient was established in the foreland/footwall granite consequent to the thrusting of the EGMB (hanging wall) granulites (Gupta et al. 2000; Bhadra et al. 2003, 2004).

11.6.2 Nature of reworking of the basement–cover interface

Despite several reports of tectonic activity within the Khariar basin, up until now there is a dearth of information on the deformation history of the Khariar metasediments, in general, and in particular, about the nature of basement–cover reworking. Recently, Ratre et al. (2010) proposed an evolutionary model of the Khariar basin based on geochronological studies of the Lakhna dyke swarm that truncates against the basal part of the Khariar metasediments.

Here, we document the deformation history of the Khariar metasediments immediately overlying the basement granites and correlate the structural fabrics across the interface. Several lines of evidences such as: (a) conformable attitude of the penetrative fabric in the basement granite and slaty cleavage in the shale unit of the overlying Khariar meta-sedimentary cover (Fig. 11.6b,c); (b) west-verging asymmetric folds in the meta-quartzite unit (Fig. 11.4b) and S-C structures in the slaty cleavage with a top-to-west shear sense (Fig. 4c); (c) intense deformation of the cover compared to feeble or little deformation in the underlying basement (Fig. 11.4 versus Fig. 11.7i); and (d) waning deformation intensity/strain gradient in the metasediments away from the basement cover interface, manifest by subvertical dips of the sedimentary beds near the interface and subhorizontal dips away from it–suggest that the metasediments deformed by far-field stresses that prevailed during westward thrusting of the EGMB over the cratonic basement. Thus, the metasediments deformed when the basement–cover interface was reworked during Terrane Boundary shearing (ca. 517 Ma, Ratre et al. 2010) across the EGMB–Bastar craton contact, that is, along the frontal foreland zone. Partitioning of deformation across the basement–cover interface (point (c)) possibly implies that the interface acted as a decòllement horizon during westward transport of the EGMB thrust sheet over the cratonic footwall (basement granites). We further envisage that, barring competency contrast between the basement granites and overlying sedimentary cover, low prevailing temperatures of deformation (greenschist facies conditions) in the interior of the foreland zone possibly facilitates deformation partitioning across the basement–cover interface. Consequently, the observations documented above demand that nature of reworking of the basement–cover interface as well as tectonic evolution of the Khariar basin needs to be evaluated in the light of granulite facies metamorphism and attendant deformation in the adjacent Eastern Ghats Province.

11.6.3 Tectonic evolution of the Khariar basin: implications for contact with the EGP

Sedimentological studies on lithologic successions from the Khariar basin support an intra-cratonic origin related to ensialic rifting of the craton (Das et al. 1992). Sediments in the Khariar basin are dominantly siliciclastic (Das et al. 2001; Chakraborty et al. 2012; Saha and Ptranabis-Deb 2014), comparable with other Purana basins such as the Chhattisgarh basin (Ratre et al. 2010, and references therein). However, geochronologic records of metasediments from Purana basins in general, and the Khariar basin, in particular, are still scanty and impede holistic understanding of the spatio-temporal evolution of the basin in relation to thrust-induced reworking of the underlying basement. Recently, Ratre et al. (2010) reported zircon SHRIMP ages at ca. 1450 Ma from the Lakhna dyke swarm, which is truncated by the basal sandstone unit of the Khariar basin. Based on these observations, and due to presence of dyke pebbles in the basal conglomeratic horizon of the lower sandstone unit of the Khariar Group, Ratre et al. (2010) proposed that the age of sedimentation in the Khariar basin post-dates emplacement of the Lakhna dyke swarm(i.e. 1450 Ma). Das et al. (2009) also reported a similar monazite age (1455 Ma) from the tuff unit in the lower part of the Khariar basin. In contrast to these Mesoproterozoic ages, Kreuzer et al. (1977) reported K-Ar Neoproterozoic ages (ca. 750–700 Ma) from authigenic glauconites in sandstone from the basal part of the Chhattisgarh basin, which is supposedly contemporaneous with the Khariar basin. The upper age limit of sedimentation in the Khariar basin was reported to be ca. 1000 Ma (Patranabis-Deb et al. 2007) on the basis of dating of pyroclastic sequences from the Chhattisgarh basin. However, based on the presence of mylonite pebbles in the conglomerate of the Khariar basin, syn-thrusting deformation of the middle shale unit and undeformed nature of the upper sandstone unit, Ratre et al. (2010) proposed that the sedimentation in the Khariar basin was contemporaneous with and continued up to the thrusting along the Terrane Boundary Shear Zone at ca. 517 Ma. This contradicts the earlier view that basin fills in Khariar and Chhattisgarh basins were dominantly during Mesoproterozoic, i.e. 1500–1000 Ma (Patranabis-Deb et al. 2007; Das et al. 2009, 2011b; Bickford et al. 2011a, b, c).

This study establishes that deformation in the metasediments was concomitant with westward thrusting of EGMB granulites onto the craton, as well as with the Terrane Boundary shear. Geochronologic data from the upper part of the sedimentary succession of the Khariar Group would be critical to constrain the timing of the Terrane Boundary shear. However, since such information is lacking, the evolution the basin can only be speculated on in the context of already published geochronologic information and models. Recent studies on the paleogeographic reconstruction of Indian craton suggest that the evolutionary history of the proto-Indian continent can be traced back to the supercontinent Columbia (1800–1500 Ma: Rogers and Santosh 2002; Zhao et al. 2004; Santosh et al. 2009). It has been proposed that the Mesoproterozoic Period was marked by widespread rifting. If the Mesoproterozoic ages reported from the base of the Khariar basin are correct, then the intra-cratonic rifting can be attributed to the initiation of the break-up of Columbia. While the SHRIMP ages of the Lakhna dyke swarm and monazite ages of the tuffaceous layer constrain the maximum possible age of sedimentation of the lower part of the Khariar basin (~1400 Ma: Ratre et al. 2010), we suspect that the Neoproterozoic age of authigenic glauconite (~750–700 Ma, Kreuzer et al. 1977) more closely approximates the age of sedimentation in the Khariar basin. The later age is significant because the final disintegration of Columbia within a time frame between 1200–1000 Ma was followed by the assembly of the supercontinent Rodinia during the globally correlatable Grenvillian orogeny (1000–900 Ma). Evidence of this tectonic reworking is widespread in the Eastern Ghats Province lithologies in the form of granulite facies metamorphism, associated migmatization and emplacement of intrusive rocks along the craton-mobile belt contact. Tectonic reworking has also been established to coincide with the accretion of the EGP with Antarctica and part of the proto-Indian craton (Nasipuri and Bhadra 2013, and references therein). Crustal-flexuring due to loading by thrust nappes is one of the popular models for the development of foreland basin (Beaumont 1981), and indeed, has been postulated as a possible mechanism of origin of late Neoproterozoic

sediments for the Chhatisgarh-, Indravati- and Khariar basins (Gupta 2012). By analogy, the development of the Khariar basin with Neoproterozoic lower sedimentary succession can be linked with the first accretion phase of the EGP with the Bastar craton during the Grenvillian tectonics (Fig. 11.10). However, the proposed crustal flexure model that has already been suggested to have resulted in a foreland basin by some workers (e.g. Biswal et al. 2003), and the earlier rift model for the origin of the Purana basins may not necessarily be contradict. Basu and Bickford (2014) suggested that though a Mesoproterozoic (ca. 1500 Ma) time frame can be established for the opening of the Purana basins (e.g. the Cuddapah basin: ca. 1900 Ma, Khariar basin: ca. 1500 Ma, Chhattisgarh basin: 1400 Ma), the time of inversion and closure of many Purana basins are still not adequately constrained. The authors also opined that inversion and closure of the Purana basins are possibly linked to the far-field stresses associated with the Grenvillian tectonothermal event that affected parts of the Indian shield. In the light of the above evidence, and as revealed from the present study, it is therefore possible, as suggested by Gupta (2012), that these Purana basins originated as rift-related basins in the Mesoproterozoic time (Phase 1 stage A, Fig. 11.10a), but the uppermost parts of the basin

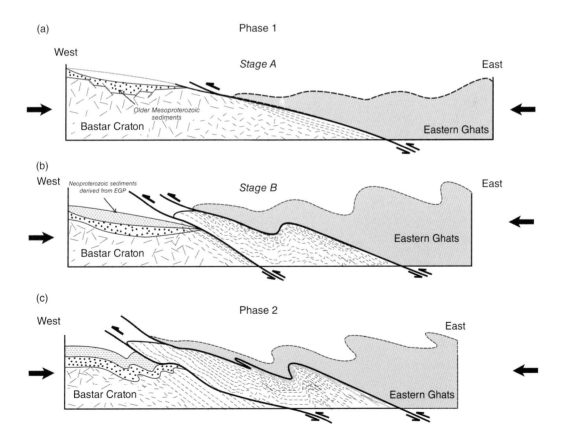

Fig. 11.10. Cartoon depicting the tectonic evolution of the Khariar basin. Phase 1 relates to the origin of the basin as an intracratonic rift (stage A) with deposition of Mesoproterozoic sediments (a), and subsequent flexuring (stage B) of the basin with deposition of Neoproterozoic sediments concomitant with westward thrusting of the Eastern Ghats Province onto the Bastar craton during the Grenvillian orogeny. Phase 2 relates to the final architecture of the Khariar foreland basin during the Pan-African orogeny.

succession may well be products of infilling of Neoproterozoic flexural depressions that originated prior to the closure of the basins themselves (Phase 1 stage B, Fig. 11.10b). Deposition of sediments, up to possibly the middle shale unit in the Khariar foreland basin continued prior to the final assembly of the EGMB with the Indian craton during the Pan-African orogeny (~550–500 Ma).

Following basin closure, tectonic reworking of the basement–cover interface in the cratonic foreland was initiated with further westward propagation of the EGMB thrust sheet. During this phase of movement (Phase 2 in Fig. 11.10c), the EGMB thrust sheet incorporated a part of the cratonic basement by successive thrusting through a sequence of progressive footwall collapse. Thus, an in-sequence thrust system formed in the interior foreland (Fig. 11.10b, c) following a piggy-back model, with older thrust being successively carried passively over newly developed footwall thrusts. Recall that Bhadra et al. (2004) recognized at least two-phases of thrusting based on structural investigation along the craton-mobile belt contact. This appears to support the piggy-back model of thrust propagation. During the final phase of the orogeny, the Khariar basin closed and developed as a foreland fold-thrust belt (accretionary wedge) by westward propagation of the thrust nappe (Fig. 11.10c). Since the dip of the accretionary wedges gradually decreases away from the thrust front (Davis et al. 1983; Westbrook et al. 1988; Dahlen, 1990), the gradual decrease in dip of the sedimentary strata away from the thrust contact (decòllement surface) can be explained suitably by the proposed model. Further, the proposed model suggests that the older thrusts will always lie at the upper structural level, as in case of piggy back model. The foreland granites in contact with the old thrust contact lie directly in contact with and overlain by the EGMB thrust sheet alone. In contrast, granites towards the interior foreland zones are likely to be overlain by a composite thrust sheet composed of EGMB granulites and also deformed cratonic granites. Therefore, the cratonic granites will register high T at relatively low P near the contact with the EGMB (i.e. frontal zone). On the other hand, low-T at relatively high P will prevail away from EGMB-Bastar craton contact (i.e. in the interior foreland zone). Thus, an inverted metamorphic field gradient is created, that tallies with the thermobarometric results (Fig. 11.8) documented in this study. The final phase of movement of the granulites (and accompanying cratonic foreland granites) is therefore essentially intracontinental. In effect, this study provides further support for a final phase of intracontinental orogenesis that involves the granulites of the EGP, postulated earlier by Nanda and Gupta (2012) in the Koraput region.

11.7   CONCLUSIONS

This study demonstrates that the easternmost part of the Khariar sedimentary basin, which lies in the foreland of a thrust contact between the Neoproterozoic granulites of the Eastern Ghats Province and the Bastar craton, is deformed. The strain pattern can be correlated with that in the adjacent deformed and metamorphosed cratonic granites. Deformation in the granites can be correlated with that in the Terrane Boundary Shear Zone. We suggest at least two phases of movement occurred along the Terrane Boundary Shear Zone: the first phase being associated with continent-continent collision that generated a flexural depression within a pre-existing riftogenic basin (the Khariar basin), and a second phase that translated much of the earlier thrust sheet, and a part of the earlier cratonic foreland, westward over the Khariar basin foreland that is consequently deformed. The Grenvillian Indo-Antarctica collision was therefore followed by a phase of intracontinental shortening in response to far-field stresses elsewhere in the supercontinent Rodinia in the Pan-African (Neoproterozoic to Cambrian) time, remobilizing and shearing the basement–cover contact in the frontal part of the Khariar basin.

ACKNOWLEDGMENT

The results presented here were part of PhD dissertation work of S.B. S.B. is grateful to Prof. Abhijit Bhattacharya for actively participating in the field and rendering valuable suggestions. We acknowledge the Elsevier for providing permission to reproduce three illustrations mentioned in the text. We thank Soumyajit Mukherjee for a detail critical review and editorial handling. We also thank two anonymous reviewers for their valuable suggestions. The Council of Scientific and Industrial Research (CSIR), India is acknowledged for the financial support (grant-in-aid no. 24/243/98/EMR-II).

REFERENCES

Basu A, Bickford E. 2014. Contributions of zircon U–Pb geochronology to understanding the volcanic and sedimentary history of some Purana basins, India. Journal of Asian Earth Sciences 91, 252–262.

Beaumont C. 1981. Foreland basins. Geophysical Journal of the Royal Astronomical Society 65, 291–329.

Bhadra S, Bhattacharya A. 2007. The barometer tremolite+tschermackite+2albite = 2pargasite+8quartz: constraints from experimental data at unit silica activity with application to garnet-free natural assemblages. American Mineralogist 92, 491–502.

Bhadra S, Gupta S, Banerjee M. 2004. Structural evolution across the Eastern Ghats Mobile Belt – Bastar craton boundary, India: Hot over cold thrusting in an ancient collision zone. Journal of Structural Geology 26, 233–245.

Bhadra S, Banerjee M, Bhattacharya A. 2003. Tectonic restoration of a polychronous mobile belt – craton assembly: constraints from corridor study across the western margin of the Eastern Ghats Belt, India. Memoirs Geological Society of India (Milestones in Petrology) 52, 109–130.

Bhadra S. 2003. Tectonometamorphic evolution of a craton-mobile belt assembly: the Bhawanipatna-Deobhog transect, Orissa, India. Unpublished PhD Thesis, IIT Kharagpur.

Bickford ME, Basu A, Patranabis-Deb S, Dhang PC, Schieber J. 2011a. Depositional history of the Chhattisgarh basin, central India: constraints from new SHRIMP zircon ages. Journal of Geology 119, 33–50.

Bickford ME, Basu A, Patranabis-Deb S, Dhang PC, Schieber J. 2011b. Depositional history of the Chhattisgarh basin, central India: constraints from new SHRIMP zircon ages; a reply. Journal of Geology, 119, 553–556.

Bickford ME, Basu A, Mukherjee A, et al. 2011c. New U-Pb SHRIMP zircon ages of the Dhamada Tuff in the Mesoproterozoic Chhattisgarh basin, Peninsular India: Stratigraphic implications and significance of a 1-Ga thermal-magmatic event. Journal of Geology 119, 535–548.

Biswal TK, Jena SK, Datta S, Das R, Khan K. 2000. Deformation of the terrain Boundary Shear zone (Lakhna shear zone) between the Eastern Ghats Mobile Belt and the Bastar craton, in the Balangir and Kalahandi district of Orissa. Journal of the Geological Society of India 55, 367–380.

Biswal TK, Sinha S, Mandal A, Ahuja H, Das MK. 2003. Deformation pattern of Bastar Craton adjoining Eastern Ghat mobile belt, NW Orissa. Gondwana Geological Magazine, Special Publication 7, 101–108.

Blumenfeld P, Mainprice D, Bouchez JL. 1986. C-slip in quartz from subsolidus deformed granite. Tectonophysics 127, 97–115.

Bose S, Dunkley DJ, Dasgupta S, Das K, Arima M. 2011. India-Antarctica-Australia-Laurentia connection in the Paleoproterozoic-Mesoproterozoic revisited: evidence from new zircon U-Pb and monazite chemical age data from the Eastern Ghats Belt, India. Bulletin of the Geological Society of America, 123, 2031–2049.

Bouchez JL. 1977. Plastic deformation of quartzites at low temperature in areas of natural strain gradient. Tectonophysics 39, 25–50.

Bouchez JL, Pecher A. 1981. The Himalayan Main Central Thrust pile and its quartz-rich tectonites in Central Nepal. Tectonophysics 78, 23–50.

Bunge HJ, Wenk HR. 1977. Three dimensional texture analysis of quartzite (trigonal crystal and triclinic specimen symmetry). Tectonophysics 40, 257–285.

Chakraborty PP, Das P, Saha S, Das K, Mishra SR, Paul P. 2012. Microbial mat related structures (MRS) from Mesoproterooic Chhattisgarh and Khriar basins, central India and their bearing on shallow marine sedimentation. Episodes 35, 513–523.

Chaudhuri AK, Chanda SK. 1991. The Proterozoic basin of Pranhita-Godavari valley: an overview. In Sedimentary Basins of India: Tectonic Context, edited by S.K. Tandon, C.C. Pant, and S.B. Casshyap, Ganodaya Prakashan, Nainital, pp. 13–30.

Chaudhuri AK, Howard JD. 1985. Ramgundam Sandstone—a middle Proterozoic shoal-bar sequence. Journal of Sedimentary Petrology 55, 392–397.

Chaudhuri AK, Mukhopadhyay J, Patranabis-Deb S, Chanda SK. 1999. The Neoproterozoic cratonic successions of Peninsular India. Gondwana Research 2, 213–225.

Crawford AR. 1969. Reconnaissance Rb-Sr dating of the Precambrian rocks of southern Peninsular India. Journal of the Geological Society of India 10, 117–166.

Crawford AR, Crompston W. 1970. The age of the Vindhyan system of peninsular India. Quaternary Journal of the Geological Society of London 125, 351–371.

Crawford AR, Compston W. 1973. The age of the Cuddapah and Kurnool systems, Southern India. Journal of the Geological Society of Australia 19, 453–464.

Culshaw N. 1987. Microstructure, c-axis pattern, microstrain and kinematics of some S-C mylonites in Grenville gneiss. Journal of Structural Geology 9, 299–311.

Dahlen FA. 1990. Critical taper model of fold-and-thrust belt and accretionary wedges. Annual Review of Earth and Planetary Sciences 18, 55–90.

Das DP, Kundu A, Das N, et al. 1992. Lithostratigraphy and sedimentation of Chhattisgarh basin. Indian Minerals 46, 271–288.

Das N, Dutta DR, Das DP. 2001. Proterozoic cover sediments of southeastern Chhattisgarh state and adjoining parts of Orissa. Geological Survey of India, Special Publication 55, 237–262.

Das K, Yokoyama K, Chakraborty PP, Sarkar A. 2009. Basal tuffs and contemporaneity of the Chattisgarh and Khariar basins based on new dates and geochemistry. Journal of Geology 117, 88–102.

Das K, Bose S, Karmakar S, Dunkley DJ, Dasgupta S. 2011a. Multiple tectonometamorphic imprints in the lower crust: first evidence of ca. 950Ma (zircon U-Pb SHRIMP) compressional reworking of UHT aluminous granulites from the Eastern Ghats Belt, India. Geological Journal 46, 217–239.

Das P, Das K, Chakraborty PP, Balakrishnan S. 2011b. 1420 Ma diabasic intrusives from the Mesoproterozoic Singhora Group, Chattisgarh Supergroup, India: Implications toward non-plume intrusive activity. Journal of Earth System Science 120, 1–14.

Davis D, Suppe J, Dahlen FA. 1983. Mechanics of fold-and-thrust belts and accretionary wedges. Journal of Geophysical Research 88, 1153–1172.

Dewey JF, Holdsworth RE, Strachan RA. 1998. Transpression and transtension zones. Geological Society of London, Special Publications 135, 1–14.

Dobmeier CJ, Raith M. 2003. Crustal architecture and evolution of the Eastern Ghats Belt and adjacent regions of India. In Proterozoic East Gondwana: supercontinent assembly and breakup, edited by M Yoshida, BF Windley, and S Dasgupta, Geological Society, London, Special Publications 206, pp. 145–168.

Fossen H, Tikoff B. 1993. The deformation matrix for simultaneous simple shearing, pure shearing and volume change, and its application to transpression-transtension tectonics. Journal of Structural Geology 15, 413–422.

Gupta S. 2012. Strain localization, granulite formation and geodynamic setting of 'hot orogens': a case study from the Eastern Ghats Province, India. Geological Journal 47, 334–351.

Gupta S, Bhattacharya A, Raith M, Nanda JK. 2000. Contrasting pressure-temperature-deformation history across a vestigial craton-mobile belt boundary: the western margin of the Eastern Ghats belt at Deobhog, India. Journal of Metamorphic Geology 18, 683–697.

Fowler CMR. 1990. The Solid Earth: An Introduction to Global Geophysics. Cambridge University Press, Cambridge, UK.

Hippert JF. 1993. Microstructures and c-axis fabrics indicative of quartz dissolution in sheared quatrzites and phyllonites. Tectonophysics 229, 141–163.

Hippert J, Rocha A, Lana C, Egydio-Silva M, Takeshita T. 2001. Quartz plastic segregation and ribbon development in high-grade striped gneisses. Journal of Structural Geology 23, 67–80.

Holland TH. 1906. Classification of the Indian strata. Presidential Address, Transaction Mining and Geological Institute, India 1.

Holland TJB, Blundy J. 1994. Non-ideal interactions in calcic amphiboles and their bearing on amphibole-plagioclase thermometry. Contribution to Mineralogy and Petrology 116, 433–447.

Jayaprakash AV, Sundaram V, Hans SK, Mishra SN. 1987. Geology of the Kaldgi–Badami basins, Karnataka. In Purana basins of Peninsular India (Middle to Late Proterozoic). Memoir Geological Society of India, 6, 201–225.

Jessel MW, Lister GS. 1990. A simulation of temperature dependence of quartz fabric. InDeformation Mechanism, Rheology and Tectonics, edited by RJ Knipe and EH Rutter, Geological Society Special Publication, 54, pp. 353–362.

Joy S, Saha D. 2000. Dynamically recrystallised quartz c-axis fabrics in greenschist facies quartzites, Singhbhum shear zone and its footwall, eastern India – influence of high fluid activity. Journal of Structural Geology 22, 777–793.

Kale VS. 1991. Constraints on the evolution of the Purana basins of Peninsular India. Journal of the Geological Society of India 38, 231–252.

Kelly NM, Clarke GL, Fanning CAM. 2002. A two-stage evolution of the Neoproterozoic Rayner Structural Episode: new U–Pb sensitive high resolution ion microprobe constraints from the Oygarden Group, Kemp Land, East Antarctica. Precambrian Research 116, 307–330.

Kreuzer H, Karre W, Karsten M, Scnitzer WA, Murti KS, Srivastava NK. 1977. K-Ar dates of two glauconites from the Chandrapur series (Chattisgarh/India): On the stratigraphic status of the late

Precambrian basins in central India. Jahrbuch Der Geologischen Bundesanstalt 28, 23–36.

Law RD. 1987. Heterogeneous deformation and quartz crystallographic fabric transitions: natural examples from the Stack of Glencoul, northern Assynt. Journal of Structural Geology 9, 819–833.

Lisker S, Fachmann S. 2001. The Phanerozoic history of Mahanadi region, India. Journal of Geophysical Research: Solid Earth 106, 22027–22050. DOI: 10.1029/2001JB000295.

Lister GS, Dornsiepen UF. 1982. Fabric transition in the Saxony granulite terrain. Journal of Structural Geology 1, 283–297.

Mainprice D, Bouchez JL, Blumenfeld P, Tubia JM. 1986. Dominant c slip in naturally deformed quartz: implication for dramatic plastic softening at high temperature. Geology 14, 819–822.

McKenzie DP. 1978. Some remarks on the development of sedimentary basins. Earth and Planetary Science Letters 40, 25–32.

Meijerink AMJ, Rao DP, Rupke J. 1984. Stratigraphic and structural development of the Precambrian Cuddapah Basin, S.E. India. Precambrian Research 26, 57–104.

Mezger K, Cosca MA. 1999. The thermal history of the Eastern Ghats Belt (India) as revealed by U-Pb and $^{40}Ar/^{39}Ar$ dating of metamorphic and magmatic minerals: implications for the SWEAT correlation. Precambrian Research 94, 251–271.

Misra S, Gupta S. 2014. Superposed deformation and inherited structures in an ancient dilational step-over zone: post-mortem of the Rengali Province, India. Journal of Structural Geology 59, 1–17.

Mukherjee S. 2010. V-pull apart structure in garnet in macro scale. Journal of Structural Geology 32, 605.

Mukherjee S. 2011. Mineral fish: their morphological classification, usefulness as shear sense indicators and genesis. International Journal of Earth Sciences100 (6), 1303–1314.

Mukherjee S. 2013. Deformation Microstructures in Rocks. Springer, Berlin, pp. 97–111.

Mukherjee S, Koyi HA. 2010. Higher Himalayan Shear Zone, Zanskar Indian Himalaya - microstructural studies and extrusion mechanism by a combination of simple shear and channel flow. International Journal of Earth Sciences 99, 1083–1100.

Mukherjee S, Mulchrone KF. 2013. Viscous dissipation pattern in incompressible Newtonian simple shear zones: an analytical model. International Journal of Earth Sciences 102, 1165–1170.

Murti KS. 1987. Stratigraphy and sedimentation in Chattisgarh basin, In Purana Basins of Peninsular India, edited by BP Radhakrishna, Memoirs of the Geological Society of India, vol. 6, pp. 239–261.

Nanda J, Gupta S. 2012. Intracontinental orogenesis in an ancient continent-continent collision zone: Evidence from structure, metamorphism and P-T paths across a suspected suture zone within the Eastern Ghats Belt, India. Journal of Asian Earth Sciences 49, 376–395.

Nanda J, Panigrahi MK, Gupta S. 2014. Fluid inclusion studies on the Koraput Alkaline Complex, Eastern Ghats Province, India: implications for granulite facies metamorphism and exhumation. Journal of Asian Earth Sciences82, 10–20.

Nasipuri P, Bhadra S. 2013. Structural framework for the emplacement of Proterozoic anorthosite massif in the Eastern Ghats Granulite Belt, India: Implications for post-Rodinia – pre Gondwana tectonics. Mineralogy and Petrology107,861–880.

Naylor M, Sinclair HD. 2008. Pro- vs. retro-foreland basins. Basin Research 20, 285–303.

Passchier CW, Trouw RAJ. 2005. Microtectonics. Springer, Berlin, New York.

Patranabis-Deb S, Chaudhuri AK. 2002. Stratigraphic architecture of the Proterozoic succession in the eastern Chattisgarth Basin, India: tectonic implications. Sedimentary Geology, 147, 105–125.

Patranabis-Deb S, Bickford ME, Hill B, Chaudhuri AK, Basu A. 2007. SHRIMP ages of zircon in the uppermost tuff in Chattisgarh

Basin in central India require 500Ma adjustment in Indian Proterozoic stratigraphy. Journal of Geology 115, 407–415.

Pryer LL. 1993. Microstructures in feldspars from a major crustal thrust zone: the Grenville Front, Ontario, Canada. Journal of Structural Geology 15, 21–36.

Ramakrishnan M, Vaidyanadhan R. 2008. Geology of India. Geological Society of India, Bangalore, vol 1, 556 pp.

Ramakrishnan M. 1987. Stratigraphy, sedimentary environment and evolution of the Late Proterozoic Indravati basin, central India. In Purana Basins of Peninsula India, edited by BP Radhakrishna, Geological Society of India Memoir vol. 6, pp. 139–160.

Ratre K, Waele BD, Biswal TK, Sinha S. 2010. SHRIMP geochronology for the 1450 Ma Lakhna dyke swarm: Its implication for the presence of Eoarchaean crust in the Bastar Craton and 1450–517 Ma depositional age for Purana basin (Khariar), Eastern Indian Peninsula. Journal of Asian Earth Sciences 39, 565–577.

Roberts DG, Bally AW. 2012. Regional Geology and Tectonics: phanerozoic rift systems and sedimentary basins. Elsevier, Oxford.

Rogers JJW, Santosh M. 2002. Supercontinents in earth history. Gondwana Research 6, 357–368.

Saha D, Patranabis-Deb S. 2014. Proterozoic evolution of Eastern Dharwar and Bastar cratons, India – An overview of the intracratonic basins, craton margins and mobile belts. Journal of Asian Earth Sciences 91, 230–251.

Santosh M, Maruyama S, Yamamoto S. 2009. The making and breakingof supercontinents: Some speculations based on superplumes, superdownwelling and the role of tectosphere. Gondwana Research 15, 324–341.

Sarkar G, Corfu F, Paul DK, McNaughton NJ, Gupta SN, Bishui PK. 1993. Early Archean crust in Bastar craton, central India – a geochemical and isotopic study. Precambrian Research 62, 127–137.

Schmid SM, Casey M. 1986. Complete fabric analysis of some commonly observed quartz c-axis patterns. Mineral and rock deformation: laboratory studies. The Paterson Volume, American Geophysical union, Geophysical Monograph 36, pp. 263–286.

Simmat R, Raith MM. 2008. U-Th-Pb monazite geochronometry of the Eastern Ghats Belt, India: Timing and spatial disposition of poly-metamorphism. Precambrian Research 162, 16–39.

Srivastava P, Mitra G.1996. Deformation mechanisms and inverted thermal profile in the North Almora thrust mylonite zone, Kumaon Lesser Himalaya. Journal of Structural Geology 18, 27–39.

Stipp M, Stunitz H, Heilbronner R, Schimd SM. 2002a. The eastern Tonale fault zone: a 'natural laboratory' for crystal plastic deformation of quartz over a temperature range from 250 to 700 °C. Journal of Structural Geology 24, 1861–1884.

Stipp M, Stunitz H, Heilbronner R,Schimd SM. 2002b. Dynamic recrystallization of quartz: correlation between natural and experimental conditions. In Deformation Mechanisms, Rheology and Tectonics: current status and future perspectives, edited by S De Meer, MR Brury, JHP De Brasser, and GM Pennock, Geological Society of London, Special Publication, vol. 200, pp. 171–190.

Tullis JA. 1977. Preferred orientation of quartz produced slip during plane strain. Tectonophysics 39, 87–102.

Tullis J. 1983. Deformation in feldspars. In Feldspar Mineralogy, edited by P.H. Ribbe, Mineralogical Society of America Reviews in Mineralogy vol. 13, pp. 297–323.

Valdiya KS. 1982. Tectonic perspective of the Vindhyanchal region. In Geology of Vindhyachal, Prof. RC Mishra volume, edited by KS Valdiya, SB Bhatia, and VK Gaur, Hindustan Publication Corporation, Delhi, pp. 23–29.

Vernon RH. 2004. A Practical Guide to Rock Microstructure. Cambridge University Press, Cambridge, UK.

Vinogradov AP, Tugarinov AI, Zhykov CI, Stapricova NI, Bibicova EV, Khorre KG. 1964. Geochronology of the Indian Precambrian.

In Proceedings of the 2nd International Geological Congress, Pt. 10, pp.553–567.

Westbrook GK, Ladd JW, Buhl P, Bangs N, Tiley GJ. 1988. Cross section of an accretionary wedge: Barbados Ridge complex. Geology, 16 631–635.

Watts AB. 1992. The formation of sedimentary basins. In Understanding the Earth, edited by G. Brown, C. Hawkesworth, and C. Wilson, Cambridge University Press, Cambridge, UK, pp. 301–326.

Wenk HR, Canova G, Molinari A, Cocks UF. 1989. Viscoplastic modelling of texture development in quartzite. Journal of Geophysical Research 94, 17895–17906.

Wilson CJL. 1975. Preferred orientation in quartz ribbon mylonites. Bulletin of the Geological Society of America 86, 968–974.

Xypolias P. 2009. Some new aspects of kinematic vorticity analysis in naturally deformed quartzites. Journal of Structural Geology 31, 3–10.

Zadins ZZ, Mitra G. 1986. Brittle-ductile deformation along thrust faults, an example from the Hudson Valley thrust Belt. Geological Society of America Annual Meeting Abstracts 18, 799.

Zhao G, Sun M, Wilde SA, Li S. 2004. A Paleo-Mesoproterozoic supercontinent: assembly, growth and breakup. Earth Science Review 67, 91–123.

Chapter 12

Intrafolial folds: Review and examples from the western Indian Higher Himalaya

SOUMYAJIT MUKHERJEE[1], JAHNAVI NARAYAN PUNEKAR[2], TANUSHREE MAHADANI[1], and RUPSA MUKHERJEE[1]

[1] *Department of Earth Sciences, Indian Institute of Technology Bombay, Powai, Mumbai 400076, Maharashtra, India*
[2] *Department of Geosciences, Princeton University, Princeton, NJ, USA*

12.1 INTRODUCTION

Folds are perhaps the most intensively studied structures in geology (for example Ramsay 1967; Ez 2000; Harris et al. 2002, 2003, 2012a,b; Alsop and Holdsworth 2004; Mandal et al. 2004; Carreras et al. 2005; Bell 2010; Hudleston and Treagus 2010; Godin et al. 2011). Depending on morphologies and orientations, folds can be classified using several schemes (reviews by Ghosh 1993; Davis et al. 2012, etc.). Besides their rheological aspects, deciphering whether folds inside any shear zones are produced by shear has been emphasized (e.g. Mandal et al. 2004; Carreras et al. 2005; Bell et al. 2010). A couple of shear zone models altogether neglected fold formation within them, for example Koyi et al. (2013), Mukherjee and Biswas (2016, Chapter 5), Mulchrone and Mukherjee (in press). Mukherjee (2012a, 2014a) investigated the issue in terms of deformation of inactive markers in inclined shear zones undergoing extrusion and subduction. Folds related to shear zones are broadly of two types: (i) those with low interlimb angles and with significantly curved hinge lines developed before shear, some of which are sheath folds; and (ii) flow perturbed syn-shear folds that may be overturned and "intrafolial" (Alsop and Holdsworth 2004). In shear zones, locally overturned isoclinally folded foliations bound by straight foliation planes are most commonly called "intrafolial folds" (*intra* = inside; *folia* = foliation) (Dennis 1987; Allaby 2013). Intrafolial folds are found most commonly in mylonites (Trouw et al. 2000). Such folds have also been reported from cataclasites and obsidian (Higgins 1971), deformed soft sediments (Jirsa and Green 2011), slump structures (Woodcock 1976) and debris flows (Gawthrope and Clemmey 1985). The vergence of these folds is in conformity with shear sense of the shear zones they occur in. Intrafolial folds are disrupted to rootless folds if shear is more intense than in the adjacent layers even on a local scale. The adjacent rocks might be undeformed as well (Neuendorf et al. 2005). These folds

tightened as shear continued (Longridge et al. 2011). Early references to classical intrafolial folds as "drag folds" (e.g. fig. IX-38 in Hills 1965) were subsequently not followed. Carreras et al. (2007) viewed intrafolial folds both as "syn-shear folds", and "shear-related late folds" (their fig. 1c). Depending on other mechanisms perceived for intrafolial folds, they have also been described as "intrafolial strain-slip folds" (Ratcliffe and Harwood 1975) and "intrafolial shear folds" (Keiter et al. 2011). Intrafolial folds can tear apart by pronounced shear into rootless folds showing opposite closure (as in fig. 4.6B of Park 1997). Intrafolial folds have been referred to mainly as byproducts of some other studies, either as one of the ductile shear sense indicators in the field (Gangopadhyay 1995 but also others) or thin-sections (Trouw et al. 2010); or for their progressive evolution and geneses. Most of these were deduced on field observations (e.g. Passchier et al. 1991), with very few analytical- (Hara and Shimamoto 1984) and analog models (Bons and Jessel 1998). Notice that well before the concept of ductile shear dominated structural geology literature, intrafolial folds were explained in the same way in terms of "fluxion plane" and "fluxion layers" (Higgins 1971). Studying intrafolial folds is of practical importance, since a few ore bodies have been deciphered to be intrafolially folded (e.g. Laznicka 1985).

This chapter reviews morphologies and geneses of intrafolial folds. We also discuss use of such folds to determine shear senses from particular Himalayan shear zones.

12.2 GEOLOGY AND TECTONICS OF THE STUDY AREAS

The Higher Himalayan Shear Zone (HHSZ) consists of gneisses and schists of Precambrian and Proterozoic ages at dominantly greenschist to amphibolite facies (Grasemann et al. 1999; Grasemann and Vannay, 1999; Vannay et al. 1999; Vannay and Grasemann 2001;

Ductile Shear Zones: From Micro- to Macro-scales, First Edition. Edited by Soumyajit Mukherjee and Kieran F. Mulchrone.

Jain et al. 2005; Mukherjee 2015; Mukherjee et al. 2015). The southern margin of the HHSZ is the "Main Central Thrust" (MCT = Main Central Thrust-Lower: MCT_L of Godin et al. 2006) and the northern boundary the South Tibetan Detachment System (or the South Tibetan Detachment System-Upper: $STDS_U$ of Godin et al. 2006). The lower boundary of the HHSZ is now conventionally viewed as the MCT-zone, which is a mixture of Higher Himalayan and the underlying Lesser Himalayan rocks. The northern boundary of this zone is designated as the MCT-Upper (MCT_U) (review by Godin et al. 2006). The HHSZ is characterized by pre-Himalayan D_1 folding, a D_2 top-to-S/SW ductile shear, and post-Himalayan D_3 folding (review by Jain et al. 2002). Throughout the HHSZ, the D_2 phase of top-to-S/SW ductile shear was documented where N/NE dipping "main foliations" acted as the primary shear C-planes. In addition, a late top-to-NE ductile shear developed inside the $STDS_U$ along the same NE dipping main foliations (Burchfiel et al. 1992; Godin et al. 2006). Inside the HHSZ, one more ductile shear zone with normal shear sense – the South Tibetan Detachment System Lower ($STDS_L$) was delineated from both the eastern (review by Godin et al. 2006) and the western Himalaya (Mukherjee and Koyi 2010a). The $STDS_U$ and the $STDS_L$ are spatially separated. The timing and the magnitude of slip of the tectonic boundaries/zones- the MCT, the $STDS_U$ and the $STDS_L$ varies along the entire Himalayan chain. Slip of the $STDS_U$ varies from 42–255 km along the Himalayan chain (review by Leloup et al. 2010). In general, the timing of extensional shear within the $STDS_U$ was ~19–94 Ma, and the $STDS_L$ was ~24–42 Ma (Godin et al. 2006; Yin 2006). The extensional shear of the $STDS_U$ stopped ~5 Ma earlier in the western than in the eastern Himalaya (Leloup et al. 2010). Unlike the previously held view that the STDS developed by continuous deformation of a low-angle normal fault system, its genesis is now linked to both channel flow and extrusion (Kellett and Grujic 2012; also see Mukherjee 2005).

An inverted metamorphic gradient is indicated inside the HHSZ by isograds of high-grade minerals at the base (that is in the S), and considerable melting at or near the top (in the N). Three metamorphic episodes of the HHSZ have been deciphered: local M_1 phase around granite plutons; prograde M_2 metamorphism during the top-to-SW shear, and finally M_3 retrogression (see review by Jain et al. 2002) during crustal unroofing. Alternately, HHSZ metamorphosed either in two events- in the Eo-Himalayan (>44 – 33 Ma) and the Neo-Himalayan periods (~18 Ma), or through a protracted single phase (reviewed by Yakymchuk and Godin 2012).

The HHSZ extruded initially by a top-to-SW shear (the D_2 phase) from ~25 Ma (Yin 2006). This was followed by a crustal channel flow along with a top-to-SW shear since ~18 Ma (Mukherjee 2005, 2009, 2012a,b, 2013a,b, 2014b, for reviews; Godin et al. 2006; Hodges 2006; Yin 2006; Harris 2007; Mukherjee and Koyi 2010a). The $STDS_U$ and the $STDS_L$ were considered to be produced during channel flow affecting different portions of the HHSZ

(see Hollister and Grujic 2006; Mukherjee and Koyi 2010a for details). The relative role of tectonics and erosion in sustaining the channel flow is debated (reviews by Mukherjee 2005; Godin et al. 2006; Hodges 2006; Jones et al. 2006; Yin 2006; Harris 2007; Mukherjee and Koyi 2010a; Godard and Burbank 2011). A more recent postulate is that the HHSZ extruded by channel flow flip-flop alternating with intervals of critical taper mechanisms (Beaumont and Jamieson 2010; Chambers et al. 2011).

The $STDS_U$ is locally known as the Zanskar Shear Zone (ZSZ) in Kashmir (India). The ZSZ is a ~1 km thick zone of mylonites (Dèzes et al. 1999), has a throw of 15–50 km (Herren 1987), and exhibits normal way up metamorphic isograds (review by Walker et al. 2001). The ZSZ shows three dominant deformations: (i) a top-to-SW compressional shear, (ii) subsequent top-to-NE extensional shear, and (iii) a latest top-to-NE (down) extensional ductile shear (Herren 1987; Patel et al. 1993; Dèzes et al. 1999; Jain and Patel 1999; Mukherjee 2007, 2010b; Mukherjee and Koyi 2010b). The $STDS_L$ has not yet been reported from the Zanskar section.

In the Sutlej section (Himachal Pradesh, India), schists constitute the southern part of the HHSZ, and gneisses, migmatites and granites the upper part (Mukherjee and Koyi 2010a). A component of pure shear was quantified from this section as having a kinematic vorticity number of 0.86 (Grasemann et al. 1999), or 0.73–3.81 (Law et al. 2010). S to N, up to the MCT_U, the peak metamorphic temperature rose from 610 to 700°C and the peak pressure fell from 900 to 700 MPa (Vannay et al. 1999) indicating metamorphism at ~30 km depth (Vannay and Grasemann 2001).

This work documents intrafolial folds in the HHSZ of the Sutlej river section (Fig. 12.1a): both from inside and outside the $STDS_U$ and the $STDS_L$; and from the XZ oriented thin-sections of the ZSZ (Fig. 12.1b). Based on pre-existing geochronological data and a number of trial tectonic models, Mukherjee and Koyi (2010a) argued that in Sutlej section, the most plausible timing of extensional shear in the $STDS_U$ (normal shear sense) had a shorter span of ~15–54 Ma than ~14–42 Ma within the $STDS_L$. Godin et al. (2006) considered that the ZSZ underwent extensional shear during 18–86 Ma (data originally by Inger 1998). The intrafolial folds of the Zanskar Shear Zone underwent a top-to-N/NE ductile shear during the same time span (Mukherjee and Koyi 2010b; Mukherjee 2010b).

12.3 REVIEW OF INTRAFOLIAL FOLDS

12.3.1 Intrafolial folds *sensu stricto*

12.3.1.1 Morphologies

Where the intrafolial folds occur in trains, these folds approximate periodic asymmetric waves (Ramsay 1967) that are tight to isoclinal (Park 1997) with thick hinges and thin limbs. These folds seldom have straight limbs or sharp hinges. Most have curved limbs and round hinges

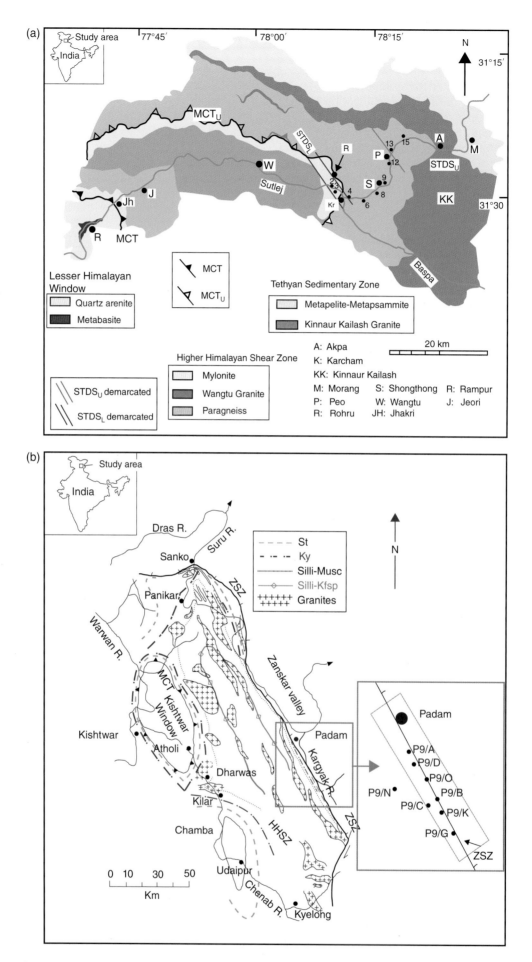

Fig. 12.1. The study areas. Numbers indicate sample locations. (a) Higher Himalayan Shear Zone, Sutlej river section (Singh, 1993; Srikantia and Bhargava, 1998; Vannay and Grasemann, 1998). The location of the "Main Central Thrust" (MCT) that bounds the Lesser Himalayan rocks is as per Singh (1993). The Vaikrita Thrust of Srikantia and Bhargava (1998) is designated as the "MCT-Upper" (MCT$_U$) of Godin et al. (2006). (b) Higher Himalayan Shear Zone, Zanskar section. Source: Searle et al. 1988. Reproduced with permission of The Royal Society.

in axial profiles. In single trains, one set of limbs is usually longer than the other. Such morphologies are also reported from intrafolial folds developed within debris flow (Gawthorpe and Clemmey 1985).

12.3.1.2 Genesis

Intrafolial folds can form in four ways: (i) folding a part of a foliation during shear along the later (Fig. 12.2a; from Passchier and Trouw 2005), or whole of the foliation planes (Winter 2012); (ii) modification of pre-existing structures/features, for example, cross-beddings and dykes; (iii) folding of foliations formed during simple shear; and (iv) remnant folded layers by flattening/unfolding that are more competent than the surrounding straight and completely unfolded layers (Llorens et al. 2013). For case (iii) above, the fold hinge rotated along with shear (Duebendorfer et al. 1990). For case (iv), the competent bed could be limestone where intrafolial folds are confined and are bound by incompetent shale dominated rock (Linn et al. 2002). Ez (2000) concluded that a shear perfectly parallel to the foliation plane cannot produce any folds. Such folds can be initiated by rotation of foliations around any harder inclusion (Fossen 2010). In other words, some hindrance to simple shear is needed for intrafolial folds to develop (Swanson 1999). However, a hindrance can also produce kinks and crenulations (Swanson 1999). He demonstrated how a pre-existing symmetric fold shears into an "overturned fold" (= folds with limbs of nearly the same dip direction). Limbs vary greatly in thickness on either side of a thick hinge (Fig. 12.2b; fig. 11.23 of Fossen 2010). Best (2006) proposed that simple shear of straight foliation

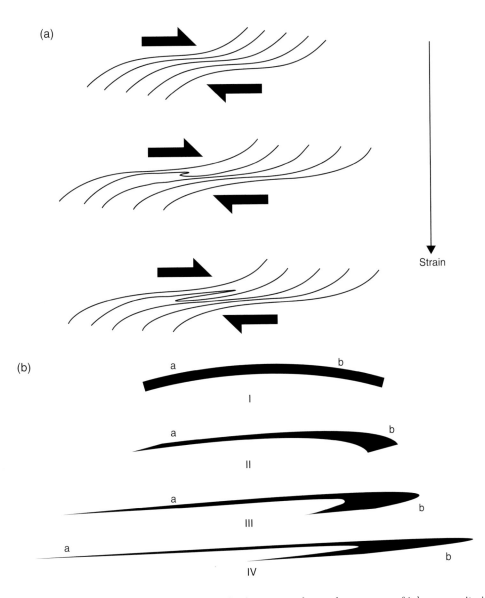

Fig. 12.2. (a) Ductile shear gives rise to intrafolial folds in a single shear event due to the presence of inhomogeneity in the rock. Source: Passchier and Trouw 2005. Reproduced with permission from Springer Science + Business Media. (b) A symmetric fold becomes asymmetric and overturned due to simple shear. The hinge is thickened and the two limbs vary in thickness and length. Source: Fossen 2010. Reproduced with permission of Cambridge University Press. Limbs 'a' and 'b' are marked and plotted in Ramsay's (1967) scheme in Fig. 12.16.

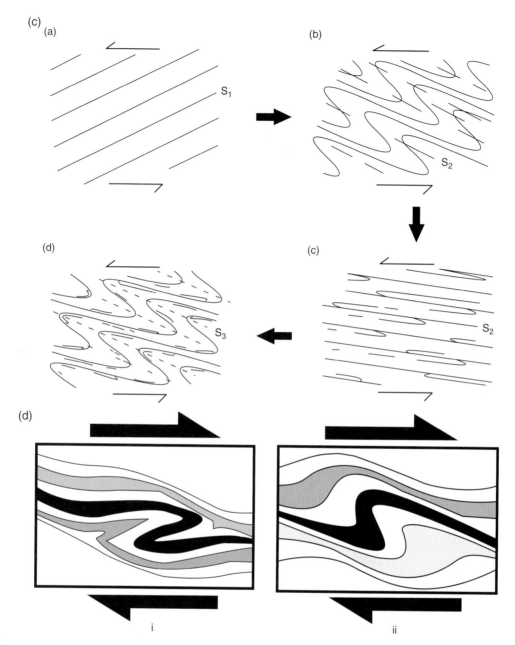

Fig. 12.2. (*Continued*)   (c) Progressive simple shear (A) on straight foliations (S_0) depicted. (B) S_0 overturned folded and axial planar foliations (S_1) develop. (C) S_1 are slipped, where the S_2 act as brittle fault planes. (D) S_2 gets overturned folded and straight S_3 axial planar foliations develop. Source: Best 2006. Reproduced with permission of John Wiley & Sons. (d) i. Synthetic fold, ii. Antithetic or back-rotated fold produced by secondary shear associated with top-to-right shear. Source: Lebit et al. 2005. Reproduced with permission from Elsevier.

planes can develop progressively overturned folds on a number of subsequent foliation planes (Fig. 12.2c). Shear magnifies initial curvatures of planar markers and their hinges to develop intrafolial folds (fig. 3.21a of Passchier et al. 1991). Taking a mixture of octachloropropane and camphor as the model material in a Taylor–Couette flow with the shear strain varying from 111 to 122, Bons and Jessel (1998) modeled development of intrafolial folds in a train. Initial irregularities evolved into fold hinges as the pre-developed fold axes attained gentler plunges.

The common perception of a fold inside an apparently undeformed matrix is that the folded material is of much lower viscosity than the matrix (fig. 3–34 of Billings 2008; also Nevin 1957). Ez (2000) questioned whether straight foliations that bound intrafolial folds are really undeformed zones. Any (competent) quartzite or pegmatite vein within a schistose rock-mass can also facilitate the competence difference and develop intrafolial folds (Harris et al. 2002). However, Whitten (1966) referred (in his fig. 175) to a number of intrafolial folds in marbles where the folded material and the matrix appear to be composed of the same mineral. Hara and Shimamoto (1984) inferred from such examples that viscosity contrast between the fold and the matrix does not seem to be a significant genetic factor for such folds. This conjecture is also found in the postulation of Carrerras et al. (2005) that these folds develop from an isotropic unfoliated rock body by ductile shear, whereby foliations form and fold simultaneously at local scales.

The disparity in thickness between different parts of a single fold becomes more acute with progressive deformation. This indicates that as shearing proceeds, materials flow from both limbs towards the hinges attenuating the limbs until they break (also fig. 3-30 of Ramsay, 1967). The limbs and axial planes are usually at low-angles to foliations that are regionally straight. Extreme shear may lead to tearing of the fold trains into disconnected folds of same overall geometry and orientation (fig. 4.19 of Passchier et al. 1991). On the other hand, shear can reduce the angle between the axial plane and the regional foliation so that they sub-parallel (fig. 12.27 of van der Pluijm and Marshak 2004). In this case, the sense of shear becomes indistinct. However, continued shear can redevelop an intrafolial fold with the previous folded fabric preserved within it (fig. 12.27c of van der Pluijm and Marshak 2004).

12.3.1.3 Shear sense

Passchier and Trouw (2005) cautioned that unless the three-dimensional shapes of the folds before they became intrafolial are known, and unless the plane of observation perpendiculars the main foliation and parallels stretching lineation, intrafolial folds are not reliable shear sense indicators (also see Fossen 2010). However, whereas the second constraint of selecting the proper plane was used to deduce shear senses in the field- and thin-sections (e.g. Mukherjee 2007, 2010a,b, 2011a; Mukherjee and Koyi 2010a,b), the first condition was not considered since the ductile shear sense indicated by those folds matched well with other shear sense indicators, most notably with mineral fish (Mukherjee 2011b) and sigmoidal quartz veins. Another point of caution is that Carrerras et al. (2007) documented back rotated folds due to ductile secondary shear. Though these folds resemble intrafolial folds, they indicate an opposed shear sense (Fig. 12.2d). Intrafolial folds are reliable ductile shear sense indicators if they are regionally common and if unassociated with any regional folds (Davis et al. 2012).

12.3.2 Intrafolial folds *sensu lato*

A number of other situations also give rise to intrafolial folds *sensu lato*. We call them *sensu lato* since these were not described so far as "intrafolial folds". For example, Harris et al. (2002) and Fossen (2010) demonstrated how a linear marker initially at a high angle to the shear direction reorients, shortens and folds to a "hook" shape (see Fig. 12.3a) showing an apparent ductile shear sense. Note that Hooper and Hatcher Jr. (1988) used the word "hook" to describe sheared out and isolated hinges of microfolded quartz ribbons. An important observation from Fig. 12.3a

is that the long limbs of such folds do not parallel the primary shear C-planes. Retro-shear on a sigmoid S-fabric (Fig. 12.3b; fig. 9 of Wennberg, 1996; also fig. 3a of Aller et al. 2011) also yields foliation bound hooks. Such hooks of leucosomes and quartz veins were documented in meta-sedimentary rocks on mesoscopic scales in other shear zones (figs. 6.38 and 6.44 of Vernon and Clarke 2008), and in mylonites on micro-scales (Mukherjee 2007; fig. 8 of Mukherjee and Koyi 2010b). At the two corners of microscopic quarter structures and symmetric phi-objects, the foliation planes locally attains overturned folds in accordance with the shear sense (Fig. 12.3c; fig. 15.33 of Bobyarchick, 1998; fig. 43B of Davis et al. 1998; fig. 15.33 of Fossen, 2010; figs. 9.7.9–9.7.12 of Trouw et al. 2010). Notice that Alves (2015) included quarter folds developed in soft sediments, especially in slumps (Jirsa and Green 2011), within intrafolial folds category. Besides soft sediments, intrafolial folds have also been recognized from lava flows as primary structures by many (e.g. Andrews and Branney 2011). Grasemann and Dabrowski (2015) considered winged inclusions can give rise to intrafolial folds surrounding them (also see Arbaret et al. 2001). Occasionally, one may encounter microscopic single mineral grains displaying intrafolial folds (Fig. 12.3d; figs. 7b,-c of ten Grotenhuis et al. 2003; same as fig. 5.32d Passchier and Trouw, 2005; and fig. 9.5.16 of Trouw et al. 2010; also folded garnet in Mukherjee, 2010a). In meso- and microscopic flanking structures, host fabric elements locally fold and drag near cross-cutting elements (Fig. 12.3e; figs. 2a-d of Mukherjee 2011; also see Becker 1995). While the cross-cutting element could be dykes, fractures, joints, faults, secondary shear planes or zones, veins, melt such as leucosomes, burrows, inclusions, minerals, or boudins; the host fabric element can be bedding planes, foliations, lineations, mineral cleavages, and grain margins (Mukherjee and Koyi. 2009, and references therein). Intrafolial parasitic folds also develop by ductile shear over first generation flexure slip folds (fig. 54 of Nevin, 1957). Parasitic S- and Z- shaped folds were considered intrafolial when they are bound by a pair of foliation planes (Maass et al. 1980). Where multiple generations of foliations survived, the nth generation of straight foliation might cut across and confine the $(n-1)$ generation of folded foliation (Fig. 12.3f). These folded foliations need not be produced by ductile shear, and could merely be asymmetric folds (fig. 4 of Bell 2010). Compression (pure shear) perpendicular to foliations that are axial planar to symmetric folds transposes/crenulates folds bound by the newly formed foliation. This holds both for vertical and inclined axial planes of the initial folds (Davis et al. 2012; figs. 12.27a-c of van der Pluijm and Marshak 2004; fig. 12.22 of Fossen 2010). On pronounced compression, however, these folds straighten (fig. 4.18-b of Passchier and Trouw 2005). If the C-plane of simple shear are compressed orthogonally, the prior shear-induced foliations can develop intrafolial folds with a vergence perpendicular to the direction of the preceding shear (Fig. 12.3g; fig. 3.21a of Passchier et al. 1991).

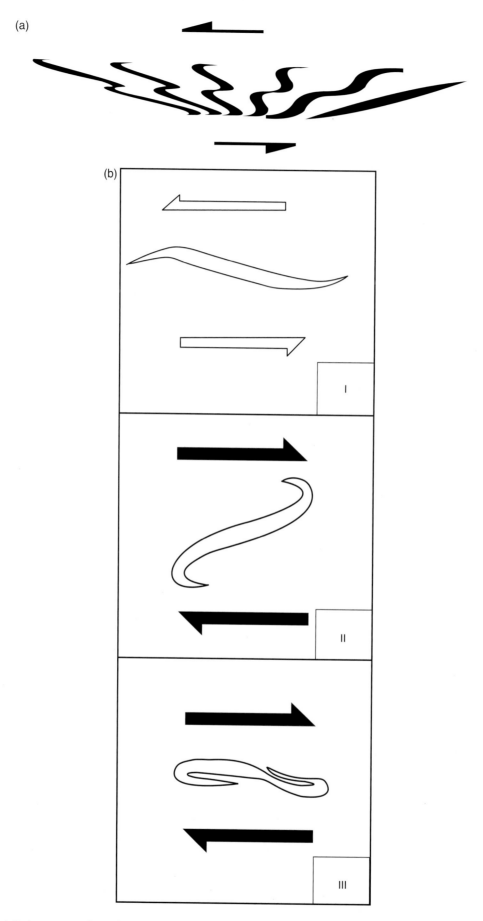

Fig. 12.3. (a) Top-to-left shear on a rightward inclined marker finally leads to intrafolial folds with longer limb dipping towards right and the shorter one towards left. Source: Fossen 2010. Reproduced with permission of Cambridge University Press; also see fig. 12.26 of van der Pluijm and Marshak 2004; fig. 7–72 of Ramsay 1967; fig. 1m of Harris 2003). (b) Reactivation of a top-to-left sheared fabric (I) into a later top-to-right one (II) leads ultimately to folded fabric at low angle to the plane of shear (III). Source: Wennberg, 1996. Reproduced with permission from Elsevier.

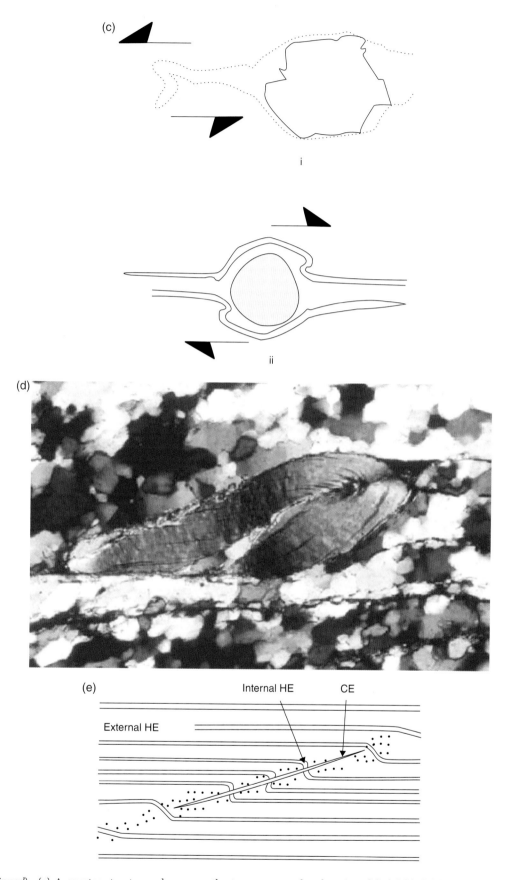

Fig. 12.3. (*Continued*) (c) A quarter structure where near the two corners of a clast, intrafolial folded foliation planes occur. Source: Passchier and Trouw 2005. Reproduced with permission from Springer Science + Business Media. (d) An isoclinally and intrafolial folded single grain of mica under microscope. Source: ten Grotenhuis et al. 2003. Reproduced with permission from Elsevier. (e) Flanking structure: dragged portion of the host fabric element (HE) close to the cross-cutting element (CE) is the "internal HE". Straight and undisturbed "external HE". Source: Passchier, 2001. Reproduced with permission from Elsevier.

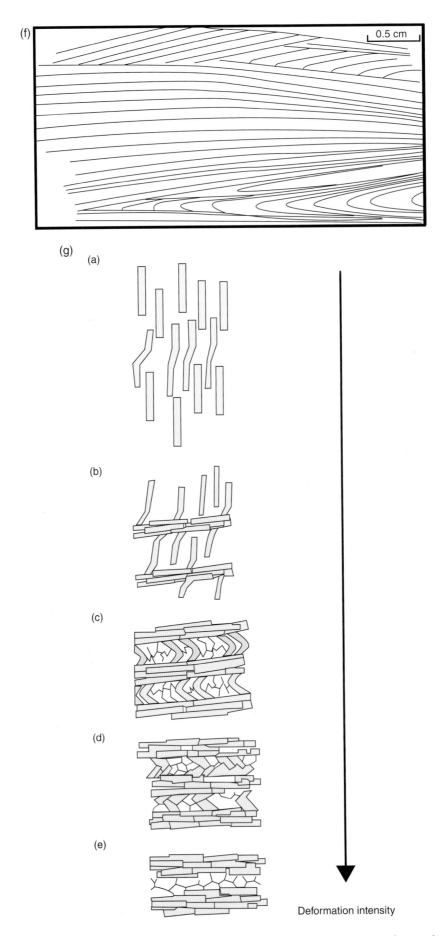

(f)

0.5 cm

(g)

(a)

(b)

(c)

(d)

(e)

Deformation intensity

Fig. 12.3. (*Continued*) (f) A number of generations of foliations are present: some of these truncate others and one is folded. The folded foliation is bound by other foliations, hence the fold is "intrafolial". Source: Bell 2010. (g) How pronounced compression flattens folded lath shaped grains is shown. Source: Passchier and Trouw 2005. Reproduced with permission from Springer Science + Business Media.

12.4   PRESENT STUDY

12.4.1   Do the studied intrafolial folds fall into *sensu stricto*, or *sensu lato category*?

Quartz rich layers and/or leucosomes define some of the intrafolial folds in the Sutlej section. They occur in a train outside the two detachments, the $STDS_U$ and the $STDS_L$, and verge SW (Figs 12.4a–d, 12.5b, c, 12.6a–c). On the other hand, folds verge NE inside those two detachments (Figs 12.5a, d; 12.7a–d). Inside the ZSZ intrafolial folds also verge NE (Fig. 12.8). These folds are intrafolial *sensu stricto*, and are unrelated to different types of intrafolial folds *sensu lato* as mentioned in Section 12.3. The reasons are as follows. Many of the folded quartz aggregates documented here have significantly continuous limbs

along the primary shear C-planes (Figs 12.5a–d, 12.6b, d, 12.7a, c, d, 12.9c). So unlike Fig. 12.3(a, b), they probably did not originate from initially antithetic layers. As deciphered on meso- and microscales, these folds confine within primary shear C-planes. Mukherjee (2010a) and Mukherjee and Koyi (2010a) distinguished secondary shear planes in the Sutlej section at 15–55° to the C-planes. Thus the studied folds contradict counter-rotated back-folds within secondary shear planes. The documented folds do not constitute any porphyroblast/clast system, and hence differ from mineral grains as in Fig. 12.3c. None of the studied folds are the tails of mineral grains, which are in contrast to the cases (i) and (ii) of Fig. 12.3c. The folds presented here are unassociated with any nearby structural element that cross-cut them.

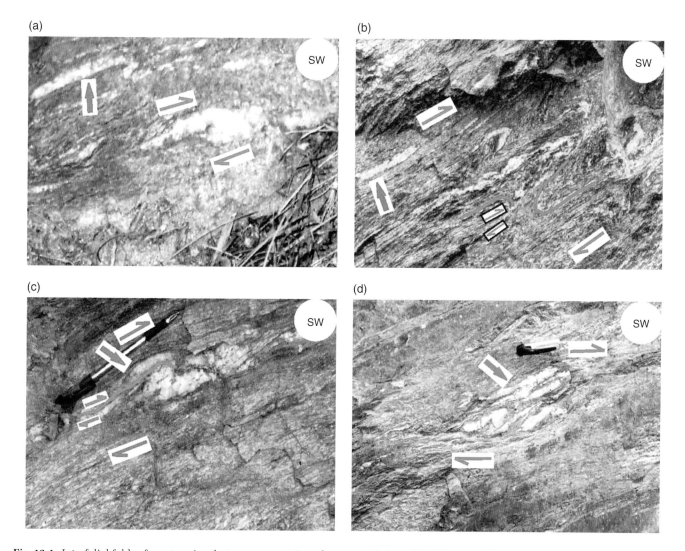

Fig. 12.4. Intrafolial folds of quartz veins that are asymmetric and overturned, from the Higher Himalayan Shear Zone, Sutlej section. Shear sense: top-to-SW (up). (a) Limbs of different thicknesses, thicker and round hinge. An intrafolial fold of different style is shown by a green arrow. Location: 6. (b) Ductile shear disrupted intrafolial folds into a number of rootless fold hinges. Brittle shear also acted along the main foliation. Thicker quartz vein along the main foliation (green arrow). Source: Mukherjee 2010a. Reproduced with permission from S. Mukherjee. Location: 15. (c) Ductile shear disrupted limb of an intrafolial fold (green arrow). The disrupted part overrode in a piggy-back manner (pair of smaller half arrows) over another fold along the main foliation. The plane of override also acted as a brittle shear plane (blue line). Location: 12. (d) Irregular adjacent folds of different morphologies, one of which is a "flame fold" with a sharp hinge (orange arrow). Source: Mukherjee 2010a. Reproduced with permission from S. Mukherjee. Location: 9.

(a)

(b)

(c)

(d)

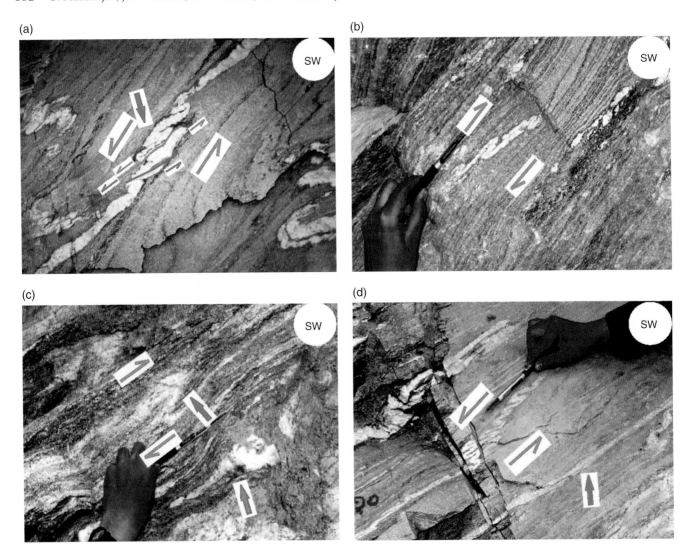

Fig. 12.5. Asymmetric overturned non-periodic intrafolial folds in a train, from the Higher Himalayan Shear Zone, Sutlej section (Fig. 12.2a). (a) Brittle shear sub-parallel to the axial trace disrupted the limbs (green and blue lines and their half arrows). The hinge zone is remarkably straight. Shear sense: top-to-NE (down). Location: 3. (b) Isoclinal fold and almost a box fold in the same train. The vergence of the isoclinal fold indicates the shear sense (top-to-SW, up) but not the box fold. Source: Mukherjee 2010a. Reproduced with permission from S. Mukherjee. Location: 13. This fold is plotted in Ramsay's (1967) scheme in Fig. 12.13. (c) The folded quartz vein cuts across the foliation planes (green full arrow). Shear sense: top-to-SW (up). Source: Mukherjee 2010a. Reproduced with permission from S. Mukherjee. Location: 3. (d) Thicknesses of limbs vary along the fold train. Uniform shear sense is displayed by every fold in the train. Shear sense: top-to-NE (down). Another quartz layer is nearly straight and not folded (orange arrow) Location: 2.

Therefore, these folds should not be correlated with flanking structures of Fig. 12.3e. Large-scale folds exist neither in the Sutlej section of the HHSZ (Vannay et al. 2004; Mukherjee 2007, 2010a – especially its fig. 14d; Mukherjee and Koyi 2010a) nor exclusively within the ZSZ (Patel et al. 1993; Mukherjee 2007, 2010b; Mukherjee and Koyi 2010b). Notice that the main foliations (or the primary shear C-planes) dip N/NE moderately. Therefore, these intrafolial folds do not categorize as parasitic folds. A few hook folded leucosomes do occur within the STDS$_L$, Sutlej section (Fig. 12.7b), and on microscale within the ZSZ defined by sillimanite and quartz minerals (see Fig. 12.12a-d). They could have formed from either an undeformed marker (such as Fig. 12.3a) or a shear fabric (e.g. Fig. 12.3b) that was

initially oriented antithetically. The final shear sense of top-to-NE sense as indicated by intrafolial folds *sensu stricto* matches with that interpreted from hooks (Fig. 12.7b). Three foliation sets occur repeatedly in all the study areas- two sets of the S-fabrics indicative of an earlier top-to-SW (Figs 12.4a–d, 12.5c, 12.6d) and a late phase top-to-NE ductile shear (Figs 12.5a, b, d, 12.6a–c, 12.7a–d, 12.8a–c, 12.9a–d, 12.10a–d, 12.11a–c, also Fig. 12.12). Both these sets of S-fabrics are bound by a common set of C-planes. Besides, a synthetic shear sense indicated by an S- plane and its bounding nearly straight C′-plane persist in the Sutlej section (figs. 4b, 5d, 6c, d, 8a of Mukherjee and Koyi 2010a). Thus, although multiple sets of foliations exist in the study areas, all of them have been categorized into specific deformation patterns,

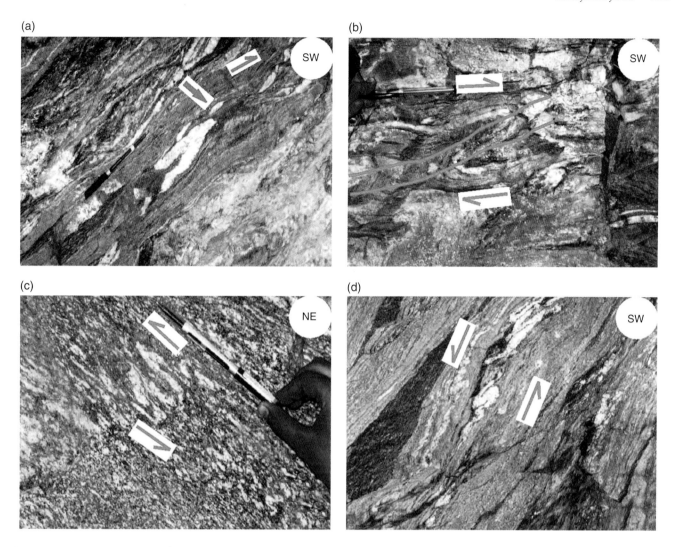

Fig. 12.6. Asymmetric overturned intrafolial folds, from the Higher Himalayan Shear Zone, Sutlej section. (a) The rootless flame fold of quartz with sharp hinge and fold axis sub-parallel to the foliation. Does not give shear sense alone. The tiny sigmoid quartz (green arrow) reveals a top-to-SW shear sense. Source: Mukherjee, 2010a. Reproduced with permission from S. Mukherjee. Location: 3. (b) Axial traces (blue lines) are somewhat curved. One of the limbs is sub-horizontal. Top-to-SW (up) shear. Location: 8. (c) S-shaped fold that truncates foliations. Top-to-SW (up) shear. Source: Mukherjee, 2010a. Reproduced with permission from S. Mukherjee. Location: 12. (d) Hinge area much thicker than the limbs. Top-to-NE (down) shear. Location: 3.

and they do not "randomly" cut across. It was found that the intrafolial folds are confined within the C-plane that is common to the northeast- and southwestward ductile shear. The studied intrafolial folds occur close to many other shear sense indicators such as mineral fish and sigmoid quartz veins (Mukherjee 2007, 2010a,b; Mukherjee and Koyi 2010a,b). This indicates simple shear produced those shear structures, and were not merely transposed foliations. As a NE–SW compression since ~55 Ma persisted in the India–Eurasia collisional regime (Keary et al. 2009), compression induced SW vergent intrafolial folding of the NE-dipping main foliation planes is one possibility, similar to Fig. 3h. However this is improbable. As argued in the previous point, it cannot explain a variety of unambiguous shear sense indicators on meso- and microscales that qualified the Higher Himalaya as a shear zone (see Jain and Anand 1988).

12.4.2 Morphology and structures

1 The two limbs of the studied individual intrafolial folds dip either SW (Figs 12.4a–d, 12.5b, c) or NE (Figs 12.5a, d, 12.6a–c, 12.7a–d, 12.8a, c, 12.9a–d, 12.10a–d, 12.11a–d). Therefore, these folds are overturned folds. Such folds are bound by primary shear C-planes (main foliations) that in the field are defined by thin planar leucosomes (Figs 12.4b, c, 12.5b–d, 12.6d, 12.7a–d), and on microscale by micas (Fig. 12.8). In the field, these fold trains can be traced for a maximum of about a meter, and in thin-section, for about a millimeter. Lengths and thicknesses of limbs of few intrafolial folds are quite dissimilar (Figs 12.4a, 12.9d, 12.10c, d, 12.11d). Most of their axial planes are straight (Figs 12.4a, d, 12.5a–d, 12.6a), but some are curved (Figs 12.6b, 12.7a, c, 12.8a–c, 12.9d; 12.10b,

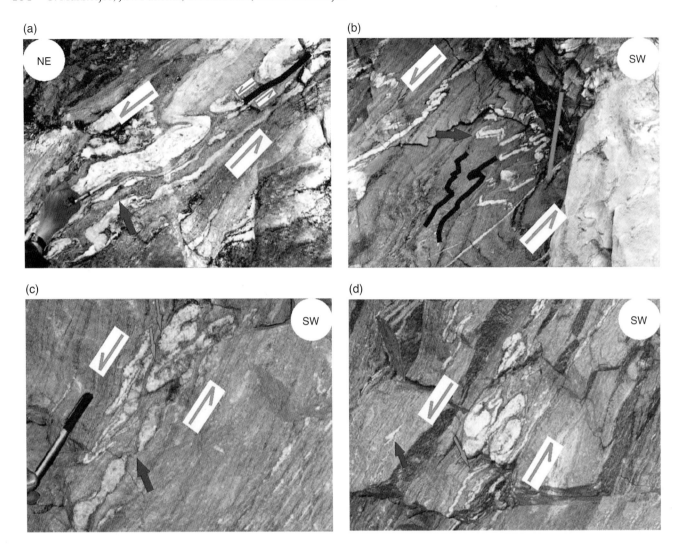

Fig. 12.7. Asymmetric overturned intrafolial folds, from the Higher Himalayan Shear Zone, Sutlej section. (a) The thicker layer of quartz shows more prominent folds. Along the main foliation, brittle faulting (green line and half arrows) disrupted the fold. Top-to-NE (down) shear sense. A top-to-SW shear is indicated by sigmoid quartz veins (orange arrow). Source: Mukherjee, 2010a. Reproduced with permission from S. Mukherjee. Location: 3. (b) Several rootless folds around the middle of the view, but only one of them have parasitic folds (orange arrow). At top-left side a continuous layer of quartz got intrafolial folded, its initial orientation with respect the shear C-plane is approximately represented by a green line. The matrix of schistose psammite is also intrafolial folded in sympathy with the quartz vein. The blue full arrow points at a fold of different style. Location: 4. (c) Folds in quartz layers with different thickness. Fold hinges of different thicknesses (blue and green full arrows). Layer parallel boudinaging took place (orange arrow) Top-to-NE (down) shear. Location: 3. (d) Side by side folds of contrasting thickness. A tiny rootless fold with well-developed hinge and limbs (orange arrow), and one more tiny intrafolial folds in a train (blue full arrow). Top-to-NE (down) shear. Location: 3.

12.11c). Microscopic intrafolial folds of single minerals seldom exhibit the geometries of sheared boxes and are rootless (Fig. 12.11b). A few rare observations are as follows. Shear planes can sharply truncate intrafolial folds with round hinges (Fig. 12.5a). Sometimes hinges of rootless intrafolial folds shears into flame shapes (Figs 12.5d and 12.7a; also designated as "intrafolial hinges" and "rootless lenses": Nicolas 1987). Microscopic sheath folds and intrafolial folds (Fig. 10c) co-existing in the same field of view indicate strain partitioning over a few mm distance. This is because sheath folds are products of extreme ductile shear and can originate through intrafolial folds (Davis et al. 2012). Note that sheath folds too can be intrafolial

(Andrews et al. 2005). An intrafolial fold with a hinge much thinner than its limbs (Fig. 12.10c) is unusual than the opposite case (Figs 12.4a, 12.6c, 12.6d, 12.7d, 12.10d).

2 Intrafolial folds of contrasting geometries (and sizes) sometimes occur adjacently (Figs 12.4a, 12.5c, 12.7d, 12.9c). Such changes in style and/or size can be across the trains (Figs 12.4a, 12.5c, 12.7d). Style can vary even in the same train – for example, folds with U-shaped and sheared box geometries can coexist (Fig. 12.9c).

3 Intrafolial folded quartz grains show wavy extinction under an optical microscope indicating that their optic axes rotated by folding (Fig. 12.10d). Individual recrystallized quartz grains too deform and assume shape-

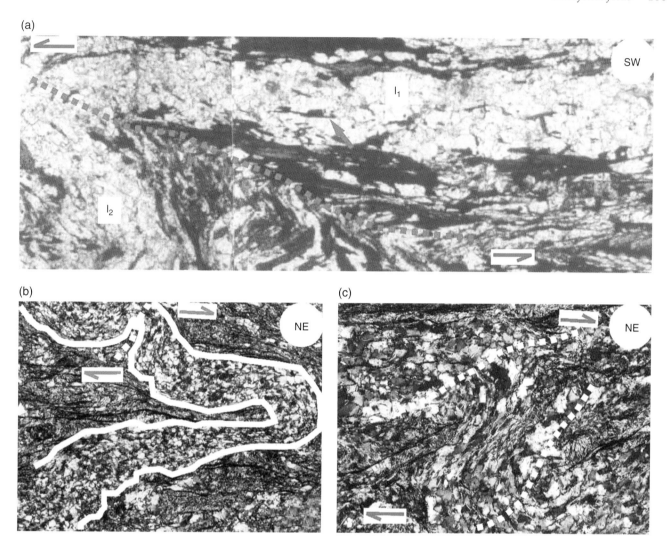

Fig. 12.8. Top-to-NE shear revealed by microscopic intrafolial folds with curved axial traces, Zanskar Shear Zone. (a) A quartz vein sub-parallel to the main foliation got round hinged folded, and maintained its longer limb (l_1) of nearly the same orientation and the shorter (l_2) at an angle. Very thin and disjointed biotites within the fold define subtle layers (blue full arrow). Thinner biotite and other quartz layers are tightly folded with sharper hinges developed in the core of the folded quartz layer. Plane polarized light. Location: P9/K. Photo length: 2 mm. Source: Mukherjee and Koyi 2010b. Reproduced with permission from Springer Science + Business Media. (b, c) The magnification of Fig. 12.8b is Fig. 12.8c: polyclinal irregular folded quartz vein. Quartz grains are elongated along folded layer. Thinner and disconnected micas also defined layers. Curved axial traces (white dash lines) define the shear sense. Location: P9/D. Photo lengths: 1 mm. Source: Mukherjee 2013. Reproduced with permission from Springer Science + Business Media.

preferred orientation (= "oblique foliation" of Vernon 2004; Trouw et al. 2010) when the aggregate intrafolial folds. In plane polarized light, such patterns appear subtle but are decipherable (compare Fig. 12.9a and Fig. 12.9b). The oblique foliation and the fold vergence show the same ductile shear sense connoting coeval folding and the obliquity of the foliation. The following effects are deciphered when intrafolial folds undergo brittle deformation. Thinning limbs thicken adjacent hinge zones so that the limbs snap. This resembles the genesis of lenticular boudins through "pinch and swell" structures (Twiss and Moores 2007). Hinge areas thicker than the limbs (especially Figs 12.4a and 12.5c) indicate that materials flowed from the limbs towards the hinges during folding. The disrupted portion of an

intrafolial fold can thrust over the hinge of an adjacent fold along a local brittle plane. This brittle plane in such a case parallels the ductile shear plane that bounds the fold (Fig. 12.4c). Brittle shear of limbs of intrafolial folds occur at an obtuse angle to the shear direction and sub-parallel the axial traces (Fig. 12.5a). These shears confine solely within the folded layer and not within the matrix. This indicates that the local brittle shears acted during the progressive evolution of the intrafolial folds, and did not relate to regional tectonics. Likewise, fractures nearly orthogonal to the limbs confine entirely to the limbs. These fractures are not perfectly straight (Figs 12.9a and 12.10a), and are unrelated to the mesoscopic deformation structures/fabrics. Some of the fractures developed only at one of

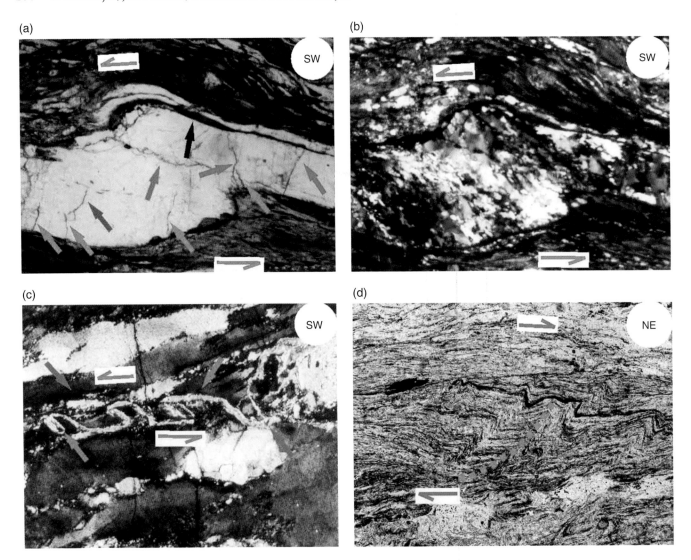

Fig. 12.9. (a.b) Intrafolial folds under microscope from the Zanskar Shear Zone. Few fractures only at the margins (blue full arrows), a few of which cross the complete limb (green full arrows). A Y-shaped fracture network formed inside the fold (purple full arrow). A much thinner layer of biotite occurs only at one of the margins of the folded quartz grain (black full arrow). Location: P9/K. Photo length: 3 mm. (a) Plane polarized light. (b) Cross-polarized light. Source: Mukherjee and Koyi 2010b. Reproduced with permission from Springer Science + Business Media. (c) Mylonitic foliation defined by train of micas is intrafolial folded into asymmetric overturned U- and box shaped folds (blue full arrows). Shear planes are defined by elongated recrystallized quartz grain and straight grain boundary (orange arrows). Cross-polarized light. Photo length: 2 mm; Location: P9/G. Source: Mukherjee 2013. Reproduced with permission from Springer Science + Business Media. (d) Kinked mica layers with SW dipping axial traces. Plane polarized light. Photo length: 1 mm. Source: Mukherjee 2013. Reproduced with permission from Springer Science + Business Media.

the margins of the folds (Fig. 12.9a). In a very rare case, a Y-shaped fracture formed inside the fold but did not reach any margin (Fig. 12.9a). A fracture can subparallel axial trace (Fig. 12.11d).

4 On mesoscopic scales, intrafolial folds of quartz aggregates are devoid of internal mica layers (Fig. 12.4a–d). On the other hand, on microscales, the folded quartz aggregates develop foliations defined by micas in one of the following ways: (i) as separated grains but of same orientation (Fig. 12.8a); (ii) as grains in close continuation and in several layers (Fig. 12.8b, c); and (iii) an aggregate of grains occurring at a part of the fold (Fig. 12.9a). In these cases, foliation planes inside the

folds mimic the later (like "F" in Fig. 12.2f). This indicates that the studied intrafolial folds with the same shear sense are not recycled, unlike Fig. 12.2a.

5 Our study of 42 intrafolial folds, both isolated/rootless and in trains in the Sutlej section, and another 18 in thin-section from the ZSZ reveal that the vast majority (e.g. Figs 12.13 and 12.14) plot within the Class 1C of Ramsay's (1967) scheme of fold classification. Ramsay (1967) defined t' as the orthogonal thickness at a point on the fold limb divided by that at its hinge. Similarly, T' is the ratio of the axial planar thicknesses. α is the acute angle between the tangent drawn at the point on the fold limb where the dip isogon was drawn. The plots

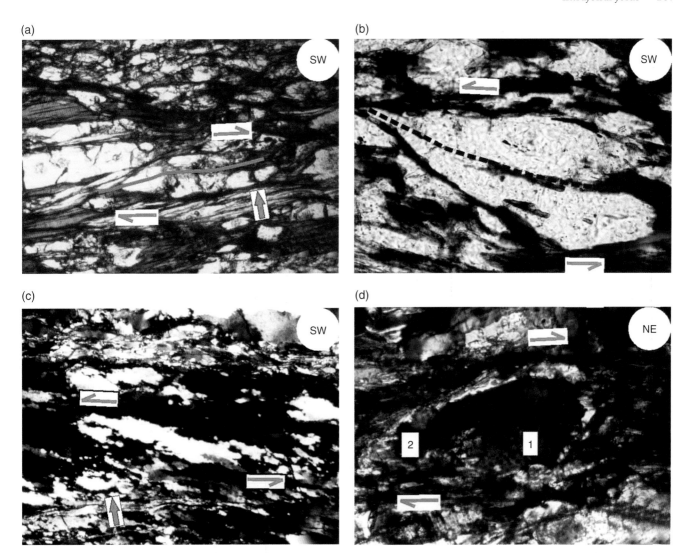

Fig. 12.10. Intrafolial folds of quartz under microscope, Zanskar Shear Zone. Shear sense: top-to-NE. (a) Axial trace at low angle to the C-foliation (green line). Boudinaged limbs (blue full arrow). Plane polarized light. Photo length: 2 mm; Location: P9/A. (b) Rootless fold. A few biotite grains inside it define a curved axial trace (blue dash line). Plane polarized light. Location: P9/O. (c) Remarkable difference in length of limbs of the folded recrystallized quartz grains. A sheath fold of quartz also present (blue full arrow). Cross-polarized light. Location: P9/G. Photo length: 3 mm. (d) Rootless fold with irregular boundaries. Wavy extinction of folded quartz grain. Limbs 1 and 2 are marked and are plotted in Ramsay's (1967) scheme in Fig. 12.14. Cross-polarized light. Photo length: 1 mm. Location: P9/B.

sub-parallel the Class 2 line in the graphs of t' vs α (Figs 12.13a, 12.14a), and Class 1B lines on T' vs α (Figs 12.13b, 12.14b). These plots, however, cannot compare with figs. 7–79 and 7.80 of Ramsay (1967). These two figures consist of plots that parallel Class 2 and the Class 1B lines and were used to estimate fold related flattening (also see caption of fig. 1c of Hudleston and Lan 1993). This is because Ramsay's (1967) exercises apply only to parallel geometries of the initial folds. Whether the studied intrafolial folds were at their initial stages of parallel geometries, however, is indeterminate.

Hook folds fall in Class 3 in t' vs α and T' vs α graphs (Fig. 12.15a,b). This means that retro shear on a pre-existing intrafolial fold can significantly alter its geometry into a stronger curvature of the outer arc than the inner one (of

every Class 3 folds). Irregular margins of some microscopic intrafolial folds (Figs 12.4a, 12.9a, 12.10c, d) and hooks (Fig. 12.12a, c) probably arose due to migration of adjacent grains (Passchier and Trouw 2005) towards the former. We manually smoothed grain-scale irregularities on tracings sheets before measuring for Ramsay's (1967) graphs. Had we not done this, the irregularities would have rendered the plots haphazard. Such smoothening has been a standard process of fold analyses (e.g. Singh 2010). Many of the mesoscopic folds had rather even margins (Figs 12.6d and 12.7a). We did not require best-fit curves for them.

The plots for intrafolial folds (Figs 12.13, 12.14) and hooks (Fig. 12.15) show the final geometries and cannot decipher fold evolution as shearing progressed. However, analyses of the few figs. published on the progressive

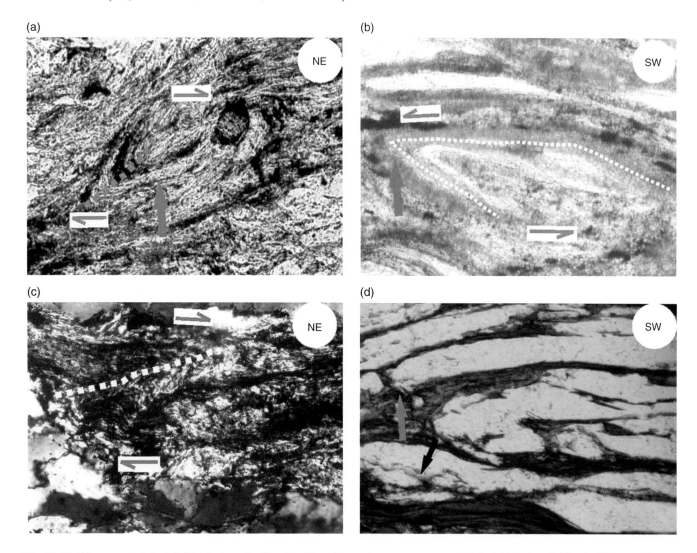

Fig. 12.11. Microscopic intrafolial folds from the Zanskar Shear Zone, shear sense: top-to-NE. (a, b) Rootless folds of sillimanite (orange arrows). That on the left has a sharp hinge whereas that on the right is more like a 'sheared box fold'. Photos in plane polarized light. Photo lengths: 1 mm; Location: P9/N. Source: Mukherjee and Koyi 2010b. Reproduced with permission from Springer Science + Business Media. (c) Crenulated mica aggregate reveal the shear sense. Axial trace (white dash line) is curved. Photo in cross-polarized light. Photo lengths: 1 mm; Location: P9/D. Source: Mukherjee 2013. Reproduced with permission from Springer Science + Business Media. (d) Fracture developed along the axial trace in a few folds (black full arrow). One of the limbs is tiny (orange arrow). Photo in plane polarized light. Photo lengths: 2 mm; Location: P9/C.

development of intrafolial folds from irregular layers by Bons and Jessel (1998) reveal that at an overturned fold (limbs "c" and "d" in Fig. 12.2f) plots within Classes 1C and 2. On the other hand, folds in the same train with limbs initially dipping in opposite directions (limbs "e" and "f" in Fig. 12.2f) more closely follow a Class 2 pattern (Fig. 12.16). Initially a Class 1C fold of Fossen (2010; here Fig. 12.2c) maintained a Class 2 geometry as simple shear progressed (Fig. 12.17). Apart from the hooks, we avoided analysis of the intrafolial folds *sensu lato* mentioned in Section 12.3.2.

12.4.3   Shear sense

Intrafolial folds with opposite vergence in the same rocks or shear zones were not encountered. Furthermore, the vergence of intrafolial folds always matched with nearby

shear sense indicators such as S-C fabrics, mineral fish and asymmetric quartz veins (Figs 12.5c, 12.7a, but many others outside the field of view of the photos). We therefore consider the studied intrafolial folds to be reliable ductile shear indicators. In a single train, shear sense can be deduced unambiguously from folds that are overturned or are asymmetric with axial traces at moderate angles to the shear planes (especially Figs 12.4a, d, 12.5a, 12.7a, c, 12.8a, b, 12.9a–d, 12.10a–d, 12.11a–c). In a polyclinal fold, some of the axial traces subparallel the main foliation – and so cannot indicate the sense (Fig. 12.8b). Instead, only those subfolds with the inclined axial traces within the fold give unequivocal shear sense (Fig. 12.8b). Interestingly, from the present orientation of hook-shaped intrafolial folds (*sensu lato*, as in Fig. 12.12a, c) of single minerals, in microscopic examples from the

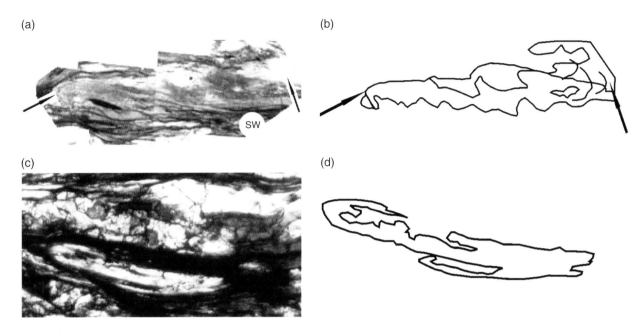

Fig. 12.12. (a) A rootless intrafolial folded sillimanite, from Zanskar Shear Zone. Note curvature of the grain at the two ends (arrows). Plane polarized light. Photo length: 1 mm; Location: P9/N. (b) Sillimanite of Fig. 12.11 is outlined. (c) A quartz hook. Cross-polarized light. (d) Outline of quartz hook. Photo length: 1 mm. Location: P9/B. It has been plotted in Ramsay's (1967) scheme in Fig. 12.15. (a, c) Source: Mukherjee and Koyi 2010b. Reproduced with permission from Springer Science + Business Media.

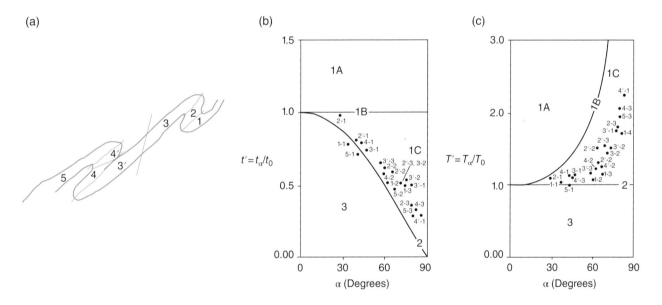

Fig. 12.13. Plot of intrafolial folds of the Higher Himalayan Shear Zone, Sutlej section in a single train into Ramsay's (1967) classification scheme. (a) The fold that was plotted is redrawn from fig. 113.2 of Bons and Jessel (1998). Limbs c, d, f, and g are marked and are plotted in Ramsay's (1967) scheme. (b) Plot of t' vs. α. (c) Plot of T' vs. α. Numbers such as 2′-1 indicate data for limb-2′ and data number 1.

ZSZ, the relative time relation of ductile shear – first a top-to-SW and then a top-to-NE – could be interpreted. However, noting that hooks can also form from initially antithetically oriented markers (as in Fig. 12.3b), the interpretation cannot be confirmed. Further, the curved axial traces of intrafolial folds resemble S-fabrics (Fig. 12.6b,–c, 12.7a, c, 12.8a–c, 12.9d, 12.10b, 12.11c) and can also give the true shear sense.

12.5 DISCUSSION AND CONCLUSIONS

Locally overturned folds in ductile shear zones formed during shear and confined within primary shear planes are commonly referred as intrafolial folds. These folds usually have round hinges and thinner limbs of unequal lengths. They can either cut the adjacent foliations or parallel them. Intrafolial folding by ductile shear either

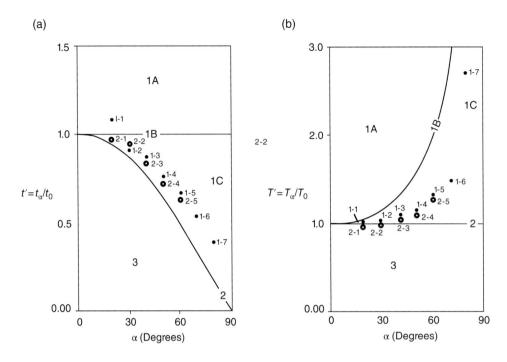

Fig. 12.14. Plot of the microscopic fold in Fig. 12.10d from the Zanskar Shear Zone into Ramsay's (1967) classification scheme. (a) Plot of t' vs. α. (b). Plot of T' vs. α. Symbols such as 1–1 indicate data from limb '1' and data number 1.

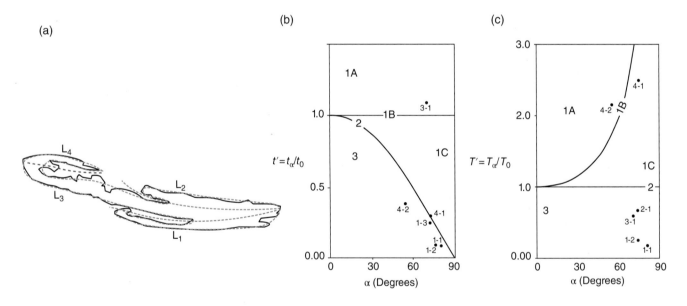

Fig. 12.15. Plot of hook shaped folds into Ramsay's (1967) classification scheme. Dashes indicate smoothening made before measurements. (a) The fold that was plotted (outlined from Fig. 12.12c). (b) Plot of t' vs. α. (c) Plot of T' vs. α. Numbers such as '4–4' indicate data for limb-4 (L_4) and data number 1.

reorient pre-existing planar foliations, or fold foliations around rigid inclusions. Pronounced shear leads materials in the folded layer to flow from limbs to the hinges until the limbs rupture into rootless folds. Axial planes progressively rotate to gentler inclinations to the enveloping foliation. In the plane perpendicular to the main foliation and parallel to the stretching lineation (the XZ section), intrafolial folds show the same shear sense as other shear indicators. This holds if these folds are unrelated to any major folds.

The other varieties of intrafolial folds *sensu lato* are: (i) those produced by shear of (un)deformed foliations antithetic to the shear planes; (ii) co- and counter rotated fabrics inside sheared lenses affected by secondary shear; (iii) folding of tails of some porphyroblasts; (iv) folded minerals; (v) drag folds of host fabric elements near cross-cutting elements; (vi) parasitic folds of first generation flexure slip folds; (vii) folded foliation cut by a straight foliation; (viii) folds trans-

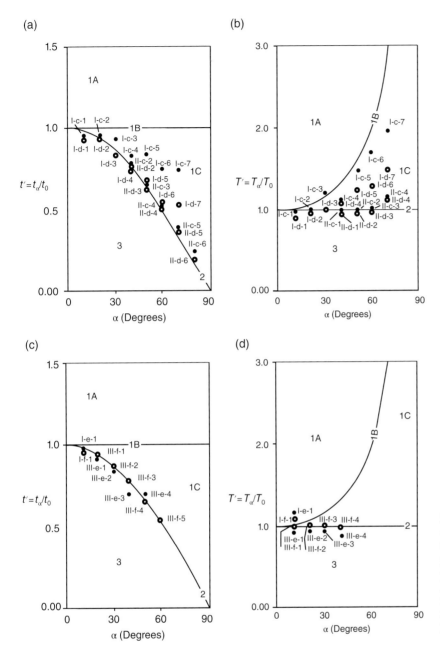

Fig. 12.16. Plot of the fold of Fossen (2010; reproduced here in Fig. 12.2(c)) into Ramsay's (1967) classification scheme. Symbols such as I-a-1 indicate data from fold I, of limb 'a' and data number 1. (a, c) Plot of t' vs. α. (b, d) Plot of T' vs. α. For each graph, two diagrams are used to avoid superposition of large number of plots. Plot of IV-a-1 to IV-a-IV, and IV-b-1 to IV-b-4 coincides with III-a-1 to III-a-IV, and IV-b-1 to IV-b-4, respectively, and are not shown in the graph to avoid crowding.

posed by layer parallel compression of a pre-existing fold; and (ix) folds produced by compression stronger than a perpendicular shear.

We argue that intrafolial folds within NE dipping primary shear C-planes of the $STDS_U$ and the $STDS_L$, and outside them, of the Sutlej section of the HHSZ (Himachal Pradesh) and in the ZSZ – an extension of the $STDS_U$ in Kashmir, do not belong to any of the categories of folds mentioned in the previous paragraph. Of regional significance are that- these folds indicate a consistent top-to-NE ductile shear, and are neither parasitic to any large scale folds, nor were they produced by the NE-SW compression related to India–Eurasia collision.

Round hinges and usually unequal limbs characterize our intrafolial folds. Some hinges are flame-shaped.

Ductile or brittle shear planes can cut round hinges along axial surfaces. Both rootless folds and sheath folds indicate intense shear. Differently shaped intrafolial folds, along and across the same train, indicate deformation partitioning. Obliquity of foliation during shear, and wavy extinction of individual mineral grains characterize microscopic intrafolial folds. Pronounced shear of these folds snap the thinning limbs. Secondary shears inside the folded layers have no regional implication.

Intrafolial folds do not indicate a consistent shear sense when, either their axial traces sub-parallel the C-planes, or are polyclinal. Our observed hook-shaped intrafolial folds of a quartz- and a sillimanite grain from the ZSZ may indicate a top-to-SW followed by a top-to-NE ductile shear.

(a)   (b)

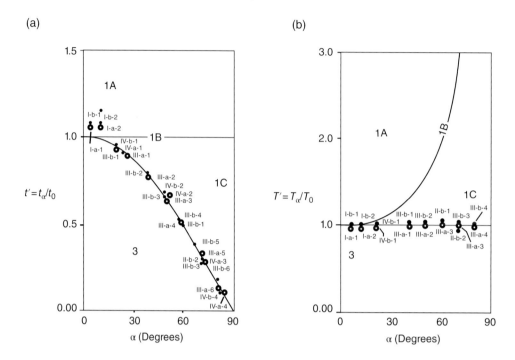

Fig. 12.17. Plot of few of the folds of Bons and Jessel (1998) into Ramsay's (1967) classification scheme. Symbols such as I-c-1 indicate data from fold I, of limb 'c' and data number 1. (a) Plot of t′ vs. α. (b) Plot of T′ vs. α. Plot II-c-1 and II-c-2 coincide with I-c-2 and I-d-2, respectively, III-g-1 and III-h-1 coincide with I-g-1 and I-h-1; therefore the former set of points are not shown.

In a series of centrifuge analogue models, Godin et al. (2011), Yakymchuk et al. (2012), Harris et al. (2012a,b) studied how channel flow extrusion modifies fold geometry by dragging adjacent layers. However, these drag folds formed outside the channel flow regime and so cannot correlate with the intrafolial folds present inside the detachments (the $STDS_U$ and the $STDS_L$ in the Sutlej section, and the ZSZ). This is because folds in this study formed inside the putative channel of the HHSZ. Second, based on U-Pb geochronology of zircons and structural geology, Larson et al. (2010) proposed that by ~35.4 Ma, crustal thickening developed upright folds near the upper part of the HHSZ in the Nepal Himalaya. From ~22.8 Ma onwards, a top-to-SW ductile shear associated with crustal-scale channel flow rotated those folds to verge SW. Thus, Larson et al.'s (2010) model cannot explain most of the NE vergeing intrafolial folds in the study areas.

Shear fabrics of leucosomes in migmatites developed during shear (Marchildon and Brown 2003; Hasalova´ et al. 2011), which is here of a top-to-NE sense inside the Higher Himalayan detachments. Following this, intrafolial folded leucosomes in the $STDS_U$ and the $STDS_L$ of the Sutlej section (Figs 12.5a–d, 12.6d, 12.7a–d) might emplaced during ductile shearing at ~15–54 and ~14–42 Ma, respectively. This matches with Mukherjee and Koyi (2010a), who suggested extensional ductile shear (normal shear sense) in detachments acted during those time periods.

Pure sheared Newtonian viscous layer within a non-Newtonian matrix develop folds with limbs dipping in opposite directions (Ord and Hobbs 2013). Thus, those folds are not overturned. On the other hand, simple shear applied at an angle to a series of non-Newtonian layers

too develop symmetric folds (Schmalholz and Schmid 2012). In none of these cases of deformation other than ductile shear, the typical overturned geometry of intrafolial folds develops.

ACKNOWLEDGMENTS

Supported by Department of Science and Technology's (New Delhi) grant: IR/S4/ESF-16/2009(G). J.N.P. and R.M. worked as Teaching Assistants. Figs 12.8b,c and 12.11c come from Roberto Weinbergs' (Monash University) thin-section. Internal and external review by several persons is acknowledged.

REFERENCES

Allaby M. 2013. A Dictionary of Geology & Earth Sciences, 4th edition. Oxford University Press, Oxford.

Aller J, Bobillo-Ares NC, Bastida F, Lisle RJ, Menéndez CO. 2010. Kinematic analysis of asymmetric folds in competent layers using mathematical modeling. Journal of Structural Geology 32, 1170–0184.

Alsop GI, Holdsworth RE. 2004. Shear zone folds: records of flow perturbation or structural inheritance? In Flow Processes in Faults and Shear Zones, edited by G.I. Alsop, R.E. Holdsworth, K.J.W. McCaffey, and M. Hand, Geological Society of London, Special Publication, vol. 224, pp. 177–799.

Alves 2015. Sub-marine slide blocks and associated soft-sediment deformation in deep-water basins: A review. Marine and Petroleum Geology 67, 262–285.

Andrews GDM, Branney MJ. 2005. Folds, fabrics and kinematic criteria in rheomorphic ignimbrites of the Snake River Plain, Idaho: Insights into emplacements and flow. In Interior United

Western States: Geological Society of America Field Guide, edited by J. Pederson and C.M. Dehler, Geological Society of America, vol. 6, pp. 311–127.

Andrews GDM, Branney MJ. 2011. Emplacement and rheomorphic deformation of a large, lava-like rhyolitic ignimbrite: Grey's Landing, southern Idaho. Geological Society of America Bulletin 123, 725–743.

Arbaret L, Mancktelow NS, Burg J-P. 2001. Effect of shape and orientation on rigid particle rotation and matrix deformation in simple shear flow. Journal of Structural Geology 23, 113–125.

Beaumont C, Jamieson RA. 2010. Himalayan-Tibetan orogeny: channel flow versus (critical) wedge models, a false dichotomy. In Proceedings for the 25th Himalaya-Karakoram-Tibet Workshop: US Geological Survey, edited by M.L. Leech et al., Open-File Report 2010–0099, 2 p. [Online] http://pubs.usgs.gov/of/2010/1099/beaumont/ (accessed 27 May 2015).

Becker A. 1995. Conical drag folds as kinematic indicators for strike-slip fault motion. Journal of Structural Geology 17, 1497–7506.

Bell TH. 2010. Deformation partitioning, foliation successions and their significance for orogenesis: hiding lengthy deformation histories in mylonites. In Continental Tectonics and Mountain Building: The Legacy of Peach and Horne, edited by R.D. Law, R.W.H. Butler, R.E. Holdsworth, M. Krabbendam, and R.A. Strachan, Geological Society of London, Special Publication, vol. 335, pp. 275–592.

Best M. 2006. Igneous and Metamorphic Petrology, 2nd edition. Blackwell, Oxford, pp. 423, 552.

Billings MP. 2008. Structural Geology, 3rd edition. Prentice-Hall of India Pvt Ltd., Delhi, pp. 54–45.

Bobyarchick AR. 1998. Foliation heterogeneities in mylonites. In Fault Related Rocks, edited by A. Snoke, J. Tullis, and V.R. Todd, Princeton University Press, Princeton, pp. 302–203.

Bons PD, Jessel MW. 1998. Folding in experimental mylonites. In Fault-Related Rocks, A Photographic Atlas, edited by A. Snoke, J. Tullis, and V.R. Todd, Princeton University Press, Princeton, pp. 366–667.

Burchfiel BC, Chen Z, Hodges KV, et al. 1992. The South Tibetan Detachment System, Himalayan orogen: extension contemporaneous with and parallel to shortening in a collisional mountain belt. Geological Society of America Special Paper 269, 1–41.

Carreras J, Druguet E, Griera A. 2005. Shear zone-related folds. Journal of Structural Geology 27, 1229–9251.

Chambers J, Parrish R, Argles T, Harris N, Mattew H. 2011. A short duration pulse of ductile normal shear on the outer South Tibetan detachment in Bhutan: alternating channel flow and critical taper mechanics of the eastern Himalaya. Tectonics 30, TC2005.

Davis GH, Cox LJ, Ornelas R. 1998. Protomylonite, augen mylonite and protocataclasite in granitic footwall of a detachment fault. In Fault-Related Rocks, A Photographic Atlas, edited by A. Snoke, J. Tullis, and V.R. Todd, Princeton University Press, Princeton, pp. 160–061.

Davis GH, Reynolds SJ, Kluth CF. 2012. Structural Geology of Rocks and Regions, 3rd edn. John Wiley & Sons, Inc., New York.

Dennis JG. 1987. Structural Geology: An Introduction. Wm. C. Brown Publishers, Dubuque, pp. 182, 196.

Dèzes PJ, Vannay JC, Steck A, Bussy F, Cosca M. 1999. Synorogenic extension: Quantitative constraints on the age and displacement of the Zanskar shear zone. Geological Society of America Bulletin 111, 364–474.

Duebendorfer EM, Sewall AJ, Smith EI. 1990. An evolving shear zone in the Lake Mead area, Nevada. In The Saddle Island Detachment, edited by B.P. Wernicke, Geological Society of America Memoir 176, 77–77.

Ez V. 2000. When shearing is a cause of folding. Earth Science Reviews 51, 155–572.

Fossen H. 2010. Structural Geology. Cambridge University Press, Cambridge, UK, pp. 230, 255, 303, 304.

Gangopadhyay PK. 1995. Intrafolial folds and associated structures in a progressive strain environment of Darjeeling-Sikkim Himalaya. Proceedings of the Indian Academy of Science 104, 523–337.

Gawthrope RL, Clemmey H. 1985. Geometry of submarine slides in the Bowland (Dinantian) and their relation to debris flow. Journal of the Geological Society of London 142, 555–565.

Ghosh SK. 1993. Structural Geology – Fundamentals and Modern Developments. Pergamon, Oxford.

Godard V, Burbank DW. 2011. Mechanical analysis of controls on strain partitioning in the Himalayas of Central Nepal. Journal of Geophysical Research: Solid Earth 116, B10402.

Godin L, Grujic D, Law RD, Searle MP. 2006. Channel flow, extrusion and exhumation incontinental collision zones: an introduction. In Channel Flow, Extrusion and Exhumation in Continental Collision Zones, edited by R.D. Law, M.P. Searle, L. Godin, Geological Society of London, Special Publication, vol. 268, pp. 1–23.

Godin L, Yakymchuk C, Harris LB. 2011. Himalayan hinterland-verging superstructure folds related to foreland-directed infrastructure ductile flow: Insights from centrifuge analogue modeling. Journal of Structural Geology 33, 39–942.

Grasemann B, Dabrowski M. 2015. Winged inclusions: Pinch-and-swell objects during high-strain simple shear. Journal of Structural Geology 70, 78–94.

Grasemann B, Vanney J-C. 1999. Flow controlled inverted metamorphism in shear zones. Journal of Structural Geology 21, 743–750.

Grasemann B, Fritz H, Vannay J-C. 1999. Quantitative kinematic flow analysis from the Main Central Thrust Zone (NW-Himalaya, India): implications for a decelerating strain path and extrusion of orogenic wedges. Journal of Structural Geology 21, 837–853.

Hara I, Shimamoto T. 1984. Geological Structures. In Folds and Folding, edited by T. Uemura and S. Mizutani, John Wiley & Sons, Chichester, pp. 199–944.

Harris LB. 2003. Folding in high-grade rocks due to back-rotation between shear zones. Journal of Structural Geology 25, 223–340.

Harris LB, Koyi HA, Fossen H. 2002. Mechanisms for folding of high-grade rocks in extensional tectonic settings. Earth Science Reviews 59, 163–310.

Harris LB, Godin L, Yakymchuk C. 2012a. Regional shortening followed by channel flow induced collapse: a new mechanism for dome and keel geometries in Neoarchaean granite-greenstone terrains. Precambrian Research 212–213, 139–954.

Harris LB, Yakymchuk C, Godin L. 2012b. Implications of centrifuge simulations of channel flow for opening out or destruction of folds. Tectonophysics 526–529, 67–87.

Harris N. 2007. Channel flow and the Himalayan-Tibetan orogen: a critical review. Journal of the Geological Society of London 164, 511–523.

Hasalova´ P, Weinberg RF, Macrae C. 2011. Microstructural evidence for magma confluence and reusage of magma pathways: implications for magma hybridization, Karakoram Shear Zone in NW India. Journal of Metamorphic Geology 25, 875–500.

Herren E. 1987. Zanskar Shear Zone: northeast southwest extension within the Higher Himalaya (Ladakh, India). Geology 15, 409–913.

Higgins MW. 1971. Cataclastic Rocks. United States Government Printing Office, Washington, pp. 1–187.

Hills ES. 1965. Elements of Structural Geology. Asia Publishing House, Bombay, pp. 88–89, 167, 285.

Hodges KV. 2006. A synthesis of the channel flow-extrusion hypothesis as developed for the Himalayan-Tibetan orogenic system. In Channel Flow, Extrusion and Exhumation in Continental Collision Zones, edited by R.D. Law, M.P. Searle, and L. Godin, Geological Society of London Special Publication vol. 268, pp. 71–90.

Hollister LS, Grujic D. 2006. Himalaya Tibet Plateau. Pulsed channel flow in Bhutan. In Channel Flow, Extrusion and Exhumation in Continental Collision Zones, edited by R.D. Law, M.P. Searle, and L. Godin, Geological Society of London Special Publication vol. 268, pp. 415–423.

Hopper RJ, Hatcher Jr. RD. 1988. Mylonites from the Towalinga fault zone, central Georgia: products of heterogeneous non-coaxial deformation. Tectonophysics 152, 1–17.

Hudleston PJ, Lan L. 1993. Information from fold shapes. Journal of Structural Geology 15, 253–364.

Hudleston PJ, Treagus SH. 2010. Information from folds. a review. Journal of Structural Geology 32, 2042–2071.

Inger S. 1998. Timing of the extensional detachment during convergent orogeny: new Rb–Sr geochronological data from the Zanskar shear zone, northwestern Himalaya. Geology 26, 223–226.

Jain AK, Anand A. 1988. Deformational and strain patterns of an intracontinental ductile shear zone- an example from the Higher Garhwal Himalaya. Journal of Structural Geology 10, 717–734.

Jain AK, Patel RC. 1999. Structure of the Higher Himalayan crystallines along the Suru-Doda valleys (Zanskar), NW Himalaya. In Geodynamics of the NW Himalaya, edited by A.K. Jain and R.M. Manickavasagam, Gondwana Research Group Memoirs No 6. Field Science, Osaka, pp. 91–110.

Jain AK, Manickavasagam RM, Singh S, Mukherjee S. 2005. Himalayan collision zone: new perspectives – its tectonic evolution in a combined ductile shear zone and channel flow model. Himalayan Geology 26, 1–18.

Jain AK, Singh S, Manickavasagam RM. 2002. Himalayan collisional tectonics. Gondwana Research Group Memoir No. 7. Field Science, Hashimoto.

Jirsa MA, Green JC. 2011. Classic Precambrian geology of northeast Minnesota. In Archean to Anthropocene: Field Guides to the Geology of the Mid-Continent of North America, edited by J.D. Miller, G.J. Hudac, and C. Wittkop, The Geological Society of America (Field Guide) vol. 24, pp. 25–55.

Jones RR, Holdsworth RE, Hand M, Goscombe B. 2006. Ductile extrusion in continental collision zones: ambiguities in the definition of channel flow and its identification in ancient orogens. In Channel Flow, Extrusion and Exhumation in Continental Collision Zones, edited by R.D. Law, M.P. Searle, and L. Godin, Geological Society of London Special Publication vol. 268, pp. 201–119.

Keary P, Klepeis KA, Vine FJ. 2009. Global Tectonics, 3rd edition. John Wiley & Sons, Chichester.

Keiter M, Ballhaus C, Tomaschek F. 2011. A new geological map of the Island of Syros (Aegean Sea, Greece): Implications for lithostratigraphy and structural history of the Cycladic Blueschist Unit. The Geological Society of America Special Paper 481.

Kellett DA, Grujic D. 2012. New insight into the South Tibetan detachment system: Not a single progressive deformation. Tectonics 31, TC2007.

Koyi H, Schmeling H, Burchardt S, et al. 2013. Shear zones between rock units with no relative movement. Journal of Structural Geology 50, 82–20.

Larson KP, Godin L, Davis WJ, Davis DW. 2010. Out-of-sequence deformation and expansion of the Himalayan orogenic wedge: insight from Changgo cumulation, south central Tibet. Tectonics 29, TC4013.

Law R, Stahr III DW, Ahmad T, Kumar S. 2010. Deformation Temperatures and flow vorticities near the base of the Greater Himalayan Crystalline Sequence, Sutlej Valley and Shimla Klippe, NW India. In Proceedings of the 25th Himalaya-Karakoram-Tibet Workshop, edited by M.L. Leech et al. US Geological Survey, Open-File Report. [Online] Available at http://pubs.usgs.gov/of/2010/1099/law/ (accessed 1 June 2015).

Laznicka P. 1985. Empirical Metallogeny: Depositional Environments, Lithological Associations and Metallic Ores. Vol. 1. Phanerozoic Environments, Associations and Deposits. Part B. Elsevier, Oxford.

Lebit H, Hudleston P, Luneberg C, Carreras J, Druguet E, Griera A. 2005. Shear zone-related folds. Journal of Structural Geology 27, 1229–1251.

Linn JK, Walker JD, Bartley JM. 2002. Late Cenozoic crustal contraction in the Kramer Hills, west-central Mojavac Desert, California. In Geologic evolution of the Mojave Desert and Southwestern Basin and Range, Boulder, Colorado, edited by A.F. Glazner, J.D. Walker, and J.M. Bartley, Geological Society of America Memoirs 195, 161–172.

Leloup PH, Mahéo G, Arnaud E, et al. 2010. The South Tibet detachment shear zone in the Dinggye area Time constraints on extrusion models of the Himalayas. Earth and Planetary Science Letters 92, 1–16.

Llorens M-G, Bons PD, Gomez-Rivas E. 2013. When do folds unfold during progressive shear? Geology 41, 563–366.

Longridge L, Gibson RL, Kinnaird JA, Armstrong RA. 2011. Constraining the timing of deformation in the southwestern Central Zone of the Damara Belt, Namibia. In The Formation and Evolution of Africa: A Synopsis of 3.8 Ga of Earth History, edited by D.J.J. Van Hinsbergen, S.J.H. Buiter, T.H. Torshvik, C. Gaina, and S.J. Webb, Geological Society of America Special Publication vol. 357, pp. 107–735.

Mandal N, Samanta SK, Chakraborty C. 2004. Problem of folding in ductile shear zones: a theoretical and experimental investigations. Journal of Structural Geology 26, 475–589.

Marchildon N, Brown M. 2003. Spatial distribution of melt-bearing structures in anatectic rocks from Southern Brittany, France: implications for melt transfer at grain- to orogen-scale. Tectonophysics 364, 215–235.

Maass RS, Medaris LG Jr, Van Schmus WR. 1980. Penokian deformation in central Wisconsin, *in* Morey, G.B., and Hanson, G.N., eds., Selected Studies of Archean Gneiss and Lower Proterozoic Rocks, Southern Canadian Shield. The Geological Society of America Memoirs 160, 85–55.

Mukherjee S. 2005. Channel flow, ductile extrusion and exhumation of lower-mid crust in continental collision zones. Current Science 89, 435–436.

Mukherjee S. 2007. Geodynamics, deformation and mathematical analysis of metamorphic belts of the NW Himalaya. PhD thesis. Indian Institute of Technology Roorkee, pp. 1–267.

Mukherjee S. 2009. Channel flow model of extrusion of the Higher Himalaya-successes & limitations. EGU General Assembly Conference Abstracts 11.

Mukherjee S. 2010a. Structures in meso- and micro-scales in the Sutlej section of the Higher Himalayan Shear Zone, Indian Himalaya. e-Terra 7, 1–17.

Mukherjee S. 2010b. Microstructures of the Zanskar Shear Zone. Earth Science India 3, 9–27.

Mukherjee S. 2011a. Flanking microstructures from the Zanskar Shear Zone, NW Indian Himalaya. YES Network Bulletin 1, 21–19.

Mukherjee S. 2011b. Mineral fish: their morphological classification, usefulness as shear sense indicators and genesis. International Journal of Earth Sciences 100, 1303–1314.

Mukherjee, S. (2012a) Tectonic implications and morphology of trapezoidal mica grains from the Sutlej section of the Higher Himalayan Shear Zone, Indian Himalaya. The Journal of Geology, v.120, pp. 575–590.

Mukherjee S. (2012b) Simple shear is not so simple! Kinematics and shear senses in Newtonian viscous simple shear zones. Geol. Mag., v.149, pp.819–926.

Mukherjee S. 2013a. Channel flow extrusion model to constrain dynamic viscosity and Prandtl number of the Higher Himalayan Shear Zone. International Journal of Earth Sciences 102, 1811–1835.

Mukherjee S. 2013b. Higher Himalaya in the Bhagirathi section (NW Himalaya, India): its structures, backthrusts and extrusion mechanism by both channel flow and critical taper mechanisms. International Journal of Earth Sciences 102, 1851–1870.

Mukherjee S. 2014. Kinematics of "top -to-down" simple shear in a Newtonian rheology. The Journal of Indian Geophysical Union 18, 273–376.

Mukherjee S. 2014. Mica inclusions inside host mica grains from the Sutlej Section of the Higher Himalayan Crystallines,

India – Morphology and Constrains in Genesis. Acta Geologica Sinica 88, 1729–9741.

Mukherjee S. 2015. A review on out-of-sequence deformation in the Himalaya. In Tectonics of the Himalaya, edited by S. Mukherjee, R. Carosi, P. van der Beek, B.K. Mukherjee, and D.M. Robinson, Geological Society, London, Special Publication 412. http://doi.org/10.1144/SP412.13.

Mukherjee S, Biswas R. 2014. Kinematics of horizontal simple shear zones of concentric arcs (Taylor–Couette flow) with incompressible Newtonian rheology. International Journal of Earth Sciences 103, 597–702.

Mukherjee S, Biswas R. 2016. Biviscous horizontal simple shear zones of concentric arcs (Taylor Couette flow) with incompressible Newtonian rheology. In Ductile Shear Zones: From Micro- to Macro-scales, edited by S. Mukherjee and K.F. Mulchrone. John Wiley & Sons, Chichester.

Mukherjee S, Carosi R, van der Beek PA, Mukherjee BK, Robinson DM. (eds) 2015. Tectonics of the Himalaya: an introduction. Geological Society, London, Special Publications, 412, 1–3.

Mukherjee S, Koyi HA. 2009. Flanking microstructures. Geological Magazine 146, 517–726.

Mukherjee S, Koyi HA. 2010a Higher Himalayan Shear Zone, Sutlej section: structural geology and extrusion mechanism by various combinations of simple shear, pure shear and channel flow in shifting modes. International Journal of Earth Sciences 99, 1267–7303.

Mukherjee S, Koyi HA. 2010b. Higher Himalayan Shear Zone, Zanskar Indian Himalaya: microstructural studies and extrusion mechanism by a combination of simple shear & channel flow. International Journal of Earth Science 99, 1083–3110.

Mulchrone KF, Mukherjee S (in press) Shear senses and viscous dissipation of layered ductile simple shear zones. Pure and Applied Geophysics.

Nevin CM. 1957. Principles of Structural Geology, 4th edition. John Wiley & Sons, New York.

Neuendorf KKE, Mehl JM Jr, Jackson JA. 2005. Glossary of Geology, 5th edition. American Geological Institute, Washington DC.

Nicolas A. 1987. Principles of Rock Deformation. D Reidel Publishing Company, Dordrecht.

Ord A, Hobbs B. 2013. Localized folding in general deformations. Tectonophysics 583, 30–05.

Park RG. 1997. Foundations of Structural Geology, 3rd edition. Routledge, London.

Passchier CW. 2001. Flanking Structures. Journal of Structural Geology 23, 951–962.

Passchier CW, Trouw RAJ. 2005. Microtectonics, 2nd edition, Springer, Heidelberg, pp. 146, 150.

Passchier CW, Myers JS, Kröner A. 1991. Field Geology of High-Grade Gneiss Terrains. Narosa Publishing House, New Delhi, pp. 40, 68.

Patel RC, Singh S, Asokan A, Manickavasagam RM, Jain AK. 1993. Extensional tectonics in the Himalayan orogen, Zanskar, NW India. In Himalayan Tectonics, edited by P.J. Treloar and M.P. Searle, Geological Society of London Special Publication, vol. 74, pp. 445–559.

Ramsay JG. 1967. Folding and Fracturing of Rocks. McGraw Hill, New York, pp. 117, 352, 390, 413, 414.

Ratcliffe NM, Harwood DS. 1975. Blastomylonites associated with recumbent folds and overthrusts at the western edge of the Berkshire massif, Connecticut and Massachusetts; a preliminary report. In Post-Carboniferous Stratigraphy, Northeastern Alaska, edited by R.L. Detterman, H.N. Reiser, W.P. Brosge, and J.T. Dutro Jr, Geological Survey Professional Paper 886. United States Government Printing Office, Waashington DC, pp. 1–19.

Rhodes S, Gayer RA. 1978. Non-cylindrical folds, linear structures in the X direction and mylonite developed during transtension of the Caledonian Kalak Nappe Complex of Finnmark. Geological Magazine 114, 329–908.

Schmalholz SM, Schmid DW. 2012. Folding in power-law viscous multi-layers. Philosophical Transactions of the Royal Society A: Mathematics, Physics & Engineering Science 370, 1798–8826.

Searle MP, Cooper DJW, Rex AJ. 1988. Collision tectonics of the Ladakh-Zanskar Himalaya. In Tectonic evolution of the Himalayas and Tibet, edited by R.M. Shackleton, J.F. Dewey, and B.F. Windley, Philosophical Transactions of the Royal Society of London. A326, 117–150.

Singh S. 1993. Collision tectonics: metamorphic and geochronological constraints from parts of Himachal Pradesh, NW-Himalaya. Unpublished PhD Thesis. University of Roorkee, India, pp. 1– 289.

Singh YK. 2010. Deformation and tectonic history of Delhi Supergroup of rocks, around Uplagarh and Sagna, SW Rajasthan, India. Unpublished PhD Thesis. Indian Institute of Technology, Bombay, pp. 87–75.

Srikantia SV, Bhargava ON. 1998. Geology of Himachal Pradesh. Geological Society of India, Bangalore, pp. 1–406.

Swanson MT. 1999. Kinematic indicators for regional dextral shear, Norumbega fault system, Casco Bay area. Geological Society of America Special Publication, vol. 331, pp. 1–13.

ten Grotenhuis SM, Trouw RAJ, Passchier CW. 2003. Evolution of mica fish in mylonitic rocks. Tectonophysics 372, 1–11.

Trouw RAJ, Passchier CW, Wiersma DJ. 2010. Atlas of Mylonites and Related Microstructures. Springer, Heidelberg.

Twiss RJ, Moores EM. 2007. Structural Geology, 2nd edition. WH Freeman and Company. New York.

van der Pluijm BA, Marshak S. 2004. Earth Structure: An Introduction to Structural Geology and Tectonics, 2nd edition. WW Norton & Company, New York, pp. 311, 312, 313.

Vannay J-C, Grasemann B, Rahn M, et al. (2004) Miocene to Holocene exhumation of metamorphic crustal wedge in the NW Himalaya: Evidence for tectonic extrusion coupled to fluvial erosion. Tectonics 23, TC1014.

Vannay J-C, Grasemann B. 2011. Himalayan inverted metamorphism and synconvergence extrusion as a consequence of a general shear extrusion. Geological Magazine 138, 253–276.Vannay J-C, Sharp DZ, Grasemann B. 1999. Himalayan inverted metamorphism constrained by oxygen thermometry. Contributions to Mineralogy and Petrology 137, 90–101.

Vernon RH. 2004. A Practical Guide to Rock Microstructure. Cambridge University Press, Cambridge UK.

Vernon RH, Clarke GL. 2008. Principles of Metamorphic Petrology. Cambridge University Press, Cambridge, UK, pp.292–293.

Walker JD, Searle MP, Waters DJ. 2001. An integrated tectonothermal model for the evolution of the Higher Himalaya in western Zanskar with constraints from thermobarometry and metamorphic modeling. Tectonics 20, 810–033.

Wennberg OP. 1996. Superimposed fabric due to reversal of shear sense: an example from the Bergen Arc Shear Zone, western Norway. Journal of Structural Geology 18, 871–889.

Whitten EHT. 1966. Structural geology of folded rocks. Rand McNally & Company, Chicago.

Winter JD. 2012. Principles of Igneous and Metamorphic Petrology, 2nd edition. Eastern Economy Edition, PHI Learning Pvt Ltd, New Delhi, pp. 37, 507.

Woodcock NH. 1976. Structural styles in slump sheets: Ludlow Series, Powys, Wales. Journal of the Geological Society, London 32, 399–915.

Yakymchuk C, Godin L. 2012. Coupled role of deformation and metamorphism in the construction of inverted metamorphic sequences: an example from far-northwest Nepal. Journal of Metamorphic Geology 30, 513–335.

Yakymchuk C, Harris LB, Godin L. 2012. Centrifuge modelling of deformation of a multi-layered sequence over a ductile substrate: 1. Style and 4D geometry of active cover folds during layer-parallel shortening. International Journal of Earth Sciences 101, 463–382.

Yin A. 2006. Cenozoic tectonic evolution of the Himalayan orogen as constrained by along-strike variation of structural geometry, exhumation history, and foreland sedimentation. Earth Science Reviews 76, 1–131.

Chapter 13
Structure and Variscan evolution of Malpica–Lamego ductile shear zone (NW of Iberian Peninsula)

JORGE PAMPLONA[1], BENEDITO C. RODRIGUES[1], SERGIO LLANA-FÚNEZ[2],
PEDRO PIMENTA SIMÕES[1], NARCISO FERREIRA[3], CARLOS COKE[4],
EURICO PEREIRA[3], PAULO CASTRO[3], and JOSÉ RODRIGUES[3]

[1] Institute of Earth Sciences (ICT)/University of Minho Pole, Universidade do Minho, Campus de Gualtar, 4710-057 Braga, Portugal
[2] Departamento de Geología, Universidad de Oviedo, Arias de Velasco s/n, 33005 Oviedo, Spain
[3] Laboratório Nacional de Energia e Geologia, Rua da Amieira, Apartado 1089, 4466 901 S. Mamede de Infesta, Portugal
[4] Departamento de Geologia, Universidade de Trás-os-Montes e Alto Douro, Apartado 1013, 5001-801 Vila Real, Portugal

13.1 INTRODUCTION

The Malpica–Lamego Ductile Shear Zone (MLDSZ) is a polyphasic structure developed in the NW of the Iberian Peninsula. The northern sector (NW Spain) of this shear zone is associated with a complex geological history that ranges from earlier Variscan to the emplacement of the Variscan nappes (allochthonous). Variscan intracontinental deformation after the allochthonous nappes emplaced, recognizably strike–slip, explains its spatial and structural continuity from the NW (Malpica, Spain) to the South (Lamego, Portugal). Here we focus on the Variscan development of the MLDSZ and characterize, longitudinally, this crustal mega-structure.

At the NW tip of the MLDSZ was considered an area (segment 1) with very good exposure of the Variscan intracontinental strike–slip component on outcrops of the coastline (e.g. Llana-Fúnez 2001). In sequence, we address segment 2, with excellent left-hand kinematic markers (shear band boudins and folds), on a metapelitic rock series (Pamplona and Rodrigues 2011). In the southernmost part of the MLDSZ, strike–slip deformation, emplacement of granitic rocks (segment 3 – based on Simões 2000; and segment 5), regional structures (segment 4: based on Coke et al. 2003), and secondary manifestations of the southern tip of MLDSZ (segment 6) were studied.

Detailed geological mapping was done. Kinematic interpretation of structures in micro- (Passchier and Trouw 2005; Mukherjee 2011) and mesoscale (e.g. Spry 1969; Goscombe and Passchier 2003; Mukherjee and Koyi 2010), quartz c-axis analysis (e.g. Schmid and Casey 1986), thermodynamic conditions inferred from textural equilibrium (e.g. Winkler 1979), magmatic fabrics (Fernández 1982), and U-Pb geochronology on zircons and monazites (Heaman and Parrish 1991), are integrated.

In order to chronologically define the activity of the MLDSZ and explain the tectonic episodes recorded during the functioning of this structure, the Variscan episodes were viewed as per Dias and Ribeiro (1995) and Dias et al. (2013). This classification was established for the northern sectors of the Variscan orogen of Portugal. However there are alternates (e.g. Noronha et al. 1979; Martínez-Catalán et al. 2008) and uncertainties on the limits of the different phases. Dias and Ribeiro (1995) and Dias et al. (2013) classified this orogeny in four phases: (i) D_1: continental collision and obduction of allochthonous terrains during the Devonian Period; (ii) D_2: structuring of the allochthonous terrains during the Early Carboniferous; (iii) D_3: intracontinental deformation forming crustal shear zones during the Later Carboniferous; (iv) D_4: brittle–ductile to brittle intracontinental deformation and complex fluid activity linked to mineralized shear zones during the Permian.

13.2 GEOLOGICAL SETTING

The MLDSZ is a major crustal structure that has a minimum length of 275 km running NW–SE, parallel to the Variscan belt in the Iberian Peninsula (Fig. 13.1) (Llana-Fúnez and Marcos 2001). Previously, the structure was described in two segments: in the northern part the Malpica–Vigo shear zone (Iglésias and Choukroune

Ductile Shear Zones: From Micro- to Macro-scales, First Edition. Edited by Soumyajit Mukherjee and Kieran F. Mulchrone.

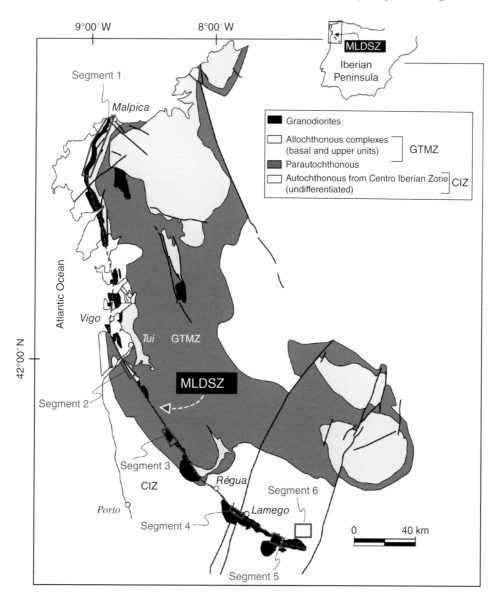

Fig. 13.1. Variscan Belt of the NW of Iberian Peninsula with representation MLDSZ and studied segments. Main geologic units related to Variscan deformation and MLDSZ are shown. CIZ: Central Iberian Zone; GTMZ: Galicia-Trás-os-Montes Zone. Adapted from Ribeiro et al. 1990 and Llana-Fúnez and Marcos 2001.

1980; Iglésias and Ribeiro 1981), and in the southern part the Vigo–Régua ductile shear zone (Ferreira et al. 1987a).

The current level of erosion shows structures generated from the middle crust. The MLDSZ has sub-vertical, or dipping to the W, penetrative foliations and sub-horizontal stretching lineations.

The kinematics in the northern segment are well characterized by multiphase fault reworking, from 350 Ma to around 310 Ma (Llana-Fúnez 2011). This evolution begins with a dip–slip movement, which began after the emplacement of allochthounous thrust sheets in NW Iberia. A second stage ~310 Ma led to dextral strike–slip shear related to intracontinental deformation during the Variscan orogenesis (Llana-Fúnez and Marcos 2001, 2007).

In the southern segment a multiphase strike–slip kinematics was identified (e.g. Fernandes 1961, Ferreira et al. 1987a; Pereira et al. 1993; Coke et al. 2000): sinistral in few deformation phases (370–310/315 Ma); and also dextral (310/315–300 Ma). However, Holtz (1987) identifies a initial thrusting, followed by a second event in the MLDSZ, that made structures vertical, which occurs before the major sinistral strike–slip event. Gomes (1984) describes the major strike–slip event to be dextral initially and subsequently becoming sinistral.

Rocks on either side of the shear zone belong to distinct sequences. For the most part of the shear zone, the rocks to the W in the hanging wall belong to parauthochthonous of Galicia–Trás-os-Montes Zone (GTMZ) of the Iberian Variscan Belt. The footwall of the shear zone is constituted by Malpica–Tui Unit (MTU), in the northernmost part, while the southern part of the shear zone are authochthonous, parauthochthonous and

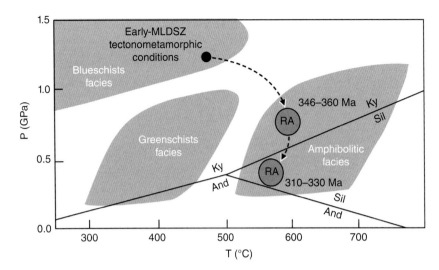

Fig. 13.2. Petrogenetic evolution of metamorphic host rocks of MLDSZ (RA). RA: relative autochthonous. Highlights thermobarometric pathway defined to northern segment, with the relative thermobarometric conditions of the terranes before the MLDSZ activated. Adapted from Llana-Fúnez 2001 and Rodríguez-Aller et al. 2003.

allochthonous sequences of the Central Iberian Zone (CIZ) and related granitic rocks.

Porphyroid granodioritic rocks, structurally controlled during their emplacement by the crustal anisotropy generated by MLDSZ, organized the lineament pattern of the ductile shear zone, particularly in the southern segment of the structure (Ferreira et al. 1987a).

The MLDSZ stands on the inner zones of Variscan orogen in Iberia. The pressure–temperature (P–T) conditions of the MLDSZ sector, achieved in each Variscan deformation phase, are related to the exposed structural level. The D_1 structures on root zone have a metamorphic peak of amphibolitic facies – kyanite zone (Llana-Fúnez and Marcos 2001), while on upper crustal levels the metamorphism conditions does not exceed greenschists facies. The D_2 is not recognized on the studied sector; however, its metamorphic conditions do not exceed the greenschists facies and have been considered as intermediate- to low-pressure type, and local shear heating (Mukherjee and Mulchrone 2013) effects are responsible for attaining amphibolitic facies (Ribeiro 1988). The D_3 metamorphic peak is achieved along structures parallel to the MLDSZ, recording amphibolitic facies metamorphism (e.g. Llana-Fúnez and Marcos 2001; Pamplona and Rodrigues 2011).

The P–T conditions of the deformation the MLDSZ was active was constrained by the thermobarometric evolution of wall rocks. The main deformation (D_3) metamorphic peak reached amphibolitic facies in regions with extensive partial melting in the hanging wall block in the northern segment (Fig. 13.2). The wall rocks metamorphism range from greenschist to amphibolitic facies.

13.3 STRUCTURAL DESCRIPTION AND INTERPRETATION

The geological and structural description of the MLDSZ is performed by presenting six selected segments along the structure (Fig. 13.1).

13.3.1 Segment 1 (Malpica, Spain)

The MLDSZ is defined along much of its length in the Spanish segment (Fig. 13.3) by sub-vertical and highly deformed schists, on map view not exceeding in general the 1 km width. In the western vicinity of Malpica phyllonites, mylonitized schists, and paragneisses define the core of the shear zone.

The coastline around Malpica is the northernmost exposure of the MLDSZ and perhaps the best one to observe deformation structures associated with the activity of this major crustal-scale shear zone during Variscan D_3 (Fig. 13.4). Due to late NW-directed faults two sections through the axial deformation zone of the MLDSZ are exposed in the Xeiruga beach and in the Seaia beach. In <1 km across the deformation zone, it outcrops the footwall (here, the Malpica–Tui unit) and the hanging wall of the shear zone (here, the autochthonous) in the Rias–Razo beach.

Metasediments of the Malpica–Tui Unit are albite-rich quartz-feldspathic paragneisses. Significant amount of albite blasts enclose a mineral assemblage that, in other similar units of the allochthonous complexes of NW Iberia, indicates equilibrium conditions at high pressure and low temperature (Arenas et al. 1995; Martínez-Catalán et al. 1996; Rodríguez-Aller et al. 2003; López-Carmona et al. 2010).

The section at the Xeiruga beach exposes the main part of the shear zone (Fig. 13.4). To the west, it occurs the glandular gneisses of San Adrian and the two mica granites; to the east, after few hundreds of meters of continuous outcrop, is the albite-rich metassediments of the Malpica–Tui unit. The contact is sharp, since the paragneisses stand out from the mostly schistose rocks of the axial deformation zone. However, no obvious change in strain is seen across the contact. In this segment can be found dykes and irregularly shaped intrusions of granodiorites (Fig. 13.5a) that are characteristic along all the MLDSZ (Llana-Fúnez and Marcos 2001; Llana-Fúnez 2011).

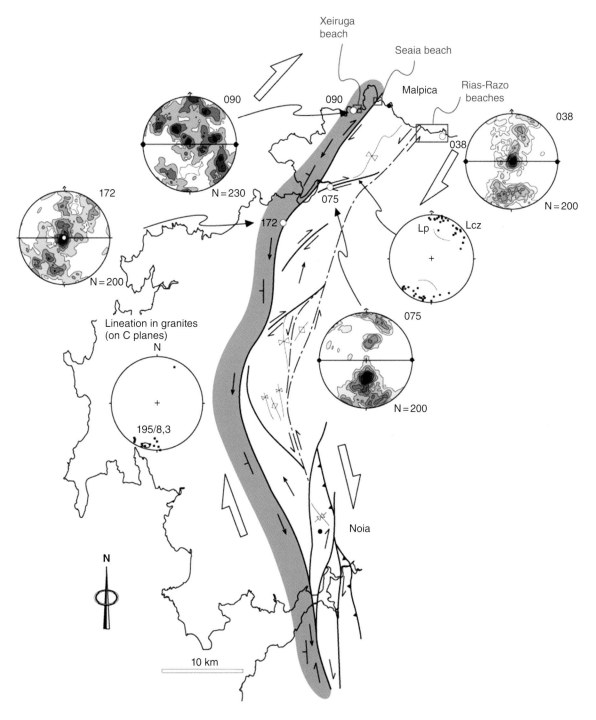

Fig. 13.3. Structures associated with the MLDSZ in northern Spain. Points in the stereo-nets show the trend of mineral lineation in granites (left) and in oblique shear zones (right). Contoured stereo-nets show distribution of c-axis in quartz from samples in the MLDSZ (090 and 172) or associated to MLDSZ structures (075). Adapted from Llana-Fúnez 2001.

Some of the shear zones affecting the schists have a left-lateral shear, opposite to the main shear sense for the MLDSZ (Fig. 8 in Llana-Fúnez and Marcos 2001, p. 1025). These are ductile shear-related minor structures at the core of the shear zone, but become predominant further east, in other structures in the footwall that develop later and at shallower crustal levels (Variscan D$_4$). Some of these later

structures are mineralized (Llana-Fúnez 2011), and a few have economic implications. In any case, all the mineralization implies large amounts of fluid circulating simultaneously to the deformation of the host rocks. The presence and the role of fluids during deformation had been commonly overlooked and is becoming imperative recently to understand their interactions (Llana-Fúnez 2011).

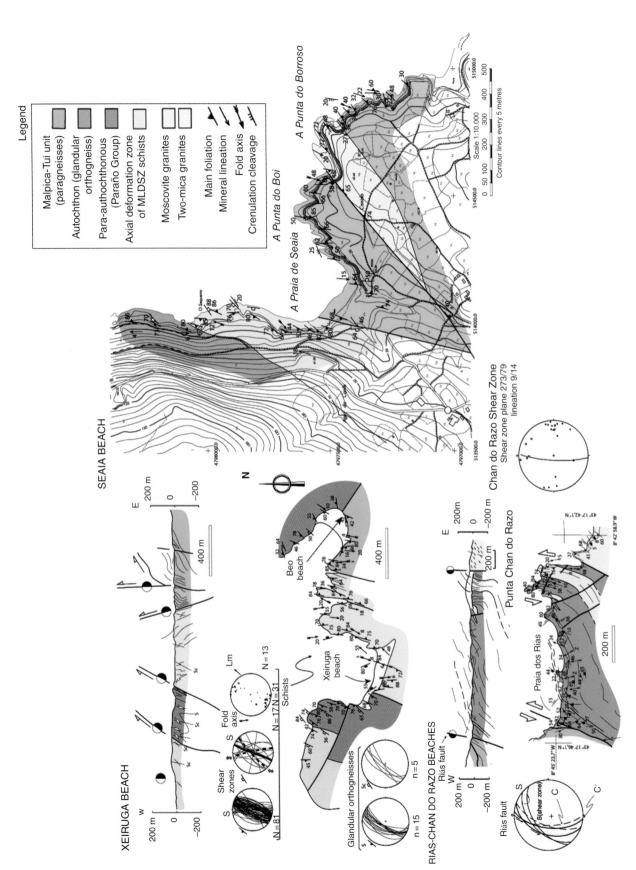

Fig. 13.4. Geological maps of selected outcrops along the coastline: Xeiruga beach (adapted from Llana-Fúnez and Marcos 2001), Seaia beach, and Praia dos Rias-Punta Chan do Razo (adapted from Llana-Fúnez 2001).

Fig. 13.5. Segment 1 is represented by dykes and irregular intrusions of granodiorite. (a) The schists exposed at the contact between the Malpica–Tui unit and the rocks further to the W, belonging to the authochthonous, show a significant amount of quartz veins. Veining may indicate the deformation zone associated with the MLDSZ had profuse fluid activity during deformation. (b) The core of the deformation zone is characterised by phyllonites – where veins thin to few millimeters and when grain size reduces significantly – the part of the high strain core of the MLDSZ, in which extremely fine-grained phyllonites alter, along with more quartz-feldspathic veins. Dextral kinematics is indicated by shear bands and sigmoid shear sense indicators (Mukherjee 2011) (c) Cataclasites in the core of the MLDSZ deformation zone. (d) A cataclastic band developed in rocks rich in quartz and feldspar.

In the Seaia beach section, the core of the MLDSZ is hundreds of meters thick and is fairly well exposed (Fig. 13.4). The core of the shear zone is defined by phyllonites and strongly deformed schists and paragneisses. On the eastern side of the beach, albite-rich quartz-feldspathic paragneisses from the basal units, belonging to the Malpica–Tui unit, are exposed. To the western side of the beach, sub-vertical well-foliated and strongly deformed schists exist, together with some fine-grained quartz-feldspathic gneisses. These rocks show abundant veins, which in cases constitute a significant fraction of the rock (Fig. 13.5b). Adjacent to the schists with the veins, there are a few meters of phyllonites, where vein thins up to few mm simultaneous to the grain size reduction (Fig. 13.5c). The rocks with finer grain size and with higher content in quartz and feldspar do show brittle fracturing and several

strands of cataclasites (Llana-Fúnez 2001) have been found (Fig. 13.5d).

In the footwall of the MLDSZ and to the East of the Malpica–Tui unit, there are two significant structures. The fault at Praia dos Rias separates the rocks from the Malpica–Tui unit, to the west, from the sequence of the para-authochthonous sequence to the east, during D_4 (Fig. 13.4). There are fault-related cataclastic rocks associated with the faulting. Further to the east, once in the para-authoctonous rock sequence, there is a left lateral shear zone hundreds of meters wide and more than a kilometer long. The structure is developed on schists intruded by muscovite granites (D_3, 317 ± 3 Ma, Rodríguez-Aller et al. 2003). A left lateral shear can be inferred from the deflection of the tectonic fabric, consistent with minor structure in the outcrop, namely C-S and C$'$-S structures (see Mukherjee 2013, 2014 for examples from different terrains).

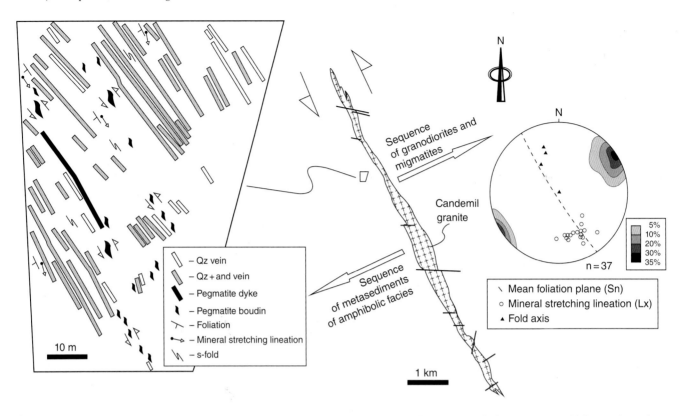

Fig. 13.6. Segment 2 of MLDSZ (V. N. Cerveira, NW Portugal) including the structural diagram with the projection of foliation (Sn) planes (represented by contours of poles and the plane of projection), mineral stretching lineation (Lx) and the cm-scale wavelength folds axes, coeval with Sn. (Lambert-Schmidt Projection, lower hemisphere).

13.3.2   Segment 2 (Vila Nova de Cerveira, Portugal)

This sector is marked by the occurrence of a deformed equigranular leucogranite, fine to medium grained, foliated with protomylonitic texture, known as the "Candemil Granite". The general orientation of the deformed granite is NW–SE, around 9 km in length and with a maximum width of 400 m. In detail, several bodies resembling mega-boudins build up this cartographic body. This granite is classically used to separate two distinct tectono-metamorphic domains: the eastern (MS-E) and western (MS-W) metasedimentary sequences (Fig. 13.6).

The western domain is composed of a monotonous pelitic sequence with a few alternations of centimeter-thick psammitic composition. It is metamorphosed in amphibolite facies and cut across by a post-tectonic granitic massif.

The eastern domain of the Candemil Granite is formed by a migmatitic sequence where the granodioritic and granite bodies emplaced into structurally controlled corridors. The granodiorites represent igneous injections – locally linked with its probable migmatitic source – exhibiting a magmatic texture structurally conditioned, whereas the granites are the mesosome, which can be described as nebulitic migmatites resembling the "Candemil Granite".

This sector recorded an event of heterogeneous progressive and high temperature shear strain, with sinistral strike–slip kinematics achieved during the Variscan D$_3$ (Fig. 13.6).

Biotite, muscovite, and sillimanite define regional foliation (Sn) in the micashists. The foliation has an associated mineral stretching lineation, defined by aggregates of quartz and sillimanite fibers, plunging 10–30° to N158° (Fig. 13.6).

The foliation (Sn) has on average an orientation N330°/85°W (strike, dip, quadrant), subparallel to the layering (S$_0$). This parallelism is accentuated on the most deformed sectors by the reactivation of Sn as a plan of movement associated with the functioning of the shear zone during D$_3$. On less deformed zones, the psammitic levels showed a refraction of Sn (Fig. 13.7a,b,c).

The main foliation surrounds pods where earlier structures are preserved such as hinges of folds with centimetric amplitude and sub-horizontal fold axes.

Aplite-pegmatite bodies of granitic composition (Qtz+Fk+Ms, Qtz+Fk+Ms+Tur, Qtz+Fk+Ms±Tur±Grt) and metamorphic segregation veins (Qtz, Qtz+Ms, Qtz+And+Ms, Qtz+And+Sil±Ms) of different ages, found in the Salgosa sector show solid-state structures such as pinch-and-swell bodies and shearband boudins (Fig. 13.7d,e) (Pamplona and Rodrigues 2011).

Throughout the deformation zone, shearband boudins appear in spatial association with intensely folded quartz veins. The geometry of the folding in quartz veins indicates a progressive deformation, recording a synthetic

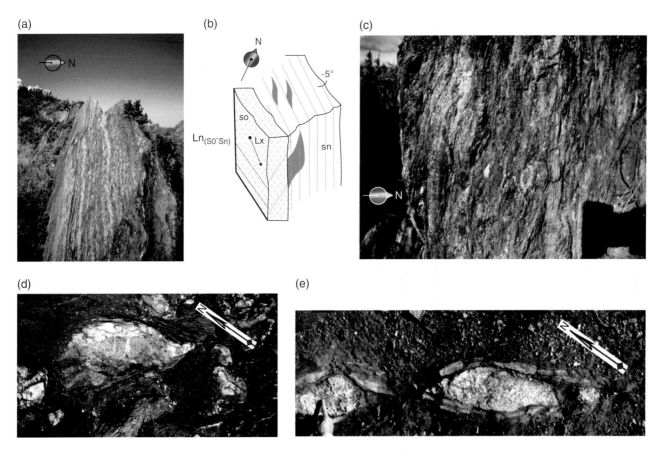

Fig. 13.7. Relationship between S_0 and Sn in metamorphosed psamitic-pelitic rock under high-grade (amphibolite facies). (a) General feature. (b) 3D interpretative sketch. (c) Detail showing the refraction of Sn in the psamitic levels. Shearband boudins in HT shear zone. (d) Quartz veins with andalusite showing well-defined sigmoidal shape. The massive andalusite exhibits a fracture pattern geometrically similar to a "torn boudin", inside the boudin block. (e) Granitic aplite-pegmatite dike deformed as a shearband boudin morphology (host-rock infilling interboudin zone).

rotation of the axial plane to a quasi-parallelism with the shear plane/C-surface (Fig. 13.8). Analyses of kinematics criteria both in the shearband boudins and folded veins indicate that they are sheared sinistrally, synthetic with kinematics of the shear zone (Pamplona et al. 2014).

13.3.3   Segment 3 (Braga, Portugal)

Here, granitic rocks of the Sameiro massif (Braga) predominate greatly over the other types of rocks. The majority of granitic rocks are syn- to post-tectonic biotite granodiorite-monzogranites, with a dominantly porphyritic texture (Ferreira et al. 1987a). They emplaced after two mica granites that cross-cut the earlier Variscan structures (Fig. 13.9). Hectometric size bodies of gabbro-granodioritic composition (gabbro, monzodiorite, quartz-monzodiorite, and granodiorite) occur, associated with the Braga composite massif. Devonian and Silurian metamorphic rocks of this region are mainly phyllites, micaschists, quartzphyllites, and greywackes with minor migmatites and basic vulcanites. The regional metamorphic rocks was affected by all deformation phases of the Variscan orogeny, with the D_1 structures being less

expressive. The Variscan D_3 shows folds of vertical axial planes striking N140° and sub-horizontal axes of about 20°S. These Variscan deformation transposed a previous foliation (striking N140°–150° and horizontal to vertical dipping) generating a D_3 foliation (S_3). The regional metamorphism in host rocks is of low pressure and high temperature.

The Sameiro massif presents a porphyritic medium-grained texture, characterized by large megacrystals of perthitic potassium feldspar, some of which are 15 cm long. The granite presents mafic microgranular and metasedimentary enclaves, and rare leucocratic ones. The majority of the mafic microgranular enclaves are rounded having (15–20 cm diameter), while the metasedimentary enclaves have decimeter to meter dimensions, some of them being roof pendants with cartographic expression.

A U–Pb geochronological study performed on the Sameiro massif reveal a minimum age of 314 ± 4 Ma for zircon fractions and an age of 318 ± 2 Ma for a monazite fraction. Monazite and zircon fractions define a reverse discordia with a lower intercept at 316 ± 2 Ma (Dias et al. 1998; Simões 2000), which matches the range 313–319

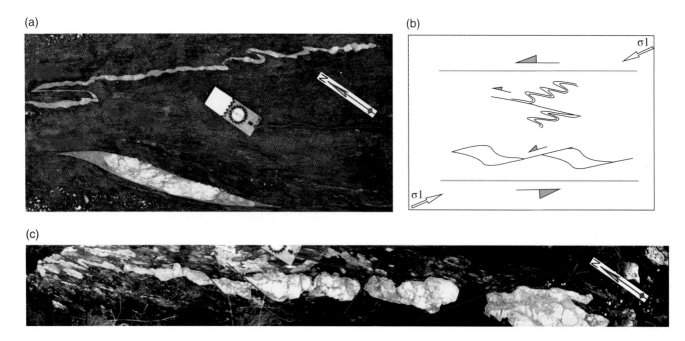

Fig. 13.8. Coeval folding and boudinage in a HT simple shear zone: (a) in side-by-side bodies; (b) interpretative kinematic model of structures like the one represented in (a); (c) in the same body. Adapted from Pamplona et al. 2014.

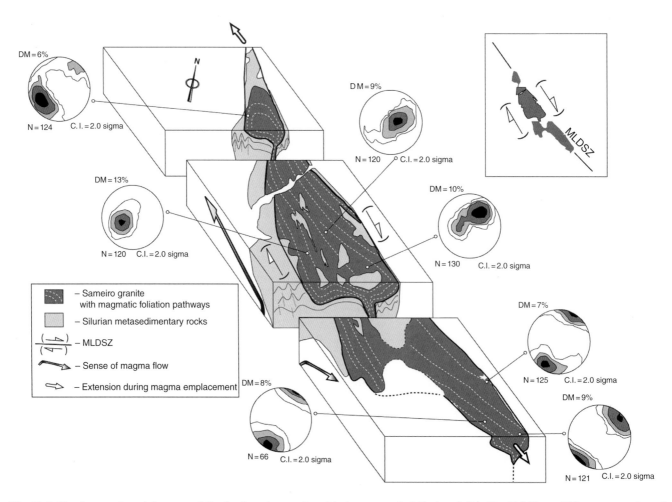

Fig. 13.9. Emplacement model proposed for the Sameiro granite with the magmatic foliation defined by K-feldspar (Kf) megacrystals and stereograms showing the distribution of (010) the Kf megacrystals poles faces density, measured at the sites indicated in this figure. The projections are Kamb contours and DM corresponds to the maximum density distribution percentage (from Simões 2000). The inset depicts D_3 kinematics within the Malpica–Lamego Ductile Shear Zone.

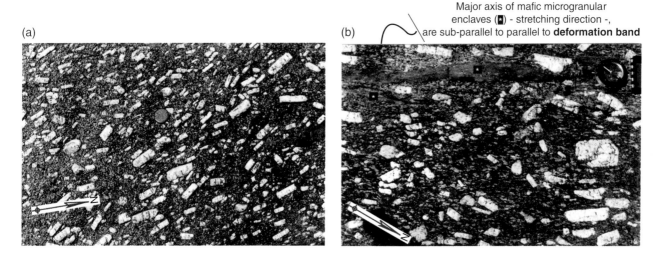

Fig. 13.10. Magmatic foliation (a) is given by the preferred orientation of potassium feldspar megacrystals. Deformation bands (b) are characterized by the flattening of quartz grains and the stretching of mafic microgranular enclaves.

Ma defined for the emplacement of similar syntectonic (Variscan D₃) granitic rocks in northern Portugal (Dias et al. 1998).

The elongated Sameiro massif is 20 km long and 4 km wide at the central sector. It has a long axis along N150, parallel to the main structures in host rocks. This massif also presents magmatic structures, deformation bands, intragranitic mylonites, ductile–brittle shears, and brittle structures.

Planar and linear magmatic structures are given by the oriented K-feldspar megacrystals, biotite and ellipsoidal mafic microgranular and metasedimentary enclaves. The main fabric in the rocks is defined by the preferred orientation of euhedral K-feldspar megacrystals (Fig. 13.10a). The magmatic foliation strikes N140–160°, with variable dip and bears a NNW–SSE lineation plunging to south and exceptionally, in the south sector, it plunges towards N. The orientations of the magmatic structures resemble those of the metasedimentary structures occurring W of the massif.

Deformation bands are m-thick and are characterized by the flattening of quartz grains within polycrystalline aggregates, the stretching of mafic microgranular enclaves (Fig. 13.10b), the development of vertical fabric of biotite and the deformation of K-feldspar megacrystals by fracturing and rounding. The bands are concordant with the magmatic foliation of the Sameiro granite, approximately N140°. Some deformations bands show N140° dextral shears based on K-feldspar megacrystal rotation criteria. The other bands indicate equal number of dextral and sinistral kinematics. In other words, no consistent rotation pattern of the K-feldspar megacrystals are found. These observations indicate that the development of syn-magmatic deformation responds to predominant flattening component. Reverse ductile shear senses from the same shear zone has been documented by many from different terrains (e.g. Mukherjee and Koyi 2010;

Chattopadhyay et al. 2016, Chapter 10; Pace et al. 2016, Chapter 8; Sengupta and Chatterjee 2016, Chapter 9).

Intragranitic mylonites are situated in the extreme north of the Sameiro massif and are characterised by the presence of C-S structures. They form sinuous metric bands trending ENE–WSW with sub-vertical dip. The bands develop C planes show sub-vertical dip to the south and strike N120–140°. C-S structures indicated a reverse movement, from S to N.

The re-orientation of granite biotite gives N–S dextral and N60–70° sinister ductile–brittle shear zones.

Measurement of strike and dip of (010) planes of K-feldspar megacrystals (magmatic foliation) indicates that the preferred orientation in the whole Sameiro massif falls in the angular foliation interval, with the northern sector exception, while dipping is variable (Fig. 13.9). This magmatic orientation parallels foliation and axial planes of open upright folds in the metasedimentary rocks. Lineation defined by (010) planes zone axes and <c> axes of K-feldspar megacrystals show slight dip to SSE and are also coincide with the D₃ folds axes. However, in the S sector, NNW lineation has an opposite orientation, plunging strongly towards N. In the central sector of the massif, the magmatic fabric shows variable dip, either to the SW or to the NE, revealing an overal curved foliation pattern for the massif (Fig. 13.10).

Contoured diagrams of K-feldspar megacrystals (010) planes show the existence of high symmetry sub-fabrics (axial to orthorhombic symmetry) in all the area of the massif (Fig. 13.9), with density maximum values (DM) reaching 13%. This type of fabric is interpreted by flattening with a reduced rotational component (Fernández 1982; Fernández and Laboue 1983) during the final stage of granite emplacement. The evaluation of the bulk strain obtained by the K-feldspar megacrystals fabric indicates a flattening of 50–65 % for the granitic magma (Simões 2000).

Two biotitic fabrics were recorded: a magmatic and a late-magmatic one. The magmatic biotite reveals an arrangement that does not always mimic that of K-feldspar megacrystals. Thus, the foliation of biotite may be parallel or subparallel to the foliation of K-feldspar megacrystals or may be vertical, with the direction N140°–160°, cutting the foliation of the megacrystals, as seen in the metric deformation bands. The late-magmatic biotite defines a vertical foliation due to a ductile–brittle episode, corresponding ~N–S dextral and N70° sinistral shears.

The field observations and structural data indicate: (i) The granite and its mafic microgranular enclaves do not record a more strained pattern near the MLDSZ than in the other parts of the massif. There is even a waning deformation towards the eastern border of the shear zone; (ii) The high intensity fabric of K-feldspar megacrystals is achieved in the western part of the massif and far from the eastern border of the shear zone. The high symmetry recorded by the K-feldspar megacrystal fabric indicates an essentially compressive deformation regime, with a weak dextral rotational component.

The magmatic structures for the Sameiro massif are the result of deformation by flattening with less rotation in a transpressive tectonic regime as dominated by a NE–SW sub-horizontal compressive component. This is compatible with the major compressive stress component ($\sigma 1$) in hosting metasediments and without significant strike–slip movement along the MLDSZ, during Variscan D_3 episode. around 316 ± 2 Ma.

13.3.4 Segment 4 (from Mesão Frio to Lamego, Portugal)

This segment is located at the southern branch of MLDSZ, developed in the lower Paleozoic formations of Autochthonous of Central Iberian Zone (Fig. 13.11).

Mapping of the southern sector of the "Serra do Marão" revealed two blocks with distinct structural signatures, separated by a sinistral (D_3) strike–slip fault called "Ferrarias Fault", a local structure included in the MLDSZ (Fig. 13.11).

The northern block, where the D_1 is penetrative, shows NE verging folds, fold-axis dipping about 12° to WNW, axial plane cleavage (S_1), S_0/S_1 intersection lineation sub-parallel to the fold-axes and sub-horizontal stretching lineation (Fig. 13.12).

In the southern block, the D_3 fabric is dominant and penetrative, marked by folds with axes dipping usually >20° to WNW, steep S_3 axial plane cleavage and S_0/S_3 intersection lineation dipping ~20° to NE (Fig. 13.11). The S_3 cleavage cuts S_1 – it is only visible in some pelitic layers. Consequently, there is a strong partition of the deformation as deduced from the parallelism between the axial planes of major folds and shear zones and from the absence of en-echelon and transect folds. The major tectonic style in this sector is described as a flower structure (Fig. 13.12).

The center of the flower structure developed during main deformation episode related tp teh activation of MLDSZ (Fig. 13.12). In the autochthonous sequence of the CIZ D_3 is more penetrative and D_1 is practically absent.

The Ferrarias Fault has been induced by one of the pre-existing anisotropies of the Pre-Cambrian basement, as it could be the case for other structures, precursors to MLDSZ, which were active during the basin infilling process on the Lower Paleozoic. This kind of faults has controlled the regional deformation during the Variscan cycle (Coke et al. 2003).

The axial zone of MLDSZ is characterized by a vertical axial plane penetrative cleavage (S_3) overwriting previous structures. On these high strained narrow zones, a sub-horizontal stretching lineation produced below greenschist facies condition, during the main deformation episode (D_3 sinistral strike–slip).

13.3.5 Segment 5 (from Ucanha to Vilar, Portugal)

The Ucanha–Vilar granite is a magmatic body striking NW–SE, parallel to the MLDSZ lineament (Fig. 13.13). It is a biotite granite with medium-grained porphyritic texture, where the K-feldspar megacrystals are ~5–6 cm long, sometimes reaching up to 10 cm. In the extreme NW of the massif, the granite is more leucocratic and shows some muscovite. The Ucanha–Vilar granite is surrounded mainly by two-mica granites. To the NW and E, the massif intrudes Cambrian metasediments of the Schist–Greywacke Complex (CXG), and in the S post-tectonic granites.

In relation to this granite, there are in some areas meter-thick bands of K-feldspar in the granite and mafic microgranular spherical and ellipsoidal enclaves of ±10 cm to 1.5 m mean diameter, especially in the Vilar region. This area also has dykes in granite (observable in the Vilar quarry) with cooling borders and lobate contacts, indicating coeval emplacement. These decimeter-thick syn-plutonic dykes are generally sub-vertical and show K-feldspar cross-cutting the contact with the hosting granite. In the massif SW tip (Vilar area), the granite has a darker color with abundant enclaves, while at the far NW (Ucanha area), the granite reveals a lighter color, with muscovite and less frequent microgranular enclaves. Note that the Ucanha–Vilar granite has more mafic microgranular enclaves than other granites.

To the NW area of the massif, the main granite appears in association with two-mica granites. Small bodies of the two-mica granite occur as enclaves within the Ucanha granite, and vice versa, and also, the contacts between these two granites have an interdigitated pattern, which support the model of contemporaneous emplacement.

The granite shows an internal structure, essentially magmatic, which is given by the orientation of K-feldspar and biotite of the matrix, and also by the parallelism of the long axis of ellipsoidal microgranular enclaves. In the Ucanha body the magmatic foliation trends ~N120°,

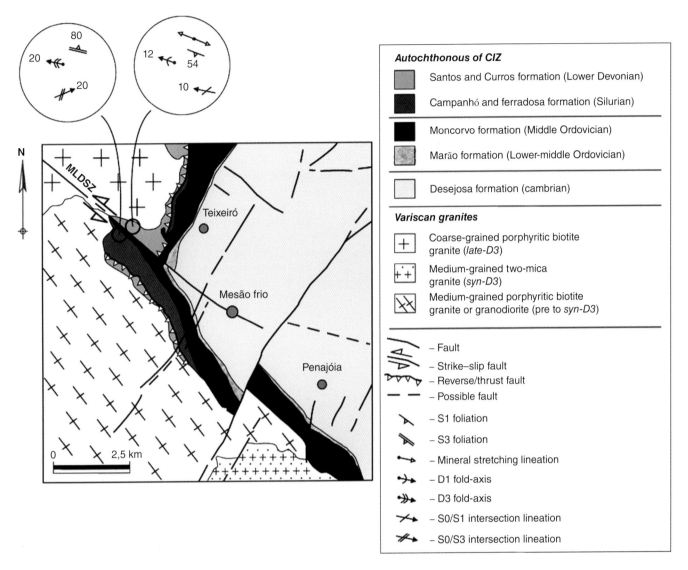

Fig. 13.11. Lower Paleozoic formations of Autochthonous of Central Iberian Zone (CIZ): Desejosa Formation (Cambrian) – Sousa (1982); Marão Formation (Lower–Middle Ordovícian, Middle–Upper(?) Arenigian) – Sá et al. (2005); Moncorvo Formation (Middle Ordovician, Lower Oretanian-Lower Dobrotivian) – Sá et al. (2005); Campanhó and Ferradosa Formations (Silurian, Llandovery-Pridolian) – Pereira (1987b) and Pereira (Coord) (2006): Santos and Curros Formations (Lower Devonian) – Pereira (1987a, b) and Pereira (Coord) (2006). Adapted from Pereira 2000 and Sá et al. 2005.

whereas in the Vilar body this direction is N–S. Some microgranular enclaves also show an internal foliation, given by the orientation of biotite, which parallels the magmatic foliation of the granite. The N70° sinistral ductile shears that affects the granite, sometimes give deformation bands. They relate with regional N70° shear zone that pass between Ucanha and Vilar granite bodies, which affected the granite when it was not crystallized fully, and also synchronously rotated the magmatic foliation near the contact.

A study of the fabric of shape-preferred orientation (SPO) of euhedral K-feldspar megacrystals in the Ucanha–Vilar granite massif enabled to compare the deformation pattern of granitic massifs situated along the MLDSZ. In the Ucanha body the preferred orientation of K-feldspar

is ~N120°, parallel to the D_3 structures of the metasedimentary country rocks, while in Vilar body it is ~N–S (Fig. 13.13). The mean dip of the SPO in the massif is relatively steep and the opposite dip directions are less frequent, which is associated with the reduced number of panels/flattened mega-enclaves of metasedimentary country rocks within the massif. This feature indicates a deeper structural level for the Ucanha–Vilar granite, not showing the "fold" pattern imposed by the metasedimentary roof of the Sameiro massif (segment 2). A change in the strike of foliation of the K-feldspar in the northern area of Vilar and SE edge of Ucanha area is attributed to a sinistral regional ductile shear zone, trending N70°, between those two areas of the massif. Near the shear zone there is flattening of the quartz

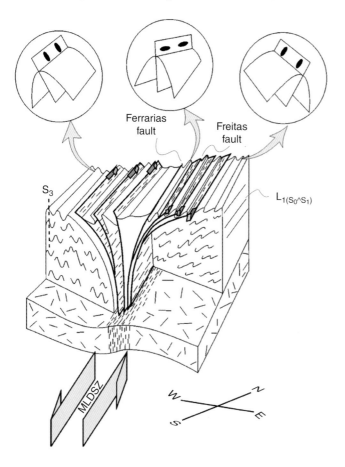

Fig. 13.12. 3D sketch of the MLDSZ and of the flower structure of the Autochthonous of CIZ in this segment. Adapted from Coke et al. 2000.

grains and C–S structures, with the direction of the S plane almost parallel to the C planes. Besides the N70° smaller ductile shears that affect Ucanha–Vilar granite, ~N–S dextral ductile–brittle shear zones were also observed.

Stereonets of K-feldspar (010) planes reveal symmetric fabrics in Ucanha–Vilar granite, with maximum density (D_M) between 7 and 11% (Fig. 13.13). That kind of symmetry indicates that the magmatic structures of Ucanha–Vilar massif, acquired during emplacement, are by magma deformation by flattening with a reduced rotational component (Fernández 1982; Fernández and Laboue 1983).

The MLDSZ, on its southern edge, could not have been so active when the Ucanha–Vilar granite emplaced ~313 ± 2 Ma (Dias et al. 1998), because its magmatic structure produced from compressive regime during the D_3. However, during the final stage of the massif emplacement, there was a sinistral N70° shear, which can explain the rotation of structures in the far northern area of the Vilar body and SE tip of the Ucanha body.

The structural study carried out in Sameiro and Ucanha–Vilar massifs indicate that the granites spatially associated with the MLDSZ emplaced in early D_3 stage;

thus, they are syn-D_3 (see also Ferreira et al. 1987a). The structural data indicates that the granite fabric is mainly magmatic and has resulted from the deformation of the magma essentially by flattening. This deformation shows a strong compressive tectonic regime, with a minor rotational component, which has acted during the final magma emplacement. Assuming a bulk transpressive regime, the sub-horizontal compressive component have been dominant with no significant movement along the MLDSZ. The shear zone recorded minor strike–slip movement, but only at the end and after the emplacement of the magmas, given the existence of N140° dextral ductile shear zones.

Magmatic structures, concordant with the D_3 regional structures, were the result of compressive tectonic regime, from ~NE–SW in the northernmost sector of the MLDSZ (e.g. Sameiro massif) and NNE–SSW on its southern sector (e.g. Ucanha–Vilar massif). This rotation probably records the sinistral rotation of σ_1 during the final episodes of the Variscan orogeny.

13.3.6 Segment 6 (Penedono, Portugal)

The Penedono area matches the southern tip of MLDSZ. By definition it is expectable that an attenuation in the main Variscan structure may be observed in this sector. The identification of the tip structures is achieved by the geometric and kinematic patterns of late reactivation that, in the present case study, is linked to gold mineralizations.

In this segment three groups of gold mines (Fig. 13.14) are disposed along NW–SE, ~5 km spaced from each other. The mineralization is mainly arsenopyrite in quartz veins, associated with second order shear zones developed in the granites. The quartz veins are deformed and exhibit en-echelon pattern (Fig. 13.14) recording sinistral shear (e.g. Blenkinsop 2000) that is conjugated with the regional dextral kinematics of the MLDSZ. It is possible to identify several generations of arsenopyrite associated with this deformation. These mineral occurrences trend N150° that parallels the MLDSZ and also to main elongation of the granitic massifs.

The rock sequence that hosts the mineralizations belongs to the Variscan autochthonous of the Douro Group, Lower Cambrian age (Sousa 1982) of the Schist–Greywacke Complex (CXG). These rocks show deformation structures identifiable regionally as D_1 and D_3 Variscan phases. D_1 manifests as the regional structure of the metasediments, developing meso- to megascopic folds of sub-horizontal axis and an axial plane cleavage S_1 striking N150°. The D_3 phase folds are coaxial with D_1 folds. A crenulation cleavage S_3 is well developed, also with recrystallization and orientation of biotite (Sousa 1982).

The granitic massif (Penedono Massif, 320–315 Ma (Ferreira et al. 1987a) that hosts the mineralization emplaces itself in the core of the D_3 antiform defined in the Cambrian metasediments of the Douro Group. The internal structures defined in these granites, such as

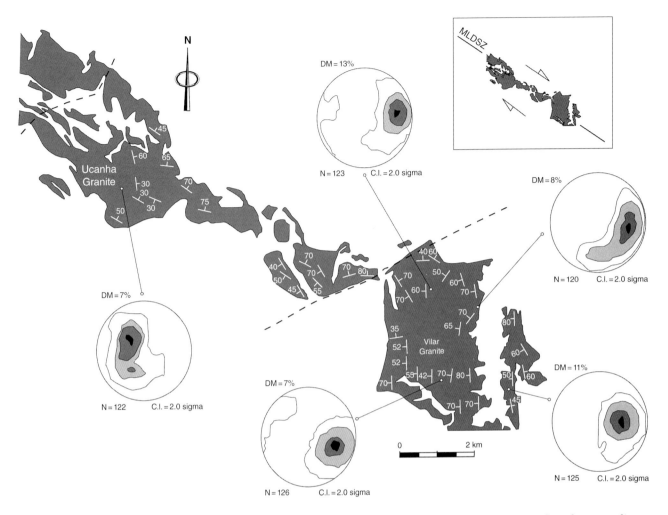

Fig. 13.13. Schematic map of Ucanha–Vilar granite with the magmatic foliation defined by K-feldspar megacrystals and contour diagrams showing the distribution of poles corresponding to planes (010) of megacrystals. The projections are Kamba contours and DM values correspond to maximum density distribution. The inset depicts D$_3$ kinematics within the Malpica-Lamego Ductile Shear Zone. Adapted from Simões 2000.

orientation of the different cartographic facies and the internal granitic foliation, parallel the D$_3$ structures, which characterizes it as syn-tectonic, relative to D$_3$.

The ~E–W sinistral ductile–brittle shear zones affect the granitic rocks of the massif. These shear zones are related spatially with the MLDSZ and developed late in the tectonic evolution of this crustal scale structure.

The gold mineralization is controlled by minor sinistral shear zones oriented E–W to ENE–WSW. These structures are related with the regional major sub-vertical MLDSZ, dextral during D$_3$ oriented N115°, and located some kilometers to the W (Fig. 13.14).

Each shear set controlling gold mineralization displays a ductile/ductile–brittle/brittle behavior. The deformation within these bands is polyphasic (Fig. 13.15). The amount of movement along the shear zones and their ductility increased towards W, i.e. towards the MLDSZ.

A sinistral E–W subvertical minor ductile shear zone develops tension gashes (T1) that were infilled with a first generation of veins (Fig. 13.15). This minor shear zone reactivated successively with sinistral movement after the opening gashes and formation of minor fractures Riedel and also R–R$'$ fractures (Fig. 13.15) on a bulk sinistral shear under the regional dextral kinematics.

The ductile–brittle process induces a second mineralization phase with fracturing of the preliminary one and deposition of a second generation of vein materials (Fig. 13.15).

A late Variscan brittle phase (D$_4$) affected rocks in Schist–Greywacke Complex (Douro Group, Lower Cambrian). This last deformation phase is due to the anti-clockwise rotation of the maximum stress σ_1 to ~N–S, which created N20° trending sub-vertical sinistral transcurrent faults. The rotation of the σ_1 to N–S direction (Fig. 13.15) either reactivated all earlier fractures with sinistral shearing when in the east quadrant, or sheared dextrally when situated towards west. Coevally, the associated transcurrent faults rotated fault-bound blocks within the granitic rocks.

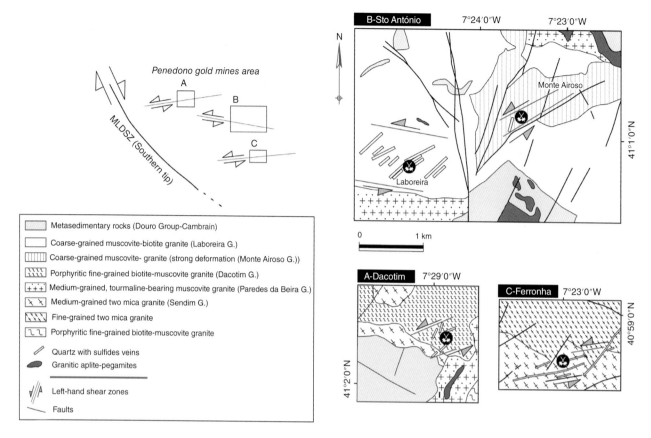

Fig. 13.14. Geological sketch map of Penedono mining area (Penedono Massif). Regional geological data. Adapted from Ferreira et al. 1987c.

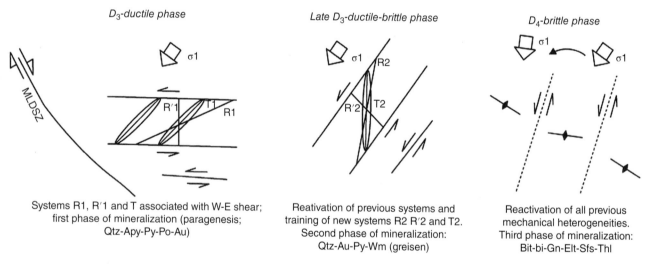

Fig. 13.15. Penedono gold mines distribution of major mineral paragenesis during variscan deformation as recorded as Sto António mine. Quartz (Qtz), arsenopyrite (Apy), pyrite (Py), pyrrotite (Po), wolframite (Wm), bismutinite (Bit), bismouth (Bi), galena (Gn), electrum (Elt), gold (Au), sulfossalts (Sfs) and tellurides (Thl).

13.4   CONCLUSIONS

The MLDSZ is a major crustal-scale tectonic structure that extends for at least 275 km from its northernmost exposure in Malpica, Spain, to its southern tip in Lamego/Penedono areas, Portugal. It manifests by a deformation zone with ~4 km maximum width and the alignment of granodioritic rocks along it.

The MLDSZ has a complex tectonic history during the Variscan orogeny. This orogeny is generally described in NW of Iberia with four main phases: D₁ – continental collision and obduction of allochthonous

Table 13.1. Characterization of the Variscan episodes (D$_3$ and later ones) recorded on MLDSZ activity: timing of deformation, *P–T* and kinematic conditions, lithologies, magmatism, and mineralizations

Description	Sequence of structures over time in relation to the MLDSZ		
	>320 Ma	320–300 Ma	<300 Ma
Malpica, Spain Northernmost tip	>350–340 Ma Dip-slip reverse fault related with overthrusting Age of earliest magmatism, e.g 341 ± 6 Ma, ^{40}Ar/^{39}Ar (Rodríguez et al. 2003)	310–300 Ma Dextral strike–slip Amphibolite facies Highly deformed schists, phyllonites Granodiorite intrusions (Llana-Fúnez and Marcos 2007) Syn-tectonic Variscan granites, e.g. 303 ± 6 Ma, ^{40}Ar/^{39}Ar (Rodríguez-Aller et al. 2003)	<300 Ma Sinistral strike–slip (late mineralized sinistral shear zones) (Llana-Fúnez and Marcos 2007) Thermal overprinting, e.g. 280 ± 5 Ma, ^{40}Ar/^{39}Ar (Rodríguez et al. 2003)
Salgosa, Portugal Central sector		Sinistral strike–slip Amphibolite facies Micaschist with intense quartz and pegmatite vein boudinage (Pamplona and Rodrigues 2011)	
Sameiro, Portugal Central sector		313–319 Ma Flattening with minor dextral strike–slip Amphibolite facies Internal deformation of granodiorite intrusion of Sameiro granite, e.g. 316 ± 2.5 Ma, U-Pb (Dias et al. 1998)	
Mesão Frio - Lamego, Portugal Central-southern sector		Sinistral strike–slip Sub-greenschist facies Ferrarias fault Pelites and psammites with axial plane cleavage (Coke et al. 2003)	
Ucanha - Vilar, Portugal Southern sector		Flattening with minor dextral strike–slip Amphibolite facies Internal deformation of granodiorite intrusions of Ucanha and Vilar granites, e.g. 313 ± 2 Ma, U-Pb (Dias et al. 1998)	<313 Ma Dextral strike–slip (late ductile dextral shear zones) (Simões 2000)
Penedono, Portugal Southern tip		Dextral strike–slip Sub-greenschist facies Ductile deformation with Au mineralization on quartz veins Syn-tectonic Variscan granites, e.g. 312 ± 6 Ma, K-Ar (Ferreira et al. 1987b)	Dextral strike–slip Ductile–brittle deformation with Au-Wm mineralization on quartz veins, followed by brittle deformation with electrum mineralization on quartz veins

terrains; D$_2$ – structuring of the allochthonous terrains; D$_3$ – intracontinental deformation that produced crustal shear zone; D$_4$ – later ductile–brittle to brittle intracontinental deformation. The MLDSZ, with its actual configuration, developed after the allochthonous complexes over Iberian plate rocks emplaced, or after plate duplication (crustal thickening resulting from the nappes emplacement of allochthonous terranes) during the Variscan collision. Most significant shear zone-related structures developed during the Variscan D$_3$ deformation phase and later episodes. Structures in six segments along the shear zone are summarized in Table 13.1.

The selected segments of MLDSZ record a major strike–slip deformation event during the D$_3$ Variscan deformation phase. The recorded data do not show a clear diachronism. If the gap of 310–300 Ma, recorded on the northernmost tip, could represent the trigger age of D$_3$ movement, the age of granite emplacement along the MLDSZ (318 and 313 Ma, respectively on segment 3 and segment 5) means that it was controlled by an earlier crustal anisotropy that reworked only on final stages of granite crystallization, later on D$_3$.

The strong dextral movement, well preserved on the northernmost and on the southern tips, aparrently contrast with sinistral kinematics that characterize the central segments. On the northern segment the initial dextral movement is followed by a later sinistral movement that indicates that the sinistral movement in the central segment could be later. These sinistral segments could record a later Variscan episode not yet well known.

The exposed outcrops show Variscan deformation conditions evolving from amphibolite facies to under greenschists facies on upper crustal blocks, closely related to magmatic activity in the early stages (D$_3$) and hydrothermal activity in the later D$_4$ stage.

Along the MLDSZ lineament, and mainly on its southern tip, brittle–ductile and brittle structures (e.g. Penedono segment) were described that are not directly related with Variscan activity of MLDSZ. However, they are possibly due to the late Variscan or Alpine deformation.

REFERENCES

Arenas R, Rubio Pascual FJ, Díaz García F, Martínez-Catalán J. 1995. High-pressure micro-inclusions and development of an inverted metamorphic gradient in the Santiago Schists (Ordones Complex, NW Iberian Massif, Spain): evidence of subduction and collisional decompression. Journal of Metamorphic Geology 13(2), 141–164.

Blenkinsop T. 2000. Deformation Microstructures and Mechanisms in Minerals and Rocks. Kluwer Academic Plubishers, Dordrecht.

Chattopadhyay N, Ray S, Sanyal S, Sengupta P. 2016. Mineralogical, textural and chemical reconstitution of granitic rock in ductile shear zone: A study from a part of the South Purulia Shear Zone, West Bengal, India. In Ductile Shear Zones: From Micro- to Macro-scales, edited by S. Mukherjee and K.F. Mulchrone, John Wiley & Sons, Chichester.

Coke C, Dias R, Ribeiro A. 2000. Malpica-Lamego shear zone: a major crustal discontinuity in the Iberian Variscan Fold Belt. Basement Tectonic 15, A Coruña, Spain, Program and Abstracts, pp. 208–210.

Coke C, Dias R, Ribeiro, A. 2003. Partição da deformação varisca induzida por anisotropias do soco Precâmbrico: o exemplo da falha de Ferrarias na Serra do Marão. Ciências da Terra (UNL), n° especial V CD-ROM, D21–D24.

Dias R, Ribeiro A. 1995. The Ibero-Armorican Arc: a collision effect against an irregular continent? Tectonophysics 246, 113–128.

Dias R, Ribeiro A, Coke C, et al. 2013. II.1.1. Evolução estrutural dos sectores setentrionais do Autóctone da Zona Centro-Ibérica. In Geologia de Portugal, edited by R. Dias, A. Araújo, P. Terrinha, and J.C. Kulberg, Vol. I, Geologia Pré-mesozóica de Portugal, Escolar Editora, Lisboa.

Dias G, Leterrier J, Mendes A, Simões PP, Bertrand JM. 1998. U-Pb zircon and monazite geochronology of post-collisional Hercynian granitoids from Central Iberian Zone (Northern Portugal). Lithos 45, 349–369.

Fernandes AP. 1961. O vale de fractura de Rio Fornelo-Padronelo-Amarante. Boletim do Museu do Laboratório de Mineralogia da Universidade de Lisboa 9, 138–147.

Fernández A, Laboue M. 1983. Développement de l'orientation préférentielle de marqueurs rigides lors d'une déformation par aplatissement de révolution. Étude théorique et application aux structures de mise en place du granite de la Margeride au voisinage du bassin du Malzieu (Massif Central français). Bulletin de la Societe Géologiqe de France XXV(3), 327–334.

Fernández A. 1982. Signification des symétries de fabrique monocliniques dans les roches magmatiques. Comptes Rendu Academie des Sciences Paris 294, 995–998.

Ferreira N, Iglésias M, Noronha F, Pereira E, Ribeiro A, Ribeiro ML. 1987a. Granitóides da Zona Centro-Ibérica e seu enquadramento geodinâmico. In Libro Homenage a L. C. Garcia Figuerola. Geologia de los granitoides e rocas associadas del Macizo Hesperico, pp. 37–52.

Ferreira N, Sousa B, Macedo R. 1987b. Cronostratigrafia dos Granitos da Região de Moimenta da Beira-Tabuaço-Penedono. Memórias do Museu e Laboratório de Mineralogia e Geologia da Faculdade de Ciências da Universidade do Porto 1, 287–301.

Ferreira N, Sousa B, Romão JC. 1987c. Carta Geológica de Portugal na escala 1/50 000, folha 14-B Moimenta da Beira. Lisboa: Serviços Geológicos de Portugal.

Gomes CL. 1994. Estudo Estrutural e Paragenético de um Sistema Pegmatóide Granítico - O Campo Aplito-pegmatítico de Arga, Minho – Portugal. PhD Thesis, Universidade do Minho, Braga.

Goscombe BD, Passchier CW. 2003. Asymmetric boudins as shear sense indicators – an assessment from field data. Journal of Structural Geology 25, 575–589.

Heaman L, Parrish RR. 1991. U-Pb geochronology of acessory minerals. In Short Course Handbook on Applications of Radiogenic Isotope Systems to Problems in Geology, edited by L. Heaman and J.N. Ludden, Mineralogical Association of Canada, Ottawa, pp. 59–102.

Holtz F. 1987. Étude Structurale, Metamorphique et Geochimique des Granitoides Hercyniens et Leur Encaissant dans la Region de Montalegre (Trás-os-Montes, Nord Portugal). Theses de Doctorat, Université de Nancy I, Nancy.

Iglésias MI, Choucrounne P. 1980. Shear zones in the Iberian Arc. Journal of Structural Geology 2, 63–68.

Iglésias MI, Ribeiro A. 1981. La zone de cisaillement ductile de Juzbado (Salamanca) – Penalva do Castelo (Viseu): Um linéament ancien réactivé pendant lórogénèse hercyienne? Comunicações Serviços Geológicos de Portugal 67, 89–93.

Llana-Fúnez S. 2001. La Estructura de la Unidad de Malpica-Tui (Cordillera Varisca en Iberia). Publicaciónes del Instituto Geológico y Minero de España, Serie Tesis Doctoral, 1, p. 1–295.

Llana-Fúnez S. (2011) Evidence of fluid circulatuion during deformation along the Malpica-Lamego Line, a major crustal scale deformation zone in the Variscan Orogen in Iberia. American Geophysical Union, Fall Meeting 2011, Abstract T53B-05.

Llana-Fúnez S, Marcos A. 2007. Convergence in a thermally softened thick crust: Variscan intracontinental tectonics in Iberian plate rocks. Terra Nova 19(6), 393–400.

Llana-Fúnez S, Marcos A. 2001. The Malpica-Lamego Line: a major crustal-scale shear zone in the Variscan belt of Iberia. Journal of Structural Geology 23(6–7), 1015–1030.

López-Carmona A, Abati J, Reche J. 2010. Petrologic modeling of chloritoid-glaucophane schists from the NW Iberian Massif. Gondwana Research 17, 377–391.

Martínez-Catalán J, Fernández-Suárez J, Meireles C, González-Clavijo E, Belousova E, Saeed A. 2008. U-Pb detrital zircon ages in synorogenic deposits of the NW Iberian Massif: interplay of syntectonic sedimentation and thrust tectónics. Journal of the Geological Society of London 165, 687–698.

Martínez-Catalán JR, Arenas R, Díaz García F, Rubio Pascal FJ, Abati J, Marquínez J. 1996. Variscan exhumation of a subducted Paleozoic continental margin; the basal units of the Ordenes Complex, Galicia, NW Spain. Tectonics 15 (1), 106–121.

Mukherjee S. 2011. Mineral fish: their morphological classification uselfuness as shear sense indicators and genesis. International Journal of Earth Science 100, 1303–1314.

Mukherjee S. 2013. Deformation Microstructures in Rocks. Springer, Berlin.

Mukherjee S. 2014. Atlas of Shear Zone Structures in Meso-scale. Springer, Cham.

Mukherjee S, Mulchrone KF. 2013. Viscous dissipation pattern in incompressible Newtonian simple shear zones: an analytical model. International Journal of Earh Sciences 102, 1165–1170.

Mukherjee S, Koyi HA. 2010. High Himalayan Shear Zone, Zanskar Indian Himalaya: microstructural studies and extrusion mechanism by a combination of simple shear and channel flow. International Journal of Earth Science 99(5), 1083–1110.

Noronha F, Ramos JMF, Rebelo JM, Ribeiro A, Ribeiro ML. 1979. Essai de corrélation des phases de déformation hercynienne dans le Nord-Ouest Péninsulaire. Boletim da Sociedade Geológica de Portugal XXI, 227–237.

Pace P, Calamita F, Tarvanelli E. 2016. Brittle-ductile shear zones along inversion-related frontal and oblique thrust ramps: insights from the Central-Northern Apennines curved thrust system (Italy). In Ductile Shear Zones: From Micro- to Macro-scales, edited by S. Mukherjee and K.F. Mulchrone, John Wiley & Sons, Chichester.

Pamplona J, Rodrigues BC. 2011. Kinematic interpretation of shear-band boudins: New parameters and ratios useful in HT simple shear zones. Journal of Structural Geology 33, 38–50.

Pamplona J, Rodrigues BC, Fernandez C. 2014. Folding as precursor of asymmetric boudinage in shear zones affecting migmatitic terranes. Geogaceta 55, 15–18.

Passchier CW, Trouw RAJ. 2005. Microtectonics, 2nd edition, Springer, Heidelberg, pp. 146-150.

Pereira E. 1987a. Carta Geológica de Portugal na Escala 1:50000 e Notícia Explicativa da Folha 10-A (Celorico de Basto). Serviços Geológicos de Portugal, Lisboa.

Pereira E. 1987b. Estudo Geológico Estrutural da Região de Celorico de Bastos e sua Interpretação Geodinâmica. PhD Thesis, Faculdade de Ciências Univiversidade de Lisboa, Lisboa.

Pereira E. (Coord) 2000. Carta Geológica de Portugal na Escala 1:200000, Folha 2. Instituto Geológico e Mineiro de Portugal, Lisboa.

Pereira E. (Coord) 2006. Notícia Explicativa da Carta Geológica de Portugal na Escala 1:200000, Folha 2. Instituto Geológico e Mineiro de Portugal, Lisboa.

Pereira E, Ribeiro A, Meireles C. 1993. Cisalhamentos hercínicos e controlo das mineralizações de Sn-W, Au e U na Zona Centro-Ibérica, em Portugal. Cuadernos do Laboratório Xeológico de Laxe 18, 89–119.

Ribeiro A, Pereira E, Dias R. 1990. Part IV Central-Iberian Zone, allochthonous sequences, structure in the northwest of the Iberian Peninsula. In Pre-Mesozoic Geology of Iberia, edited by R.D. Dallmeyer and E. Martinez-Garcia, Springer-Verlag, Berlin, pp. 220–236.

Ribeiro ML. 1988. Shear heating effects on the evolution of metamorphic gradients at Macedo de Cavaleiros Region (north eastern Portugal). Comunicações dos Serviços Geológicos de Portugal 74, 35-29.

Rodríguez-Aller J, Cosca CMA, Ibarguchi JG, Dallmeyer RD. 2003. Strain partitioning and preservation of 40Ar/39Ar ages during Variscan exhumation of a subducted crust (Malpica-Tui Complex, NW Spain). Lithos 70, 111–139.

Sá AA, Meireles C, Coke C, Gutiérrez-Marco JC. 2005. Unidades litoestratigráficas do Ordovícico da região de Trás-os-Montes (Zona Centro Ibérica, Portugal). Comunicações Geológicas 92, 31–74.

Schmid SM, Casey M. 1986. Complete fabric analysis of some commonly observed quartz c-axis patterns. In Mineral and Rock Deformation: Laboratory Studies. The Paterson Volume, edited by B.E. Hobbs and H.C. Heard, Geophysical Monographs, American Geophysics Union, vol. 36, pp. 263–286.

Sengupta S, Chatterjee SM. 2016. Microstructural variations in quartzofeldspathic mylonites and the problem of vorticity analysis using rotating porphyroclasts in the Phulad Shear Zone, Rajasthan, India. In Ductile Shear Zones: From Micro- to Macro-scales, edited by S. Mukherjee and K.F. Mulchrone, John Wiley & Sons, Chichester.

Simões PP. 2000. Instalação, geocronologia e petrogénese de granitóides biotíticos hercínicos associados ao cisalhamento Vigo-Régua (ZCI, Norte de Portugal). PhD Thesis, Universidade do Minho/Institut National Polytechnique de Lorraine, Braga/Nancy.

Sousa MB. 1982. Litoestratigrafia e estrutura do Complexo Xisto-Grauváquico Ante-Ordovícico - Grupo do Douro (Nordeste de Portugal). PhD Thesis, Universidade de Coimbra, Coimbra.

Spry A. 1969. Metamorphic Textures. Pergamon Press, Oxford.

Winkler HF. 1979. Petrogenesis of Metamorfic Rocks, 5th edition, Springer-Verlag, New York.

Chapter 14

Microstructural development in ductile deformed metapelitic–metapsamitic rocks: A case study from the greenschist to granulite facies megashear zone of the Pringles Metamorphic Complex, Argentina

SERGIO DELPINO[1,2], MARINA RUEDA[2], IVANA URRAZA[2], and BERNHARD GRASEMANN[3]

[1] INGEOSUR, (CONICET-UNS), Departamento de Geología (UNS), San Juan 670 (B8000ICN), Bahía Blanca, Argentina
[2] Departamento de Geología, (Universidad Nacional del Sur), Bahía Blanca, Argentina
[3] Department of Geodynamics and Sedimentology, Structural Processes Group, University of Vienna, Austria

14.1 INTRODUCTION

Deformation microstructures in minerals represent an extremely valuable tool to determine the physical conditions of ductile deformation of rocks (e.g. Vernon 2004; Passchier and Trouw 2005; and references therein). Nevertheless, most of the existing experimental studies and natural examples are based on the rheologic behavior of monomineral aggregates, eminently of quartz (Hirth and Tullis 1992 and references therein; Stipp et al. 2002 and references therein). On the contrary, the available information for the interpretation of polymineral aggregates -where the interaction between phases with different compositional and chrystallographic characteristics can influence its rheological behavior considerably- is scarce (e.g. Renard et al. 2001; Herwegh et al. 2005, Huet et al. 2014).

Another major drawback when performing a comparative analysis between deformation microstructures observed in natural samples and those obtained in experimental studies is that most of them involve coarse-grained monomineral aggregates as starting materials. Consequently, recrystallized grains smaller than the parent crystals are produced, which increase its size with the increasing of temperature and/or fall in strain rate. This is the most frequently used reference when deformation mechanisms in minerals and associated microstructures, are used to estimate the thermal conditions of ductile deformation of natural felsic and intermediate rocks (see for example, Spear 1993; Vernon 2004; Passchier and Trouw 2005; Trouw et al. 2010, and references therein).

But what happens to the rocks generated from fine-grained polymineralic sediments undergoing prograde metamorphism? The characteristics of the microstructural development of fine-grained polymineralic sediments undergoing progressive metamorphism and deformation, is less well understood. It is well known that ductile behavior of previously formed coarse-grained quartz aggregates begins at ~250–300°C, and that the dominant deformation mechanism is dislocation glide and low-temperature grain boundary migration recrystallization or bulging recrystallization (Bailey and Hirsch 1962; Drury et al. 1985; Stipp et al. 2002; Passchier and Trouw 2005), corresponding to Regime 1 of Hirth and Tullis (1992). However, at these thermal conditions a rock derived from the metamorphism of fine-grained sediments lies in the very low-grade metamorphic field, in which phases are usually very intermingled and grains are still small-sized. Therefore, the rheological behavior of these minerals will surely be very different to that of coarse-grained monomineralic aggregates (e.g. Herwegh et al. 2005 and references therein).

Therefore, it is important to highlight the absence of the typical reference microstructures in metasedimentary rocks undergoing ductile deformation within low to medium metamorphic grades. This does not mean that they were not subject to shear stresses of enough magnitude that, in the presence of coarse-grained aggregates, would generate such microstructures.

Thus, a lack of correlation exists between microstructures originated by ductile deformation of previously formed coarse-grained monomineral aggregates (most frequently, a retrograde transformation) and those generated from fine-grained polymineral aggregates, at equivalent physical conditions. The Pringles Metamorphic Complex, consisting of a metasedimentary sequence with intercalated coarse-grained rocks, altogether ductile deformed along a geothermal gradient, constitutes a very suitable area for the establishment of the above mentioned relationships.

This study described in this chapter aims to provide simple and rapid tools to correlate the observed microstructures in the essential minerals of metapsamitic–metapelitic

Ductile Shear Zones: From Micro- to Macro-scales, First Edition. Edited by Soumyajit Mukherjee and Kieran F. Mulchrone.

rocks, with the thermal conditions of deformation prevailing during the development of mylonitic shear zones. The present study emphasizes the use of methodologies of broad and easy access, as standard methods of microfabric analysis and petrographic microscopy.

14.2 GEOLOGICAL SETTING

The Sierras Pampeanas Range, located in central-western Argentina (Fig. 14.1), has been divided based on lithology and magmatic–metamorphic evolution into two contrasting units: the Western and Eastern Sierras Pampeanas (Caminos 1973, 1979; Dalla Salda 1987). This range formed part of the western margin of Gondwana during the Late Proterozoic–Early Paleozoic. Four geological cycles (see Pankhurst and Rapela 1998) intervened in its genesis: (1) Pampean (Neoproterozoic–Late Cambrian), (2) Famatinian (Early Ordovician–Early Carboniferous), (3) Gondwanian (Early Carboniferous–Early Cretaceous), and (4) Andean (Early Cretaceous–present).

Geochronological studies (Sims et al. 1997, 1998; Rapela et al. 1998; Pankhurst et al. 1998, 2000; von Gosen et al. 2002; Steenken, et al. 2006, 2011; Castro de Machuca et al. 2008, 2010, 2012; Dalquist et al. 2008; Gallien et al. 2010, Siegesmund et al. 2010; Larrovere et al. 2011; Iannizzotto et al. 2013), indicate that the main Sierras Pampeanas metamorphic- and igneous events were related to the evolution of SW Gondwana margin during the Early-Middle Paleozoic Pampean- and Famatinian orogenies.

Located between latitudes 32–34°S and longitudes 66–68°W, the Sierra de San Luis represents one of the southernmost exposures of the Eastern Sierras Pampeanas (Fig. 14.1). In the complex context described above, the Sierra de San Luis constitute a key piece and a valuable source of information for the development of the Famatinian Orogenic Cycle.

According to Sims et al. (1997) and Hauzenberger et al. (2001), the crystalline basement of the sierra (Fig. 14.1) consists of three main units, which from E to W, are: (1) the Conlara Metamorphic Complex, composed mainly of high-grade gneisses and migmatites; (2) the Pringles Metamorphic Complex, which varies in metamorphic grade from greenschist to granulite facies and comprises slates, phyllites, micaschists, staurolitic schists, gneisses, migmatites, intercalated mafic–ultramafic rocks, and tonalitic–granodioritic bodies and pegmatites; and (3) the Nogolí Metamorphic Complex, consisting mainly of migmatitic and high-grade gneisses with amphibolite lenses.

The Pringles Metamorphic Complex differs substantially from the neighboring blocks, in relation to metamorphic grade, the occurrence of metabasites within a sequence of quartzofeldspathic and pelitic metasediments, and ubiquitous mylonitization affecting all lithologies and imprinting the dominant structural pattern in this region. These contrasting characteristics, in addition to differences in the structural configurations (von Gosen,

1998; Sims et al. 1998; Delpino et al. 2001, 2007; González et al. 2002), indicate that tectonometamorphic evolution of the Pringles Metamorphic Complex was, at least temporarily, independent of the evolution of the Conlara- and Nogolí metamorphic complexes that delimit it towards the E and W, respectively.

Within the Pringles Metamorphic Complex, mafic–ultramafic bodies occur as discontinuous lenses with maximum size in outcrop of up to 3.5 × 1.5 km², distributed along a 50 km long and 2.5–3 km wide central belt concordant with the general NNE–SSW structural trend. They are composed of orthopyroxene + clinopyroxene + hornblende + plagioclase + sulfides ± olivine ± phlogopite ± Cr-spinel ± platinum group minerals, with accessory minerals such as apatite (Mogessie et al. 1994, 1995, 1998; Brogioni and Ribot 1994; Cruciani et al. 2011, 2012). On both sides of the mafic–ultramafic bodies, a metamorphic gradient from granulite to greenschist facies is apparent (Delpino et al. 2002, 2007, 2012; Rueda et al. 2013). Granulite facies country-rock assemblages (garnet + cordierite + sillimanite + biotite + K-feldspar + plagioclase + quartz + rutile + ilmenite ± orthopyroxene) near mafic–ultramafic bodies grade toward both extremes to the amphibolite facies assemblage (garnet + biotite + muscovite + plagioclase + quartz ± staurolite ± sillimanite ± chlorite) and towards the easternmost part of the Pringles Metamorphic Complex to the greenschist facies assemblage: chlorite + biotite + muscovite + quartz + plagioclase ± K-feldspar (Hauzenberger et al. 2001; Delpino et al. 2007; Rueda et al. 2013). The granulite facies metamorphism reached by the adjacent country-rocks has been attributed to the intrusion of mafic–ultramafic bodies into a greenschist or low-amphibolite facies metamorphic sequence (González Bonorino 1961; Hauzenberger et al. 1997; Sims et al. 1998; Hauzenberger et al. 1998; von Gosen and Prozzi 1998). According to Hauzenberger et al. (2001), the volume of mafic lenses estimated through geophysical methods by Kostadinoff et al. (1998a, b) suffices to produce the observed granulite facies rocks. On the other hand, some contribution from viscous dissipation/shear heating during mylonitization (e.g. Brun and Cobbold 1980; Stüwe 2002; Burg and Gerya 2005; Mukherjee and Mulchrone 2013; Mulchrone and Mukherjee, in press) could be possible. However, viscous heating is a local perturbation around shear zones and faults that require fast movement and high shear stress over long periods (Scholz 1980), but can be negligible on a regional scale. Also, assuming weakness of deeply buried rocks (e.g. Ranalli 1995), temperature increase by viscous heating is less than other temperature perturbations (Burg and Gerya 2005).

The structure of the Pringles Metamorphic Complex has been discussed in detail in von Gosen and Prozzi (1996, 1998) and Delpino et al. (2001, 2007). Pressure–temperature (*P–T*) paths for the tectonometamorphic evolution of the Pringles Metamorphic Complex, have been proposed by Hauzenberger et al. (2001), Delpino et al. (2001, 2007), and Cruciani et al. (2008, 2011, 2012).

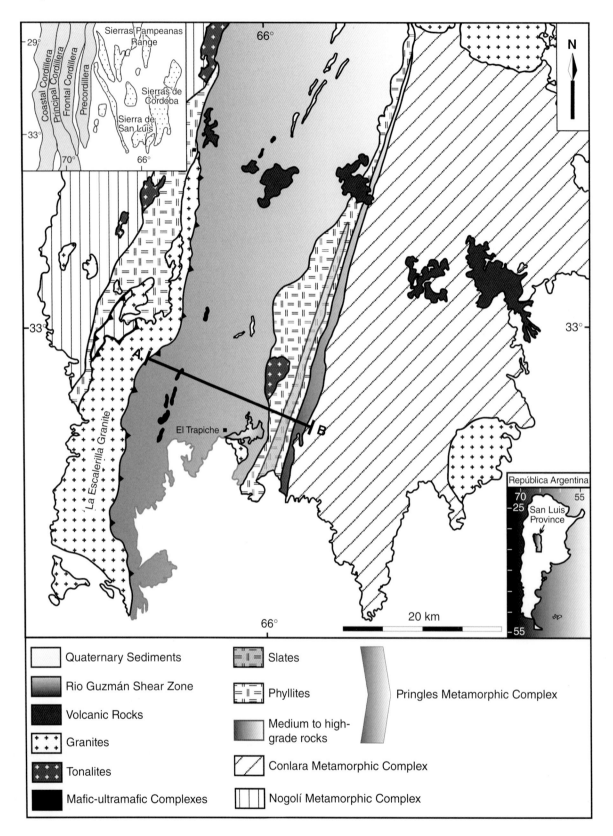

Fig. 14.1. Geological map of the Sierra de San Luis. Inset shows location of Sierra de San Luis in the regional context of Sierras Pampeanas range, and younger Precordillera and Andes Cordillera units. A–B is the location and length of the integrated profile studied in the present contribution. Details in the text.

14.3 FIELD ASPECT AND PETROGRAPHY

We built a representative profile sub-perpendicular to the NNE dominant structural trend characterizing the region (Figs 14.1 and 14.2). Based on field, petrographic, and geochemical characteristics, two groups of rocks were differentiated along the cross section: (i) subsolidus metamorphic rocks (slates, phyllites, Ms-Bt-micaschists, St-micaschists and Grt-Bt-gneisses); and (ii) suprasolidus anatectic rocks (nebulitic and stromatic migmatites with leucosome segregates).

14.3.1 Subsolidus metamorphic rocks

14.3.1.1 Slates

These rocks are located at the eastern edge of the Pringles Metamorphic Complex. At outcrop scale, they appear as bluish-green very fine-grained with a non-domainal continuous foliation. Microscopically, very fine elongated irregular quartz and albite grains and aligned phengite flakes define foliation (Fig. 14.3a,b). Phengite rich-bands alternate with quartz–feldspar-rich bands. Bigger sized quartz crystals are only observed in isolated lenses dispersed in the matrix, in asymmetric strain shadows developed on sub-idiomorphic ilmenite clasts (Fig. 14.3a) or in boudinaged necks between two fragments of oxide. Quartz originally grew as fibers elongated subparallel to foliation, but sometimes the external sectors or the totality of the tails recrystallized to aggregates of polygonal crystals with ~120° triple junctions (Fig. 14.3b and inset). Fine reddish brown irregular aggregates of organic matter are distributed in the sample. Very thin pressure dissolution planes highlighted by the presence of undissolved organic matter, cut the grains forming coarser quartz lenses and extend along the whole sample (Fig. 14.3b).

Chlorite appears as fine flakes parallel to foliation, or as bigger crystals in the low-pressure sites where it is oriented randomly and is undeformed. The rock shows intense microcrenulation affecting matrix, coarser quartz lenses and ilmenite porphyroclasts and their strain shadows (Fig. 14.3a, b). Close to western contact with the Río Guzmán Shear Zone (Fig. 14.1), very dark slates occur. These rocks are constituted by quartz, feldspars and micas with abundant dark brown to black graphitized organic matter. The habit, size and distribution of silicates resemble with those described previously. Few coarser quartz recrystallized lenses also exist. But, in addition to abundance and greater graphitization of organic matter, the more noticeable difference with the previously described slates is the absence of euhedral oxides and the presence of large lensoidal structures that seem to be the result of pseudomorphosis of some previous microstructure. The center of these structures is occupied by leucoxene and/or a granular aggregates of dark, semi-translucent organic matter, surrounded by quartz with larger size than in the matrix (Fig. 14.3b, inset). In ilmenite tails, quartz crystals near the center are elongated, while those in contact with the matrix are pseudopolygonal and of smaller size. Abundant fine aggregates of pentlandite associated to leucoxene and graphitized aggregates of organic matter have been recognized. Scattered in the matrix, there are small crystals of chlorite, oriented parallel to foliation.

14.3.1.2 Phyllites

In the eastern sector of the Pringles Metamorphic Complex, phyllites occur at both sides of slates. These rocks also occur along the northwestern edge of the complex, where slates are absent (Fig. 14.1). In the field, phyllites show

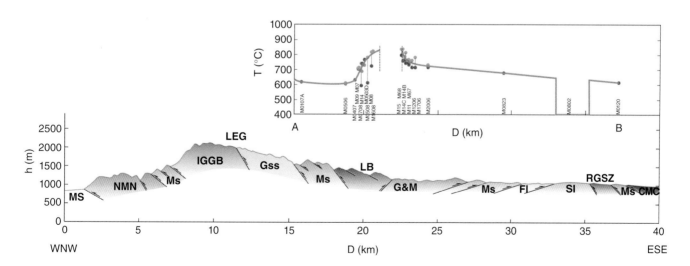

Fig. 14.2. Representative cross-section showing main lithological units and simplified macrostructures. Also shown, thermal curve along A-B cross-section, obtained through geothermobarometry. MS. Modern sediments; NMC: Nogolí Metamorphic Complex; CMC: Conlara Metamorphic Complex; IGGB: gabbro/granodiorite-granite-basement intercalations; GLE: La Escalerilla granite; Gss: granite *sensu stricto*; Ms: micaschists; G&M: gneisses and migmatites; Fl: phyllites; Sl: slates; LB: La Bolsa mafic-ultramafic body outcrop; RGSZ: Río Guzmán Shear Zone. Details in the text.

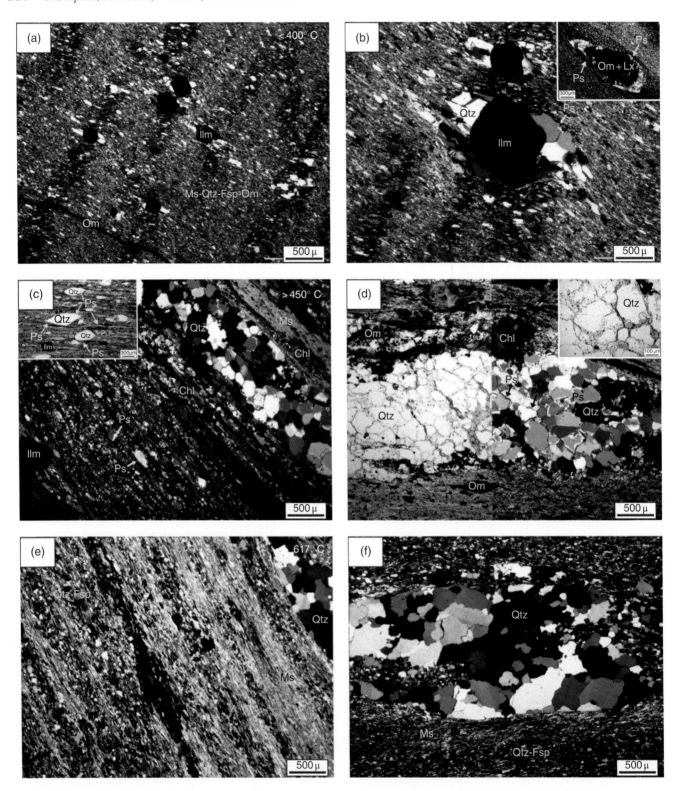

Fig. 14.3. Microstructural development in metapsamitic–metapelitic rocks of the Pringles Metamorphic Complex with the increase of temperature. Also shown, contrasting rheological behavior of originally fine and coarse-grained rocks (compare left and right columns). (a,b) Very fine-grained slates from the eastern extreme of the Pringles Metamorphic Complex. (c,d) Fine grained rock, transitional between slates and phyllites showing intercalations of cuasi-monomineralic coarse-grained quartz layers. Arrows and Ps: pressure solution surfaces. (e,f) Phyllites with a well-defined domainal foliation and evidence of grain boundary migration recrystallization in matrix and coarse-grained quartz layers. Note continuity of phyllosilicate-rich layers.

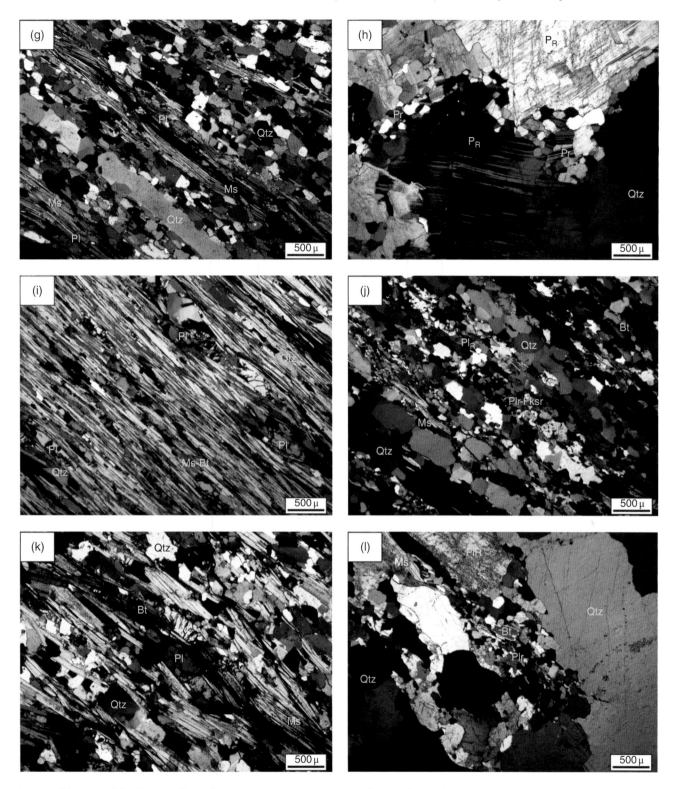

Fig. 14.3. (*Continued*) (g,i,k) Micaschists showing progressive coarsening of plagioclase and quartz grains with the increase of temperature. Note blocky plagioclase grains pinned by micas and without evidence of ductile deformation, and development of quartz ribbons indicating high-mobility of grain boundaries. (h,j,l) Coarse-grained neighboring rocks (granites and pegmatites), in which intense recrystallization to polygonal new-grains clearly evidence ductile deformation;

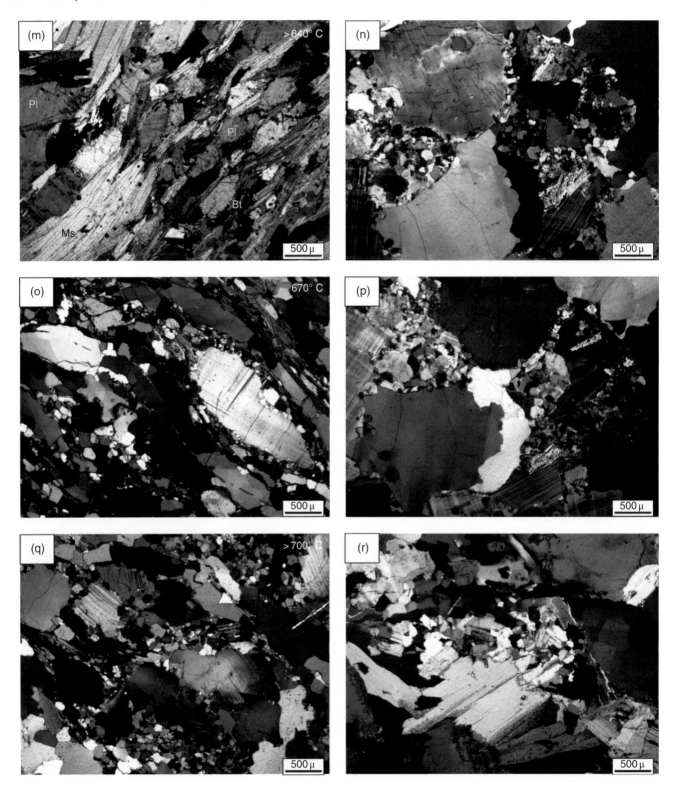

Fig. 14.3. (*Continued*) (m) Coarse grained Ms-Bt micaschist with coarse-subhedral plagioclase crystals still limited in its growth by aligned mica flakes and without intracrystalline deformation. Note the continuity of micaceous cleavage domains, which inhibit mutual contact between plagioclase grains. (o–q) Strongly recrystallized Qtz-Fsp-Bt gneisses. Note decrease of phyllosilicate content and discontinuity of biotite folia, which allow mutual contact between the other silicates. (n–r) Note the contrasting rheological response showed by the micaschist (n, left) with mica-rich continuous cleavage planes and a contiguous pegmatite (m, right) with original coarse granular textures. However, coarse grained gneisses (o–q, left) show the same rheological behavior that neigboring pegmatites (p–r, right). All photographs with crossed nicols, except left side of (d) and insets in (c) and (d). Abbreviations after Bucher and Grapes (2011). Details in the text.

alternation of mica- and quartz-rich lithologies. They show greenish or yellowish greenish colors and a planar disjunctive foliation. Microscopically, rocks intermediate between slates and phyllites are characterized by grain size increase and loss of the homogeneity of the matrix essentially by the differential growth of some quartz crystals (Fig. 14.3c). These quartz crystals show clear evidence of pressure solution parallel to foliation (Fig. 14.3c). Increasing anisotropy mainly due to segregation into alternating quartz or mica-dominated layers, results in a smooth disjunctive foliation. A feature not observed in slates, is the presence of monomineralic coarse-grained quartz layers (Fig. 14.3c, d). These layers are often folded and pinched and swelled. In the matrix, quartz-rich bands contain quartz crystals of sizes of up to 50 µm and form aggregates of polygonal grains meeting at 120° triple junctions (Fig. 14.6a$_1$). Finer-grained micaceous bands contain rectangular quartz crystals due to mica pinning, and equidimensional grains with smoothly lobate grain boundaries. Rectangular crystals reach up to 50 µm parallel to the foliation, while equidimensional crystals reach sizes of ~25 µm. Monomineralic coarse-grained quartz layers preserve sedimentary textures (Fig. 14.3d), consisting of irregular, angular quartz fragments cemented with a reddish brown very fine grained matrix (Fig. 14.3d, see Vernon 2004). A close inspection shows that siliceous cement overgrows the sedimentary preexisting grains. The siliceous overgrowth can be discriminated from the detrital clear quartz grains by abundant inclusions/impurities (essentially aggregates of organic matter) and fluid inclusions (Fig. 14.3d, inset). Toward the margins of these layers, quartz microstructures indicate static growth.

Unlike what is observed in the matrix, pressure solution planes are developed oblique to the rock foliation (Fig. 14.3d). These rocks contain significant amount of chlorite, oriented parallel to foliation. Towards the W of the cross-section, the increase of the metamorphic grade give place to phyllites with a lepidoblastic texture given by the alignment of muscovite and subordinate amounts of pale brown biotite. Together with elongated quartz-feldspar rich-bands and sparsely distributed monomineralic quartz layers, originate a moderately rough disjunctive foliation (Fig. 14.3e). Matrix quartz-rich bands are characterized by grains with gently lobate or pseudopolygonal outlines, which reach up to ~75 µm and form microribbons. Some rectangular crystals fill the full width of the ribbon and reach up to ~100 µm in the direction of foliation (Fig. 14.6a$_2$). Within almost pure coarse quartz layers, crystals show microstructures indicating mobile grain boundaries with the presence of large grains with lobate boundaries and/or unequally sized polygonal crystals meeting at 120° triple junctions (Fig. 14.3f). Lobate crystals reach up to 500 µm and show evidence of internal deformation as sweeping undulatory extinction, incipient development of deformation bands with diffuse boundaries and moderate preferential crystallographic orientation. Polygonal crystals reach up to 150 µm and most of them do not display internal deformation and have straight or slightly curved edges, although a moderately crystallographic orientation

is present. The feldspar present in phyllites is plagioclase. It can be seen in the interior of the quartzose bands forming either irregularly shaped crystals identifiable by its characteristic polysynthetic twinning (Fig. 14.6b$_2$), or interstitial grains with grain boundaries defined by the surrounding quartz crystals. These grains grow parallel to foliation including micas from the matrix. In the micaceous bands, plagioclase shows very irregular cuspate geometry that extends apices towards the cleavage planes of muscovites, which entirely surround them. These individual grains reach up to 100 µm along their major axes parallel to the foliation (Fig. 14.6b$_1$). Chlorite partially replaces biotite and muscovite. Organic matter, when present, is entirely graphitized. Ilmenite and apatite are the accessory minerals. Apatite forms small interstitial grains or inclusions in muscovite and plagioclase crystals.

14.3.1.3 Micaschists

Lower-grade basement rocks of the W extreme of the profile, at the contact with La Escalerilla granite (Figs 14.1 and 14.2), are constituted by centimeter-scale intercalated metapsamitic–metapelitic layers. They show the same domainal foliation as in phyllites, but are coarser grained and sligthly more anastomosed (Fig. 14.3g). These rocks represent the transition between phyllites and micaschists. Like the phyllites of the eastern sector, these rocks show asymmetric microfolds (Fig. 14.6d). Muscovite represents the dominant mica in the cleavage domains, in which subordinated amounts of biotite can be observed. Microlithons are composed mainly of quartz with lower proportions of plagioclase. Quartz in microlithons appear either as elongated tabular crystals parallel to foliation due to mica pinning or as approximately equidimensional grains with smoothly lobated to straight grain boundaries. These grains reach up to 250 µm and sometimes show 120° triple junctions. Furthermore, quartz segregates as monomineral ribbons characterized by polygonal to highly elongated tabular crystals (Type 2 and 3 of Boullier and Bouchez 1978), which fill the entire width of the ribbon and reach lengths of up to 1500 µm (Fig. 14.3g). Plagioclase crystals are scarcer than quartz and appear as elongated grains parallel to foliation due to mica pinning in mica-rich domains, or as interstitial smaller grains in the quartz-rich domains where its external shape is defined by the surrounding quartz grains. Towards the center of the profile these rocks change progressively to coarser-grained micaschists, which preserve the alternation of metapsamitic–metapelitic layers. Micaschists show a more prominent anastomosed disjunctive foliation with cleavage domains composed mainly of medium-grained muscovite with subordinated amounts of biotite and quartz-rich microlithons (Figs 14.3i,k and 14.6c,d). The only compositional difference between metapsamites and metapelites is the relative proportion of quartz- and mica-rich bands. In metapelitic layers, quartz in microlithons forms ribbons composed of polygonal or tabular crystals limited by fine-grained muscovite-biotite flakes. Polygonal grains reach up to 250 µm, whereas blocky quartz grains reach lengths of up 400 μm in the direction of foliation. In the metapsamites with higher prevalence of quartzose bands, this mineral shows similar textures to those observed in the monomineralic coarse-grained quartz layers observed in the phyllites of the eastern sector, i.e. large amoeboid crystals and polygonal new-grains. The formers reach up to 600 μm and show undulose extinctions, incipient development of deformation bands and high-amplitude bulging. Polygonal recrystallized grains of up to 300 μm, contact each other through triple junctions at 120°. Plagioclase crystals are now easily identified by their larger crystal size (up to 1000 μm). It is more abundant in metapelitic than in metapsamitic layers, and is associated mainly with mica-rich domains. Elongated grains parallel to foliation show straight faces due to pinning by micas in the direction normal to foliation, and irregular to cuspate outlines in the direction of foliation (Figs 14.3 l, k, 14.6c, d). Most crystals lack of the characteristic polysynthetic twinning. More equidimensional crystals, can be observed in fold hinges. Plagioclase crystals show inclusion of matrix quartz and micas, these later usually preserving the preferred orientation parallel to foliation. Higher-temperature Ms-Bt micaschists show an increment in the grain size of all its components. These rocks show a coarser anastomose disjunctive foliation. Still continuous cleavage domains constituted by coarse muscovite and biotite crystals, wrap around tabular to lenticular plagioclase crystals still pinned by micas. The proportion of biotite increase in accordance with a decrease in the muscovite content. Plagioclase crystals reach lengths in the direction parallel to foliation >1500 μm (Fig. 14.3m). Some grains also develop straight faces in directions not parallel to foliation, forming subhedral crystals. Some individuals develop polysynthetic twinning and often show fracturing and fragmentary extinctions, but do not show evidence of recrystallization. Accessory minerals are ilmenite and apatite.

14.3.1.4   Staurolitic micaschists

These rocks outcrop at approximately the half of the distance between La Escalerilla granitoid (Fig. 14.1) and the mafic-ultramafic bodies emplaced in the center of the Pringles Metamorphic Complex. The alternation between quartzose (lighter) and micaceous (darker) bands is still recognized, and the increase of the average grain size is clear. Staurolitic micaschists located further away from mafic-ultramafic bodies, show folded and kinked muscovite porphyroclasts surrounded by coarse reddish biotite and disrupted muscovite fragments, both defining an anastomosing foliation. Biotite also crystallizes in low differential stress sites, and replaces muscovite at their margins and along cleavage planes and kink bands. Quartz and plagioclase form coarse lenticular or ribbon-like microlithons partially delimited by mica folia. In monomineralic ribbons, individual blocky quartz crystals fill the entire width of the ribbon, whereas in polymineralic lenses quartz develops grains with dissimilar sizes and lobate boundaries. Plagioclase shows irregular, sometimes cuspate, relict crystals, and towards the borders of the lenses some small recrystallized pseudopolygonal grains can be observed. Staurolite appears as pleochroic yellow to pale brownish-yellow large prismatic crystals and low-birefringence pseudo-hexagonal basal sections. Prismatic crystals preferentially parallel the foliation. The crystals show undulose extinctions or fragmentary extinctions associated to subgrain development. In the higher-grade extreme of the staurolitic micaschist outcrop (closer to the mafic-ultramafic bodies), coarse irregular garnet grains and prismatic sillimanite crystals partially replace staurolite (Fig. 14.6e and f). Quartz and plagioclase in microlithons are of larger sizes and show either quartz–quartz or quartz–plagioclase strongly lobated contacts. Relict plagioclase crystals show bending and partial homogenization of the primary twinning and development of secondary twinning. Bulging of moderate amplitude evidence important mobility of plagioclase grain boundaries, which associate to local recrystallization forming pseudopolygonal new grains meeting at 120° triple junctions (Fig. 14.6e, f). Garnet appears as large irregular porphyroclasts with numerous fractures filled by biotite. Biotite is the dominant phyllosilicate and wraps porphyroclasts defining a coarse anastomosing foliation, and also crystallizes in low differential stress sites like pressure shadows. Sillimanite forms long prismatic crystals parallel to foliation. Locally, secondary muscovite and chlorite replace partially the previous described minerals along foliation planes or massively without any preferred orientation. Accessory minerals are apatite and ilmenite.

14.3.1.5   Bt-Grt-gneisses

Gray coarse-grained rocks with an anastomosed banded compositional foliation. Microscopically they show strongly aligned fine biotite flakes and coarse quartz ribbons intercalated with thinner ribbons composed of recrystallized plagioclase grains intermixed with fine biotite, all wrapping around plagioclase and garnet porphyroclasts (Fig. 14.3q). Quartz forms ribbons of type 3 or 4 (Boullier and Bouchez 1978). Biotite distributed along cleavage planes and intermixed with plagioclase in ribbons is fine grained, but greater-sized biotite crystallizes within strain shadows and low differential stress sites. Plagioclase porphyroclasts are equidimensional or elliptical and show wavy extinction and recrystallization to aggregates of polygonal new grains at the margins of relict porphyroclasts or extending into the thin ribbons described above. Some highly irregular, skeletal garnet porphyroclasts seem to have formed from pseudomorphic replacement of a previous mineral possibly staurolite. Accessory mineral is ilmenite.

14.3.2   Suprasolidus anatectic rocks

14.3.2.1   Nebulitic migmatites

Two types of nebulitic migmatites were recognized: a leucocratic muscovite-garnet variety and a mesocratic biotite–epidote variety. In the field, both appear as coarse grained "granitoid" rocks, with a granular fabric and a poorly developed rough anastomosing foliation. Under the microscope, the leucocratic variety shows large plagioclase, quartz, muscovite and garnet crystals arranged in a granular fabric (Fig. 14.3r). Fine biotite appears only along thin foliation planes, in pressure shadows or replacing large muscovite crystals at their edges and cleavage planes. Quartz forms very large ameboidal crystals with very feeble internal deformation, which enclose remnants of plagioclase crystals and show preferential crystallographic orientation. Plagioclase appear as irregular patches with intense internal deformation and recrystallized at their margins to pseudopolygonal aggregates, sometimes constituting core-and-mantle structures. The mesocratic variety, collected near the mafic-ultramafic bodies, is composed of quartz and plagioclase with similar characteristics to that described for the leucocratic variety, but they have large biotite crystals in place of muscovite and totally lack garnet. Abundant epidote and opaque minerals are also present.

14.3.2.2   Stromatic migmatites

These rocks are characterized by dark melanosomes and leucosomes of variable sizes ranging between millimetric lit-par-lit sheets, to centimeter-scale veins predominantly distributed along foliation planes. Mesosomes well preserve the fabric of the non-melted rocks, as the compositional banded foliation and granulometry. At the microscopic scale, these rocks show sectors with a well preserved anastomosing discontinuous foliation marked by biotite flakes, quartz ribbons and prismatic sillimanite crystals, and by sectors enclose lenticular leucosome patches with a granular appearance. Quartz forms ribbons or large ameboid individual grains, which sometimes show rectangular subgrains with "chessboard" extinction. Most grains within ribbons show a marked preferred crystallographic orientation. Plagioclase grains and biotite are highly irregular and both present resorbed edges. K-feldspar is interstitial and usually shows cuspate boundaries. Garnet porphyroclasts are very often elongated parallel to foliation, and show poikilitic inclusions of quartz and plagioclase. Sometimes, fibrolitic sillimanite is observed inside garnet crystals. Cordierite appears invariably as porphyroclasts intensely altered to prismatic/fibrolitic sillimanite and green biotite, and/or show the characteristic pinitization. Some tonalitic mesosomes without K-feldspar, sillimanite and garnet are also present. Ilmenite occurs as an accessory mineral.

14.3.2.3   Leucosome segregates

Leucosomes were differentiated in three groups based on field appearance and petrographic-geochemical characteristics.

Leucosome I   In the field, they appear as thin veins of millimeter- to centimeter-scale in stromatic migmatites. They can be classified as in-source leucosomes (this and all types of leucosome in the following text, classified after Sawyer 2008). Under the microscope, they show coarse-grained granular texture formed by subhedral plagioclase crystals, ameboidal quartz grains, interstitial K-feldspar and large muscovite flakes lacking preferential orientation. Quartz and K-feldspar usually enclose relics of plagioclase crystals. K-feldspar grains show elliptical shapes elongated parallel to foliation when they are included in quartz. A rough disjunctive foliation is marked by fine biotite flakes. At both sides of the thin cleavage planes, quartz, plagioclase and K-feldspar show recrystallization to polygonal new-grains.

Leucosome II   Within this group, three sub-varieties can be differentiated. The first one is an *in-situ* leucosome, which forms a patch within a migmatitic gneiss located close to the mafic-ultramafic bodies. This rock has coarser grain size than the enclosing gneiss, but preserves a foliated appearance. It is formed by quartz, plagioclase and biotite forming a coarse granular texture with a rough disjunctive foliation. The second corresponds to a pegmatitic vein sampled close to the external migmatitic front, and can be classified as an in-source vein. It is a decimeter-scale vein composed of abundant K-feldspar and subordinated plagioclase, muscovite and garnet. The third variety is out of the migmatitic zone, and it is classified as a leucocratic vein. This vein is similar to the previous one, but does not contain K-feldspar and has greater amounts of muscovite.

Leucosome III   They occur as big veins or sills of meter-scale sizes. They are composed of very coarse quartz, plagioclase, and muscovite crystals without preferential orientation. Chessboard extinction in quartz and plagioclase recrystallized to polygonal grains meeting at $120°$ triple junctions indicates high-temperature ductile deformation. Some of the pegmatites of this subgroup contain garnet and large tourmaline crystals.

14.4   PHASE RELATIONS

Phase relations were evaluated through detailed petrographic studies and a suitable pseudosection based on a representative whole rock chemical composition. Pseudosection, shown in Fig. 14.4, was constructed with the program Perplex (Connolly, 1990, updated 2014), with internally consistent dataset of Holland and Powell (1998, 2002).

Figure 14.4 shows the counterclockwise trajectory for the tectonometamorphic evolution of the Pringles Metamorphic Complex, previously proposed by Delpino et al. (2007).

In previous studies (Delpino et al. 2002, 2007, 2012), it has been stated that the mylonitic event (T_3–M_3: S_3, L_3, b_3, Delpino, et al. 2001, 2007) should have begun after, but at conditions close to, the metamorphic peak marked by the intrusion of the mafic–ultramafic bodies, which generates the geothermal gradient over which the ductile shear zone develops.

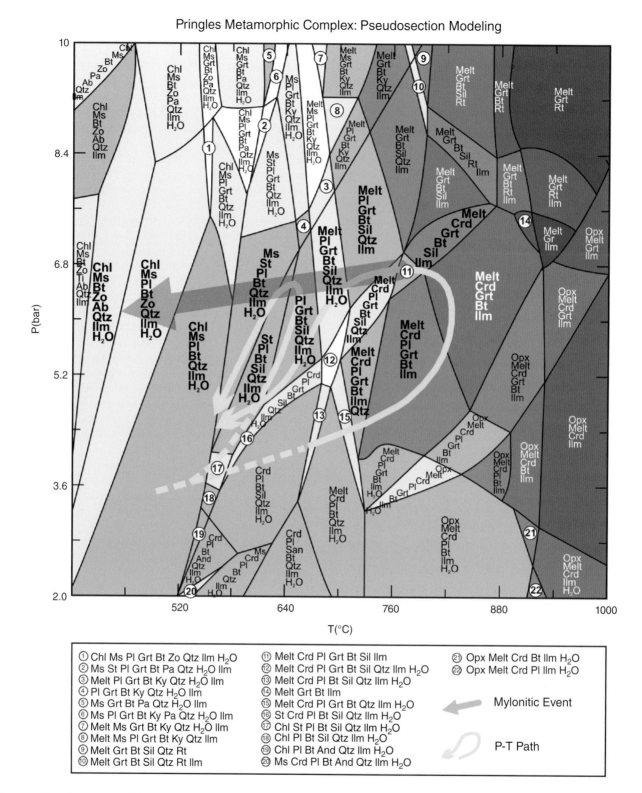

Fig. 14.4. Pseudosection modeling of metapsamitic–metapelitic rocks of the Pringles Metamorphic Complex. *P–T* path taken from Delpino et al. (2007). Arrow indicates the *P–T* range of development of the mylonitic event and the path followed for the construction of Fig. 14.5. Associations of phases predicted by the model (enhanced in bold), agree very well with those observed petrographically. Diagram was constructed with the following conditions: System: NKCFMSAT-H; Whole rock chemical composition (St-schist): SiO$_2$=52.99, Al$_2$O$_3$=19.15, TiO$_2$=1.52, FeO=11.06, MgO=5.04, Na$_2$O=1.10, K$_2$O=4.46, H$_2$O=5.00; Solution models: melt(HP), Opx(HP), hCrd, Gt(HP), St(HP), TiBio(HP), Pheng(HP), Chl(HP), Pl(h), San, IlHm(A), Ctd(SGH) (solution_models, Perplex program package, Conolly 1990, updated 2014). Abbreviations after Bucher and Grapes (2011). Details in the text.

Table 14.1. Synoptic table of phase relationships. Left column: summary of field and petrographically observed mineral associations in each lithological type. Right column: predicted low to high-T succession of equilibrium associations and stability temperature range from pseudosection modeling, along the path shown by the arrow in Fig. 14.4. In *italic*, main continuous reactions between established lithological types and temperature range of development calculated through Fig. 14.5.

Subsolidus rocks			
Rock name	Mineral associations observed in the field and petrographically	Equilibrium associations, main reactions and temperature (°C) from pseudosection modeling	
Slates	Qtz+Ab+Ms+Ilm+Chl+Om±Ap±Gr±Lx±Pn	Qtz+Ab+Ms+Bt+Ilm+Chl+Zo^+H_2O	415–458
		Ms+Chl+Ab+Zo=Bt+Qtz+Pl+H_2O	*446–458*
Phyllites	Qtz+Pl+Ms+Bt+Ilm+Chl+Ap±Gr	Qtz+Pl+Ms+Bt+Ilm+Chl+Zo^+H_2O	458–519
		Ms+Chl+Zo=Bt+Qtz+Pl+H_2O	*458–519*
Bt-Ms-Schists	Qtz+Pl+Ms+Bt+Ilm+Chl+Ap	Qtz+Pl+Ms+Bt+Ilm+Chl+H_2O	519–580
		Ms+Chl+Ilm=Qtz+Bt+Pl+St+Bt+H_2O	*573–580*
St-Schists	Qtz+Pl+Bt+Ms+St+Ilm+Ap	Qtz+Pl+Ms+Bt+Ilm+St+H_2O	>580
		Ms+St=Qtz+Bt+H_2O	*573–640*
	Qtz+Pl+Bt+Ms+St+Grt+Sil+Ilm+Ap*	Qtz+Pl+Sil+Bt+Ilm+St+H_2O	<640
		Pl+St+Qtz=Bt+Sil+Grt+H_2O	*640–651*
Gneisses	Qtz+Pl+Bt+Grtl+Ilm	Qtz+Pl+Sil+Bt+Grt+Ilml+H_2O	651–673
Suprasolidus rocks			
		Pl+Bt+Sil+Qtz+H_2O=Melt+Grt+Ilm	*673–709*
		Qtz+Pl+Sil+Bt+Grt+Ilm+Melt+H_2O	673–709
		Qtz+Pl+Sil+Bt+Grt+Ilm+Melt	709–750
Stromatic Migmatites	Melanosome	Qtz+Pl+Kfs+Bt+Grt+Sil+Ilm±Crd	
		Pl+Bt+Sil+Qtz=Melt+Grt+Crd+Ilm	*>750*
		Qtz+Pl+Sil+Bt+Grt+Ilm+Crd+Melt	750–769
		Pl+Bt+Sil+Qtz=Melt+Grt+Crd+Ilm	*>769*
		Pl+Sil+Bt+Grt+Ilm+Crd+Melt	769–794
	Neosome — Leucosome I — Qtz+Pl+Kfs		
	Neosome — Leucosome II — Qtz+Pl+Bt		
	Neosome — Qtz+Pl+Kfs+Ms+Grt		
	Neosome — Qtz+Pl+Ms+Grt		
	Neosome — Leucosome III — Qtz+Pl+Ms±Grt±Tur		
Nebulitic Migmatites	Qtz+Pl+Ms+Grt		
	Qtz+Pl+Bt+Ilm+Ep		

Abbreviations after Bucher and Grapes 2011, except Om: organic matter and Lx: leucoxene. Details in the text.
Zo^: not observed in these rocks. See text for details.
*Rock mode, which chemical composition was used for pseudosection modeling.

The arrow in the pseudosection in Fig 14.4 represents the variation of mineral paragenesis along the geothermal gradient, at the pressure range (6–7 kbar) for the mylonitic event estimated through geothermobarometry.

A summary of phase associations observed in the field and petrographically for each rock type is shown in the left column of Table 14.1. Predicted paragenesis from pseudosection modeling along the geothermal gradient are highlighted in bold in Fig. 14.4. The comparison between the left and right columns of Table 14.1 shows the very good correspondence between the equilibrium associations observed in mylonites along the considered profile and those generated by the model. The main discrepancy is observed in the lower grade rocks. Zoisite was not recognized in the lower-grade rocks. This phase probably represents the presence of apatite non-representable in the pseudosection, given that Vertex does not support fluoride as a component. The presence of a calcium phase is indispensable for the generation of plagioclase above the albite stability field. Concordant with previous geochemical studies (Rueda et al. 2013),

pseudosection modeling also indicate that there were no significant chemical variations along the cross-section. Thus, if not significant chemical variations can be inferred from the calculated pseudosection, the variation of equilibrium associations should be attributed essentially to the progressive change of temperature along the profile. Temperature range for each lithological type (right column, Table 14.1), agrees well with the temperatures obtained through geothermobarometry (Fig. 14.2, Table 14.5).

Variation of the volumetric fractions of the main mineral phases (Fig. 14.5) along a path parallel to the arrow in Fig. 14.4, together with textural relationships, determine the most probable reactants and products of reactions producing the different mineral associations characterizing each stability field (Inset, Fig. 14.5). Selected reactions which produce total consumption or first appearance of index minerals, used in conjunction with the textural and structural characteristics to define the different type of rocks, are shown in Table 14.1. Given that most reactions are continuous due to the

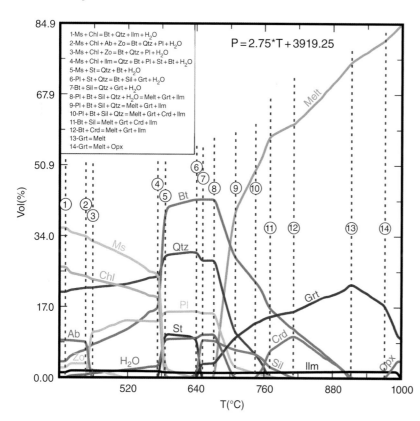

1-Ms + Chl = Bt + Qtz + Ilm + H$_2$O
2-Ms + Chl + Ab + Zo = Bt + Qtz + Pl + H$_2$O
3-Ms + Chl + Zo = Bt + Qtz + Pl + H$_2$O
4-Ms + Chl + Ilm = Qtz + Bt + Pl + St + Bt + H$_2$O
5-Ms + St = Qtz + Bt + H$_2$O
6-Pl + St + Qtz = Bt + Sil + Grt + H$_2$O
7-Bt + Sil = Qtz + Grt + H$_2$O
8-Pl + Bt + Sil + Qtz + H$_2$O = Melt + Grt + Ilm
9-Pl + Bt + Sil + Qtz = Melt + Grt + Ilm
10-Pl + Bt + Sil + Qtz = Melt + Grt + Crd + Ilm
11-Bt + Sil = Melt + Grt + Crd + Ilm
12-Bt + Crd = Melt + Grt + Ilm
13-Grt = Melt
14-Grt = Melt + Opx

P = 2.75*T + 3919.25

Fig. 14.5. Variation of volumetric proportions of phases along the path shown by the arrow in Figure 14.5. Variations clearly indicate the possible reactions taking place along the path (Inset) and the extent of the thermal range in which they occur. Reactions in Inset are not balanced. Abbreviations after Bucher and Grapes (2011). Details in the text.

participation of solid solution phases, this method estimates the approximate temperature range for each reaction (Fig. 14.5 and right column in Table 14.1).

14.5   GEOTHERMOBAROMETRY

Several geothermobarometers were applied to different lithological types to constrain the variations in P–T conditions along the representative profile. In medium to high-grade rocks containing garnet and biotite, temperatures were constrained through three independent Grt-Bt geothermometers (Holdaway and Lee 1977; Perchuk and Lavrent'eva 1983; Bhattacharya et al. 1992), Ti-in-Bt geothermometer (Henry et al 2005) and multiequilibrium calculation through TWEEQU program (Berman 1991), updated version WinTWQ 2004. In medium-grade rocks without garnet, Ti-in-Bt geothermometer was used. In low-grade rocks located in the eastern extreme of the cross-section, which do not contain garnet and biotite, temperature was estimated through graphite crystallinity (Landis 1971; Shengelia et al. 1979).

The difference in P–T conditions between metamorphic peak and the ductile event M_3–T_3, was reflected in the re-equilibrium of the main phases (see Delpino et al. 2007, for detailed analysis).

Throughout the studied profile, biotite shows clear textural evidence of having been the stable phyllosilicate during mylonitization, except in the eastern end of the cross-section where this phase is absent. Biotite is distributed along the mylonitic folia and crystallized within

pressure shadows and internal fractures of relict porphyroclasts. Biotite also constitutes the main component of the matrix in mylonitic–ultramylonitic bands. These biotites are underformed and compositionally homogeneous in each lithological type.

Despite that the former growth of garnet may have preceded the starting of the mylonitization, there is evidence that its composition does not represent peak metamorphic conditions, but re-equilibrium during the shear event. Notable among them are the greater abundance and larger sizes reached by this phase within ultramylonitic bands with predominant biotite-rich matrix, compared with garnet observed in the least sheared adjacent mylonitic and protomylonitic rocks. This feature suggests mylonitic garnet growth, at least in the first stage of the shear event. On the other hand, almandine-rich garnet appears in the profile at >600°C. Several profiles performed on crystals of rocks used for geothermobarometry show homogeneous compositions. This characteristic is consistent with rapid diffusion of elements at very-high temperatures (e.g. Yardley 1989; Spear 1995). For these reasons, it is assumed that both phases, biotite and garnet, were in equilibrium during mylonitization, and thus garnet-biotite and Ti-in-biotite are the appropriated geothermometers to estimate the thermal conditions during the development of this last event (details in Delpino et al. 2007).

Microprobe analysis of minerals used for geothermobarometry are presented in Tables 14.2–14.4. Geothermobarometric results are shown in Table 14.5. The geothermal curve for the studied profile is shown in Fig. 14.2.

Table 14.2. Representative chemical analysis of plagioclases used for geothermobarometry. The location of the samples along the profile is shown in Fig. 14.2

Sample	01070313 A	01070313 B	01090313 C	01090313 D	05060804	08070804	05080804	16080804	M07	05031209 Bo	05031209 C	05031209 D
SiO_2	64.00	63.92	64.07	66.46	60.51	61.01	58.51	59.41	56.97	58.87	58.59	59.37
TiO_2	0.00	0.00	0.00	0.00	0.00	0.03	0.01	0.04	0.04	0.02	0.01	0.01
Al_2O_3	22.37	22.25	22.42	20.48	25.31	27.46	25.45	24.65	27.32	25.87	25.98	25.51
FeOt	0.04	0.10	0.02	0.13	0.08	0.04	0.05	0.04	0.15	0.12	0.18	0.12
MnO	0.00	0.00	0.00	0.00	0.00	0.00	0.00	0.03	0.04	0.00	0.00	0.00
MgO	0.00	0.00	0.00	0.00	0.00	0.00	0.00	0.16	0.00	0.00	0.00	0.00
CaO	3.54	3.23	3.69	1.28	6.49	8.59	7.72	6.13	9.28	7.61	7.79	7.11
Na_2O	9.32	9.52	9.30	10.71	7.96	6.59	7.10	7.72	5.81	7.10	7.09	7.44
K_2O	0.07	0.11	0.14	0.10	0.08	0.17	0.04	0.13	0.20	0.10	0.09	0.11
Total	99.34	99.13	99.66	99.16	100.23	103.89	98.89	98.31	99.81	99.69	99.73	99.67

Numbers of ions on the basis of 8 Ox. and 5 Cat.

	01070313 A	01070313 B	01090313 C	01090313 D	05060804	08070804	05080804	16080804	M07	05031209 Bo	05031209 C	05031209 D
Si	2.849	2.846	2.843	2.941	2.685	2.637	2.644	2.688	2.558	2.640	2.625	2.657
Ti	0.000	0.000	0.000	0.000	0.000	0.001	0.000	0.001	0.001	0.001	0.000	0.000
Al	1.174	1.168	1.172	1.068	1.314	1.399	1.355	1.314	1.445	1.367	1.372	1.345
Fe^{2+}	0.001	0.004	0.001	0.005	0.003	0.001	0.002	0.002	0.006	0.005	0.007	0.005
Mn	0.000	0.000	0.000	0.000	0.000	0.000	0.000	0.001	0.002	0.000	0.000	0.000
Mg	0.000	0.000	0.000	0.000	0.000	0.000	0.000	0.010	0.000	0.000	0.000	0.000
Ca	0.169	0.154	0.176	0.061	0.309	0.398	0.374	0.297	0.446	0.366	0.374	0.341
Na	0.804	0.822	0.800	0.919	0.685	0.552	0.622	0.677	0.506	0.617	0.616	0.645
K	0.004	0.006	0.008	0.006	0.004	0.010	0.002	0.007	0.011	0.006	0.005	0.006
Sum	5.001	5.000	5.000	5.000	5.000	4.998	4.999	4.997	4.975	5.002	4.999	4.999

End-members (mol %)

	01070313 A	01070313 B	01090313 C	01090313 D	05060804	08070804	05080804	16080804	M07	05031209 Bo	05031209 C	05031209 D
An	17.29	15.67	17.85	6.15	30.96	41.49	37.43	30.29	46.30	36.98	37.60	34.36
Ab	82.32	83.70	81.35	93.27	68.61	57.52	62.32	68.95	52.50	62.42	61.86	65.00
Or	0.39	0.63	0.81	0.58	0.43	0.99	0.25	0.76	1.10	0.59	0.53	0.64

Sample	05031209 E	M09	M08	M03	M15	M14	M11	M67	M68	12060405 B	17060405	20060405
SiO_2	59.29	56.79	58.71	59.92	59.81	60.46	60.87	47.23	46.94	60.57	57.29	61.01
TiO_2	0.01	0.05	0.06	0.02	0.04	0.00	0.03	0.04	0.01	0.00	0.01	0.01
Al_2O_3	25.61	25.49	25.42	24.99	25.75	24.89	23.91	33.40	33.45	24.61	25.99	24.42
FeOt	0.03	0.11	0.12	0.05	0.01	0.23	0.10	0.28	0.09	0.05	0.12	0.03
MnO	0.00	0.00	0.02	0.04	0.03	0.01	0.03	0.01	0.02	0.00	0.00	0.00
MgO	0.00	1.24	0.02	0.00	0.41	0.72	0.00	0.15	0.08	0.00	0.00	0.00
CaO	7.30	7.63	7.81	6.90	7.02	6.14	6.06	16.74	17.29	6.15	8.79	5.88
Na_2O	7.44	7.58	7.05	7.43	7.38	8.09	7.83	2.05	1.93	7.91	6.40	7.93
K_2O	0.06	0.12	0.18	0.14	0.39	0.11	0.19	0.13	0.02	0.18	0.11	0.23
Total	99.72	99.01	99.37	99.49	100.84	100.65	99.02	100.03	99.83	99.47	98.70	99.52

Table 14.2. (Continued)

Sample	01070313 A	01070313 B	01090313 C	01090313 D	05060804	08070804	05080804	16080804	M07	05031209 Bo	05031209 C	05031209 D
					Numbers of ions on the basis of 8 Ox. and 5 Cat.							
Si	2.651	2.576	2.640	2.681	2.646	2.676	2.730	2.171	2.162	2.709	2.604	2.729
Ti	0.000	0.002	0.002	0.001	0.001	0.000	0.001	0.001	0.000	0.000	0.000	0.000
Al	1.350	1.362	1.346	1.317	1.342	1.297	1.263	1.808	1.815	1.298	1.392	1.287
Fe^{2+}	0.001	0.004	0.005	0.002	0.000	0.009	0.004	0.011	0.003	0.002	0.005	0.001
Mn	0.000	0.000	0.001	0.002	0.001	0.000	0.001	0.000	0.001	0.000	0.000	0.000
Mg	0.000	0.084	0.001	0.000	0.027	0.048	0.000	0.010	0.005	0.000	0.000	0.000
Ca	0.350	0.371	0.376	0.331	0.333	0.291	0.291	0.824	0.853	0.295	0.428	0.282
Na	0.645	0.667	0.615	0.645	0.633	0.694	0.681	0.183	0.172	0.686	0.564	0.688
K	0.003	0.007	0.010	0.008	0.022	0.006	0.011	0.008	0.001	0.010	0.006	0.013
Sum	5.000	5.073	4.996	4.987	5.005	5.021	4.982	5.016	5.012	5.000	4.999	5.000
					End-members (mol %)							
An	35.05	35.50	37.60	33.60	33.70	29.40	29.60	81.20	83.10	29.73	42.88	28.69
Ab	64.63	63.80	61.40	65.50	64.10	70.00	69.30	18.00	16.80	69.23	56.47	69.99
Or	0.33	0.70	1.00	0.80	2.20	0.60	1.10	0.80	0.10	1.05	0.65	1.32

Table 14.3. Representative chemical analysis of biotites used for geothermobarometry. The location of the samples along the profile is shown in Fig. 14.2

Sample	01070313A	01070313B	01090313C	01090313D	05060804	04070804	08070804	05080804	1.6E+07	M07	05031209Bo	05031209C	05031209D	05031209 E
SiO_2	36.17	36.10	35.81	36.65	35.87	36.16	36.81	36.42	39.37	35.46	35.31	35.37	35.78	35.37
TiO_2	2.06	1.77	3.22	2.35	1.79	2.04	1.93	1.64	3.20	3.16	3.58	3.61	3.44	2.24
Al_2O_3	17.54	17.39	17.86	17.52	18.36	18.77	17.57	17.57	13.08	17.76	16.72	16.68	17.72	17.61
FeOt	19.44	19.77	18.30	18.56	17.78	17.21	20.85	15.39	4.20	16.50	20.62	20.45	17.72	20.46
MnO	0.27	0.30	0.36	0.16	0.13	0.07	0.25	0.05	0.10	0.02	0.16	0.10	0.12	0.23
MgO	9.12	8.63	8.88	7.73	10.95	10.37	8.94	13.06	21.62	10.51	8.30	8.93	10.27	9.14
CaO	0.07	0.07	0.12	0.17	0.04	0.17	0.13	0.05	0.10	0.00	0.13	0.07	0.05	0.04
Na_2O	0.14	0.22	0.04	0.17	0.11	0.27	0.08	0.31	0.31	0.17	0.22	0.18	0.08	0.20
K_2O	8.96	8.85	8.99	8.42	8.56	7.78	9.02	9.24	9.86	9.96	8.98	9.43	9.60	9.33
Total	93.75	93.09	93.57	91.72	93.58	92.83	95.58	93.71	91.84	93.54	93.92	94.81	94.48	94.63
Numbers of ions on the basis of 22 Ox.														
Si	5.586	5.626	5.518	5.731	5.487	5.529	5.602	5.527	5.807	5.452	5.498	5.464	5.466	5.466
Ti	0.239	0.208	0.373	0.276	0.206	0.235	0.221	0.187	0.356	0.365	0.419	0.420	0.395	0.261
Al	3.192	3.193	3.242	3.228	3.309	3.382	3.153	3.142	2.271	3.216	3.068	3.037	3.138	3.207
Fe^{2+}	2.511	2.576	2.358	2.427	2.276	2.200	2.654	1.952	0.518	2.120	2.685	2.642	2.263	2.645
Mn	0.035	0.039	0.047	0.021	0.017	0.009	0.033	0.006	0.012	0.000	0.022	0.013	0.016	0.030
Mg	2.101	2.004	2.040	1.802	2.498	2.364	2.029	2.954	4.751	2.409	1.927	2.056	2.336	2.106
Ca	0.011	0.012	0.019	0.028	0.006	0.028	0.022	0.008	0.015	0.000	0.021	0.012	0.008	0.006
Na	0.042	0.065	0.013	0.050	0.034	0.081	0.024	0.090	0.089	0.051	0.035	0.053	0.024	0.061
K	1.765	1.760	1.767	1.680	1.669	1.517	1.751	1.788	1.856	1.954	1.781	1.858	1.870	1.840

Sample	M09	M08	M03	M15	M14	M11	M67	M68	17060405	12060405 B	20060405	08230306	1E+07
SiO_2	35.55	35.88	36.35	35.74	35.24	35.74	37.10	38.49	35.34	35.20	35.07	35.74	36.34
TiO_2	2.70	3.77	4.81	4.48	4.09	4.47	3.40	3.30	3.52	4.02	3.82	2.69	1.94
Al_2O_3	17.64	17.13	16.59	16.68	16.72	16.50	16.45	16.08	17.62	17.51	17.57	18.01	16.03
FeOt	13.75	17.53	14.15	15.47	16.21	16.27	14.15	7.55	18.46	18.47	20.77	16.16	19.19
MnO	0.07	0.08	0.06	0.06	0.05	0.03	0.03	0.03	0.08	0.06	0.22	0.21	0.13
MgO	12.25	9.79	12.76	12.04	10.44	11.20	14.83	19.73	9.68	9.76	7.89	9.78	10.06
CaO	0.04	0.03	0.02	0.04	0.03	0.02	0.21	0.05	0.01	0.02	0.00	0.33	0.14
Na_2O	0.16	0.18	0.23	0.00	0.11	0.16	0.27	0.46	0.13	0.25	0.19	0.23	0.17
K_2O	9.29	9.77	9.68	9.77	9.81	9.93	8.90	8.90	9.61	9.68	9.73	7.95	8.63
Total	91.34	94.11	94.65	94.21	92.61	94.32	95.34	94.58	94.44	94.97	95.26	91.10	92.62
Numbers of ions on the basis of 22 Ox.													
Si	5.490	5.499	5.451	5.423	5.467	5.451	5.240	5.280	5.425	5.382	5.409	5.566	5.674
Ti	0.314	0.435	0.543	0.511	0.477	0.513	0.360	0.340	0.406	0.462	0.443	0.315	0.228
Al	3.208	3.092	2.930	2.981	3.054	2.964	2.740	2.600	3.188	3.156	3.193	3.305	2.949
Fe^{2+}	1.780	2.250	1.770	1.960	2.100	2.080	1.670	0.870	2.370	2.362	2.679	2.104	2.506
Mn	0.009	0.010	0.008	0.008	0.007	0.004	0.000	0.000	0.010	0.008	0.029	0.028	0.018
Mg	2.820	2.237	2.853	2.724	2.414	2.547	3.120	4.040	2.214	2.224	1.814	2.270	2.341
Ca	0.007	0.005	0.003	0.007	0.005	0.003	0.030	0.010	0.001	0.003	0.000	0.055	0.023
Na	0.048	0.053	0.067	0.000	0.033	0.047	0.070	0.120	0.038	0.073	0.057	0.070	0.051
K	1.830	1.910	1.852	1.891	1.941	1.932	1.610	1.560	1.882	1.888	1.914	1.579	1.719

Table 14.4. Representative chemical analysis of garnets used for geothermobarometry. The location of the samples along the profile is shown in Fig. 14.2.

Sample	01070313A	05060804	08070804	05080804	16080804	M07	05031209	05031209	05031209 D	05031209 E
SiO_2	36.69	36.66	38.00	38.04	40.93	38.24	37.42	37.57	37.50	37.73
TiO_2	0.00	0.03	0.00	0.00	0.47	0.01	0.00	0.00	0.00	0.00
Al_2O_3	20.86	21.75	21.36	21.24	21.33	21.09	19.34	21.34	21.49	21.53
FeOt	22.63	28.27	34.03	28.98	11.31	32.93	31.27	32.61	31.07	31.37
MnO	15.12	7.78	2.25	2.64	0.31	1.24	3.01	2.81	3.35	2.07
MgO	0.84	3.41	4.56	7.33	16.32	4.48	4.84	4.62	5.06	5.99
CaO	3.40	1.71	2.36	1.15	7.41	2.21	1.55	1.55	1.66	1.41
Na_2O	0.00	0.00	0.00	0.00	0.00	0.00	0.08	0.00	0.00	0.00
Total	99.56	99.60	102.56	99.41	98.40	100.20	97.53	100.50	100.12	100.09

Numbers of ions on the basis of 12 Ox. and 8 Cat.										
Si	2.998	2.948	2.955	2.992	3.013	3.040	3.065	2.978	2.971	2.973
Ti	0.000	0.002	0.000	0.000	0.026	0.001	0.000	0.000	0.000	0.000
Al	2.009	2.061	1.958	1.965	1.850	1.975	1.857	1.994	2.007	2.000
Fe^{2+}	1.544	1.863	2.082	1.859	0.642	2.182	2.127	2.112	2.009	2.013
Mn	1.046	0.530	0.148	0.176	0.020	0.084	0.208	0.189	0.225	0.138
Mg	0.103	0.409	0.529	0.859	1.791	0.531	0.592	0.545	0.597	0.703
Ca	0.297	0.148	0.197	0.097	0.585	0.188	0.136	0.132	0.141	0.119
Sum	7.997	7.961	7.869	7.948	7.927	8.001	7.985	7.950	7.950	7.946

End-members (mol%)										
Alm	51.63	63.35	70.44	62.16	22.98	73.12	69.45	70.93	67.60	67.71
Py	3.44	13.71	17.89	28.73	56.99	17.79	19.29	18.31	20.08	23.66
Gr	9.94	5.01	6.65	3.24	19.50	6.30	4.44	4.42	4.75	4.00
Sp	34.99	17.94	5.02	5.88	0.53	2.82	6.81	6.34	7.57	4.64

Sample	M09	M08	M03	M15	M14	M11	M67	12060405	17060405	20060405
SiO_2	37.14	38.41	38.13	38.01	37.41	38.23	38.48	37.81	37.85	37.30
TiO_2	0.06	0.01	0.05	0.10	0.02	0.04	0.04	0.00	0.00	0.00
Al_2O_3	21.34	21.63	20.99	21.76	21.05	21.02	21.73	21.56	21.25	21.43
FeOt	29.85	27.54	29.99	28.89	33.51	31.42	27.07	32.36	31.27	33.43
MnO	2.88	1.69	1.18	1.69	1.48	1.86	1.61	1.54	2.05	2.98
MgO	5.90	8.78	6.62	7.99	4.44	6.16	6.57	5.74	5.99	3.96
CaO	1.50	1.30	1.30	1.36	1.53	1.19	4.88	1.29	1.54	1.21
Na_2O	0.02	0.00	0.16	0.00	0.03	0.02	0.02	0.18	0.09	0.08
Total	98.73	99.36	98.47	99.80	99.47	99.94	100.40	100.47	100.06	100.40

Numbers of ions on the basis of 12 Ox. and 8 Cat.										
Si	2.967	2.980	3.033	2.960	3.000	3.022	2.983	2.980	2.987	2.977
Ti	0.004	0.000	0.003	0.010	0.001	0.002	0.002	0.000	0.000	0.000
Al	2.009	1.980	1.967	1.990	1.989	1.957	1.983	2.002	1.976	2.016
Fe^{2+}	1.987	1.800	1.994	1.900	2.240	2.064	1.756	2.095	2.015	2.200
Mn	0.196	0.110	0.080	0.110	0.101	0.125	0.106	0.102	0.137	0.201
Mg	0.703	1.020	0.785	0.930	0.532	0.726	0.759	0.674	0.705	0.472
Ca	0.128	0.110	0.111	0.110	0.130	0.101	0.405	0.109	0.130	0.104
Sum	7.994	8.000	7.973	8.010	7.993	7.997	7.994	7.962	7.950	7.970

End-members (mol%)										
Alm	65.86	59.32	66.60	62.21	74.44	68.39	57.97	70.29	67.48	74.05
Py	23.30	33.48	26.22	30.40	17.68	24.06	25.06	22.61	23.59	15.53
Gr	4.09	3.55	3.56	3.71	4.32	3.35	13.27	3.66	4.35	3.53
Sp	6.50	3.65	2.67	3.64	3.36	4.14	3.50	3.44	4.59	6.89

Table 14.5. Results of the geothermobarometric study, corresponding to points plotted in Fig. 14.2 to draw the geothermal curve along the cross-section. Details in the text

	Pringles Metamorphic Complex geothermobarometry						
Method	TWQ		Ti-in-Bt	Grt-Bt			
	Con-91-14/HP-02		Henry-05	B92-HW	B92-GS	PL-83	HL-77
Sample	P (bar)			T (°C)			
01050804							
01070313 A	3900	460	617				
01070313 B			585				
01090313 C			693				
01090313 D			638				
05060804	4700	610	605	591	583	606	607
04070804			628				
08070804	6600	740	593	735	737	718	752
05080804	5600	780	611	702	706	698	725
16080804	6800	820	821	750	849	697	723
M7	6850	713	703	635	644	699	726
05031209 Bo	6800	800	706	772	775	758	806
05031209 C	5200	750	709	727	728	718	751
05031209 D	6100	715	713	690	693	683	706
05031209 E	7000	840	634	813	821	793	854
M9	6288	709	701	633	641	628	635
M3	6521	728	766	653	657	647	659
M15	8250	837	758	740	744	733	772
M14	6595	687	740	622	632	612	615
M11	6850	762	752	687	691	680	702
M67	6271	765	737	665	681	635	644
12060405 B	6800	760	731	738	741	725	761
17060405	6100	780	713	764	768	749	794
20060405	6600	730	714	712	710	707	736
08230306			678				
10200306 A			614				

	Graphite crystallinity						
	Sh et al-79						
08021209	380						

Obtained temperatures match with the predicted stability fields for the corresponding associations determined through pseudosection modeling based on whole rock chemical composition (Fig. 14.4 and Table 14.1).

14.6 MECHANISMS OF MICROSTRUCTURAL DEVELOPMENT

14.6.1 Slates

The described fabric indicates an important contribution of pressure dissolution-mass transfer and fluid-assisted grain boundary migration mechanisms in the microstructure development of these rocks. Matrix quartz and feldspar grains show high-mobility of grain boundaries and pinning by micas and organic matter. As a result, elongated crystals with slightly lobate boundaries with moderate preferred crystallographic orientation develop (Fig. 14.6a$_1$ and a$_2$). Quartz crystals in the matrix reach lengths of up to 50 µm parallel to foliation. Microstructures suggest that recrystallized quartz in low-differential stress sites is the result of precipitation from silica-supersaturated fluids generated by pressure dissolution. However, some grains with slightly lobate edges, smooth undulose extinctions and weak preferred crystallographic orientation in coarse-grained lenses, magnetite tails, and lensoidal pseudomorphs indicate grain boundary migration recrystallization after crystallization. Furthermore, the presence of some unstrained, up to 100 µm long polygonal crystals, with dissimilar crystallographic orientation and straight or gently curved edges forming 120° triple junctions, point to some contribution of post-deformation static recrystallization (Fig. 14.3a,b).

14.6.2 Phyllites

In phyllites transitional to slates, pressure dissolution is evidenced by straight faces parallel to foliation, developed in quartz grains with differential growth immersed in the finer-grained matrix (Fig. 14.3c). The precipitation of the siliceous cement in the adjacent coarse-monomineralic quartz layers (Fig. 14.3d), furthermore suggests a major contribution of diffusive mass transfer processes.

In the transitional rocks between slates and phyllites, the microstructures observed in matrix quartz seem to indicate,

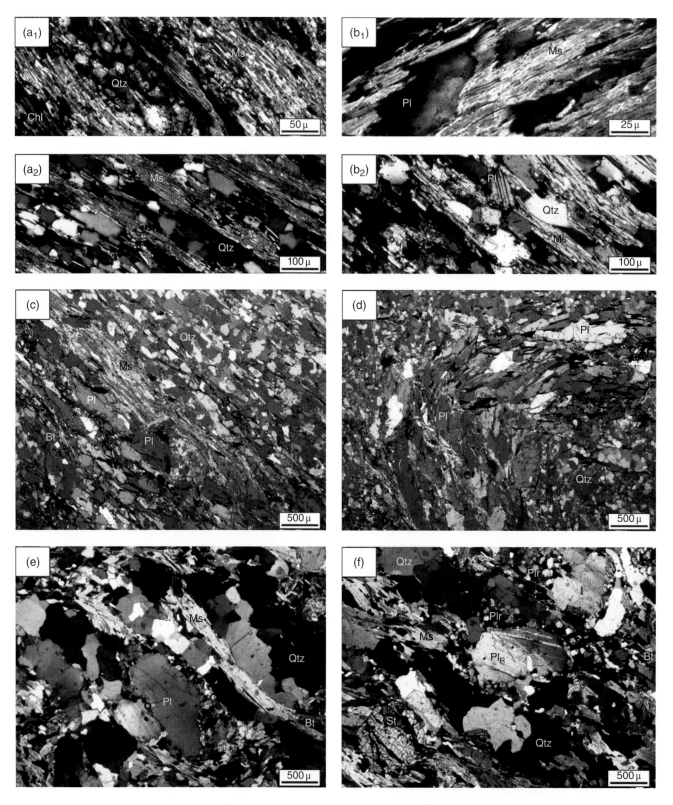

Fig. 14.6. Comparative microstructural development of quartz and plagioclase in metapsamitic–metapelitic rocks during temperature increase. Low-temperature initial growth of quartz (a) and plagioclase (b). After the transition from a non-domainal to a domainal fabric, grain boundary migration recrystallization of quartz occur. Observe quartz-rich microlithons in which this phase begins to develop more or less equidimensional grains with gently lobated boundaries (a_1). At a most advanced stage, quartz shows bigger and elongated grains parallel to foliation, forming very thin ribbons (a_2); at this stage, plagioclase show growth of irregular, cuspate grains, usually associated to mica-rich domains (b_1). Note that small subhedral grains can be also observed in quartz-rich microlithons (b_2). (c, d) Higher-temperature folded micaschist, showing increase of quartz and plagioclase crystal size. Observe polygonal or tabular quartz grains forming coarse microlithons separated by micaceous cleavage domains. Plagioclase growth preferentially in mica-rich domains. Note growth parallel to foliation due to mica pinning and cuspate outlines in the direction of growing. Observe also lack of intracrystalline ductile deformation. (e, f) Staurolitic micaschist in which growth of crystals in combination with metamorphic reactions significantly modify the rock fabric and allow starting of ductile deformation of plagioclase crystals. Note unlimited growth of quartz and plagioclase, discontinuity of micaceous folia and mutual contact between anhydrous phases.

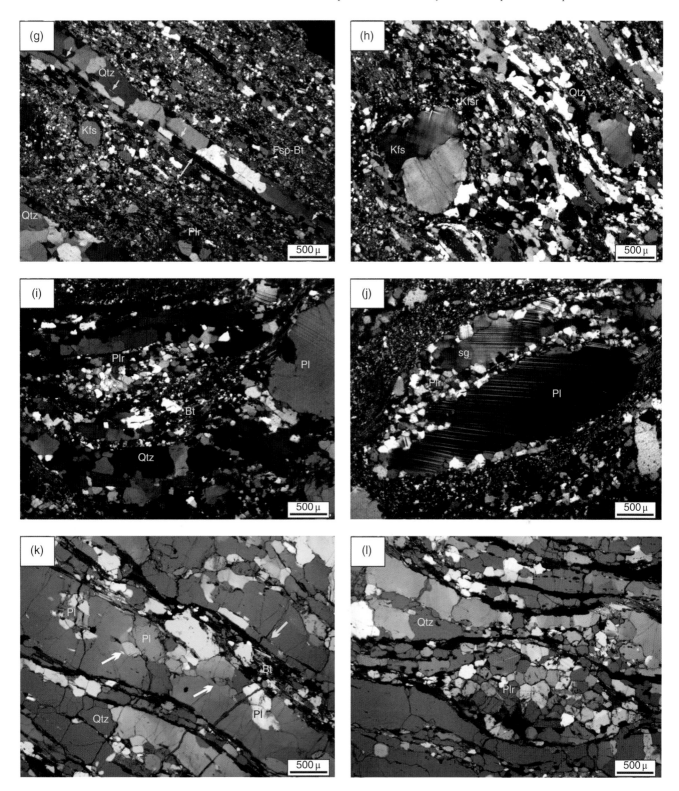

Fig. 14.6. (*Continued*) (g, h) Temperatures >600°C allow coalescence of quartz ribbons (arrows in g) and progressive coarsening of recrystallization of feldspars (h) by grain boundary recrystallization. (i, j) At temperatures >700°C, quartz forms very irregular ribbons with strongly ameboid coarse-grains (i). Plagioclase shows subgrains in some elongated relict crystals (j). (k, l) At very high-temperatures, quartz shows extreme coalescence, forming very thick ribbons and enclosing entirely other phases. Observe sigmoidal mica fishes with smooth outlines (k) and wholly recrystallized plagioclase porphyroclast with perfect polygonal new-grains meeting at triple junctions. All photographs with crossed nicols. (c), (d), (k), and (l) with gypsum plate to enhance textures. Abbreviations after Bucher and Grapes (2011). Details in the text.

as in slates, growth by fluid-assisted grain boundary migration, with a probable late adjustment by static growth.

The increase in metamorphic grade promotes the development of the domainal fabric and reduce the interspersing of phases in both quartzose and micaceous bands, changing the microstructural patterns. The generation of bands dominated by the presence of a phase or quasi-monomineralic, gives rise not only to greater crystal growth, but also to the change in the shape of recrystallized grains. The matrix shows incipient quartz ribbons where alternate elongated rectangular grains pinned by micas, with polygonal grains frequently showing 120° triple junctions (Fig. 14.3e). In quasi-monomineralic quartzose layers, typical microstructures representative of grain boundary migration recrystallization in regime 3 (Hirth and Tullis 1992; Stipp et al. 2002) develop. Bulging of large amplitude (100–150 μm) in ameboidal relic grains, is consistent with the size of recrystallized polygonal crystals (Fig. 14.3e, f). As in slates, fabric observed in polygonal quartz aggregates indicates some participation of static recrystallization during the final growth of crystals. However, the preservation of a preferential crystallographic orientation (Fig. 14.3f), suggests that crystals formed previously by dynamic grain boundary migration recrystallization. In case of plagioclase, textures observed in isolated grains in micaceous and quartz layers (Fig. 14.6b$_1$ and b$_2$), indicate metamorphic growth of feldspar without intracrystalline deformation.

14.6.3 Ms-Bt schists

In metapsamites with predominant quartzose bands, large ameboid crystals reaching up to 600 μm with undulose extinctions and incipient development of deformation bands are present. These grains show high-amplitude bulging and polygonal new-grains of up to 300 μm contacting each other through triple junctions at 120°. Similar characteristics are observed in quartz-rich microlithons of metapelitic rocks, where this phase forms elongated tabular crystals parallel to foliation due to mica pinning or more or less equidimensional grains with smoothly lobated to straight edges that reach up to 250 μm and meet each other through 120° triple junctions (Fig. 14.6c, d). The most outstanding feature in relation to the lower-grade rocks described previously is segregation of quartz in well-defined quasi-monomineralic ribbons characterized by polygonal to highly elongated tabular crystals reaching lengths up to 1500 μm (Type 2 and 3 of Boullier and Bouchez 1978) (Fig. 14.3g,k). All these characteristics documents dynamic recrystallization through grain boundary migration.

Plagioclase shows no evidence of dynamic recrystallization. The formation of tabular crystals pinned by mica, their cuspate grain boundaries in the direction of foliation and the incorporation of matrix minerals, indicates fluid-assisted metamorphic growth (Fig. 14.3g–i, m).

14.6.4 Ms-Bt-St schists

The most abrupt structural and mineralogical change occurs in staurolite schists. Muscovite diminishes drastically and chlorite disappears, whereas biotite becomes the dominant phyllosilicate.

The increase in the content of non-laminar minerals (St-Pl-Qtz) associated with the increase of the grain size, produces significant changes in the rock fabric. Quartz and plagioclase crystals overgrow the mica pinning with the consequent loss of continuity of the micaceous cleavage domains, allowing non-laminar minerals to be in contact with each other (Fig. 14.6e, f). Strain previously partitioned into interconnected weak micaceous layers, is now localized in touching non-laminar phases (see Hunter et al. 2013; Holyoke and Tullis 2006; Mariani et al. 2006; Montési 2013). This fabric change affects the rheological behavior of rock phases, triggering strain-induced grain boundary migration recrystallization of feldspars.

Ductile deformation of plagioclase crystals produces relict crystals with microfracturing, undulose extinctions, secondary twinning, bulging of moderate amplitude and recrystallization to aggregates of polygonal crystals that contact each other through 120° triple junctions. Recrystallization occurred essentially at the margins of feldspar grains and along internal microfractures (Fig. 14.6e, f).

The beginning of feldspar recrystallization coincides with the transformation of micas and staurolite to coarse garnet and sillimanite. These dehydration reactions occur at >600°C and release water to the system (reaction 6–8 in Fig. 14.5). Free water and elevated temperature enhance dynamic recrystallization. In addition, these reactions take place just before the onset of anatexis (Fig. 14.4 and 14.5), and may have also contributed significantly to partial melting in rocks located east of the staurolitic schists.

14.6.5 Gneisses and migmatites

As in higher-temperature staurolite-bearing micaschists, coarse-grained gneisses and migmatites with segregates neosomes show deformation microstructures similar to those observed in natural or experimental studies of ductile deformation of medium to high-grade rocks. However, it is interesting to highlight the features and changes observed in quartz and feldspars with the increase of temperature within the fields of high and very-high metamorphic grade.

At these thermal conditions, relict quartz crystals in remnant coarse-grained portions of rocks show big crystals with marked lobate boundaries (Fig. 14.6g). In high-strain zones where the biotitic matrix predominate, quartz forms ribbons of Type 3 to 4 of Boullier and Bouchez (1978), with straight or gently lobated grain boundaries. In the wider ribbons, evidence can be seen that these are the product of the coalescence of thinner ribbons by the unrestricted growth of quartz at high temperature (Fig. 14.6g, i). At higher temperatures, quartz coalescence leads to thick ribbons that usually include not only to biotite, but also feldspars. Some sigmoid feldspars record evidence for dynamic recrystallization (Fig. 14.6k) (see, Mukherjee 2011). At these very high temperatures, inside rocks subject to low differential stress, quartz often show chessboard extinction (Fig. 14.3r) and contact with feldspars through strongly lobate boundaries.

At moderate temperatures, feldspars show similar behavior that in the staurolitic schists. Relict crystals are more or less equidimensional, rounded or elliptical, and show microfracture–microshear zones, slightly undulatory or fragmentary extinctions, bulging and recrystallization to aggregates of polygonal grains with 120° triple junctions (Fig. 14.3o, q and Fig. 14.6h, i). With increase in temperature, the size of bulging and recrystallized new-grains increase. At very high temperature, subgrains can be observed in some relict crystals. Sub-grains are found only in some relict plagioclase crystals elongated parallel to foliation (Fig. 14.6j). In accordance with the development of subgrains, these relict crystals show sweeping undulatory extinctions, deformation bands and tapering secondary twins. The adjacent relict equidimensional-shaped crystals show no subgrains. This feature suggests that subgrains forms in grains with a suitable orientation for sliding. Another interesting characteristic observed, is the usual localization of recrystallization along irregular microfracture–microshear zones developed inside relict porphyroclasts. Often, the new-grains in contact with the relict porphyroclast show irregular shapes and one or more of their faces are limited by tiny fractures, whereas away from the contact the recrystallized grains are polygonal and meet at 120° triple junctions.

The characteristics described above, indicate that microfracturing–microshearing combined with fluid-assisted grain boundary migration is the most likely mechanism for feldspar recrystallization.

14.7   DISCUSSION

Low- to medium-grade metapelitic–metapsamitic rocks have been sometimes considered as undeformed, mainly because of the absence of the typical microstructures that indicate ductile deformation. In the Pringles Metamorphic Complex, a wide shear zone developed under a prevailing geothermal gradient previously established on a monotonous and very homogeneous sedimentary sequence. Study of this shear zone rocks demonstrates that the lack of such microstructures is not enough proof of the absence of significant ductile deformation. An alternative explanation to the above mentioned absence can be found in a different rheological response inherent to the different characteristics of polymineralic fine-grained sedimentary rocks (e.g.

Renard et al. 2001; Herwegh et al. 2005, Huet et al. 2014) and coarse-grained monomineralic rocks from which ductile deformation is usually evaluated (e.g. Tullis and Yund 1985, 1987; Hirth and Tullis 1992; Stipp et al. 2002; Passchier and Trouw 2005, and references therein).

Below 600°C, microstructures observed in metapelitic–metapsamitic rocks differ markedly from those expected for ductile deformation of coarse-grained quartz and/or feldspars aggregates at equivalent thermal conditions. Quartz subgrain rotation crystallization and bulging recrystallization in feldspars that should take place within this thermal range, do not occur. Instead, quartz crystals growth progressively by grain boundary migration, both in polymineralic microlithons and monomineralic ribbons. Plagioclase shows metamorphic growth without evidence of intracrystalline deformation. The rheologic response of both phases is a function of strain partitioning within the domainal fabric of metasedimentary rocks. The presence of continuous cleavage domains, consisting of micas predominantly oriented with their [001] cleavage planes parallel to foliation, accommodates most of the shear strain and inhibit intracrystalline deformation in the other silicates. Presence of coarse-grained rocks contiguous, or included, in basement rocks (La Escalerilla granite, veins and pegmatitic bodies), serves as proof that the necessary conditions for ductile deformation were the appropriated (compare left and right columns in Fig. 14.3a–n).

At temperatures ~600°C, a combination of factors substantially modifies the rock fabric and, hence, its mechanical response to the applied differential stress. The most significant factors include crystal growth, mineralogical changes generated by metamorphic reactions and the presence of fluids. Metamorphic dehydration reactions reduce the micaceous content and release abundant amounts of water. All these factors discontinue micaceous cleavage domains and allow the mutual contact of anhydrous phases. Thus, high differential stress along point contacts of feldspar grains triggers recrystallization processes. The presence of free water facilitates ductile deformation of quartz and feldspars (Urai 1983; Farver and Yund 1990; Kronemberg and Tullis 1994).

At temperatures >600°C, coarse-grained metapelitic–metapsamitic rocks show the typical ductile microstructures usually used as reference for ductile deformation of quartz–feldspar rocks. Ductile deformation in foliated metasedimentary rocks and the neighboring granular rocks (coarse-grained neosomes, pegmatites), shows similar characteristics (compare left and right columns in Fig. 14.3(o–r)). Gneisses and migmatites show quartz ribbons which progressively widen by the unlimited growth and coalescence of quartz crystals, whereas feldspars experience grain boundary migration recrystallization with progressive increase of new-grains size.

At temperatures above ~700°C, subgrains in elongated plagioclase crystals indicate some contribution of the rotation recrystallization mechanism in the ductile deformation of plagioclase (Fig. 14.6j). In sectors of bodies subject to low-strain like inside pegmatitic veins and bodies, there is evidence of very high-temperature deformation such as chessboard extinctions in quartz (Fig. 14.3r), "fish like" feldspar inclusions in optically continuous big quartz crystals (e.g. Mukherjee 2013) and lobate contacts between quartz and feldspars. It is considered that the first microstructure occurs only above the Qtz_α/Qtz_β transition (Kruhl 1996, Stipp et al. 2002), which at the estimated pressure of ~6 kbar imply temperatures >700°C. The second microstructure is attributable to the diffusion creep mechanism, which is characteristically observed in rocks deformed within granulite facies (e.g. Simpson and De Paor 1991; Gower and Simpson 1992; Martelat et al. 1999).

14.8   CONCLUSIONS

1   The Pringles Metamorphic Complex mega-shear zone constitutes a natural laboratory for the study of the microstructural development in metapelitic–metapsamitic rocks. Polymineralic metasedimentary rocks at temperatures <600°C show a rheological response that do not agree with typical behavior of compositionally equivalent monomineralic coarse-grained rocks. Microstructures developed by progressive transformation of fine-grained sediments with temperature rise are often considered erroneously to indicate absence of ductile deformation.

2   The distinct rheologic response of quartz and feldspar can be attributed to the development of a domainal fabric in metasedimentary rocks. Continuous mica-rich cleavage domains absorbs most of the differential stress and attenuates intracrystalline deformation in the other silicates. As a result, grain boundary migration growth of quartz and metamorphic growth of plagioclase are favored, instead of quartz subgrain rotation crystallization and bulging recrystallization in feldspars. Adjacent coarse-grained rocks with granular textures (leucosome veins and bodies), show typical microstructures of recrystallization connoting conditions for ductile deformation.

3   At ~600°C typical of upper amphibolite facies, grain growth and dehydration reactions promote significant mineralogical and microstructural changes. Reduction of phyllosilicate proportions and discontinuation of micaceous cleavage domains allow mutual contact between anhydrous phases and promote strain-induced grain boundary migration recrystallization of feldspars. Free water might enhance grain boundary migration and recrystallization of quartz and feldspars.

4   At >600°C, coarse-grained metapelitic–metapsamitic rocks behave as typically observed in natural and experimental studies of ductile deformation. Quartz forms Type 3 and 4 ribbons, whereas dynamic recrystallization of feldspars through the grain boundary migration mechanism forms progressively bigger new grains. Observed microstructures point to microfracturing–microshearing combined with fluid-assisted grain boundary migration as the most suitable deformation mechanism for feldspars recrystallization.

5   At >700°C (typical of granulite facies), quartz ribbons coalesce and enclose feldspars. At these thermal conditions, first evidence of rotation recrystallization is observed in plagioclase porphyroclasts. Low-strain sectors of coarse-grained rocks show microstructures typical of ductile deformation at very-high temperature, like diffusion creep between quartz and feldspars grains.

ACKNOWLEDGMENTS

This work was supported by ANPCyT (Argentina) Grant PICT2008–1352 and SCyT-UNS (Argentina) Grants 24/H098 and 24/H120 to SHD. We thank Soumyajit Mukherjee who edited and reviewed this work. Review by S. Verdecchia and an anonymous researcher helped a lot.

REFERENCES

Bailey J, Hirsch P. 1962. The recrystallization process in some polycrystalline metals. Proceedings of the Royal Society of London A267, 11–30.

Bhattacharya A, Mohanty L, Maji A, Sen SK, Raith M., 1992. Non-Ideal mixing in the phlogopite-annite binary: constraints from experimental data on Mg-Fe partitioning and formation of the biotite-garnet geothermometer. Contributions to Mineralogy and Petrology 111, 87–93.

Berman RG. 1991. Thermobarometry using multi-equilibrium calculations: a new technique, with petrological applications. Canadian Mineralogist 29, 833–855.

Boullier A, Bouchez J. 1978. Le quartz en rubans dans les mylonites. Bulletin de la Societé Géologique de France 7, 253–262.

Brogioni N, Ribot A. 1994. Petrología de los cuerpos La Melada y La Gruta, faja máfica-ultramáfica del borde oriental de la Sierra de San Luis. Revista de la Asociación Geológica Argentina 49, 269–283.

Brun J, Cobbold P. 1980. Strain heating and thermal softeningin continental shear zones: a review. Journal of Structural Geology 2, 149–158.

Bucher K, Grapes R. 2011. Petrogenesis of Metamorphic Rocks. Springer, Heidelberg, Dordrecht, London, New York.

Burg J, Gerya T. 2005. The role of viscous heating in Barrovian metamorphism of collisional orogens: thermomechanical models and application to the Lepontine Dome in the Central Alps. Journal of Metamorphic Geology 23, 75–95.

Caminos R. 1973. Some granites, gneisses and metamorphites of Argentina. In Symposium on Granites, Gneisses and Related Rocks, edited by L. Lister, Geological Society, South Africa, Special Publication, vol. 3, pp. 333–338.

Caminos R. 1979. Sierras Pampeanas Noroccidentales de Salta, Tucumán Catamarca, La Rioja y San Juan. Segundo Simposio de Geología Regional Argentina. Academia Nacional de Ciencias, Córdoba 1, 225–291.

Castro de Machuca B, Arancibia G, Morata D, et al. 2008. P-T-t evolution of an Early Silurian medium-grade shear zone on the west side of the Famatinian magmatic arc, Argentina: implications for the assembly of the Western Gondwana margin. Gondwana Research 13, 216–226.

Castro de Machuca B, Delpino S, Mogessie A, et al. 2010. 40Ar/39Ar age dating of ductile deformation in a shear zone from the Sierra de La Huerta, San Juan, Argentina: evidence of the Famatinian orogeny. 7° South American Symposium on Isotope Geology Extended Abstract, 186–189. Brasilia.

Castro de Machuca B, Delpino S, Previley L, et al. 2012. Evolution of a high-to medium-grade ductile shear zone from the Famatinian Orogen, Western Sierras Pampeanas, Argentina. Journal of Structural Geology 41, 1–18.

Connolly J. 1990. Multivariable phase-diagramas – an algorithm based on generalized thermodynamics. American Journal of Science 290, 666–718.

Cruciani G, Franceschelli M, Groppo et al. 2008. Formation of clinopyroxene+spinel and amphibole+spinel symplectites in coronitic gabbros from the Sierra de San Luis (Argentina): a key to post-magmatic evolution. Journal of Metamorphic Geology 26, 759–774

Cruciani G, Franceschelli M, Brogioni N. 2011. Mineral re-equilibration and P-T path of metagabbros, Sierra de San Luis, Argentina: insights into the exhumation of mafic-ultramafic belt. European Journal of Mineralogy 23, 591–608.

Cruciani G, Franceschelli M, Brogioni N. 2012. Early stage evolution of the mafic-ultramafic belt at La Melada, Sierra de San Luis, Argentina: P-T constraints from metapyroxenite pseudo-section modelling. Journal of South American Earth Sciences 37, 1–12.

Dalla Salda L. 1987. Basement tectonics of the Southern Pampean Ranges, Argentina. Tectonics 6, 249–260.

Dahlquist J, Pankhurst R, Rapela C, et al. 2008. New SHRIMP U–Pb data from the Famatina Complex: constraining Early–Mid Ordovician Famatinian magmatism in the Sierras Pampeanas, Argentina. Geologica Acta, 6, 319–333.

Delpino, S., Dimieri, L., Bjerg, E., et al. 2001. Geometrical analysis and timing of structures on mafic–ultramafic bodies and high-grade metamorphic rocks, Sierras Grandes of San Luis, Argentina. Journal of South American Earth Sciences 14 (1), 101–112.

Delpino S, Bjerg E, Ferracutti et al. 2002. Upper-amphibolite facies mylonitization of mafic-ultramafic rocks and gneissic-migmatitic country rocks, Sierras de San Luis, Argentina: implications in the remobilization of ore sulfides. In Mineralogía y Metalogenia 2002, edited by M. Brodtkorb, M. Koukharsky, and P. Leal, Universidad Nacional de Buenos Aires, Buenos Aires, pp. 123–126.

Delpino S, Bjerg E, Ferracutti et al. 2007. Counterclockwise tectono-metamorphic evolution of the Pringles Metamorphic Complex, Sierras Pampeanas of San Luis (Argentina). Journal of South American Earth Sciences 23 (2–3), 147–175.

Delpino S, Grasemann B, Bjerg et al. 2012. Thermal evolution model for the Pringles Metamorphic Complex, Sierra de San Luis, Argentina. 15° Reunión de Tectónica, Resúmenes CD: 42. San Juan, Argentina.

Drury M. Humphreys F. White S. 1985. Large strain deformation studies using polycrystalline magnesium as rock analogue. Part II: dynamic recrystallisation mechanisms at high temperatures. Physics of the Earth and Planetary Interiors 40, 208–222.

Farver J, Yund R. 1990. The effect of hydrogen, oxigen and water fugacity on oxigen diffusion in alkali feldspar. Geochimica et Cosmochimica Acta 54, 2953–2964.

Gallien F, Mogessie A, Bjerg E, et al. 2010. Timing and rate of granulite facies metamorphism and cooling from multi-mineral chronology on migmatitic gneisses, Sierras de La Huerta and Valle Fértil, NW Argentina. Lithos 114, 229–252.

González P, Sato A, Basei M, et al. 2002. Structure, metamorphism and age of the Pampean–Famatinian Orogenies in the western Sierra de San Luis, XV Congreso Geológico Argentino (CD), El Calafate.

González Bonorino F. 1961. Petrología de algunos cuerpos básicos de San Luis y las granulitas asociadas. Revista de la Asociación Geológica Argentina 16 (1–2), 61–106.

Gower R, Simpson C. 1992. Phase boundary mobility in naturally deformed, high-grade quartzofeldspathic rocks: evidence for diffusional creep. Journal of Structural Geology 14, 301–314.

Hauzenberger Ch, Mogessie A, Hoinkes G, et al. 1997. Platinum group minerals in the basic to ultrabasic complex of the Sierras de San Luis, Argentina. In Mineral Deposits: Research and Explorations – Where Do They Meet, edited by H. Papunen, A. A. Balkema, Rotterdam, pp. 439–442. .

Hauzenberger Ch, Mogessie A, Hoinkes G, et al. 1998. Metamorphic Evolution of the Southern Part of the Sierras de San Luis, Argentina. IV Reunio´n de Mineralogía y Metalogenia y IV Jornadas de Mineralogía, Petrología y Metalogénesis de Rocas Máficas y Ultramáficas, pp. 121–130.

Hauzenberger Ch, Mogessie A, Hoinkes G, et al. 2001. Metamorphic evolution of the Sierras de San Luis: Granulite facies metamorphism related to mafic intrusions. Mineralogy and Petrology 71 (1/2), 95–126.

Henry D, Guidotti Ch, Thomson J 2005. The Ti-saturation surface for low-to-medium pressure metapelitic biotites: Implications for geothermometry and Ti-substitution mechanisms. American Mineralogist 90, 316-328,

Herwegh M, Berger A, Ebert A 2005. Grain coarsening maps: A new tool to predict microfabric evolution of polymineralic rocks. Geology 33 (10), 801–804.

Hirth G, Tullis J. 1992. Dislocation creep regimes in quartz aggregates. Journal of Structural Geology 14, 145–159.

Holdaway M. Lee S. 1977. Fe-Mg cordierite stability in high-grade pelitic rocks based on experimental, theoretical and natural observations. Contribution to Mineralogy and Petrology 63, 175–98.

Holland T. Powell R. 1998. An internally consistent thermodynamic data set for phases of petrological interest. Journal of Metamorphic Geology 16, 309–343.

Holyoke C, Tullis J. 2006. Mechanisms of weak phase interconnection and the effects of phase strength contrast on fabric development. Journal of Structural Geology 28, 621–640.

Huet B, Yamato P, Grasemann B. 2014, The Minimized PowerGeometric model: An analytical mixingmodel for calculating polyphase rockviscosities consistent with experimental data, Journal of Geophysical Research: Solid Earth119, 1–28.

Hunter J, Hasalová P, Weinberg R. 2013. Strain partitioning in crustal shear zones: the effect of interconnected micaceous layers on quartz deformation. In Deformation Mechanisms, Rheology and Tectonics, Programme and abstract, International Conference, Leuven 2013, p. 35.

Iannizzotto N, Rapela C, Baldo et al. 2013. The Sierra Norte-Ambargasta batholith: late Ediacaran early Cambrian magmatism associated with Pampean transpressional tectonics. Journal of South American Earth Sciences 42, 127–143.

Kostadinoff J, Bjerg E, Delpino S, et al. 1998a. Gravimetric and magnetometric anomalies in the Sierras Pampeanas of San Luis. Revista de la Asociación Geológica Argentina 53 (4), 549–552.

Kostadinoff J, Bjerg EA, Dimieri L, et al. 1998b. Anomalías geofísicas en la faja de rocas máficas-ultramáficas de la Sierra Grande de San Luis, Argentina. IV Reunión de Mineralogía y Metalogenia y IV Jornadas de Mineralogía, Petrografía, Metalogénesis de Rocas Máficas y Ultramáficas, Actas, 139–146, Bahía Blanca.

Kronemberg A, Tullis J. 1994. Flow strengths of quartz aggregates: grain size and pressure effects due to hydrolytic weakening. Journal of Geophysical Research 89, 4281–4297.

Kruhl JH. 1996. Prism- and basal -plane parallel subgrain boundaries in quartz: a microstructural geothermobarometer. Journal of
Metamorphic Geology 14, 581–589.

Landis C. 1971. Graphitization of dispersed carbonaceous material
in metamorphic rocks. Contribution to Mineralogy and Petrology
30, 34–45.

Larrovere M, de los Hoyos C. Toselli A. et al. 2011, High T/P evolution and metamorphic ages of the migmatitic basement of northern Sierras Pampeanas, Argentina: characterization of a
mid-crustal segment of the Famatinian belt: Journal of South
American Earth Sciences 31 (2–3), 279–297.

Mariani E, Brodie K, Rutter E. 2006. Experimental deformation of
muscovite shear zones at high temperatures under hydrothermal
conditions and the strength of phyllosilicate-bearing faults in
nature. Journal of Structural Geology 28, 1569–1587.

Martelat J, Schulmann K, Lardeaux J, et al. 1999. Granulite microfabrics and deformation in southern Madagascar. Journal of
Structural Geology 21, 671–687.

Mogessie A, Hoinkes G, Stumpfl EF, et al. 1994. The petrology and
mineralization of the basement and associated mafic–ultramafic
rocks, San Luis Province, Central

Argentina. Mitteilungen der O¨ sterreichischen Mineralogischen
Gesellschafl 139, 347–348. Austria.

Mogessie A, Hoinkes G, Stumpfl et al. 1995. Occurrence of Platinum
Group Minerals in the Las Aguilas Ultramafic Unit within a
Granulite Facies Basement, San Luis Province, Central
Argentina. In Mineral Deposits: From Their Origin to Their
Environmental Impacts, edited by J. Paiava, B. Ktibek, and K.
Zak, A. A. Balkema, Rotterdam, pp. 897–900. .

Mogessie A, Hauzenberger Ch, Hoinkes et al. 1998. Origin of
Platinum Group Minerals in the Las Aguilas Mafic-Ultramafic
Intrusion, San Luis Province, Argentina. IV Reunio´n de
Mineralogı´a y Metalogenia y IV Jornadas de Mineralogía,
Petrografía, Metalogénesis de Rocas Máficas y Ultramáficas,
Actas, 285–289, Argentina.

Montési L. 2013. Fabric development as the key for forming ductile
shear zones and enabling plate tectonics. Journal of Structural
Geology 50, 254–266.

Mukherjee S. 2011. Mineral fish: their morphological classification,
usefulness as shear sense indicators and genesis. International
Journal of Earth Sciences 100(6), 1303–1314.

Mukherjee S. 2013. Deformation Microstructures in Rocks.
Springer-Verlag, Berlin.

Mukherjee S, Mulchrone K 2013. Viscous dissipation pattern in
incompressible Newtonian simple shear zones: an analytical
model. International Journal of Earth Sciences 102 (4),
1165–1170.

Mulchrone K Mukherjee S 2015. Shear senses and viscous dissipation of layered ductile simple shear zones. Pure and Applied
Geophysics, in press.

Pankhurst R, Rapela C. 1998a. The proto-Andean margin of
Gondwana: an introduction. In The Proto-Andean Margin of
Gondwana, edited by R. Pankhurst and C. Rapela, Geological
Society, London, Special Publication, vol. 142, pp. 1–9.

Pankhurst RJ, Rapela CW, Saavedra J, et al. 1998b. The Famatinian
magmatic arc in the central Sierras Pampeanas: an Early to Mid-
Ordovician arc on the Gondwana margin. In The Proto-Andean
Margin of Gondwana, edited by R. Pankhurst and C. Rapela,
Geological Society, London, Special Publication vol. 142,
pp. 343–367.

Pankhurst R, Rapela C, Fanning C. 2000. Age and origin ofcoeval TTG,
I- and S-type granites in the Famatian belt of NWArgentina.
Transaction Royal Society Edinburgh: Earth Sciences 91, 151–168.

Passchier C. Trouw R. 2005. Microtectonics. Springer-Verlag,
Berlin.

Perchuk L, Lavrent'eva L. 1983. Experimental investigation of
exchange equilibria in the system cordierite-garnet-biotite.
In Kinetics and Equilibrium in Mineral Reactions, edited by
S.K. Saxena, Springer-Verlag, New York, pp. 199–239.

Ranalli G. 1995. Rheology of the Earth, 2nd edition. Chapman &
Hall, New York

Rapela C, Pankhurst R, Casquet R, et al. 1998. The Pampean Orogeny
of the southern proto-Andes: Cambrian continental collision in
the Sierras de Córdoba. In The Proto-Andean Margin of
Gondwana, edited by R.J. Pankhurst and C.W. Rapela, Geological
Society, London, Special Publication vol. 142, pp. 181–217.

Renard F, Dysthe D, Feder et al. 2001. Enhanced pressure solution
creep rates induced by clay particles: Experimental evidence in
salt aggregates. Geophysiscal Research Letters 28 (7), 1295–1298.

Rueda M, Delpino S, Grasemann et al. 2013. From sedimentary precursor to anatectic products, field, petrographic and geochemical constraints: Pringles Metamorphic Complex, Sierra de San
Luis, Argentina. In Avances en Mineralogía, Metalogenia y
Petrología 2013, pp. 349–356.

Sawyer E. 2008. Working with migmatites: Nomenclature for the
constituent parts. In Working with Migmatites, edited by E.
Sayer and M. Brown, Mineralogical Association of Canada,
Short Course Series vol. 38, pp. 1–28.

Scholz C. 1980. Shear heating and the state of stress on faults.
Journal of Geophysical Research 85, 6174–6184.

Shengelia D, Akhvlediani R, Ketskhoveli D. 1979. The graphite geothermometer. Doklady Akademii Nauk SSSR 235, 132–134.

Siegesmund S, Steenken A, Martino R, et al. 2010. Time constraints
on the tectonic evolution of the Eastern Sierras Pampeanas (central Argentina). International Journal of Earth Science
(Geologische Rundschau) 99, 1199–1226.

Simpson C, De Paor D. 1991. Deformation and kinematics of high
strain zones. Annual GSA Meeting, Structural and Tectonics
Division, 116 p. San Diego.

Sims J, Skirrow R, Stuart-Smith P, et al. 1997. Informe geológico y met-
alogénico de las Sierras de San Luis y Comechingones (Provincias
de San Luis y Córdoba), 1:250000. Anales XXVIII, Instituto de
Geología y Recursos Minerales, SEGEMAR, Buenos Aires.

Sims J, Ireland T, Camacho A, et al. 1998. U–Pb, Th–Pb and Ar–Ar
geochronolgy from the southern Sierras Pampeanas, Argentina:
implications for the Palaeozoic tectonic evolution of the western
Gondwana margin. In The Proto-Andean Margin of Gondwana,
edited by R. Pankhurst and C. Rapela, Geological Society
London, Special Publication vol. 142, pp. 259–281.

Spear F. 1995. Metamorphic Phase Equilibria and Pressure–
Temperature-Time Paths. Mineralogical Society of America,
Washington.

Steenken A, Siegesmund S, López deLuchi et al. 2006.
Neoproterozoic to early Palaeozoic events in the Sierra de San
Luis: implications for the Famatinian geodynamics in the
Eastern Sierras Pampeanas (Argentina). Journal of the Geological
Society of London 163, 965–982.

Steenken A, López de Luchi M, Martínez Dopico C, et al. 2011. The
Neoproterozoic-early Paleozoic metamorphic and magmatic
evolution of the Eastern Sierras Pampeanas: An Overview.
International Journal of Earth Sciences 100, 465–488.

Stipp M, Stunitz H, Heilbronner R, et al. 2002. The eastern Tonale
Fault Zone: a "natural laboratory" for crystal plastic deformation
of quartz over a temperature range from 250 to 700°C. Journal of
Structural Geology 24, 1861–1884. Stüwe K. 2002, Geodynamic
of the Lithosphere: An introduction. Springer Verlag, Berlin.

Trouw R, Passchier C, Wiersma DJ. 2010. Atlas of Mylonites – and
related microstructures. Springer-Verlag, Berlin.

Tullis J, Yund R. 1985. Dynamic recrystallization of feldspars: a
mechanism for ductile shear zone formation. Geology 13,
238–241.

Tullis J, Yund R. 1987. Transition from cataclastic flow to dislocation creep of feldspar: mechanisms and microstructures. Geology
15, 606–609.

Urai J. 1983. Water assisted dynamic recrystallization and weakening in polycrystalline bischofite. Tectonophysics 96, 125–127.

Vernon R. 2004. A practical guide to Rock Microstructure.
Cambridge University Press, New York.

von Gosen W. 1998. Transpressive deformation in the southwestern part of the Sierra de San Luis (Sierrras Pampeanas, Argentina). Journal of South American Earth Sciences 11(3), 233–264.

von Gosen W, Prozzi C. 1996. Geology, Structure, and Metamorphism in the Area South of La Carolina (Sierra de San Luis, Argentina). XIII Congreso Geolo´gico Argentino y III Congreso de Exploración de Hidrocarburos, Actas 2, 301–314, Buenos Aires.

von Gosen W, Prozzi C. 1998. Structural evolution of the Sierra de San Luis (Eastern Sierras Pampeanas, Argentina): implications for the proto-Andean margin of Gondwana. In The Proto-Andean Margin of Gondwana, edited by R. Pankhurst and C. Rapela, Geological Society, London, Special Publication vol. 142, pp. 235–258.

von Gosen W, Loske W, Prozzi C. 2002. New isotopic dating of intrusive rocks in the Sierra de San Luis (Argentina): implications for the geodynamic history of the Eastern Sierras Pampeanas. Journal of South American Earth Sciences 15 (2), 237–250.

Yardley B. 1989. An Introduction to Metamorphic Petrology. Longman Scientific & Technical, New York.

Chapter 15
Strike–slip ductile shear zones in Thailand

PITSANUPONG KANJANAPAYONT

Department of Geology, Faculty of Science, Chulalongkorn University, Bangkok 10330, Thailand

15.1 INTRODUCTION

Mainland south-east Asia includes development of numerous intraplate strike–slip deformations. These intense deformations confine show in narrow and sub-parallel sided zones, which normally are termed as "shear zones" (Ramsay 1980). Shear zones can reveal brittle, brittle–ductile, and ductile deformations. Many brittle shear zones are also described as fault zone, where rocks in the shear planes could be brecciated. The brittle–ductile shear zones are associated with fault rocks with some ductile deformation, while the ductile shear zones are commonly in the high plastic deformed rocks or mylonites (Ramsay 1980; White et al. 1980).

Strike–slip ductile shear zones in Thailand are the Mae Ping shear zone, the Three Pagodas shear zone, the Ranong shear zone, and the Khlong Marui shear zone (Fig. 15.1). In the last few years, geological studies on the strike–slip shear zones in Thailand have been carried out by a number of workers (Watkinson et al. 2008, 2011; Morley et al. 2011; Kanjanapayont et al. 2012a, b; Nantasin et al. 2012; Palin et al. 2013) to understand their genesis and kinematics. The whole system of strike–slip shear zones in Thailand requires a review. First, I present an overview tectonics of Thailand. I also present the meso- and micro-structural analyses, plus previous geochronological data in the best exposed mylonite shear zones in Thailand.

15.2 TECTONIC FRAMEWORK OF THAILAND

The strike–slip shear zones in Thailand are situated in the Sibumasu terrane, which is a western part of Thailand (e.g. Ridd 1971; Tapponnier et al. 1986; Leloup et al. 1995). The name "SIBUMASU" comes from SI (Siam), BU (Burma= now Myanmar), MA (Malaya), and SU (Sumatra) (Metcalfe 1984). The terrane includes the NW and peninsular Thailand, the Shan States of Myanmar, western Malaya, and Sumatra. It is bound at E by sutures: the Changning–Menglian suture in SW China, the Chiang Mai–Inthanon suture in Thailand, and the Bentong–Raub suture in Peninsular Malaysia, and to the W by the Mogok Metamorphic belt, the Andaman Sea, and the medial Sumatra tectonic zone (Barber and Crow 2009). The Sibumasu have widely used as "Shan-Thai" in the past (e.g. Bunopas 1992; Charusiri et al. 2002; Hirsch et al. 2006). The Sibumasu terrane was derived from the Indian–Australian margin of eastern Gondwana based on: (1) Cambrian–Early Permian Gondwanaland faunas (Archbold et al. 1982; Burrett and Stait 1985; Metcalfe 1991, 1994; Shi and Waterhouse 1991); (2) Late Carboniferous–Early Permian glacial–marine diamictites (Stauffer and Lee 1989; Ampaiwan et al. 2009); and (3) southern paleolatitudes of Paleozoic paleomagnetic data (Bunopas 1981; Fang et al. 1989; Huang and Opdyke 1991). The separation of the Sibumasu terrane from Gondwana happened during the Early Permian, before the collision of the Sibumasu with the Sukhothai Arc and Indochina in the Late Triassic gave rise to the Indosinian Orogeny (Metcalfe 2011, 2013). This collision was followed by the Late Triassic to Early Jurassic S–Type granite emplacement in W and central Thailand. The intraplate strike–slip activity occurred during the Late Cretaceous to Early Palaeogene, and it was already present when the India–Asia collision affected this terrain (Morley 2012). This strike–slip system may have controlled the Cenozoic rift basin development in both onshore and the Gulf of Thailand (Polachan et al. 1991; Morley 2012).

15.3 STRUCTURAL GEOLOGY OF THESTRIKE–SLIP DUCTILE SHEAR ZONES

15.3.1 Mae Ping shear zone

The Mae Ping shear zone was described as the Wang Chao shear zone by Lacassin et al. (1993, 1997). This NW–SE trending strike–slip shear zone has a structure that parallels southward to the Ailao Shan–Red River shear zone in China and Vietnam (Fig. 15.1). The shear system seems to be the branch of the Sagaing fault in Myanmar, and transects the N–S Chiang Mai–Lincang metamorphic belt along the Mae Ping river in western Thailand. This shear zone was proposed to splay into the

Ductile Shear Zones: From Micro- to Macro-scales, First Edition. Edited by Soumyajit Mukherjee and Kieran F. Mulchrone.

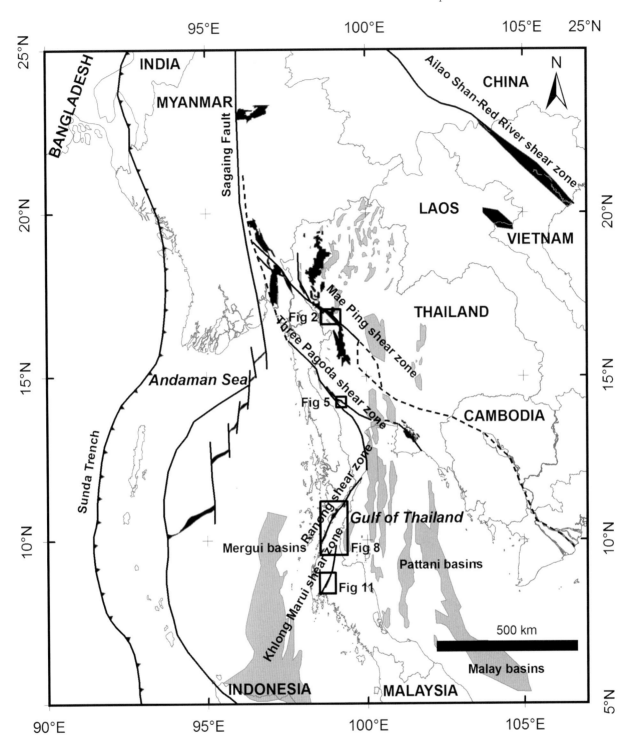

Fig. 15.1. Regional tectonic framework in Thailand showing the major strike–slip ductile shear zones in Thailand and related structures. Black, metamorphic complex; gray, Cenozoic basins. Box refers to Fig.15.2, 15.5, 15.8, and 15.11. Adapted from Macdonald et al. 2010; Mitchell et al. 2012; Morley, 2002 and Polachan and Sattayarak, 1989

Chainat Duplex in central Thailand, connect the KhaoYai Fault on the S margin of the Khorat Plateau, and passing into Cambodia (Morley 2002; Morley et al. 2007; Smith et al. 2007; Ridd and Morley 2011).

The Mae Ping shear zone contains the mylonite outcrops of the high-grade metamorphics named Lansang Gneiss, which is best preserved at the Lansang National Park, Tak province. The mylonites in the shear zone are ~5 km wide, and consists of paragneisses, orthogneisses, and mica-schist (Lacassin et al. 1993, 1997) (Fig. 15.2). The Mae Ping shear zones sheared sinistrally during the latter stages or after the greenschist to

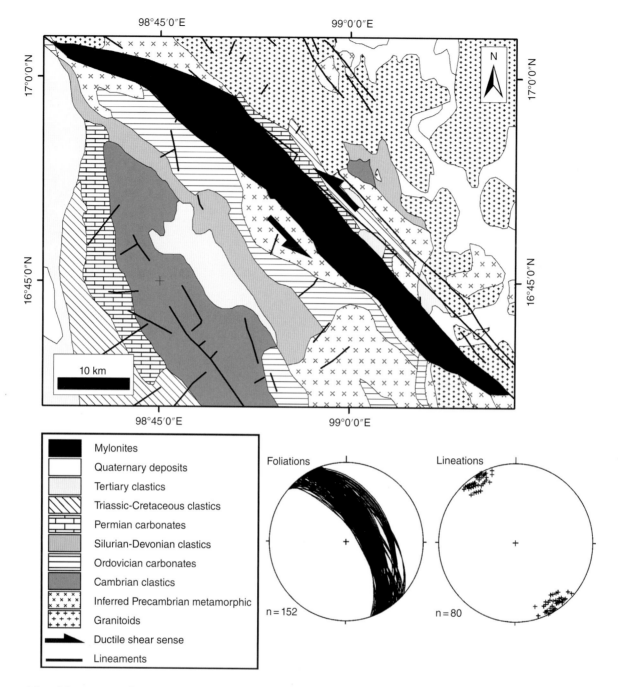

Fig. 15.2. Map of the Mae Ping shear zone in western Thailand with stereographic plots of the mylonitic foliations and stretching lineations. Adapted from Department of Mineral Resources (1982) Geological map of Thailand. Department of Mineral Resources.

amphibolite metamorphism (Lacassin et al. 1993, 1997; Palin et al. 2013). Stereographic plots illustrate that the mylonitic foliations strike NW–SE dip steeply, and NW–SE stretching lineations are most prolific (Fig. 15.2).

The kinematic indictors within the Mae Ping shear zone have a strongly defined sinistral shear. The mylonitic foliations and stretching lineations are clearly preserved in the mylonites. The σ-object (Fig. 15.3a–c), S–C (Fig. 15.3b), and S–C' fabrics are usually shown by the gneissic mylonite. Asymmetric fold (Fig. 15.3e),

δ-object (Fig. 15.3d), shear band (Fig. 15.3f,g) and domino type (Fig. 15.3h) (Passchier and Trouw 2005) can be observed in calc-silicate mylonite.

The microstructures include the σ-object (Fig. 15.4a–c,e), strain shadow (Fig. 15.4b), mica fish (Fig. 15.4d,g), stair steps (Fig. 15.4e,f), S–C (Fig. 15.4c), and S–C' fabrics. The large mica fish (Fig. 15.4d) can be found in the gneissic mylonite, which is sigmoidal (Mukherjee 2011). Undulose extinction, basal gliding, bulging, and grain boundary migration (Fig. 15.4h) are generally present along the oblique foliation.

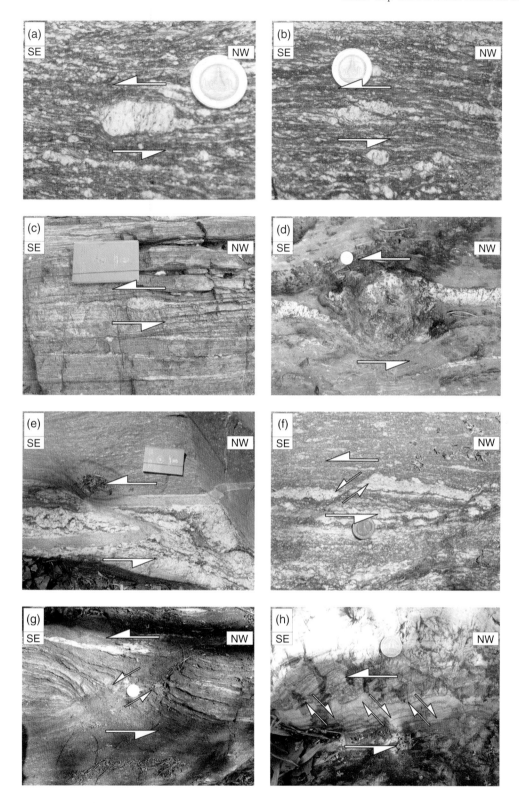

Fig. 15.3. The outcrops showing the mesostructures of the rocks within the Mae Ping shear zone; (a) σ-object of K-feldspar, (b) σ-object and SC fabric, (c) σ-object along the foliation, (d) δ-object, (e) asymmetric fold, (f) shear band type in gneissic mylonite, (g) shear band type in calc-silicate mylonite, (h) domino type or flanking structure in calc-silicate mylonite.

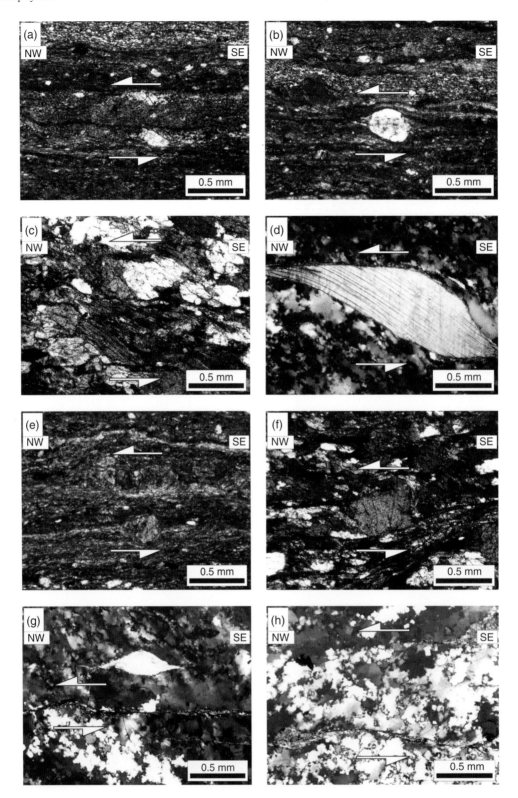

Fig. 15.4. Photomicrographs showing the microstructures of the rocks within the Mae Ping shear zone; (a) σ-object of K-feldspar, (b) K-feldspar with strain shadow; (c) SC fabric, (d) large mica fish, (e) σ-object and stair stepping, (f) stair stepping in gneissic mylonite, (g) mica fish and oblique foliation, (h) oblique foliation and grain boundary migration of quartz.

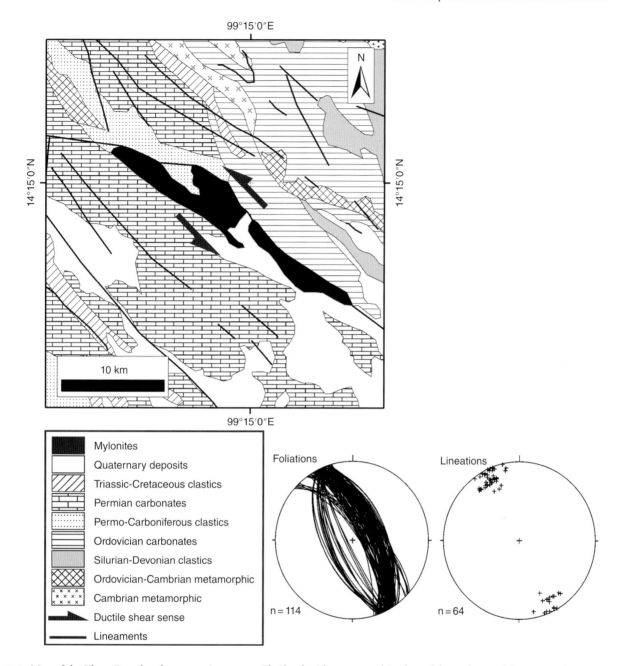

Fig. 15.5. Map of the Three Pagodas shear zone in western Thailand with stereographic plots of the mylonitic foliations and stretching lineations. Adapted from Department of Mineral Resources (1982) Geological map of Thailand. Department of Mineral Resources.

15.3.2 Three Pagodas shear zone

The Three Pagodas shear zone is a NW–SE trending strike–slip structure paralleled southward to the Mae Ping shear zone. This shear zone is located at W Thailand from the Three Pagodas Pass on the Thai–Myanmar border, towards central Thailand. This strike-slip zone was proposed to splay into an E–W zone running through the central Thailand and pass into the NW–SE Klaeng fault zone in the eastern Thailand, based on magnetic anomaly profiles (Morley 2002). Other splays run continuously to the NW–SE and N–S southward to join the NNE–SSW Ranong strike–slip zone in the Peninsular Thailand (Ridd 2012).

The mylonites of the Three Pagodas shear zone includes a narrow ~250 km long and ~25 km wide lenticular slice of the high-grade basement metamorphic rocks named the Thabsila Metamorphic Complex (Nantasin et al. 2012) (Fig. 15.5). This high-grade metamorphic complex consists of four units, based on lithology and structural features: (1) Unit A: marble, mica schist and quartzite, (2) Unit B: mylonites, (3) Unit C: calc-silicate, and (4) Unit D: various gneisses (Nantasin et al. 2012). Based on petrography and mineral chemistry, the Thabsila Metamorphic Complex was found to have a single metamorphic event under similar pressure–temperature (*P–T*) conditions ranging from upper amphibolite to lower

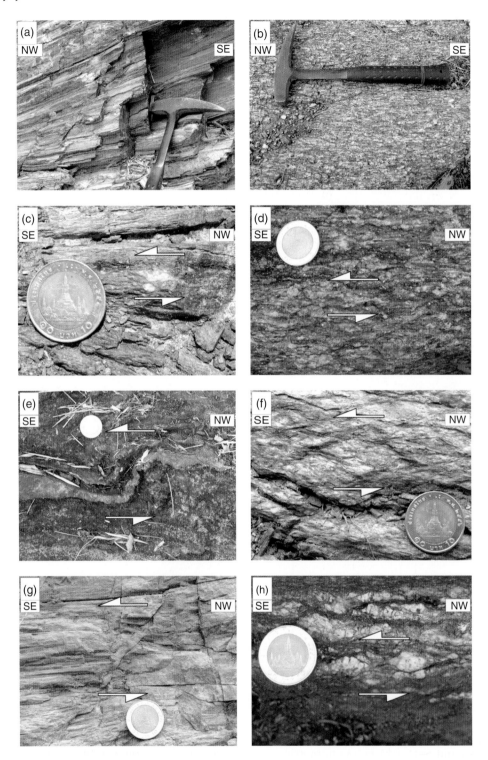

Fig. 15.6. The outcrops showing the mesostructures of the rocks within the Three Pagodas shear zone; (a) mylonitic foliations and stretching lineations in quartz mylonite, (b) stretching lineations in granite mylonite, (c) σ-object of K-feldspar, (d) σ-object and SC fabric in granite mylonite, (e) asymmetric fold, (f) SC' fabric, (g) shear band type in quartz mylonite, (h) shear band type in granite mylonite.

granulite facies (Nantasin et al. 2012). This metamorphic complex constricted, exhumed, and sheared sinistrally (Nantasin et al. 2012). The mylonitic foliations and stretching lineations in the stereographic plots indicate they trend NW–SE, trending with the steep dip (Fig. 15.5).

Mylonite outcrops within the Three Pagodas shear zone record sinistral strike–slip kinematic indictors. The mylonitic foliations and stretching lineations (Fig. 15.6a,b) are clearly presented by the quartz mylonite. The σ-object (Fig. 15.6c), S–C (Fig. 15.6d), and S–C' fabrics are noted from granite mylonite. Asymmetric fold preserved in the calc-silicate mylonite (Fig. 15.6e), while shear bands (Fig. 15.6f,g) can be observed in both quartz mylonite and granite mylonite.

Shear sense given by asymmetric microstructures matches with that indicated by mesoscale structures in the granite mylonite. The σ-object (Fig. 15.7a–c), strain shadow (Fig. 15.7a), mica fish (Fig. 15.7d–f), and stair stepping can be clearly found along the oblique foliation, S–C (Fig. 15.7f) and S–C' fabrics. Steep sigmoid fish and lenticular fish (Mukherjee 2011) are the morphological types of the mica fish in this shear zone. Undulose extinction (Fig. 15.7g), basal gliding, and grain bulging (Fig. 15.7h) connote dynamic crystallization.

15.3.3   Ranong shear zone

In contrast to the Mae Ping and Three Pagodas shear zones, the Ranong shear zone is a NNE–SSW trending strike–slip system that runs from the Gulf of Thailand through the Ranong province in southern Thailand into the Andaman Sea. The Ranong shear zone is represented by the widespread exposures of the migmatite, foliated granite, orthogneiss, and quartz–biotite mylonite, which have been metamorphosed at greenschist to amphibolite facies (Watkinson et al. 2008, 2011) (Fig. 15.8). The kinematic deformation of the Ranong shear zone involves the ductile dextral shear and the dextral transpression causing exhumation (Watkinson et al. 2008). The mylonitic foliations strike NNE–SSW and the stretching lineations trend dominantly NNE–SSW are indicated by the stereographic plots (Fig. 15.8).

The exposures of the mylonites illustrate the typical mylonitic foliations (Fig. 15.9a,b), stretching lineations, σ-object (Fig. 15.9c–e), and strain shadow (Fig. 15.9f). Some K-feldspars present high strain with long σ-object (Fig. 15.9c–e). Asymmetric fold (e.g. Aller et al. 2010) (Fig. 15.9g) and shear band (Fig. 15.9h) are found commonly within the granite mylonite. Kinematic indicators of the mylonites within the Ranong shear zone define a dextral shear.

Microstructures of the Ranong shear zone typically include σ-object (Fig. 15.10a–c) of K-feldspar and quartz, strain shadow, mica fish, and stair stepping of mylonitic foliation. The mica fish (Fig. 15.10d, e) can be found within the granite mylonite, and generally associate with S–C and S–C' fabrics. Asymmetric myrmerkite (Fig. 15.10f) of quarter structure and "V"–pull-apart structure (e.g. Hippertt 1993; Mukherjee 2010, 2014a; Mukherjee and Koyi 2010) (Fig. 15.10g) are associated with the K-feldspar grains in the granite mylonite. Undulose extinction, basal gliding, grain-bulging (Fig. 15.10h), and subgrain rotation along the oblique foliation of quartz indicated the dynamic recrystallization in this shear zone (Passchier and Trouw 2005).

15.3.4   Khlong Marui shear zone

The southernmost Khlong Marui shear zone trends NNE–SSW, next to the Ranong shear zone in the southern direction. The strike–slip system affected peninsular Thailand, trending NNE–SSW and stretching ~150 km from the Gulf of Thailand to the Andaman Sea (e.g. Garson and Mitchell 1970; Ridd 1971; Kanjanapayont et al. 2009). The mylonitic shear zone marked by the lozenge-shaped 40×6 km Khao Phanom mountain range (Charusiri 1989; Watkinson et al. 2008; Kanjanapayont et al. 2012a) (Fig. 15.11). The lithologies mainly consist of migmatites, orthognesisses, granite mylonites, quartz–biotite mylonites, and sheared sedimentary rocks (Watkinson et al. 2008), or metasedimentary mylonites associated with orthogneisses, granite mylonites, and pegmatitic veins (Kanjanapayont et al. 2012a). The rocks in this shear zone have been metamorphosed at greenschist to amphibolite facies (Watkinson et al. 2011; Kanjanapayont et al. 2012a). The ductile shearing of the Khlong Marui shear zone started with a dextral movement (Watkinson et al. 2008; Kanjanapayont et al. 2012a). Two possibilities of the later deformations, which are either Eocene sinistral brittle transpression, with exhumation followed by the dextral shear (Watkinson et al. 2008), or dextral ductile transpression causing exhumation in the Eocene (Kanjanapayont et al. 2012a). Stereographic plots show that the mylonitic foliations strike NNE–SSW and dip steeply towards either WNW or ESE, and the stretching lineations have a clearly defined maximum trend at NNE and SSW (Fig. 15.11).

Mylonites within the Khlong Marui shear zone deformed non-coaxially. Abundant spectacular monoclinic quartz lenses and porphyroclasts (Fig. 15.12a–f) indicate dextral movement, the same as those given by asymmetric fold (Fig. 15.12g). These quartz lenses and porphyroclasts frequently appear in all rocks in this shear zone. The σ-objects (Fig. 15.12b–f) of quartz and K-feldspar grains, with strong mylonitic foliations and stretching lineations (Fig. 15.12a), generally indicate high strain. The shear band (Fig. 15.12h) and domino types (Passchier and Trouw 2005) are probably relics of stretched and rotated quartz and pegmatitic veins.

Microstructures of mylonitized metasedimentary rocks show different deformation structures (Fig. 15.13a–h). Mica fish (Fig. 15.13c–d), stair steps (Fig. 15.13e) of mylonitic foliation, σ-objects (Fig. 15.13a,b) of quartz, and K-feldspar grains indicate dextral shear. Micas dominantly represent either the S–C (Fig. 15.13d) or the S–C' (Fig. 15.13c) fabrics. Asymmetric myrmerkite (Fig. 15.13f) of quarter structure and "V"-pull-apart structure (Fig. 15.13g) in the granite mylonite indicate dextral shear. Dynamic recrystallization such as undulose extinction, basal gliding, bulging (Fig. 15.13h), subgrain rotation, and grain boundary migration associated with oblique foliation, can be observed in quartz-rich layers in mylonites.

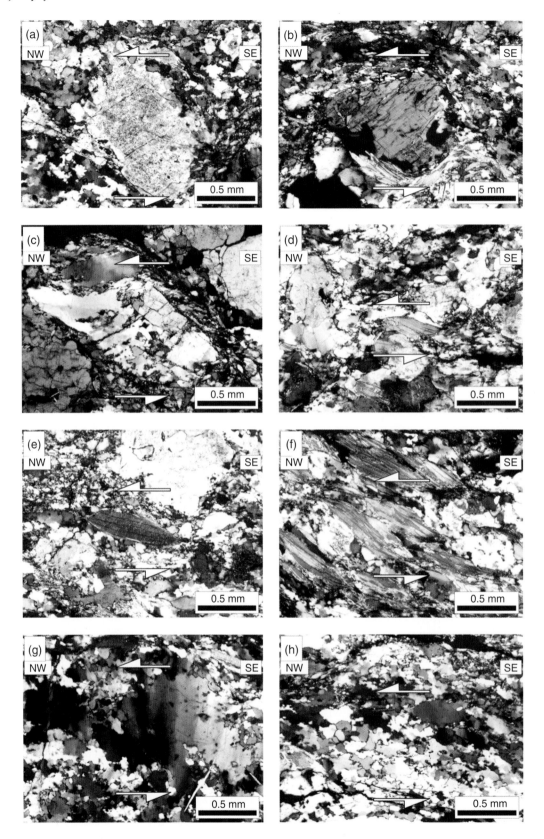

Fig. 15.7. Photomicrographs showing the microstructures of the rocks within the Three Pagodas shear zone; (a) K-feldspar with strain shadow, (b) σ-object of K-feldspar, (c) σ-object of quartz, (d) mica fish in granite mylonite, (e) lenticular mica fish, (f) mica fish and SC fabric, (g) undulose extinction and oblique foliation, (h) oblique foliation and bulging of quartz.

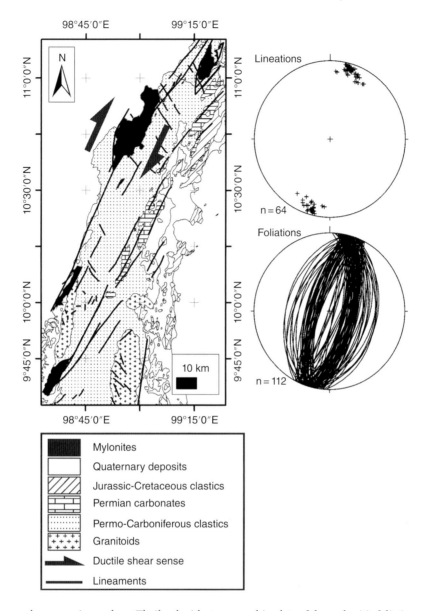

Fig. 15.8. Map of the Ranong shear zone in southern Thailand with stereographic plots of the mylonitic foliations and stretching lineations. Adapted from Department of Mineral Resources 1982 and Watkinson et al. 2011.

15.4 GEOCHRONOLOGY

Strike–slip ductile shear zones in Thailand have been dated by various methods. The K–Ar biotite and illite, ^{40}Ar/^{39}Ar biotite and K-feldspar, and U–Pb monazite methods were done with rocks from the Mae Ping shear zone (Ahrendt et al. 1993; Lacassin et al. 1997; Palin et al. 2013). Dating of K–Ar biotite, ^{40}Ar/^{39}Ar biotite, Rb–Sr biotite, and U–Pb zircon were performed within the Three Pagodas shear zone (Bunopas 1981; Lacassin et al. 1997; Nantasin et al. 2012). The ^{40}Ar/^{39}Ar muscovite, biotite, hornblende, and U–Pb zircon dating systems were applied to the Ranong shear zone (Watkinson et al. 2011), while the ^{40}Ar/^{39}Ar muscovite and biotite, Rb–Sr white mica and biotite, and U–Pb zircon dating were done within the Khlong Marui shear zone (Charusiri 1989;

Watkinson et al. 2011; Kanjanapayontet al. 2012b). The geochronology data of the mylonites within the strike–slip ductile shear zones in Thailand are summarized in Table 15.1.

15.5 DISCUSSION

15.5.1 Deformation style

The mesostructures of all shear zones contain kinematic indicators, such as asymmetric fold, σ-object, shear band, S–C and S–C' fabrics (Table 15.2). Domino type (Passchier and Trouw 2005) or flanking structures (Grasemann and Stüwe 2001; Grasemann et al. 2003; Mukherjee and Koyi 2009; Mukherjee, 2014b) and δ-objects only appear

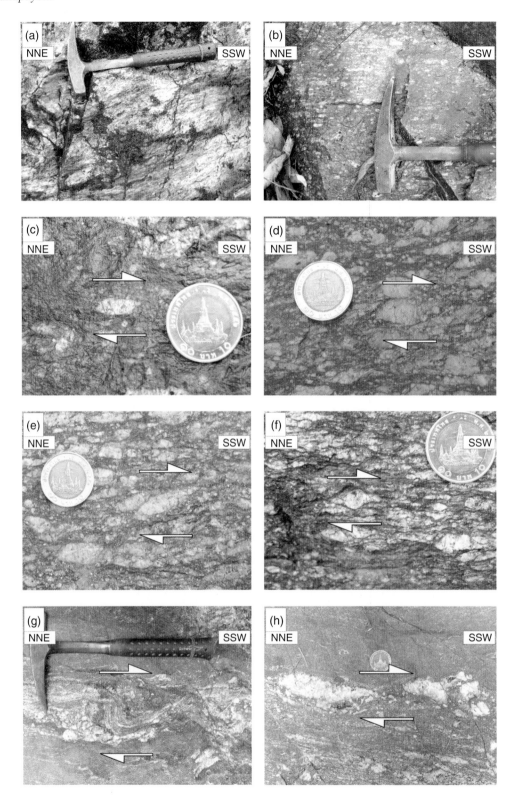

Fig. 15.9. The outcrops showing the mesostructures of the rocks within the Ranong shear zone; (a) mylonitic foliations and stretching lineations in quartz mylonite, (b) stretching lineations in granite mylonite, (c) σ-object of K-feldspar, (d) σ-object and SC fabric, (e) long σ-object in granite mylonite, (f) K-feldspar with strain shadow, (g) asymmetric fold, (h) shear band type in granite mylonite.

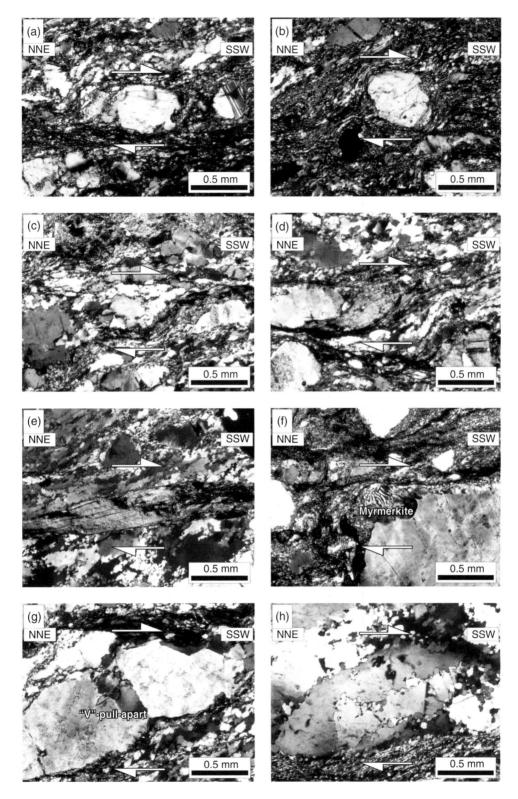

Fig. 15.10. Photomicrographs showing the microstructures of the rocks within the Ranong shear zone; (a) σ-object of K-feldspar, (b) K-feldspar with strain shadow, (c) σ-object of quartz, (d) mica fish in granite mylonite, (e) long mica fish, (f) asymmetric mymerkite, (g) "V"-pull–apart structure, (h) oblique foliation and bulging of quartz.

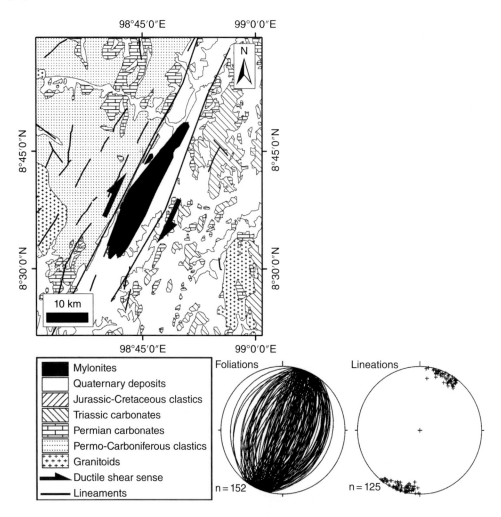

Fig. 15.11. Map of the Khlong Marui shear zone in southern Thailand with stereographic plots of the mylonitic foliations and stretching lineations. Adapted from Department of Mineral Resources (1982) Geological map of Thailand. Department of Mineral Resources.

within the MaePing shear zone. Mylonitic foliations of the Mae Ping and Three Pagodas shear zones strike NW–SE, steeply dipping, and the stretching lineations dominantly trend NW and SE. On the other hand, the Ranong and Khlong Marui shear zones present NNE–SSW stretching lineations.

The microstructures include the σ-objects, strain shadows, mica fish, stair stepping, oblique foliations, shear band type, S–C and S–C' fabrics (Table 15.2). Secondary C' shear indicates a pure shear component along with simple shear (as reviewed in Mukherjee 2013a, b, 2014b). Therefore this shear zone possibly underwent a general shear/sub-simple shear. Asymmetric folds and δ-objects were not encountered in the microscale, while asymmetric myrmerkite of quarter structure and "V"-pull-apart structures were found from the Ranong and Khlong Marui shear zones. Undulose extinction, basal gliding, bulging, subgrain rotation, and grain boundary migration were also documented.

Both meso- and microstructures indicate the sinistral movement within the Mae Ping and Three Pagodas shear zones, and the dextral shear within the Ranong and

Khlong Marui shear zones (Fig. 15.14). This shearing event possibly coincided with the sinistral strike–slip ductile movement of the Chinat Duplex in central Thailand (Prasongtham and Kanjanapayont 2014) and the Klaeng fault zone in eastern Thailand (Kanjanapayont et al. 2013). Furthermore, it may be related to the WNW–ESE to E–W contractional deformation of the Mesozoic sedimentary rocks in southern Thailand (Kanjanapayont 2014).

15.5.2 Timing of shear

The first possibility to have a timing of shear wasin the Late Cretaceous for the Ranong shear zone (Watkinson et al. 2011), and the summary of geochronology data strongly present the Paleogene activity along the strike–slip ductile shear zones in Thailand (Fig. 15.15). U–Pb zircon and monazite dating suggest the metamorphism in the Eocene (Kanjanapayont et al. 2012b; Nantasin et al. 2012; Palin et al. 2013), while the younger ages of K–Ar, ^{40}Ar/^{39}Ar, and Rb–Sr methods deduce the time of the strike–slip deformation (Ahrendt et al. 1993; Lacassin

Fig. 15.12. The outcrops showing the mesostructures of the rocks within the Khlong Marui shear zone; (a) mylonitic foliations and stretching lineations in meta-sedimentary mylonite, (b) abundant σ-objects in meta-sedimentary mylonite, (c) σ-object of quartz porphyroclast, (d) σ-object quartz lenses along the foliation, (e) σ-objectin metasedimentary mylonite, (f) long σ-object in metasedimentary mylonite, (g) asymmetric fold, (h) shear band type.

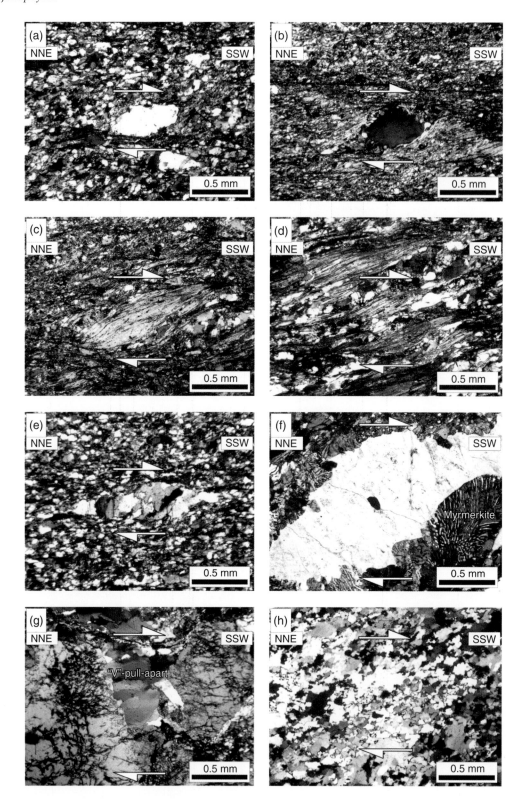

Fig. 15.13. Photomicrographs showing the microstructures of the rocks within the Khlong Marui shear zone; (a) σ-object of K-feldspar, (b) σ-object of quartz, (c) mica fish and SC' fabric, (d) abundant mica fish and SC fabric, (e) σ-object and stair stepping, (f) asymmetric mymerkite, (g) "V"-pull-apart structure, (h) oblique foliation, bulging, and subgrain rotation of quartz.

Table 15.1. Summary of the geochronology data of the rocks within the strike–slip ductile shear zones in Thailand

Shear zone	Method	Mineral	Age (Ma)	References
Mae Ping	K–Ar	biotite	29.9 ± 0.6	Ahrendt et al. 1993
			31.0 ± 0.6	
			31.9 ± 1.6	
	K–Ar	illite	29.6 ± 1.2	Ahrendt et al. 1993
			30.5 ± 1.0	
	^{40}Ar/^{39}Ar	biotite	30.6 ± 0.3	Lacassin et al. 1997
			31.3 ± 0.7	
			33.0 ± 0.2	
			33.1 ± 0.4	
	^{40}Ar/^{39}Ar	K-feldspar	30.5	Lacassin et al. 1997
	U–Pb	monazite	37.1 ± 1.5	Palin et al. 2013
			39.5 ± 3.7	
			41.0 ± 1.9	
			41.1 ± 8.8	
			41.2 ± 1.4	
			41.3 ± 1.0	
			43.5 ± 3.6	
			44.5 ± 6.1	
Three Pagodas	K–Ar	biotite	33 ± 2	Bunopas 1981
			36 ± 1	
	^{40}Ar/^{39}Ar	biotite	33.4 ± 0.4	Lacassin et al. 1997
	Rb–Sr	biotite	31.78 ± 0.32	Nantasin et al. 2012
			36.05 ± 0.36	
	U–Pb	zircon	51 ± 7	Nantasin et al. 2012
			57 ± 1	
Ranong	^{40}Ar/^{39}Ar	muscovite	42.35 ± 0.46	Watkinson et al. 2011
	^{40}Ar/^{39}Ar	biotite	41.41 ± 0.45	Watkinson et al. 2011
			41.84 ± 0.48	
			42.85 ± 0.68	
			44.88 ± 0.51	
			46.09 ± 0.55	
			49.43 ± 0.61	
			51.16 ± 0.65	
			58.74 ± 0.62	
	^{40}Ar/^{39}Ar	hornblende	43.99 ± 0.51	Watkinson et al. 2011
			88.12 ± 1.12	
	U–Pb	zircon	47.6 ± 0.8	Watkinson et al. 2011
			71.0 ± 0.7	
			80.5 ± 0.6	
Khlong Marui	^{40}Ar/^{39}Ar	muscovite	41.3	Charusiri 1989
			42.6 ± 0.5	
	^{40}Ar/^{39}Ar	muscovite	41.10 ± 0.26	Watkinson et al. 2011
			43.58 ± 0.52	
	^{40}Ar/^{39}Ar	biotite	37.11 ± 0.31	Watkinson et al. 2011
			37.47 ± 0.28	
			40.33 ± 0.47	
			41.84 ± 0.46	
			41.40 ± 0.50	
	Rb–Sr	white mica	22.6 ± 0.2	Kanjanapayont et al. 2012b
			38.3 ± 0.4	
	Rb–Sr	biotite	33 ± 0.3	Kanjanapayont et al. 2012b
			36 ± 0.4	
			35.9 ± 0.4	
	U–Pb	zircon	45.6 ± 0.7	Kanjanapayont et al. 2012b
			49 ± 1	
			50.9 ± 0.6	
			55 ± 3	

Table 15.2. Summary of the meso- and microstructures of the strike–slip ductile shear zone in Thailand.

	Shear zone			
	Mae Ping	Three Pagodas	Ranong	Khlong Marui
Mesostructures				
Asymmetric fold	X	X	X	X
σ-object	X	X	X	X
δ-object	X			
SC fabric	X	X	X	X
SC' fabric	X	X	X	X
Shear band type	X	X	X	X
Domino type	X			X
Microstructures				
Asymmetric fold				
σ-object	X	X	X	X
δ-object				
Strain shadow	X	X	X	X
Mica fish	X	X	X	X
SC fabric	X	X	X	X
SC' fabric	X	X	X	X
Oblique foliation	X	X	X	X
Stair stepping	X	X	X	X
Asymmetric mymerkite			X	X
"V"-pull-apart structure			X	X

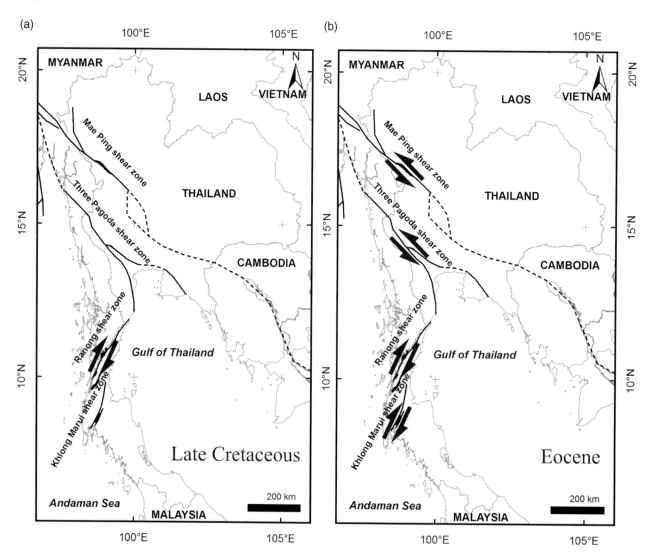

Fig. 15.14. Schematic diagram showing the deformation pattern of the strike–slip ductile shear zones in Thailand with respect to the mylonites in (a) the Late Cretaceous and (b) the Eocene.

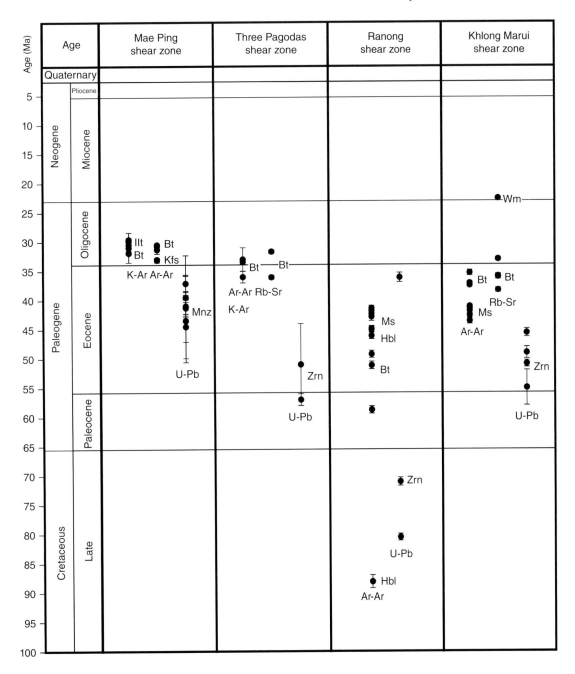

Fig. 15.15. Summary of age constraints for periods of shear along the Mae Ping, Three Pagodas, Ranong, and Khlong Marui shear zones. The data are taken from Table 15.1.

et al. 1997; Watkinson et al. 2011; Kanjanapayont et al. 2012b; Palin et al. 2013), or the exhumation (Kanjanapayont et al. 2012b; Nantasin et al. 2012). However, the major strike–slip and the apatite fission track ages give the Late Oligocene–Early Miocene cooling patterns along the western Thailand, which is unrelated to the shear zones (Upton 1999; Morley 2007). Therefore, the age constraints for periods of shear is the Late Cretaceous and the Eocene along the Ranong shear zone, and the Eocene–Oligocene along the Mae Ping, Three Pagodas, and Khlong Marui shear zones.

15.6 CONCLUSIONS

The meso- and microstructures of the mylonites within the strike–slip shear zones in Thailand include asymmetric folds, σ-objects, δ-objects, strain shadow, S–C and S–C' fabrics, shear band types, domino types, mica fish, stair stepping, asymmetric mymerkites, and "V"-pull-apart structures. The dynamic recrystallization in the shear zones were expressed by the undulose extinction, basal gliding, bulging, subgrain rotation, and grain boundary migration. Rocks deformed under the ductile strike–slip motion. The NW–SE Mae Ping and Three Pagodas shear zones deformed by sinistral movement, while the NNE–SSW Ranong and Khlong Marui shear zones have undergone dextral motion. Geochronology data indicates Late Cretaceous timing for the Ranong shear zone, and Eocene–Oligocenefor all four strike–slip zones. These ages suggest the timing of shear.

ACKNOWLEDGMENTS

This research was granted by the Ratchadaphiseksomphot Endowment Fund, Chulalongkorn University, and the Thailand Research Fund (TRF) TRG5780235. Montri Choowong and Punya Charusiri are thanked for discussions and very fruitful comments. Reviews by anonymous reviewers and editing plus reviewing by Soumyajit Mukherjee are acknowledged.

REFERENCES

Ahrendt H, Chonglakmani C, Hansen BT, Helmcke D. 1993. Geochronological cross section through northern Thailand. Journal of Southeast Asian Earth Sciences 8, 207–217.

Aller J, Bobillo-Ares NC, Bastida F, Lisle RJ, Menendez CO. 2010. Kinematic analysis of asymmetric folds in competent layers using mathematical modelling. Journal of Structural Geology 32, 1170–1184.

Ampaiwan T, Hisada K, Charusiri C. 2009. Lower Permian glacially influenced deposits in Phuket and adjacent islands, peninsular Thailand. Island Arc 18, 52–68.

Archbold NW, Pigram CJ, Ratman N, Hakim S. 1982. Indonesian Permian brachiopod fauna and Gondwana– South-East Asia relationships. Nature 296, 556–558.

Barber AJ, Crow MJ. 2009. The structure of Sumatra and its implications for the tectonic assembly of Southeast Asia and the destruction of Paleotethys. Island Arc 18, 3–20.

Bunopas S. 1981. Paleogeographic history of western Thailand and adjacent parts of Southeast Asia: A plate tectonic interpretation. PhD thesis, Victoria University of Wellington, New Zealand.

Burrett C, Stait B. 1985. South–East Asia as part of an Ordovician Gondwanaland. Earth and Planetary Science Letters 75, 184–190.

Charusiri P. 1989. Lithophile metallogenetic epochs of Thailand: A geological and geochronological investigation. PhD thesis, Queen's University, Kingston, Canada.

Charusiri P, Daorerk V, Archibald D, Hisada K, Ampaiwan T. 2002. Geotectonic evolution of Thailand: A new synthesis. Journal of the Geological Society of Thailand 1, 1–20.

Department of Mineral Resources. 1982. Geological map of Thailand. Department of Mineral Resources, Bangkok, scale 1:1,000,000.

Fang W, Van Der Voo R, Liang Q. 1989. Devonian palaeomagnetism of Yunnan province across the Shan Thai–South China suture. Tectonics 8, 939–952.

Garson MS, Mitchell AHG. 1970. Transform faulting in the Thai Peninsula. Nature 22, 45–47.

Grasemann B, Stüwe K. 2001. The development of flanking folds during simple shear and their use as kinematic indicators. Journal of Structural Geology 23, 715–724.

Grasemann B, Stüwe K, Vannay J. 2003. Sense and non-sense of shear in flanking structures. Journal of Structural Geology 25, 19–34.

Hippertt JFM. 1993. "V"-pull-apart microstructures: a new shear sense indicator. Journal of Structural Geology 15, 1393–1403.

Hirsch, F., Ishida, K., Kozai, T, Meesook, A. 2006. The welding of Shan–Thai. Geosciences Journal 10 (3), 195–204.

Huang K, Opdyke ND. 1991. Paleomagnetic results from the Upper Carboniferous of the Shan–Thai–Malay block of western Yunnan, China. Tectonophysics 192, 333–344.

Kanjanapayont P. 2014. Deformation style of the Mesozoic sedimentary rocks in southern Thailand. Journal of Asian Earth Sciences 92, 1–9.

Kanjanapayont P, Edwards MA, Grasemann B. 2009. The dextral strike-slip Khlong Marui Fault, southern Thailand.Trabajos de Geologia 29, 393–398.

Kanjanapayont P, Grasemann B, Edwards MA, Fritz H. 2012a. Quantitative kinematic analysis within the Khlong Marui shear zone, southern Thailand. Journal of Structural Geology 35, 17–27.

Kanjanapayont P, Klötzli U, Thöni M, Grasemann B, Edwards MA. 2012b. Rb–Sr, Sm–Nd, and U–Pb geochronology of the rocks within the Khlong Marui shear zone, southern Thailand. Journal of Asian Earth Sciences 56, 263–275.

Kanjanapayont P, Kieduppatum P, Klötzli U, Klötzli E, Charusiri P. 2013. Deformation history and U–Pb zircon geochronology of the high grade metamorphic rocks within the Klaeng fault zone, eastern Thailand. Journal of Asian Earth Sciences 77, 224–233.

Lacassin R, Leloup PH, Tapponnier P. 1993. Bounds on strain in large Tertiary shear zones of SE Asia from boudinage restoration. Journal of Structural Geology 15,677–692.

Lacassin R, Maluski H, Leloup PH, et al. 1997. Tertiary diachronic extrusion and deformation of western Indochina: Structure and $^{40}Ar/^{39}Ar$ evidence from NW Thailand. Journal of Geophysical Research 102 (B5), 10013–10037.

Leloup PH, Lacassin R, Tapponnier P, et al. 1995. The Ailao Shan–Red River shear zone (Yunnan, China), Tertiary transform boundary of Incochina. Tectonophysics 251, 3–84.

Macdonald AS, Barr SM, Miller BV, Reynolds PH, Rhodes BP, Yokart B. 2010. P–T–t constraints on the development of the Doi Inthanon metamorphic core complex domain and implications for the evolution of the western gneiss belt, northern Thailand. Journal of Asian Earth Sciences 37, 82–104.

Metcalfe I. 1984. Stratigraphy, palaeontology and palaeogeography of the Carboniferous of Southeast Asia. Memoires de la Societe Geologique de France 147, 107–118.

Metcalfe I. 1991. Late Palaeozoic and Mesozoic palaeogeography of Southeast Asia. Palaeogeography, Palaeoclimatology, Palaeoecology 87, 211–221.

Metcalfe I. 1994. Gondwanaland origin, dispersion, and accretion of East and Southeast Asian continental terranes. Journal of South American Earth Sciences 7, 333–347.

Metcalfe I. 2011. Tectonic framework and Phanerozoic evolution of Sundaland. Gondwana Research 19, 3–21.

Metcalfe I. 2013. Gondwana dispersion and Asian accretion: Tectonic and palaeogeographic evolution of eastern Tethys. Journal of Asian Earth Sciences 66, 1–33.

Mitchell A, Chung S, Oo T, Lin T, Hung C. 2012. Zircon U–Pb ages in Myanmar: magmatic-metamorphic events and the closure of a neo-Tethys ocean? Journal of Asian Earth Sciences 56, 1–23.

Morley CK. 2002. A tectonic model for the Tertiary evolution of strike-slip faults and rift basins in SE Asia. Tectonophysics 347, 189–215.

Morley CK. 2012. Late Cretaceous–Early Palaeogene tectonic development of SE Asia. Earth–Science Reviews 115, 37–75.

Morley CK, Charusiri P, Watkinson I. 2011. Chapter 11 Structural geology of Thailand during the Cenozoic. In The geology of Thailand, edited by MF Ridd, AJ Barber, and MJ Crow, The Geological Society of London, pp.273–334.

Morley CK, Smith M, Carter A, Charusiri P, Chantraprasert S. 2007. Evolution of deformation styles at a major restraining bend, constraints from cooling histories, Mae Ping fault zone, western Thailand. In Tectonics of Strike-slip Restraining and Releasing

Bends, edited by WD Cunningham and P Mann, The Geological Society of London, Special Publication vol. 290, pp. 325–349.

Mukherjee S. 2010. Macroscopic 'V' pull-apart structure in garnet. Journal of Structural Geology 32, 605.

Mukherjee S. 2011. Mineral fish: their morphological classification, usefulness as shear sense indicators and genesis. International Journal of Earth Sciences 100, 1303–1314.

Mukherjee S. 2013a. Channel flow extrusion model to constrain dynamic viscosity and Prandtl number of the Higher Himalayan Shear Zone. International Journal of Earth Sciences 102, 1811–1835.

Mukherjee S. 2013b. Deformation Microstructures in Rocks. Springer, Heidelberg.

Mukherjee S. 2014a Atlas of Shear Zone Structures in Meso-scale. Springer International Publishing, Cham.

Mukherjee S. 2014b. Review of flanking structures in meso-and micro-scales. Geological Magazine 151, 957–974.

Mukherjee S, Koyi HA. 2009. Flanking microstructures. Geological Magazine 146, 517–526.

Mukherjee S, Koyi HA. 2010. Higher Himalayan Shear Zone, Zanskar Indian Himalaya: microstructural studies and extrusion mechanism by a combination of simple shear and channel flow. International Journal of Earth Sciences 99, 1083–1110.

Nantasin P, Hauzenberger C, Liu X, et al. 2012. Occurrence of the high grade Thabsila metamorphic complex within the low grade Three Pagodas shear zone, Kanchanaburi Province, western Thailand: petrology and geochronology. Journal of Asian Earth Sciences 60, 68–87.

Palin RM, Searle MP, Morley CK, Charusiri P, Horstwood MSA, Roberts NMW. 2013. Timing of metamorphism of the Lansang gneiss and implications for left–lateral motion along the Mae Ping (Wang Chao) strike–slip fault, Thailand. Journal of Asian Earth Sciences 76, 120–136.

Passchier CW, Trouw RAJ. 2005. Microtectonics. Springer, Heidelberg.

Polachan S, Sattayarak N. 1989 Strike-slip tectonics and the development of Tertiary basins in Thailand. In Proceedings of the International symposium on intermontane basins: Geology and resources, edited by T Thanasuthipitak and P Ounchanum, Chiang Mai University, Chiang Mai, pp. 243–253.

Polachan S, Pradidtan S, Tongtaow C, Janmaha S, Intarawijitr K, Sunsuwan C. 1991. Development of Cenozoic basins in Thailand. Marine and Petroleum Geology 8, 84–97.

Prasongtham P, Kanjanapayont P. 2014. Deformation styles of the UthaiThani-NakhonSawan Ridge within the Chainat Duplex, Thailand. Journal of Earth Science 25 (5), 854–860.

Ramsay JG. 1980. Shear zone geometry: a review. Journal of Structural Geology 2, 83–99.

Ridd MF. 1971. Faults in Southeast Asia, and the Andaman rhombochasm. Nature 229, 51–52.

Ridd MF. 2012. The role of strike–slip faults in the displacement of the Palaeotethys suture zone in Southeast Thailand 51, 63–84.

Ridd MF, Morley CK. 2011. The KhaoYai Fault on the southern margin of the Khorat Plateau, and the pattern of faulting in Southeast Thailand. Proceedings of the Geologists' Association 122, 143–156.

Shi GR, Waterhouse JB. 1991. Early Permian brachiopods from Perak, west Malaysia. Journal of Southeast Asian Earth Sciences 6, 25–39.

Smith M, Chantraprasert S, Morley CK, Cartwright I. 2007. Structural geometry and timing of deformation in the Chainat duplex, Thailand. In Tectonics of Strike-slip Restraining and Releasing Bends, edited by W.D. Cunnningham and P. Mann, The Geological Society of London, Special Publication vol. 290, pp. 305–323.

Stauffer PH, Lee CP. 1989. Late Palaeozoic glacial marine facies in Southeast Asia and its implications.Geological Society of Malaysia Bulletin 20, 363–397.

Tapponnier P, Peltzer G, Armijo R. 1986. On the mechanism of collision between India and Asia. InCollision Tectonics, edited by M.P. Coward and A.C. Ries, Geological Society of London, Special Publication vol. 19, pp. 115–157.

Upton DR. 1999. A regional fission track study of Thailand: implications for thermal history and denudation. PhD thesis, University of London, UK.

Watkinson I, Elders C, Hall R.2008. The kinematic history of the Khlong Marui and Ranong Faults, southern Thailand. Journal of Structural Geology 30, 1554–1571.

Watkinson I, Elders C, Batt G, Jourdan F, Hall R, McNaughton NJ. 2011. The timing of strike–slip shear along the Ranong and Khlong Marui faults, Thailand. Journal of Geophysical Research 116 (B9), 1–26.

White SH, Burrows SE, Carreras J, Shaw ND, Humphreys FJ. 1980. On mylonites in ductile shear zones. Journal of Structural Geology 2, 175–187.

Chapter 16

Geotectonic evolution of the Nihonkoku Mylonite Zone of north central Japan based on geology, geochemistry, and radiometric ages of the Nihonkoku Mylonites: Implications for Cretaceous to Paleogene tectonics of the Japanese Islands

YUTAKA TAKAHASHI

Geological Survey of Japan, AIST, 1-1-1 Higashi, Tsukuba, Ibaraki 305-8567, Japan

16.1 INTRODUCTION

The Japanese Islands are situated on the eastern margin of the Asian Continent before opening of the Sea of Japan during Miocene time (Otofuji and Matsuda 1984; Maruyama et al. 1989). In the Japanese Islands, Cretaceous to Paleogene sinistral shear zones are widely distributed along the Median Tectonic Line (MTL; e.g. Takagi 1986; Takagi et al. 1989; Shimada et al. 1998), the Tanagura Tectonic Line (TTL; Koshiya 1986), the Hatakawa Tectonic Line (Sasada 1988; Takagi et al. 2000; Shigematsu and Yamagishi 2002) and in the Kitakami Mountains (Sasaki and Otoh 2000; Sasaki 2001) (Fig. 16.1). Therefore, studies on the geological and temporal relations among these shear zones are important to understand the pre-Neogene tectonics of the eastern margin of the Asian Continent.

The Nihonkoku Mylonite Zone (NMZ) is situated around Mount Nihonkoku (555 m) on the Japan Sea side of central Japan (Takahashi 1998a) (Fig. 16.1). It is located to the NNW of the Tanagura Shear Zone (Omori et al. 1953), and is regarded as one of the most likely northern extension of the TTL (Shimazu 1964a, b), which represents the boundary between the Ashio Belt (a Jurassic accretionary complex in Southwest Japan) and the Abukuma Belt (a Cretaceous high-temperature–pressure (T–P) regional metamorphic belt in Northeast Japan).

In this chapter, geology, structure, petrography, geochemistry, and radiometric ages of the Nihonkoku Mylonites and related rocks are described briefly (Takahashi 1998a, b, 1999, 2000; Takahashi et al. 2012). Timings of mylonitization within the NMZ are summarized, based on geology and radiometric ages of the Nihonkoku Mylonites and related rocks. The tectonic framework of Cretaceous to Paleogene sinistral ductile shear zones around the TTL and MTL in the Japanese Islands is also discussed. Study of ductile shear zones is important since plate boundaries are usually such zones (Mukherjee and Mulchrone 2013).

16.2 GEOLOGY OF THE NIHONKOKU MYLONITE ZONE

The NMZ strikes NW–SE, which extends to the upper reaches of the Miomote River on the SE side, forming the Nihonkoku–Miomote Milonite Zone (Takahashi 1999; Fig. 16.1). In this chapter, "mylonite" means plastically deformed and foliated rocks with porphyroclastic texture. On the other hand, plastically deformed and foliated granitic rocks *without* porphyroclast are called gneissose granitic rocks.

The NMZ is composed mainly of mylonites and gneissose rocks of hornblende–biotite granodiorite, biotite granodiorite, and biotite granite. Biotite–muscovite schist occurs at the center of the mylonite zone, which originated from the sedimentary rocks of the Ashio Belt through contact metamorphism and mylonitization (Takahashi 1998a) (Fig. 16.2). It is bound by faults to the surrounding granitic mylonites, although it is originally intruded by the granitic rocks (Takahashi 1998a). The original rocks of these granitic mylonites and gneissose granites intrude one another (Takahashi 1998a) (Fig. 16.3). The biotite granodiorite mylonite grades into gneissose biotite granodiorite and massive biotite granodiorite. Also, the biotite granite mylonite grades into gneissose biotite granite and massive biotite granite (Iwafune Granite). As a whole, the SW side of the biotite–muscovite schist layer in the NMZ is occupied by granitic rocks, and NE side is dominated by granodioritic rocks. Several lithologies of the Nihonkoku Mylonites are present, which depend on the degree of mylonitization and petrographic characteristics of the original rocks (Takahashi 1998a) (Fig. 16.2).

The Nishitagawa Granodiorite intruded the parent rocks of the Nihonkoku Mylonites, resulting contact metamorphism on the NE side of the NMZ (Fig. 16.2). Also, the pre-Neogene basement rocks are overlain unconformably by Neogene and Quaternary strata.

Ductile Shear Zones: From Micro- to Macro-scales, First Edition. Edited by Soumyajit Mukherjee and Kieran F. Mulchrone.

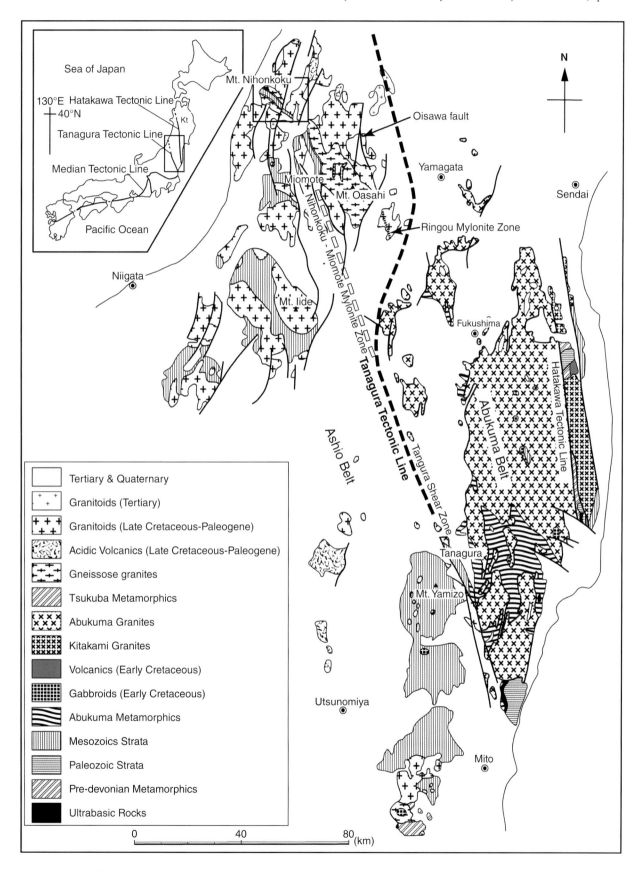

Fig. 16.1. Regional geological map of the pre-Neogene basement rocks around the Tanagura Tectonic Line. Kt: Kitakami Mountains. Adapted from Takahashi 1999. Reproduced with permission of the Geological Society of Japan.

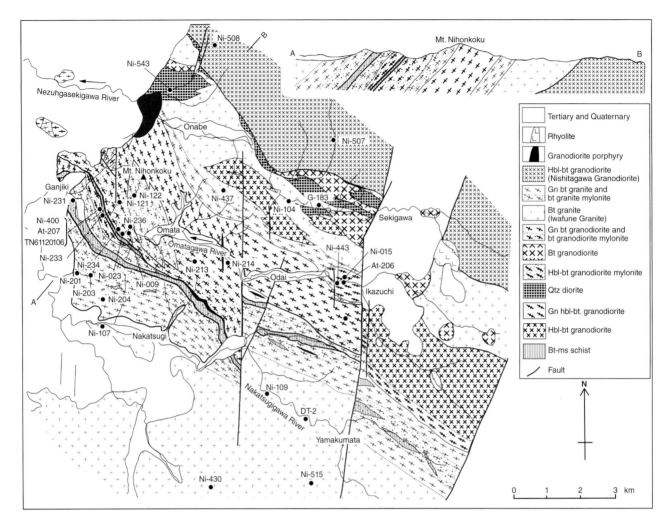

Fig. 16.2. Geological map of the Nihonkoku Mylonite Zone, showing the locations of samples for chemical analyses and age dating. Gn: gneissose, Hbl: hornblende, Bt: biotite, ms: muscovite, Qtz: quartz. Adapted from Takahashi 1998a with permission from the Geological Society of Japan.

Figure 16.3 summarizes the relationships amongst the pre-Neogene rocks.

16.3   PETROGRAPHY OF THE NIHONKOKU MYLONITES AND RELATED ROCKS

Takahashi (1998a) described the petrography of the Nihonkoku Mylonites and related rocks, and the Nishitagawa Granodiorite. In this chapter, representative specimens of the Nihonkoku Mylonites and related rocks are briefly described, referring to Takahashi (1998a, b) and Takahashi et al. (2012).

16.3.1   Biotite–muscovite schist

Biotite–muscovite schist consists mainly of quartz, plagioclase, muscovite, biotite, and carbonaceous material, with accessory alkali feldspar, tourmaline, and zircon, with rare garnet in some samples. Shape-preferred

arrangement of biotite and muscovite clearly define schistosity. Quartz appears as lenticular aggregates of fine-grained crystals or ribbons, oriented parallel to the foliation. These can be therefore called "composite lenticular quartz fish" (cf. Mukherjee 2011). S-C fabric (Lister and Snoke 1984) indicates that the mylonitization had taken place with a sinistral shear (Fig. 16.4). Some samples are massive and unfoliated and composed in part of alternate layers of sandstone and mudstone.

16.3.2   Gneissose hornblende–biotite granodiorite

The gneissose hornblende–biotite granodiorite is composed of quartz, plagioclase, alkali feldspar, hornblende, and biotite, with accessory sphene and opaque minerals (Fig. 16.5e, f; Fig. 16.6). The shape-preferred arrangement of hornblende and biotite forms a weak foliation. Quartz appears as aggregates of fine-grained crystals with serrated grain boundaries and wavy extinction. Plagioclase is euhedral, with albite twins

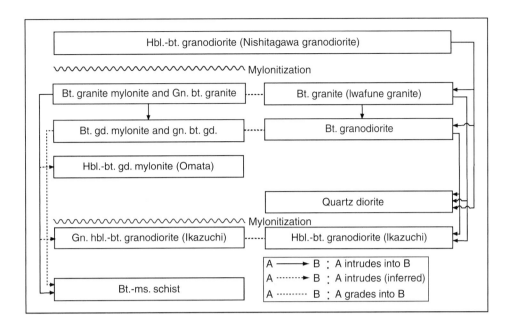

Fig. 16.3. Mutual relationships of the rock units in and around the Nihonkoku Mylonite Zone. Gn: gneissose, gd: granodiorite, Hbl.: hornblende, Bt.: biotite, ms.: muscovite. Adapted from Takahashi et al. 2012, with permission from Elsevier.

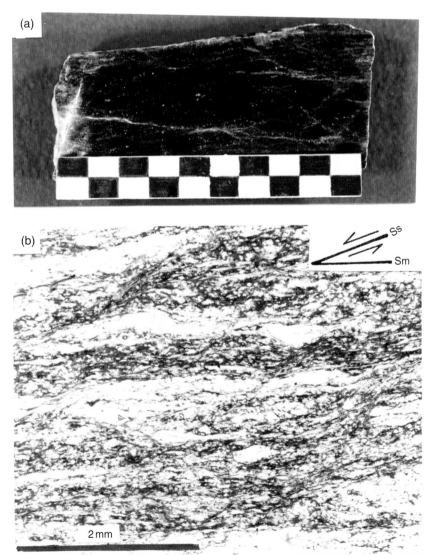

Fig. 16.4. Photograph of a polished slab (a) and photomicrograph (plane-polarized light) (b) of biotite–muscovite schist. Adapted from Takahashi et al. 2012, with permission from Elsevier. Scale bar in (a) is 10 cm across. Sm: main foliation defined by the shape-preferred orientation of biotite and muscovite. Ss: shear band.

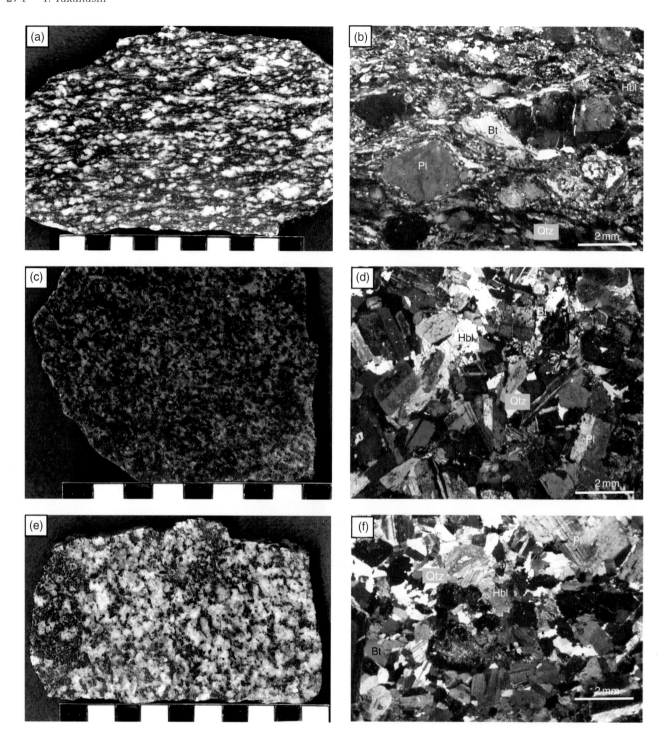

Fig. 16.5. Photographs of polished slabs (a, c, e) and photomicrographs (cross-polarized light) (b, d, f) of hornblende–biotite granodiorite mylonite (a, b), quartz diorite (c, d), and gneissose hornblende–biotite granodiorite (e, f). In the polished slabs, blackish grains are hornblende and biotite, pale gray grains are quartz, and white grains are plagioclase. Hbl: hornblende, Bt: biotite, Pl: plagioclase, Qtz: quartz. Scale bars are 10 cm across (a, c, e). Adapted from Takahashi et al. 2012, with permission from Elsevier.

and rare oscillatory zoning. Hornblende (greenish brown in color for Z) is euhedral and exhibits a weak shape-preferred arrangement. Biotite, brown in color for Y–Z, is partly altered to chlorite, exhibiting a weak shape-preferred arrangement, with hornblende. Alkali feldspar rarely appears as interstitial crystals.

16.3.3   Quartz diorite

The quartz diorite is composed of hornblende, plagioclase, and biotite, with accessory quartz and opaque minerals. Mylonitic textures are rare in the quartz diorite (Fig. 16.5c, d), whereas weakly mylonitized textures

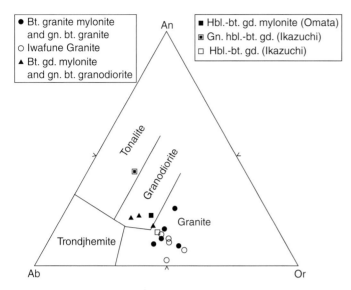

Fig. 16.6. Classification of sampled granitic rocks on a normative ternary anorthite–albite–orthoclase diagram (Baker 1979). Bt.: biotite, Hbl.: hornblende, Gn.: gneissose. Adapted from Takahashi et al. 2012, with permission from Elsevier. With additional data from Takahashi 1998b.

consisting of aggregates of fine-grained quartz with serrated grain boundaries and the shape-preferred arrangement of hornblende and quartz are present in some samples. Also, decussate textures of hornblende and biotite, possibly indicate thermal overprinting due to intrusion of younger plutonic rocks are observed in some samples.

16.3.4   Hornblende–biotite granodiorite mylonite

The hornblende–biotite granodiorite mylonite is composed of quartz, plagioclase, alkali feldspar, hornblende, and biotite, with accessory allanite and opaque minerals. On the ternary An-Ab-Or diagram, the rock is a granodiorite (Fig. 16.6). The shape-preferred arrangement of biotite and aggregates of fine-grained recrystallized crystals forms a clear foliation (Figs. 16.5a, b). Ribbons of fine-grained recrystallized quartz and porphyroclasts of normally zoned plagioclase, alkali feldspar (orthoclase with perthitic structure), and hornblende (green in color for Z) exhibit typical mylonitic textures. Biotite, brown in color for Y–Z, is partly altered to chlorite. The rock is a proto-mylonite to mylonite based on the classification proposed by Sibson (1977).

16.3.5   Biotite granodiorite mylonite and gneissose biotite granodiorite

The biotite granodiorite mylonite and gneissose biotite granodiorite consist of quartz, plagioclase, alkali feldspar, and biotite, with accessory zircon, apatite, allanite, and opaque minerals. Hornblende (green for Z) and muscovite are rarely observed. On the ternary An-Ab-Or diagram, the rock is mostly granodiorite with granite (Fig. 16.6). The biotite granodiorite mylonite is intensely

mylonitized, with porphyroclasts of normally zoned plagioclase and alkali feldspar (orthoclase) (Fig. 16.7e, f). Ribbons of fine-grained recrystallized quartz and film-like aggregates of fine-grained recrystallized biotite (brown for Y–Z) locally occurring around coarse-grained original biotite (brown for Y–Z) are present in the matrix (see Fig 16.17c). These elongate aggregates of fine-grained crystals define a prominent foliation. The rock is mylonite to protomylonite, locally ultramylonite according to Sibson (1977). The slightly mylonitized rock shows plutonic texture of subhedral granular and quartz is present as aggregates of fine, irregular crystals exhibiting wavy extinction (Fig. 16.7a and b).

16.3.6   Biotite granite mylonite and gneissose biotite granite

Biotite granite mylonite and gneissose biotite granite are composed of quartz, alkali feldspar, plagioclase, and biotite, with accesory allanite, zircon, tourmaline, and opaque minerals. On the ternary An-Ab-Or diagram (Fig. 16.6), the rock is identified as granite. Porphyroclasts of alkali feldspar (orthoclase) and normally zoned plagioclase are observed in strongly mylonitized parts (Fig. 16.8e–h). Elongate aggregates of fine-grained recrystallized quartz and biotite (brown for Y–Z) are present in the matrix surrounding the porphyroclasts and defining a prominent foliation (Fig. 16.8e and f). According to the classification of Lister and Snoke (1984), this rock is a typical mylonite with an S-C fabric. Ultramylonite (Sibson 1977) occurs locally, similar to biotite granodiorite mylonite (Fig. 16.8g, h). Weakly mylonitized rock shows plutonic texture of subhedral granular, and quartz occurs as weakly elongated aggregates of fine-grained crystals (Fig. 16.8c and d).

16.3.7   Biotite granite (Iwafune Granite)

The Iwafune Granite is a medium- to coarse-grained biotite granite consists of quartz, alkali feldspar, plagioclase, and biotite, with accessory zircon and opaque minerals (Fig. 16.8a, b). Quartz is subhedral to anhedral with wavy extinction. Alkali feldspar (orthoclase) is anhedral to subhedral with perthitic texture. Plagioclase is euhedral with normal zoning. Biotite is euhedral to subhedral with brownish axial color for Y–Z. This rock represents a typical plutonic texture, without evidence of mylonitization.

16.4   STRUCTURE OF THE NIHONKOKU MYLONITE ZONE

16.4.1   Foliation and lineation of the Nihonkoku Mylonites and related rocks

Foliation and lineation are described as structural elements of the Nihonkoku Mylonites and related rocks (Takahashi 1998a). The main foliation (Sm) is defined as

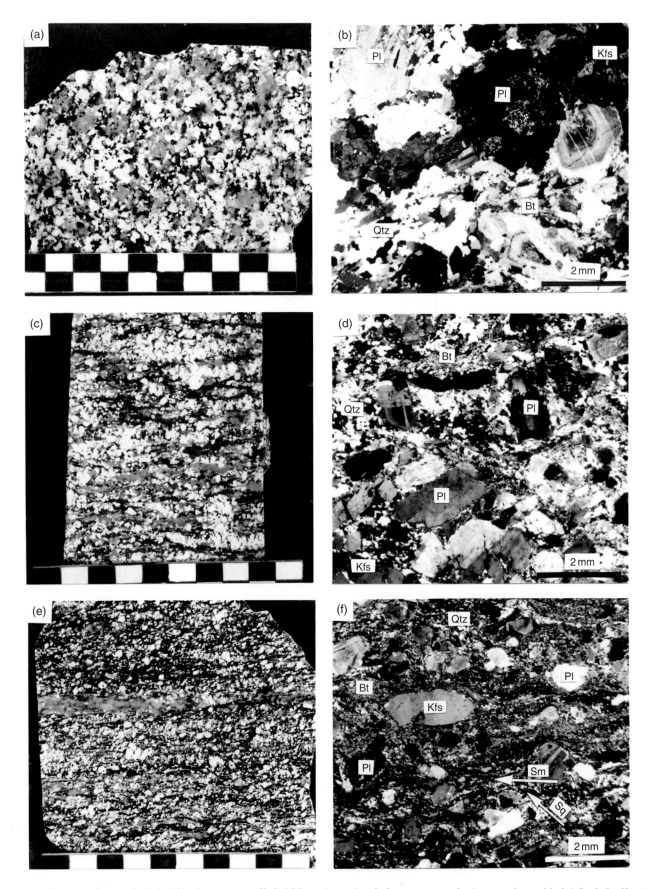

Fig. 16.7. Photographs of polished slabs after staining alkali feldspar (a, c, e) and photomicrographs (cross-polarized light) (b, d, f) of biotite granodiorite (a, b), gneissose biotite granodioritic (c, d), and biotite granodiorite mylonite (e, f). In the polished slabs, blackish grains are biotite, yellow grains are alkali feldspar, pale gray grains are quartz, and white grains are plagioclase. Qtz: quartz, Pl: plagioclase, Kfs: alkali feldspar, Bt: biotite. Sm: main foliation, Sq: foliation defined by the shape-preferred orientation of fine recrystallized quartz and lenticular aggregates of fine recrystallized quartz. Abbreviations are the same as in Fig. 16.5. Scale bars are 10 cm across (a, c, e). Adapted from Takahashi et al. 2012, with permission from Elsevier.

Fig. 16.8. Photographs of polished slabs after staining alkali feldspar (a, c, e, g) and photomicrographs (cross-polarized light) (b, d, f, h) of the Iwafune Granite (a, b), weakly mylonitized biotite granite (c, d), biotite granite mylonite (e, f), and biotite granite ultramylonite (g, h). In the polished slabs, blackish grains are biotite, yellow grains are alkali feldspar, pale gray grains are quartz, and white grains are plagioclase. Sm: main foliation defined by fluxion banding, Ss: shear band, S: foliation defined by the shape-preferred orientation of fine recrystallized quartz and lenticular aggregates of fine recrystallized quartz (Sq). Abbreviations are the same as in Fig. 16.5. Scale bars are 10 cm across (a, c, e, g). Adapted from Takahashi et al. 2012, with permission from Elsevier.

planar fabric made by preferred arrangement of mafic minerals such as biotite for granitic mylonites and preferred arrangements of biotite and muscovite for the biotite-muscovite schist. The lineation is defined as elongated linear arrangements of mineral aggregates on the main foliation. The microstructures described below are observed on the *XZ* plane, which is perpendicular to the main foliation (Sm) and parallel to the lineation.

The NMZ is delimited in the W, central and at E areas by N–S trending faults (Fig. 16.2). Structure of the W area around Omata is described here (Fig. 16.9). The foliation and lineation of the Nihonkoku Mylonites around Omata are different to each other divided by the northeastern marginal fault of the southern biotite–muscovite schist layer striking NW–SE. In the NE side of the fault, the mylonitic foliations strike mainly N–S to NW–SE, dip 50–80°W–SW, and mylonitic lineations plunge 20–40° to the S. On the other hand, in the SW side of the fault, mylonitic foliations strike mainly E–W to NW–SE, dip 30–50°S–SW, and mylonitic lineations plunge 30–35° to the S (Fig. 16.9).

16.4.2   Shear sense and direction in the Nihonkoku Mylonite Zone

Various asymmetric microstructures are observed in the NMZ (Takahashi 1998a). Based on those and stretching lineations, shear senses can be determined.

The S-C fabric is observed in the biotite–muscovite schist (Fig. 16.4), exhibiting sinistral shear. Typical asymmetric microstructures, such as asymmetric pressure shadow, mica fish (Mukherjee 2007, 2010a,b; Mukherjee and Koyi 2010a,b) and the preferred arrangement of fine recrystallized quartz, are observed in the hornblende–biotite granodiorite mylonite, biotite granodiorite mylonite and biotite granite mylonite. Mylonitic foliations, lineations and shear senses are shown in Fig. 16.9a. Also, the mylonitic foliations, stretching lineations and shear senses are plotted on the Schimit net (lower hemisphere) (Fig. 16.9b, c).

On the NE side of the biotite–muscovite schist layer, the Nihonkoku Mylonites were sinistrally sheared with extensional normal fault sense plunging 20–40° to the S, along a plane striking N–S to NW–SE and dipping 50–80° to W–SW. Also, on the SW side of the biotite-muscovite schist layer, the shear sense of the Nihonkoku Mylonites is sinistral with normal fault sense plunging 30–35° to the S, along a plane striking E–W to NW–SE and dipping 30–50° to S–SW.

16.4.3   Grain size reduction of recrystallized quartz through mylonitization

Grain size of recrystallized quartz in mylonites is an important indicator of mylonitization (Hara et al. 1977; Takagi 1984). Accordingly, maximum grain sizes (geometric mean of longest and shortest diameters) of

recrystallized quartz in the Nihonkoku Mylonites were measured on the *XZ* plane under the microscope. The results are plotted on the geological map (Fig. 16.10). The fine-grained (<0.25 mm) recrystallized crystals of quartz concentrate along the biotite–muscovite schist layer and get coarser away from the biotite–muscovite schist layer. The mylonite, with grain size of recrystallized quartz <0.25 mm, is classified as ultramylonite of Sibson (1977), as the mylonite usually contains porphyroclast of <10%. Therefore, it is obvious that the strain by mylonitization concentrates in the granitic mylonites beside the biotite–muscovite layer and the strain weakens away apart from the biotite–muscovite schist layer.

16.5   GEOCHEMISTRY OF THE NIHONKOKU MYLONITES AND RELATED ROCKS

Geochemistry of the Nihonkoku Mylonites and related rocks was reported by Takahashi (1998b). Here, geochemical features of the Nihonkoku Mylonites and related rocks are briefly described, referring to Takahashi (1998b).

16.5.1   Whole rock chemistry

The chemistry of the major elements of the Nihonkoku Mylonites and related rocks was analyzed using X-ray fluorescence spectrometer (XRF; Philips PW1404) at the Geological Survey of Japan. Analytical conditions of the XRF were 40 kV, 75 mA, using a Sc/Mo tube.

On the Harker's diagram (Fig. 16.11), chemical compositions of mylonites and gneissose rocks of hornblende–biotite granodiorite, biotite granodiorite and biotite granite of the NMZ, and massive biotite granite of the Iwafune Granite, are plotted nearly along a straight trend, except for Na_2O. The range of SiO_2 contents increases from hornblende–biotite granodiorite mylonite and gneissose hornblende–biotite granodiorite (61.18–71.52%) through biotite granodiorite mylonite and gneissose biotite granodiorite (69.76–72.18%), biotite granite mylonite and gneissose biotite granite (72.31–73.78%), to massive biotite granite (72.02–75.97%). Chemical compositions of the biotite granite mylonite and gneissose biotite granite are plotted in the range of those of massive biotite granite (Iwafune Granite).

16.5.2   Mineral chemistry

The geochemistry of the constituent minerals of the Nihonkoku Mylonites, gneissose granitic rocks, and Iwafune Granite were analyzed using an electron probe microanalyzer (EPMA; JEOL 8800) at the Geological Survey of Japan (Takahashi 1998b). Accelerating voltage, specimen current, and beam diameter were kept at 15 kV, 12 nA on Faraday cup, 2 μm, respectively.

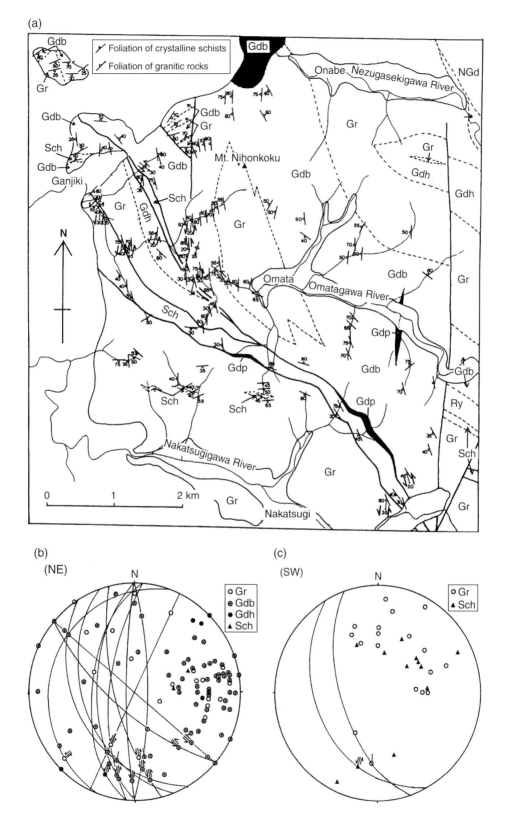

Fig. 16.9. Foliation and lineation map of the Nihonkoku Mylonite Zone around Omata (a) and Schmidt net projections (lower hemisphere) of mylonitic foliations and stretching lineations in the NE side (b) and SW side (c) of the southern biotite-muscovite schist layer. Symbols with bars indicate the lineations and arrows beside the lineation bars show the sense of shear determined under the microscope. On the lineated specimens, great circles of mylonitic foliations are also shown. Gr: biotite granite, Gdb: biotite granodiorite, Gdh: hornblende–biotite granodiorite, Sch: biotite-muscovite schist. Adapted from Takahashi 1998a, with permission from the Geological Society of Japan.

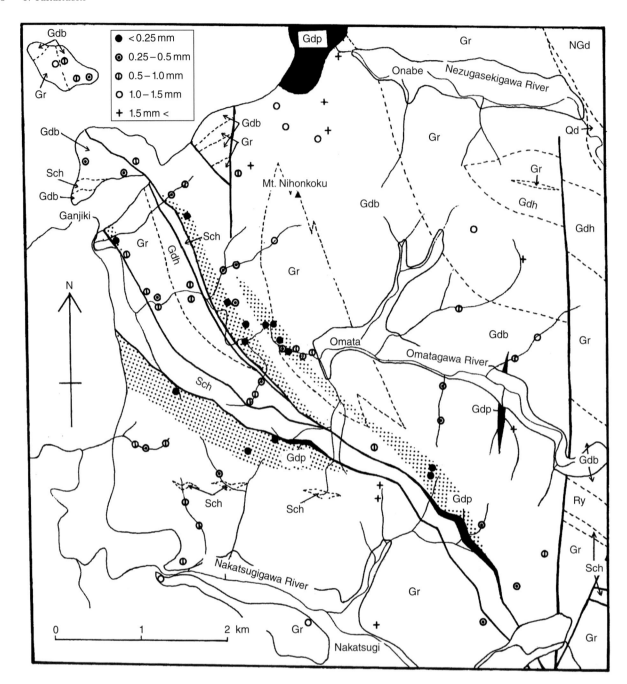

Fig. 16.10. Division of the intensity of mylonitization around Omata on the basis of maximum grain size of recrystallized quartz. Dotted fields represent the strongly sheared area in which maximum grain size of recrystallized quartz is less than 0.25 mm. Abbreviations are the same as those in Fig. 16.9. Adapted from Takahashi 1998a, with permission from the Geological Society of Japan.

16.5.2.1 Biotite

Aggregates of fine-grained biotite crystals are elongated to form foliations, and sometimes appear around original igneous coarse-grained crystals (see Fig. 16.17c). On the Ti vs. Si diagram (Fig. 16.12), chemical compositions of these fine-grained crystals and the original igneous coarse-grained crystals are connected by arrows. Most of the fine-grained crystals are poorer in Ti than the original igneous coarse-grained crystals (Fig. 16.12).

16.5.2.2 Amphibole

Amphibole is one of the main constituents in hornblende–biotite granodiorite mylonite and gneissose hornblende–biotite granodiorite. Amphibole occurs rarely in the gneissose biotite granodiorite. It appears as a porphyroclast, which is not surrounded by fine-grained recrystallized crystals of amphibole (Fig. 16.5 a, b). Chemical compositions of amphibole in the Nihonkoku Mylonite are given in Fig. 16.13. According to Deer et al. (1980),

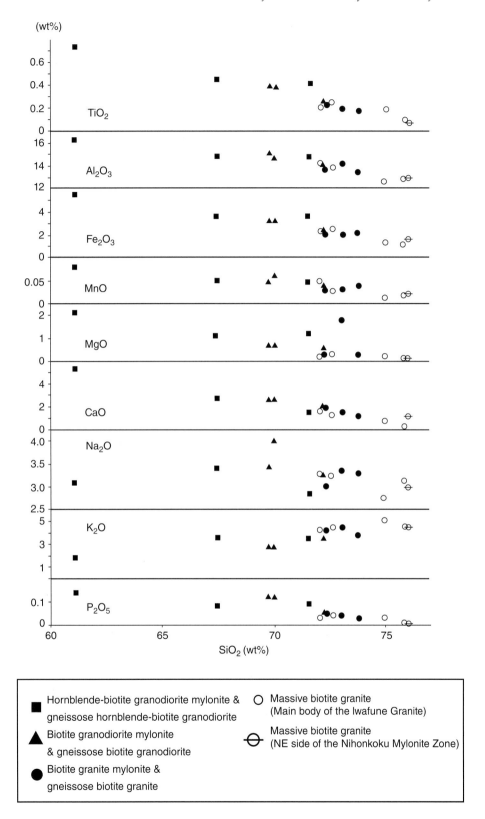

Fig. 16.11. Harker's diagram for the Nihonkoku Mylonites and related rocks. Adapted from Takahashi 1998b, with permission from Japan Association of Mineralogical Sciences.

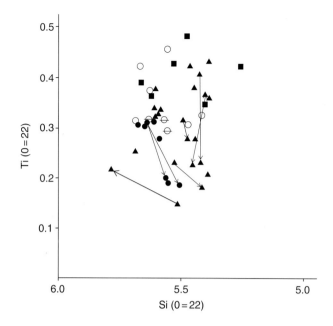

Fig. 16.12. Chemical compositions of biotite in the Nihonkoku Mylonites and related rocks, plotted on the Ti vs. Si diagram (atomic ratio, based on O=22) after Takahashi (1998b). Data for fine-grained recrystallized crystals of biotite are connected by arrows from the original coarse-grained crystals. Symbols are the same as Fig. 16.11. Adapted from Takahashi 1998b, with permission from Japan Association of Mineralogical Sciences.

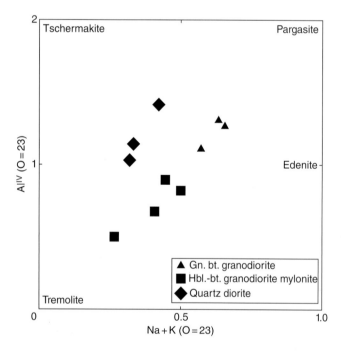

Fig. 16.13. Chemical compositions of amphibole in the Nihonkoku Mylonites and related rocks.

these amphiboles are classified into hornblende. The chemical composition of rim of the porphyroclast matches closely to that of core (Table 16.1). Based on chemical compositions and non-existence of fine-

Table 6.1. Chemical compositions of amphibole in the Nihonkoku mylonites

Rock name	Gn. bt. granodiorite			Hbl.-bt. Granodiorite mylonite		
Rock No.	Ni214 core	Ni214 core	Ni214 core	Ni400 core	Ni400 core	Ni400 rim
SiO	43.59	42.22	42.82	48.90	45.71	47.18
TiO	1.03	1.07	1.29	0.56	0.73	0.62
Al	7.79	9.15	8.99	4.52	5.92	5.20
Cr	0.02	0.00	0.01	0.07	0.05	0.01
FeOt	22.06	22.24	21.64	21.01	21.98	21.43
MnO	0.87	0.90	0.85	0.83	1.03	0.79
MgO	7.13	6.96	7.30	8.82	7.94	8.12
CaO	10.76	10.87	10.69	11.99	10.61	11.15
Na	1.28	1.23	1.41	0.57	1.20	1.03
K	0.87	1.26	1.11	0.49	0.69	0.59
Total	95.40	95.90	96.11	97.76	95.86	96.12
	Cations per 23 oxygens					
Si	6.885	6.670	6.714	7.407	7.149	7.308
Al	1.115	1.330	1.286	0.593	0.851	0.692
Al	0.334	0.374	0.374	0.215	0.239	0.257
Ti	0.122	0.127	0.153	0.064	0.085	0.072
Cr	0.002	0.000	0.001	0.008	0.007	0.001
Fe	0.000	0.000	0.000	0.000	0.000	0.000
Fe	2.913	2.938	2.837	2.660	2.873	2.776
Mn	0.117	0.120	0.112	0.106	0.137	0.104
Mg	1.677	1.639	1.706	1.990	1.850	1.873
Ca	1.820	1.839	1.795	1.946	1.777	1.849
Na	0.391	0.376	0.428	0.168	0.362	0.308
K	0.174	0.254	0.222	0.094	0.137	0.117
Total	15.550	15.666	15.628	15.252	15.467	15.357

FeOt: total Fe as FeO. Abbreviations same as Fig. 16.3.

grained recrystallized crystals, amphibole in the hornblende–biotite granodiorite mylonite is suggested not to recrystallize by mylonitization.

16.5.2.3 Plagioclase

Plagioclase appears as porphyroclast in highly mylonitized rocks in the NMZ. Chemical compositions of plagioclase in the Nihonkoku Mylonites and related rocks are plotted on the ternary An-Ab-Or diagram (Fig. 16.14). As a whole, the ranges of anorthite contents decrease systematically from hornblende–biotite granodiorite mylonite and gneissose hornblende–biotite granodiorite (An 23.6–42.9%), through biotite granodiorite mylonite and gneissose biotite granodiorite (An 15.1–35.4%), to biotite granite mylonite and gneissose biotite granite (An 6.0–33.7%), and massive biotite granite of the Iwafune Granite (An 6.4–32.3%).

Ranges of anorthite contents of plagioclase in the biotite granite mylonite and gneissose biotite granite, and the biotite granite (Iwafune Granite) are almost the same (Fig. 16.14). Anorthite contents of plagioclase in massive biotite granite located on the northeast side of the NMZ resemble with those at the SW side (Iwafune Granite).

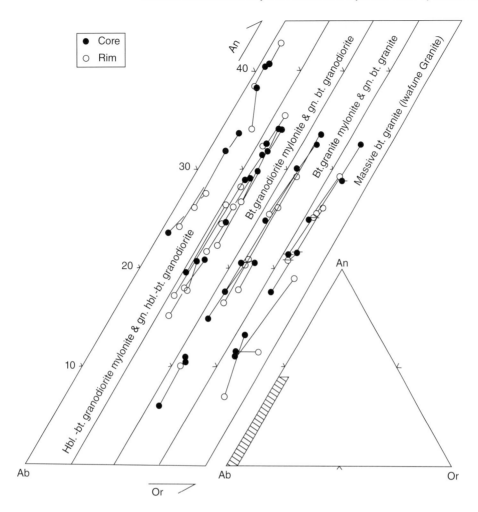

Fig. 16.14. Chemical compositions of plagioclase in the Nihonkoku Mylonites and related rocks, plotted on the ternary Anorthite (An)-Albite (Ab)-Orthoclase (Or) diagram. Compositions of core and rim in the same crystal are connected with a line. Abbreviations are the same as Fig. 16.3. Adapted from Takahashi 1998b, with permission from Japan Association of Mineralogical Sciences.

16.6 RADIOMETRIC AGES OF THE NIHONKOKU MYLONITES AND RELATED ROCKS

16.6.1 SHRIMP U–Pb Zircon ages of the Nihonkoku Mylonites and related rocks

SHRIMP U–Pb dating of zircon for the Nihonkoku Mylonites and related rocks are briefly described following Takahashi et al. (2012).

Zircon was separated at the Nanjing Institute of Geology and Mineral Resources, China. SHRIMP U-Pb analyses were performed at the Beijing SHRIMP Center, Chinese Academy of Geological Sciences, China. Analysis and data processing were performed according to those described in Williams (1998). The programs SQUID (Ludwig 2001a) and Isoplot/Ex (Ludwig 2001b) were used to reduce the raw data and for age calculations, respectively. The errors provided for weighted mean $^{206}Pb/^{238}U$ ages are 2σ (95% confidence level). The results of zircon analyses are plotted on conventional $^{206}Pb/^{238}U$ vs. $^{207}Pb/^{235}U$ concordia diagrams in Fig. 16.15.

On cathodoluminescence images, zircon grains in the Iwafune Granite and biotite granite mylonite show oscillatory zoning, and inherited cores are rare or absent. The same features are observed in zircons from the biotite granodiorite mylonite, hornblende–biotite granodiorite mylonite, and gneissose hornblende–biotite granodiorite (Takahashi et al. 2012).

16.6.2 K–Ar ages of the Nihonkoku Mylonites

The results of K–Ar age datings of biotite and hornblende in the Nihonkoku Mylonites and related rocks, and the Nishitagawa Granodiorite are described briefly, following Takahashi et al. (2012). Approximately 1–1.5 kg of each sample was crushed to powder using a jaw crusher and disk grinder. Biotite and hornblende were concentrated using a magnetic separator and heavy liquids. K–Ar age datings were carried out at Teledyne Isotopes Ltd. (Westwood, NJ, USA) and the Institute of Geological & Nuclear Sciences Ltd. (Lower Foot, New Zealand). The results of K–Ar age dating are listed in Table 16.2 and Fig. 16.16, with U–Pb data and previously published age data for comparison.

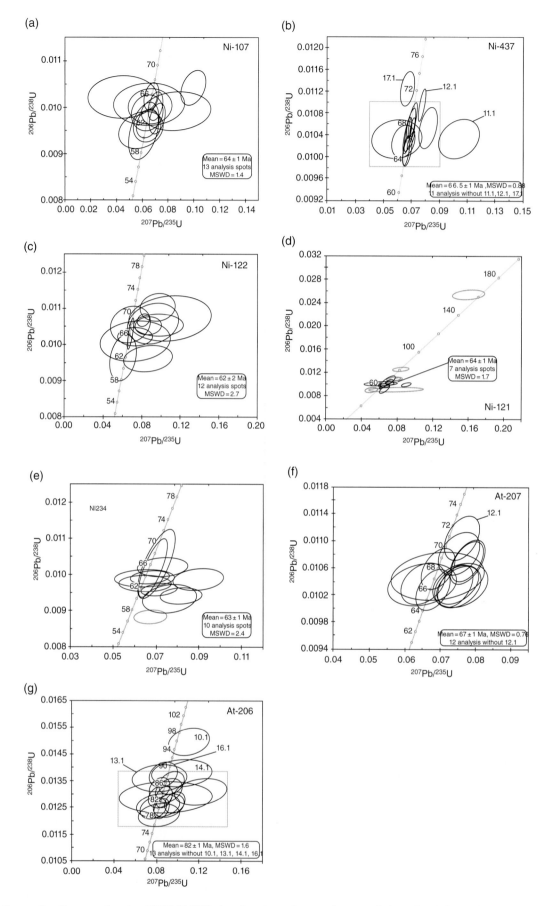

Fig. 16.15. Concordia diagrams showing SHRIMP U-Pb data for zircon from rocks around the Nihonkoku Mylonite Zone. a, b: Iwafune Granite (samples Ni-107, Ni-437), c: biotite granite mylonite (Ni-122), d, e: biotite granodiorite mylonite (Ni-121, Ni-234), (f) hornblende–biotite granodiorite mylonite (At-207), and (g) gneissose hornblende-biotite tonalite (At-206). Adapted from Takahashi et al. 2012, with permission from Elsevier.

Table 16.2. K-Ar ages of the Nihonkoku Mylonites and related rocks

Geologic Unit	Rock name	Sample No.	Sample	K (%)	^{40}Ar rad (10^{-5}ml STP/g)	^{40}Ar rad (%)	Age (Ma)
Dyke	Gd. porphyry	Ni 222	whole rock	1.58	0.170	56.7	27.4 ± 1.4
				1.59	0.173	60.6	27.9 ± 1.4
							av. 27.6 ± 1.4
	Bt. gd.	DT-3*	Biotite	4.46	0.948	78.2	53.9 ± 2.7
				4.53	0.983	79.9	55.0 ± 2.8
							av. 54.4 ± 2.7
Nishitagawa Granodiorite	Hbl.-bt. gd.	Ni 508	Biotite	2.94	0.681	71.5	58.5 ± 2.9
				2.95	0.672	84.0	57.8 ± 2.9
							av. 58.1 ± 2.9
		"	Hornblende	1.10	0.257	83.5	58.6 ± 2.9
				1.12	0.264	85.1	60.2 ± 3.0
							av. 59.4 ± 3.0
	Bt. gr.	DT-2*	Biotite	4.94	1.020	84.2	52.4 ± 2.6
				4.95	0.969	78.5	49.7 ± 2.5
							av. 50.7 ± 2.5
Iwafune Granite	"	Ni 109	Biotite	4.65	0.938	91.5	51.5 ± 2.6
				4.58	0.945	88.9	51.9 ± 2.6
							av. 51.7 ± 2.6
	"	G-183**	Biotite	5.29	1.152		55
	Bt. gr. mylonite	Ni 122	Biotite	3.93	0.790	84.0	51 ± 1
	Gn. bt. gr.	Ni 204	Biotite	3.87	0.767	83.6	50.3 ± 2.5
				3.87	0.775	82.2	50.8 ± 2.5
							av. 50.5 ± 2.5
	Bt. gd. Mylonite	Ni 121	Biotite	5.53	1.030	94.0	47 ± 1
	"	Ni 236	Biotite	4.80	0.918	92.5	48.4 ± 2.4
				4.82	0.938	89.3	49.6 ± 2.5
					0.909	91.5	48.0 ± 2.4
					0.916	92.1	48.3 ± 2.4
							av. 48.6 ± 2.4
	Gn. bt. gd.	Ni 213	Biotite	3.53	0.683	85.1	48.9 ± 2.4
				3.56	0.695	83.0	49.7 ± 2.5
					0.690	84.7	49.4 ± 2.5
					0.687	76.7	49.2 ± 2.5
							av. 49.3 ± 2.5
Nihonkoku Mylonite	Hbl.-bt. gd. mylonite (Omata)	Ni 233	Biotite	4.09	0.737	91.3	45.8 ± 2.3
				4.09	0.756	90.5	46.9 ± 2.3
					0.751	91.3	46.6 ± 2.3
					0.753	91.4	46.8 ± 2.3
							av. 45.6 ± 2.3
	"	Ni 400	Biotite	4.80	0.987	87.0	52 ± 1
	"	At 207	Biotite	4.61	0.959	93.8	52.7 ± 2.6
				4.62	0.926	96.3	50.9 ± 2.5
							av. 51.8 ± 2.6
	"	TN61120106***	Biotite	5.90	2.170		92 ± 6
	Gn. hbl.-bt. gd. (Ikazuchi)	Ni 015	Biotite	2.47	0.442	68.0	45 ± 1
		At 206	Biotite	3.33	0.999	94.5	75.4 ± 3.8
				3.35	0.981	86.8	74.0 ± 3.7
							av. 74.7 ± 3.7
		"	Hornblende	0.82	0.255	61.7	75.4 ± 3.8
				0.82	0.258	67.4	74.0 ± 3.7
							av. 78.7 ± 3.9
	Quartz diorite	Ni 543	Biotite	0.80	0.210	71.7	66.3 ± 1.7
				0.80	0.210	75.8	66.3 ± 1.7
							av. 66.3 ± 1.7
	"	"	Hornblende	0.44	0.121	50.2	68.6 ± 1.7
				0.45	0.120	52.0	68.1 ± 1.7
							av. 68.4 ± 1.7
	Bt.-ms. sch.	Ni231	Biotite	3.15	0.501	80.0	40.8 ± 0.8

λβ = 4.962 × 10^{-10}y, λe = 0.581 × 10^{-10} years
^{40}K/K = 1.167 × 10^{-2} atom%
* After Agency of Resources and Energy (1982).
** After Kawano and Ueda (1966).
*** After Shibata and Nozawa (1966).
Source: Adapted from Takahashi, et al. 2012 with permission from Elsevier.

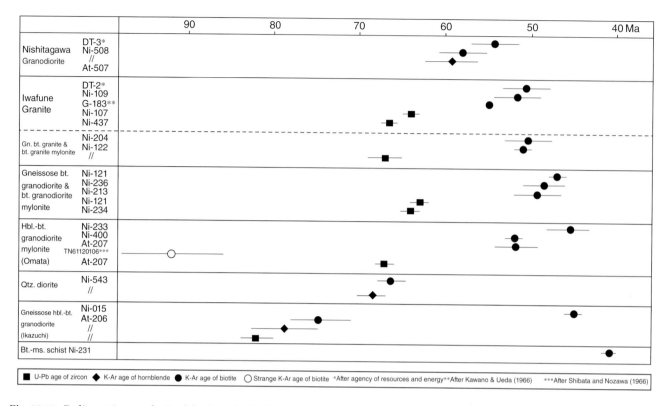

Fig. 16.16. Radiometric ages obtained for the Nihonkoku Mylonites and related rocks. Adapted from Takahashi et al. 2012, with permission from Elsevier.

16.7 DISCUSSION

16.7.1 Timing of mylonitization and cooling history of the Nihonkoku Mylonites and related rocks

The mylonitization in the NMZ happened at upper greenschist to amphibolite facies conditions (~500°C; Miyashiro 1994), based on the fact that fine-grained crystals of biotite – the product of recrystallization during mylonitization – appear around original igneous biotite in the biotite granodiorite mylonite (Takahashi 1998b, Simpson 1985, Fig. 16.17c). Moreover, myrmekite occurs around alkali feldspar porphyroclasts (Fig. 16.17a, b), and aggregates of fine-grained recrystallized alkali feldspar appear in pressure shadows around the porphyroclasts (Fig. 16.17a). Intermediate metamorphic conditions of amphibolite facies (450–600°C; Simpson 1985; Pryer 1993; Simpson and Wintsch 1989; Passchier and Trouw 2005) are suggested by these textures. Therefore, mylonitization should have happened before 47–49 Ma, which is the K–Ar age of biotite in the biotite granodiorite mylonite (Table 16.2), indicating the age of cooling ~300°C (Ar blocking temperature of biotite; Dodson and McClelland-Brown 1985).

Therefore, the hornblende–biotite granodiorite mylonite was generated between 67 Ma (SHRIMP U-Pb zircon age) and 46–52 Ma (K–Ar biotite age). Also, the biotite granodiorite mylonite was generated after 63–64 Ma (SHRIMP U–Pb zircon ages) and prior to 47–49 Ma (K–Ar biotite age). The biotite granite mylonite was after

67 Ma (SHRIMP U–Pb zircon age) and prior to 51 Ma (K–Ar biotite age). Consequently, the main mylonitization in the NMZ happened within 63–67 to 46–52 Ma.

Figure 16.18 shows cooling histories of the Nihonkoku Mylonites and related rocks. The parent magmas of the Nihonkoku Mylonites and related rocks intruded and consolidated at ca. 65 Ma (63–67 Ma) and mylonitization occurred at ca. 53–63 Ma. This is because, mylonitization in the NMZ is considered to have occurred at upper greenschist to amphibolite facies condition around 450–600°C (Fig. 16.15).

Also, an older stage of mylonitization is recognized by the radiometric ages of the gneissose hornblende-biotite granodiorite (At-206), whose U–Pb age of zircon is 82 Ma and K–Ar ages are 79 Ma for hornblende and 75 Ma for biotite (Table 16.2). The temperature for the older mylonitization is not clear but it should be >300°C as quartz deformed plastically. Therefore, the older stage mylonitization had possibly taken place at ca. 71–83 Ma (Fig. 16.18).

The quartz diorite intruded into the NE side of the NMZ at ca. 70 Ma and cooled rapidly (Fig. 16.18). However, it did not undergo mylonitization due to the position of intrusion, which is out of the mylonite zone (Fig. 16.2). Also, the Nishitagawa Granodiorite intruded into NE side of the Nihonkoku Mylonite Zone at ca. 60 Ma after the intrusion of the original rocks of the Nihonkoku Mylonites and caused contact metamorphism to them while cooling rapidly.

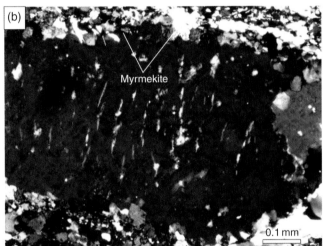

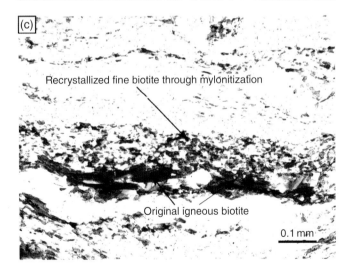

Fig. 16.17. Photomicrograph of myrmekite around alkali-feldspar porphyroclast (a, b), and fine-grained alkali-feldspar new grains in pressure shadow of the alkali-feldspar porphyroclast in the biotite granodiorite mylonite (a) and photomicrograph of recrystallized fine-grained biotite around original igneous biotite in the biotite granodiorite mylonite (c). Adapted from Takahashi et al. 2012, with permission from Elsevier.

16.7.2 Geological correlation with Cretaceous to Paleogene sinistral shear zones in the Japanese Islands

The TTL is regarded as the NE extension of the MTL based on regional geology and paleogeography before opening of the Sea of Japan (Yoshida 1977, 1981; Shibata and Takagi 1989; Takahashi 1999; Yamakita and Otoh 2000) (Fig. 16.19). On the other hand, the TTL is regarded as a pre-Neogene tectonic line that divides SW Japan from NE Japan, as the zonal arrangement of pre-Neogene geological units in SW Japan terminate at the TTL (Ichikawa 1990). The TTL divides the Ashio Belt in the W from the Abukuma Belt in the E (Fig. 16.1). The Ashio Belt, which is equivalent to the Mino–Tanba Belt of Southwest Japan, is composed mainly of a Jurassic accretionary complex in which Late Cretaceous volcano-plutonic rhyolites, welded tuffs, and granites distribute widely. The Mino (near Nagoya City), Tanba (near Osaka City) and Ashio (north of Tokyo) are local names of the areas. Also, the Asahi Belt is regarded as the northern extension of the Ryoke Belt (Sudo 1977). On the other hand, the Abukuma Belt of NE Japan is an Early Cretaceous high-temperature/pressure metamorphic belt composed of metapelites, metabasites, and voluminous granitic rocks, without volcanic rocks (Miyashiro 1973).

The timing of mylonitization along the TTL is considered to post-date 107 Ma, before phyllitization prior to 60–70 Ma and following low-temperature sinistral deformation which produced fault breccia and gouge after 60–70 Ma (Koshiya 1986). This period of low-temperature deformation is close to the timing of mylonitization in the NMZ. As for the MTL, a two-stage sinistral ductile shear is recognized along the MTL. At first, high-temperature deformation happened producing a foliation within the older Ryoke granite at ca. 90 Ma. Then, subsequent low-temperature deformation that produced the Kashio Mylonites of central Japan had taken place at ca. 65–70 Ma (Takagi, 1997). The final phase of ductile flow within the MTL happened at ca. 62–63 Ma, based on Ar–Ar whole-rock plateau ages obtained for protomylonite along the MTL (Dallmeyer and Takasu 1991). In case of the Inner Shear Zone of the Ryoke Belt (RISZ), mylonitization happened at ca. 67–74 Ma (Shimada et al. 1998), which is almost contemporaneous to the main mylonitization of the MTL. Therefore, the timing of main mylonitization within the NMZ (53–63 Ma) is correlated with: (1) low-temperature deformation after 60–70 Ma following the main mylonitization along the TTL; and possibly (2) the final phase of ductile deformation along the MTL, which had taken place at ca. 62–63 Ma (Fig. 16.19).

16.7.3 Geotectonic Evolution of the Nihonkoku Mylonite Zone

To consider the original geologic structure, paleogeography of the Japanese Islands before opening of the Sea of Japan are restored, based on paleomagnetic data (Otofuji and Matsuda 1984; Hamano and Tosha 1985) (Fig. 16.19).

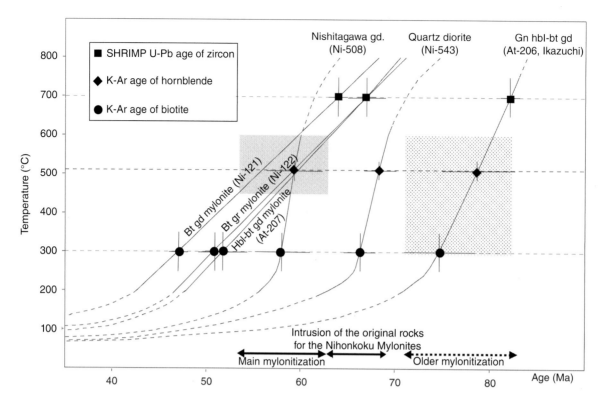

Fig. 16.18. Cooling history of the rocks around the Nihonkoku Mylonite Zone. Horizontal bars beside the symbols are error bar for analyses and the vertical bars beside the symbols are range of blocking temperatures for each system. In the shaded area, fine biotite crystals around original magmatic biotite and myrmekite around alkali-feldspar porphyroclasts were generated by recrystallization through mylonitization. In the stippled area, older stage mylonitization had possibly taken place. Bt, bt: biotite, Hbl, hbl: hornblende, gd: granodiorite, gr: granite, Gn: gneissose. Adapted from Takahashi et al. 2012, with permission from Elsevier.

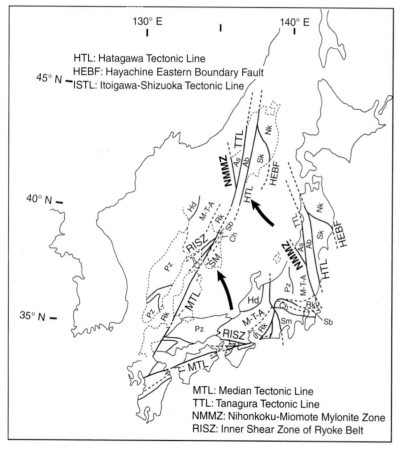

Fig. 16.19. Reconstructed paleogeography of the Japanese Islands before the opening of the Sea of Japan in Miocene time. The paleo-positions of Southwest and Northeast Japan were restored based on paleo-magnetic data reported by Otofuji and Matsuda (1984) and Hamano and Tosha (1985), respectively. Geotectonic data are after Ichikawa (1990). Hd: Hida Belt, Pz: Paleozoic terranes, M-T-A: Mino-Tanba-Ashio Belt, Rk: Ryoke Belt, As: Asahi Belt, Ab: Abukuma belt, Sk: South Kitakami Belt, Nk: North Kitakami Belt, Sb: Sanbagawa Belt, Ch: Chichibu Belt, Sm: Shimanto Belt. Adapted from Takahashi 2000.

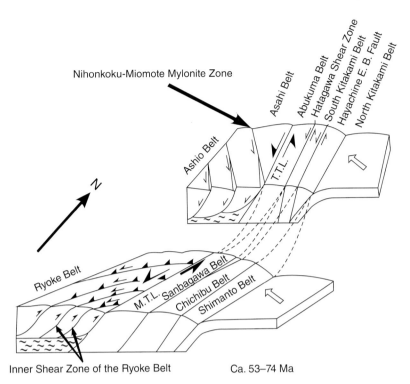

Fig. 16.20. Geotectonic model of central Japan during Late Cretaceous to Paleogene. Adapted from Takahashi 1999. Reproduced with permission of the Geological Society of Japan.

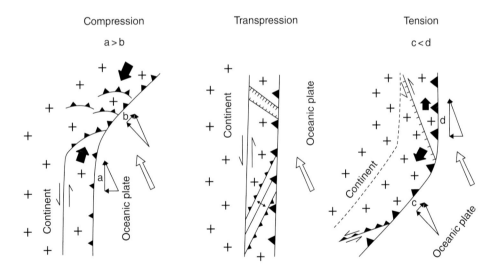

Fig. 16.21. Modes of oblique subduction of the oceanic plate beneath the continental crust and structures in the continental crust resulted due to the oblique subduction. Adapted from Takahashi 1999. Reproduced with permission of the Geological Society of Japan.

According to the restored paleogeography, the MTL smoothly extends to the TTL, although they are not straight but are convexly curved to the Pacific side. The Nihonkoku–Miomote Mylonite Zone (NMMZ) branches off from the TTL and the RISZ also branches off from the MTL. The NMMZ and the RISZ are symmetric about the convex corner. A schematic model of the structural state of the NMMZ compared with the RISZ of Late Cretaceous to Paleogene at ca. 53–74 Ma is shown in Fig. 16.20. The tectonic movement of the NMMZ was sinistral with

extensional normal fault sense plunging 20–40° to the S, along a plane striking N–S to NW–SE and dipping 50–80° to SW–W. The timing of the main mylonitization for the NMZ was constrained to 53–63 Ma. On the other hand, tectonic movement of the RISZ was top-to-W with reverse fault sense, which happened at ca. 67–74 Ma (Shimada et al. 1998) slightly prior to the main mylonitization of the NMZ. In the period of Late Cretaceous to Paleogene at ca. 53–74 Ma, the Pacific Plate was obliquely subducting beneath the Asian Continent toward NW direction with moderate speed of 10.4 cm/year (Maruyama and Seno 1986).

As a tectonic model for the NMMZ with the RISZ, the mode of oblique subduction of the oceanic plate beneath the continental plate and structures in the continental crust due to the oblique subduction is given in Fig. 16.21. If the plate boundary is straight (central figure of Fig. 16.21), the continental side is governed by state of transpression. In this condition, sinistral fault with reverse sense and its parallel fold, which is at low-angle to trench was formed (Sanderson and Marchini 1984). If accretional complex is present along the trench, reverse fault dipping to the continental side is commonly created as the accretional complex dips toward the continental side. The RISZ is regarded to form by transpression, in which sinistral (top-to-west) shear zones with reverse fault and its parallel fold branching off from the MTL is observed (Shimada et al. 1998). This condition is suitable to the transpression model (Fig. 16.21, center). If the continent is concave to the subducting oceanic plate (Fig. 16.21, left), element of plate motion parallel to the trench below the corner "a" is larger than the element above the corner "b". Consequently, continental side near the corner of this situation is under compression. In this case, reverse fault develops. If the continent is convex to the subducting oceanic plate (Fig. 16.21, right), element of plate motion parallel to the trench below the corner "c" is smaller than the element above the corner "d". Consequently, the continental side, near the corner of this situation, is under a tensional environment. In this case, normal fault with sinistral component of strike slip branches off from the convexed corner. Also, other than the convex corner area, the continental region is under transpressional environment, with a sinistral reverse fault. This situation applies to the eastern margin of the Asian continent in Late Cretaceous to Paleogene ca. 53–73 Ma. Namely, the normal fault near the corner corresponds to the NMMZ and the reverse fault branching off from the trench correspond to the RISZ. This tectonic model fits the restored paleogeogrphy of the Japanese Island (Fig. 16.19) and oblique subduction of the Pacific Plate beneath the eastern margin of the Asian Continent.

16.8 SUMMARY

1 The NW–SE trending NMZ, situated at the Sea of Japan side of central Japan, is one of the Cretaceous to Paleogene sinistral ductile shear zones in the Japanese Islands, which branches off from the TTL.

2 The NMZ is composed mainly of mylonites and gneissose rocks of the hornblende–biotite granodiorite, biotite granodiorite, and biotite granite, derived from Late Cretaceous to Paleogene granitic rocks in the Ashio Belt.

3 The Nihonkoku Mylonites sheared with normal fault sense along a plane striking N–S to NW-SE and dipping 50–80° to WSW on the NE side of the biotite–muscovite schist unit. Also, on the SW side of the biotite-muscovite schist layer, the shear sense of the Nihonkoku Mylonites is like a normal fault.

4 Whole rock chemistry of the Nihonkoku Mylonites and related rocks represent a nearly linear trend for every oxide except for Na_2O on the Harker's diagram. Anorthite content of plagioclase decrease systematically from the hornblende–biotite granodiorite mylonite and gneissose hornblende–biotite granodiorite through biotite granodiorite mylonite and gneissose biotite granodiorite, biotite granite mylonite and gneissose biotite granite, to massive biotite granite of the Iwafune Granite. These systematic chemical features suggest that each rock facies of the Nihonkoku Mylonites and related rocks is comagmatic.

5 Aggregates of fine-grained crystals of biotite are poorer in Ti than the original coarse-grained magmatic crystals. The decrease in Ti suggests lower temperature condition of recrystallization through mylonitization than that of magmatic crystallization.

6 The parent rocks of the Nihonkoku Mylonites intruded into the Jurrasic accretionary complex of the Ashio Belt and consolidated at ca. 65 Ma (based on SHRIMP U–Pb ages of 63–67 Ma) following mylonitization at ca. 53–63 Ma (based on the history of cooling to upper greenschist to amphibolite facies of ca. 450–600°C for the Nihonkoku Mylonites).

7 The main period of mylonitization along the NMZ (ca. 55–60 Ma) is correlative with (i) low-temperature deformation that occurred after 60–70 Ma along the Tanagura Tectonic Line, subsequent to the main mylonitization, and possibly (ii) the final phase of ductile deformation along the MTL (62–63 Ma).

8 The NMZ is regarded to form with the Inner Shear Zone of the Ryoke Belt (RISZ), in the framework of sinistral tectonic movement of eastern margin of the Asian Continent, due to oblique subduction of the Pacific Plate beneath the Asian Continent. The tectonic movement with normal fault sense of the NMZ is due to transtensional tectonics around the convexly curved plate boundary. On the other hand, top-to-W strike slip component with reverse fault sense of the RISZ is explained by transpressional tectonics due to the oblique subduction.

ACKNOWLEDGMENTS

This paper is based on the data of study on the geology of the Atsumi District (geological sheet mapping project at 1:50,000 of the Geological Survey of Japan, GSJ). I acknowledge Soumyajit Mukherjee for handling and detailed review of the manuscript. Also two anonymous reviewers are greatly acknowledged for important comments and improvements. Kelvin Matthews, Delia Sandford, and Ian Francis (Wiley Blackwell) are also acknowledged.

REFERENCES

Agency of Resources and Energy. 1982. Report of Regional Geological Survey, Uetsu Area (I) (in Japanese).

Barker F. 1979. Trondhjemite: definition, environment and hypotheses of origin. In: Barker, F. (Ed.), Trondhjemites, Dacites and Related Rocks. Developments in Petrology 6, 1–12.

Dallmeyer RD, Takasu A. 1991. Middle Paleocene terrane juxtaposition along the Median Tectonic Line, southwest Japan: evidence from $^{40}Ar/^{39}Ar$ mineral ages. Tectonophysics 200, 281–297.

Deer WA, Howie RA, Zussman J. 1980. An Introduction to the Rock Forming Minerals. Longman Ltd., Harlow.

Dodson MH, McClelland-Brown E. 1985. Isotopic and palaeomagnetic evidence for rates of cooling, uplift and erosion. Geological Society Memoir 10, 315–325.

Hamano Y, Tosha T. 1985. Movement of northeast Japan and paleomagnetism. Kagaku 55, 476–483 (in Japanese).

Hara I, Yamada T, Yokoyama T, Arita M, Hiraga Y. 1977. Study on the Southern Marginal Shear Belt of the Ryoke Metamorphic Terrain – initial movement picture of the Median Tectonic Line. Earth Science 31, 204–217 (in Japanese with English abstract).

Ichikawa K. 1990. Pre-Cretaceous terranes of Japan. In Pre-Cretaceous Terranes of Japan, edited by K. Ichikawa, S. Mizutani, I. Hara, S. Hada, and A. Yao, IGCP, International Geoscience Programme, UNESCO/IUGS, vol. 224, pp. 1–12.

Kawano Y, Ueda Y. 1966. K-A dating on the igneous rocks in Japan (IV)—Granitic rocks in northeastern Japan. The Journal of the Japanese Association of Mineralogists, Petrologists and Economic Geologists 56, 41–55.

Koshiya S. 1986. Tanakura Shear Zone: The deformation process of fault rocks and its kinematics. Journal of the Geological Society of Japan 92, 15–29 (in Japanese with English abstract).

Lister GS. Snoke AW. 1984. S-C mylonites. Journal of Structural Geology 6, 617–638.

Ludwig KR. 2001a. User's Manual for Isoplot/Ex, version 2.47. A Geochronological Toolkit for Microsoft Excel. Berkeley Geochronology Center Special Publication 1a, 2455. Ridge Road, Berkley, CA 94709, USA.

Ludwig KR. 2001b. SQUID 1.02. A User's Manual. Berkeley Geochronology Center Special Publication 2, 2455. Ridge Road, Berkley, CA 94709, USA.

Maruyama S. Seno T. 1986. Orogeny and relative plate motions: Example of the Japanese Islands. Tectonophysics 127, 305–329.

Maruyama S. Liou JG. Seno T. 1989. Mesozoic and Cenozoic evolution of Asia. In The Evolution of the Pacific Ocean Margins, edited by Z. Ben-Abraham, Oxford Monographs on Geology and Geophysics vol. 8, pp. 75–99.

Miyashiro A. 1973. Metamorphism and Metamorphic Belts. George Allen & Unwin, London.

Miyashiro A. 1994. Metamorphic Petrology. UCL Press, London.

Mukherjee S. 2007. Geodynamics, deformation and mathematical analysis of metamorphic belts of the NW Himalaya. Unpublished PhD thesis. Indian Institute of Technology, Roorkee.

Mukherjee S. 2010a. Structures in meso-and micro-scales in the Sutlej section of the Higher Himalayan Shear Zone, Indian Himalaya. e-Terra 7, 1–27.

Mukherjee S. 2010b. Microstructures of the Zanskar shear zone. Earth Science India 3, 9–27.

Mukherjee S. 2011. Mineral fish: their morphological classification, usefulness as shear sense indicators and genesis. International Journal of Earth Science 100, 1303–1314.

Mukherjee S, Koyi HA. 2010a. Higher Himalayan Shear Zone, Sutlej section: structural geology and extrusion mechanism by various combinations of simple shear, pure shear and channel flow in shifting modes. International Journal of Earth Sciences 99, 1267–1303.

Mukherjee S, Koyi HA. 2010b. Higher Himalayan Shear Zone, Zanskar Indian Himalaya: microstructural studies and extrusion mechanism by a combination of simple shear and channel flow. International Journal of Earth Sciences 99, 1083–1110.

Mukherjee S, Mulchrone KF. 2013. Viscous dissipation pattern in incompressible Newtonian simple shear zones: an analytical model. International Journal of Earth Sciences 102, 1165–1170.

Omori M, Horikoshi K, Suzuki K, Fujita Y. 1953. On the Tanakura Sheared Zone in the SW margin of the Abukuma Mountainland (study of the Cenozoic history of the southwestern margin of the Abukuma Plateau, Part 3). Journal of the Geological Society of Japan 59, 217–223 (in Japanese with English abstract).

Otofuji Y, Matsuda T. 1984. Timing of rotational motion of Southwest Japan inferred from paleomagnetism. Earth and Planetary Science Letters 70, 373–382.

Passchier CW, Trouw RAJ. 2005. Microtectonics, 2nd edition. Springer, Berlin.

Pryer L. 1993. Microstructures in feldspar from a major crustal thrust zone: the Grenville front, Ontario, Canada. Journal of Structural Geology 15, 21–36.

Sanderson DJ, Marchini WRD. 1984. Transpression. Journal of Structural Geology 6, 449–458.

Sasada M. 1988. Onikobe-Yuzawa Mylonite Zone. Earth Science 42, 346–353.

Sasaki M. 2001. Restoration of Early Cretaceous sinistral displacement and deformation in the South Kitakami Belt, NE Japan: an example of the Motai-Nagasaka area. Earth Science 55, 83–101.

Sasaki M, Otoh S. 2000. A cataclastic shear zone in the Motai-Nagasaka area, Southern Kitakami Belt – an example of the Tsuwamono-zawa shear zone. Journal of the Geological Society of Japan 106, 659–669 (in Japanese with English abstract)

Shibata K, Nozawa T. 1966. K-Ar age of the Nihonkoku Gneiss, Northeast Japan. Bulletin of the Geological Survey of Japan 17, 426–429.

Shibata K, Takagi H. 1989. Tectonic relationship between the Median Tectonic Line and the Tanakura Tectonic Line viewed from isotopic ages and Sr isotopes of granitic rocks in the northern Kanto Mountains. Journal of the Geological Society of Japan 95, 687–700 (in Japanese with English abstract).

Shigematsu N, Yamagishi H. 2002. Quartz microstructures and deformation conditions in the Hatagawa shear zone, northeastern Japan. The Island Arc 11, 45–60.

Shimada K, Takagi H, Osawa H. 1998. Geotectonic evolution in transpressional regime: time and space relationships between mylonitization and folding in the southern Ryoke belt, eastern Kii Peninsula, southwest Japan. Journal of the Geological Society of Japan 104, 825–844 (in Japanese with English abstract).

Shimazu M. 1964a. Cretaceous granites of Northeast Japan, I. Earth Science 71, 18–27 (in Japanese).

Shimazu M. 1964b. Cretaceous granites of Northeast Japan, II. Earth Science 72, 24–29 (in Japanese with English abstract).

Sibson RH. 1977. Fault rocks and fault mechanisms. Journal of the Geological Society London 133, 191–213.

Simpson C. 1985. Deformation of granitic rocks across the brittle-ductile transition. Journal of Structural Geology 7, 503–511.

Simpson C, Wintsch RP. 1989. Evidence for deformation-induced K-feldspar replacement by myrmekite. Journal of Metamorphic Geology 7, 261–275.

Sudo S. 1977. Some problems on the Late Cretaceous to Paleogene volcano-plutonic activity in central Japan. The Association for the Geological Collaboration in Japan, Monograph no. 20, 53–60 (in Japanese with English abstract).

Takagi H. 1984. Mylonitic rocks along the Median Tectonic Line in Takato-Ichinose area, Nagano Prefecture. Journal of the Geological Society of Japan 90, 81–100 (In Japanese with English abstract).

Takagi H. 1986. Implications of mylonitic microstructures for the geotectonic evolution of the Median Tectonic Line, central Japan. Journal of Structural Geology 8, 3–14.

Takagi H. 1997. Timing of mylonitization for the Ryoke Belt of the Chubu district. Earth Monthly 19, 111–115 (in Japanese).

Takagi H. Goto K., Shigematsu N. 2000. Ultramylonite bands derived from cataclasite and pseudotachylyte in granites, northeast Japan. Journal of Structural Geology 22, 1325–1339.

Takagi H, Shibata K, Sugiyama Y, Uchiumi S, Matsumoto A. 1989. Isotopic ages of rocks along the Median Tectonic Line in the Kayumi area, Mie Prefecture. Journal of Mineralogy, Petrology and Economic Geology 84, 75–88.

Takahashi Y. 1998a. Geology and structure of the Nihonkoku Mylonite Zone on the borders of Niigata and Yamagata Prefectures, northeast Japan. Journal of the Geological Society of Japan 104, 122–136 (in Japanese with English abstract).

Takahashi Y. 1998b. Geochemistry of the Nihonkoku Mylonite along the border between Niigata and Yamagata Prefectures, Northeast Japan. Journal of Mineralogy Petrology and Economic Geology 93, 330–343.

Takahashi Y. 1999. Reexamination of the northern extension of the Tanagura Tectonic Line, with special reference to the Nihonkoku-Miomote Mylonite Zone. Structural Geology (Journal of the Tectonic Research Group of Japan) 43, 69–78 (in Japanese).

Takahashi Y. 2000. Tectonics around the Asahi Mountains, northern Japan, based on deformation structures of granitic rocks. Earth Monthly 30, 120–126 (in Japanese).

Takahashi Y, Mao J, Zhao X. 2012. Timing of mylonitization in the Nihonkoku Mylonite Zone of north central Japan: Implications for Cretaceous to Paleogene sinistral ductile deformation in the Japanese Islands. Journal of Asian Earth Science 47, 265–280.

Williams IS. 1998. U–Th–Pb geochronology by ion microprobe. Reviews in Economic Geology 7, 1–35.

Yamakita S, Otoh S. 2000. Cretaceous rearrangement processes of pre-Cretaceous geologic units of the Japanese Islands by MTL-Kurosegawa left-lateral strike-slip fault system. Memory of the Geological Society of Japan 56, 23–38 (in Japanese with English abstract).

Yoshida T. 1977. Northeast Japan and the Median Tectonic Line – an attempt at interpretation. Monograph of the Association for the Geological Collaboration in Japan 20, 113–116 (In Japanese).

Yoshida T. 1981. Remarks on the pre-Neogene geotectonics in Northeast Japan. Structural Geology (Journal of the Tectonic Research Group of Japan) 26, 3–29 (in Japanese).

Chapter 17

Flanking structures as shear sense indicators in the Higher Himalayan gneisses near Tato, West Siang District, Arunachal Pradesh, India

TAPOS KUMAR GOSWAMI and SUKUMAR BARUAH

Department of Applied Geology, Dibrugarh University, Dibrugarh 786004, Assam, India

17.1 INTRODUCTION

Flanking structures (FS)/flanking folds are deflections of a planar layer or host fabric elements (HE) such as bedding, foliation or compositional layering around a cross cutting element (CE) such as a fault, vein, joint, a patch of melt, a mineral grain, or even a boudin (Passchier 2001; Coelho et al. 2005; Mulchrone 2007; Mukherjee and Koyi 2009; Mukherjee 2014a). The first descriptions of flanking structures were based mainly on sub-meter scale (Passchier 2001). Microscale examples of the structures came later (Mukherjee 2007, 2010a, b, 2011, 2014b; Mukherjee and Koyi 2009; Grasemann et al. 2011). Description of flanking structures is an expansion of the concept of fault drag-the deflection of layers in the vicinity of the fault (Gayer et al. 1978; Hudleston 1989; Druguet et al. 1997). Between the two end members of simple shear and pure shear, the resulting flanking structures are classified as s-type flanking folds, a-type flanking folds and shear bands (Grasemann et al. 2003). As contractional or extensional offset of central markers, both s-type and a-type FS may exhibit normal or reverse drag (Hamblin 1965) of the central markers in reference to the shear sense along the CE (Wiesmayr and Grasemann 2005). Two new schemes of classification of flanking structures have recently been proposed (Mukherjee 2014b). First, whether (i) the CE is a sharp plane of discontinuity or, (ii) it consists of rock(s)/mineral(s). In case (ii), FS is again divided in to whether HE penetrates CE or not. The second classification is based on drag and slip along the CE margins. It is already defined that along the direction of shear if a convex HE is reached, the sense of drag is "normal" and in the opposite case is "reverse" (Grasemann et al. 2003). What controls the sense of drag is a long studied issue. The latest view is that it depends on (i) angle between the HE and the CE before deformation started; and (ii) relative values of vertical separation and throw of faulting along the CE (reviewed in Mukherjee 2014b). A clear convex/concave drag does not always develop in all the mesoscopic FS. Also, the HE could be thicker near the CE.

The two latest classification schemes consider all possible slip and drag of the HE.

s-type flanking structures are regarded as reliable shear sense indicators (Exner et al. 2004) and are used here for that purpose under contractional tectonic setting.

Numerous flanking structures developed in a ductile shear zone affecting the Higher Himalayan migmatitic gneisses around the Tato area of the West Siang District, Arunachal Pradesh, India. Using the transparent overlays on the outcrop structures, we calculated the value of Φ (the deflection of the slip surface) and β (angle between the tangent to HE at the intersection with CE and the x-axis) in a WNW–ESE traverse, and found that the geometrical types depict co-shearing with both normal and reverse drags. The bulk shear thus indicates a top-to-SSE dextral sense. Horizontal shear bands with top-to-SE shear is reported from the biotite gneisses exposed W to Yapuik (Saha 2011). Feldspar porphyroclasts in the Bomdila gneisses indicate a top-to-SE shear in the western Arunachal Himalaya (Singh and Gururajan, 2011).

17.2 GEOLOGY

The Himalayan Main Central Thrust (MCT) evolved both in space and time from a deep level ductile shear zone to a shallow level brittle thrust fault from Early to Middle Miocene (Ahmed et al. 2000; Jain et al. 2005; Mukherjee, 2005, 2012a,b, 2013a,b,c, 2015; Searly et al. 2008; Mukherjee and Koyi, 2010a,b; Mukherjee et al. 2012, 2014b, 2015). Although the MCT is defined as broad shear zone in most parts of central and western Himalaya, it as a sharp contact between kyanite and garnet-bearing schist and gneisses of the Greater Himalayan Sequence (GHS) near Pene (West Siang District, Arunachal Pradesh) over the quartzites, phyllites, carbonates and metavolcanics of Lesser Himalayan Sequence (LHS) (Singh and Chowdhary 1990; Gururajan and Chowdhary 2003; Yin 2006). In the Siyom River Section of West Siang District, Arunachal Pradesh (Fig. 17.1) thick bedded quartzites, phyllites and quartz -mica-schists constitutes the Bomdilla Group of

Ductile Shear Zones: From Micro- to Macro-scales, First Edition. Edited by Soumyajit Mukherjee and Kieran F. Mulchrone.

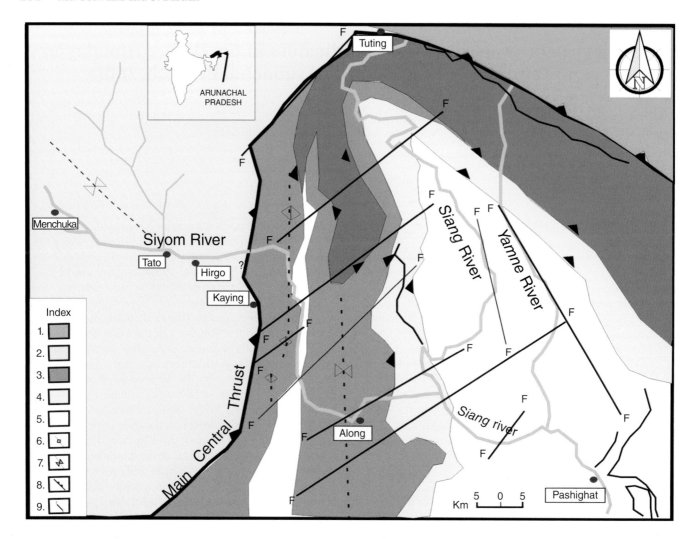

Fig. 17.1. Structural and Tectonic map of Siang Fold and Thrust belt. **1.** Tuting Thrust Sheet **2.** Siang Thrust Sheet **3.** Siyom Thrust Sheet **4.** Rikor Thrust Sheet **5.** Parautocthonous Zone **6.** Axial trace of the antiformal fold **7.** Axial Trace of the synformal fold **8.** Thrust **9.** Fault. The MCT is placed east of Hirgo in the original map is shown with a question mark. Map of India (inset). Adapted from Singh 1993.

rocks or Zero Gneisses (Saha 2011) in the MCT footwall, while the biotite–granite gneisses of Hone and Hirgo at the west of Pene, represents the base of the Higher Himalayan slab (Kumar 1997).

The gneisses grade into garnetiferous migmatitic gneisses near Tato which is traceable up to ~1 km along a WNW trend towards Tato-Menchuka road (Fig. 17.2). The migmatitic gneisses are associated with biotite granite gneisses and quartzites towards Menchuka. The MCT ductile shear zone at Tato is bound by the brittle MCT fault at Pene. The foliations trend ENE in the gneisses and has a moderate dip 20–25° towards N. The compositional layers in the gneisses are asymmetric folded and exhibit flanking structures in the fold train. Three generations of folds are preserved in the migmatitic gneisses. The mesoscopic shear fabric indicates a top-to-SSE left lateral shear. The entire sequence is folded and the large-scale regional fold axis trend ENE-WSW plunging SE at 20–30° (Singh and Malhotra 1980)

N to Kaying, in the downstream of Siyom River, thick-bedded quartzites and quartz–mica schists trend N. The mineral stretching lineation plunges ~15° northerly from Kaying to Dipu and this trend parallels the N–S trend of the regional-scale folds of the Lesser Himalaya (Singh 1993). This sequence is affected by S/SE verging meso-scopic folds of the Higher Himalaya. From Kaying north-ward, the S1 axial planar foliation of these north-south folds dips steeply to the W on the right bank of the Siyom River (Fig. 17.3a). The L1 fold axis lineation plunges gen-tly either to N or S (Fig. 17.3b).

In the left bank of the Siyom River, the limb dips at moderate angle to the E. The whole quartzite and quartz mica schists sequence is thus folded into N-S open folds plunging gently either to N or S (Fig. 17.1). In the mica-ceous schist at Duku, the new S and L fabrics are observed and represent mylonitization due to ductile shear related to the thrusting. From Bille to Pene, the S1 and S2 schistosities dip at moderate angles to the SW to NW

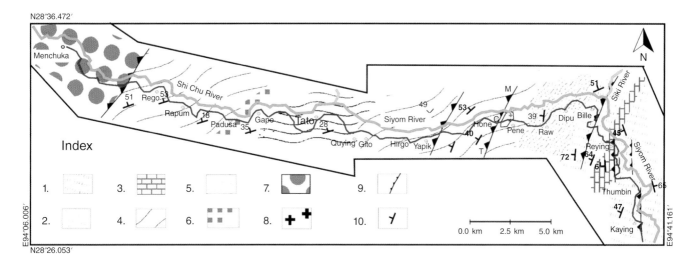

Fig. 17.2. Geological map of the study area. **1.** Quartzite and quartz mica schist. **2.** Lime stone band. **3.** biotite –muscovite schist. **4.** Biotite gneiss. **5.** Garnetiferous migmatitic gneiss. **6.** Quartzites. **7.** Tillites (glacial deposit). **8.** Tourmaline bearing leucogranites. **9.** Thrust. **10.** Foliation dip. MCT is placed at Pene, based on our study.

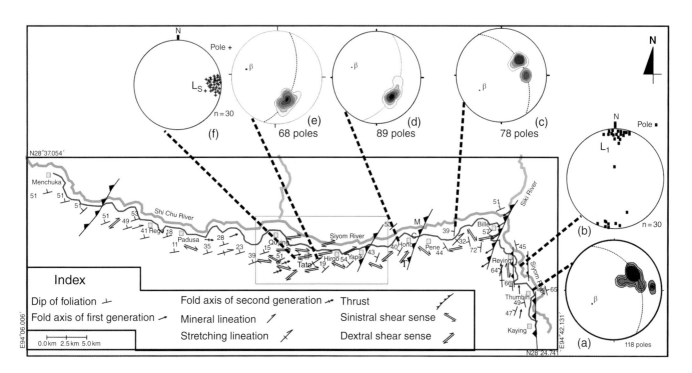

Fig. 17.3. Structural map of the area from Kaying to Menchuka. Synoptic orientations of S1 foliations in different subareas a, c, d, e. The contour intervals are: < 1.50, 3.50, 9.80, 16.10, >22% per unit area. Numbers of poles are mentioned. L1 fold axis lineation in sub area b and Ls stretching lineation in subarea f area also shown.

(Fig. 17.3c,d). Further W, around Tato, the S1 and S2 foliations dip consistently towards NW or WNW at low to moderate angles (Fig. 17.3e). The stretching lineations (Ls) have low plunge to the ENE or ESE (Fig. 17.3f). The shaded area in Fig. 17.3 shows fold axes of first and second generations dipping ENE and consistent right lateral top-to-SE shear sense.

17.3 DUCTILE SHEAR FABRICS

Ductile shear zones are characterized by concentrated non-coaxial/Couette flow in a relatively planar- (Ramsay 1980), or a curvi-planar (Mukherjee and Biswas 2014; Mukherjee and Biswas, 2016, Chapter 5) domain. The displacements in the wall rock in ductile shear zones are higher than any shortening/stretching within the wall rock (Passchier and Coelho 2006). The ductile shear zone in the migmatitic gneisses in Tato represents a domain of pervasive planar mylonitic foliation. The stretching lineations plunge 10–12° towards ENE/ESE and parallel the vorticity profile plane (VPP). Evidences of stretching parallel to the lineation are exhibited by boudinaged and elongated enclaves of mylonitic magmatic rocks (Hubert et al. 2011) (Figs 17.4 and 17.5). Asymmetric protolith boudins (Fig. 17.5a), commonly parallel to the lineation, define a sinistral shear.

The asymmetric fold verges ENE while the gneissic layer is small-scale ENE verging drag folded. The movement in fabrics of the shear zone defines a monoclinic symmetry. The symmetry plane is normal to the line of intersection between the S- and the C- and contains stretching lineations on the S-planes (Fig. 17.4 and 17.5). Thus, the shear direction parallels the orthogonal projection of the stretching lineation on the C-surface (Lin and William 1992).

17.4 FLANKING STRUCTURES FOR SHEAR SENSE

The flanking folds in the present outcrop are of centimeter-scale and are open to isoclinal. Amongst the different markers, mesoscale flanking folds indicate shear sense in the migmatitic gneisses. The compositional banding due to partial melting in the migmatitic gneisses is also observed (Fig. 17.4c). The compositional banding shows flanking structures around a planar discontinuity (fracture/vein) (Fig. 17.4a-d). A non-homogeneous flow field thus developed near fractures during deformation. We considered that the CE intruded prior to or during shear deflecting the layer. The type of flanking structures develop depend upon (i) flow field defined by the vorticity number (W_k); and (ii) the initial fracture orientation. The basic flanking structures observed are the s-type (Passchier 2001; Coelho et al. 2005). The fold train in the host fabric element (HE) perhaps developed following the shear zone formation. The CE in most cases are

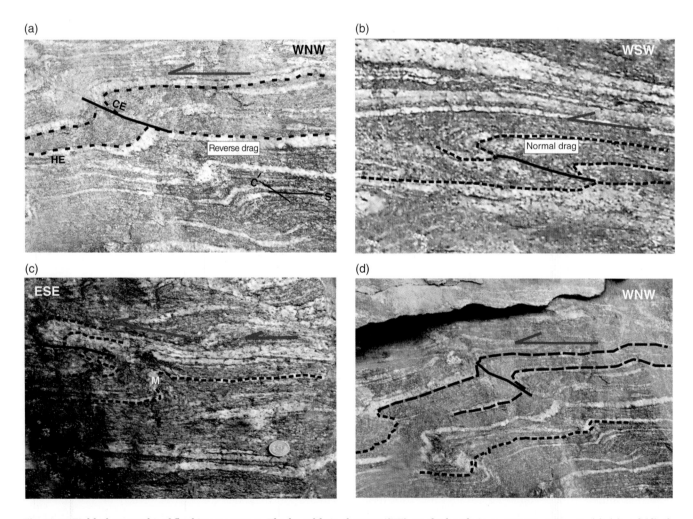

Fig. 17.4. Field photographs of flanking structures. The host fabric elements (HE) are the bands in composite gneiss. In (a), (c) and (d), the cross cutting elements (CE) are fractures. The HE concave to the direction of shear, indicate a reverse drag. In a, the extension crenulations cleavage (ecc) C and the foliation S makes a small angle. (b) The CE is patches of melt. The HE convex to the direction of shear indicates a normal drag. Figs 4a, 4b and 4d are of *a.1.1* type and 4c is of *b.1.2.2* type Flanking structure (Mukherjee 2014b).

Fig. 17.5. Field photographs of the flanking structures west of the area described in Fig. 17.4. (a) The protolith boudin of metabasite serve as CE for the composite bands to form the flanking structure with reverse drag. Normal drag is observed in photographs (b), (c), and (d). In (d), the vergence of the asymmetric folds towards ENE. (b) is of *b.1.1* type and (c) and (d) are *a.1.1.* type flanking structure (Mukherjee 2014a).

fractures (Figs 17.4a,c,d and 17.5b–d). However, in two cases flanking structures around melt patches and around boudinaged protolith are observed (Figs 17.4b and 17.5a). The folds are thicker in the hinges and thinner in the limbs similar to Class 1B of Ramsay (1967). The central HE in Fig. 17.5a is folded. These fold verge ESE. We measured the deflection/tilt angle (Φ) of the slip surface (CE) and β (angle between the tangent to HE at the intersection with CE and the *x*-axis) in an WNW to ESE traverse (Table 17.1) on transparent overlays. The *x*-axis orients WNW–ESE in most cases in the field. The measurement shows that Φ ranges 140–170° and β 50–130°. The initial and the final Φ measured are presented in Fig. 17.6b. The origin of the Cartesian coordinates is at the intersection between the CE and the HE. The *x*-axis parallels the far field HE with its positive half set considering the dip of CE (Fig. 17.6a). As the photographs are taken viewing S, the shear sense appears sinistral (Fig. 17.6b). However, viewing N, progressive flanking fold evolution on the basis of Φ value changes (Wiesmayr and Grasemann 2005), as shown in Fig. 17.7a,b. The Φ value is considered positive portion of the *x*-axis

which is parallel to the far field HE (counterclockwise) in the slip sense. We interpret the geometric structures of the outcrop as contractional s-type flanking folds with both normal and reverse drags (Figs 17.4 and 17.5) as the fault co-rotates to a higher angle. In general, the structures can be described as reverse drag s-type flanking structures (Exner et al. 2004). The bulk shear thus depicts a dextral sense viewing N (Fig. 17.7a,b). In summary, s-type contractional flanking structures are observed mostly with reverse drag. These structures mostly show positive slip; over roll and positive lift. Thus the geometry of the flanking structures of the present area describes s-type of flanking folds mostly with reverse drag with a positive slip, lift, and over roll. The s-type flanking structures indicate a synthetic displacement where leucosomes cross-cut the host fabric elements of gneissic bands. The shear slip is generally seen at the central part, which weakens to the marginal lower part.

Flanking structures vary geometrically depending on tilt, orientation of CE with respect to the main foliation. Qualitatively, *slip* of the HE along CE, *lift* or elevation of the HE above or below x-axis and *roll*- the magnitude and

Table 17.1. The quantitative parameters for the flanking folds of the area

Outcrop location: Tato, West Siang District, Arunachal Pradesh, India
Lat 28° 30′ 976″ N Long 94° 22′ 642″ E

Sl. No	Fig. No	Φ	β	CE	Slip	Lift	Roll	Drag	Type
1	17.4a	160	100	Phases of melt	Positive	Positive	Over	Reverse	S
2	17.4b	160	110	Phases of melt	Positive	Positive	Over	Normal	S
3	17.4c	170	120	Fracture	Zero	Zero	Over	Reverse	S
4	17.4d	160	110	Fracture	Positive	Positive	Over	Reverse	S
5	17.5a	160	130	Fracture	Positive	Positive	Over	Reverse	S
6	17.5b	140	100	Fracture	Positive	Positive	Over	Reverse	S
7	17.5c	168	50	Fracture	Positive	Positive	Over	Normal	S
8	17.5d	165	63	Fracture	Zero	Zero	Over	Reverse	S

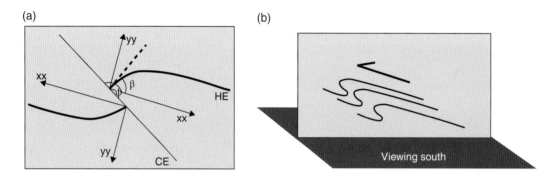

Fig. 17.6. (a) Diagram for measurement of Φ and β values with respect to the x-axis parallel to the far filed HE (Coelho et al. 2005). (b) Viewing south the shear sense is top to the ESE or ENE directions.

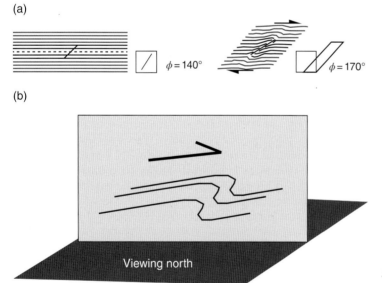

Fig. 17.7. (a, b) Viewing north, the Φ value calculated with respect to the x-axis varies from 140° to 170°. The shear sense is dextral with top to the ESE or ENE directions.

sense of curvature of HE with the cut of point on CE are also described from the overlays as summarized in Table 17.1.

The description of the flanking structures following the new scheme (Mukherjee 2014b) gives a better insight considering whether the CE as rocks and minerals or as a clear fracture and whether HE penetrates CE or not. Following the scheme, we can describe the Figs 17.4a,b,d and 17.5c,d as *a.1.1* type of flanking structure, as the drag at one side of CE is observed and HE dragged in the same direction without a slip. In Fig. 17.4c, the HE dragged in the opposite senses (see top left and near the coin) can be termed as the *b.1.2.2* type of flanking structure. The flanking structure where a slip without drag on the single side of CE can be considered as the *b.1.1* type, as in Fig. 17.5d.

17.5   DISCUSSION

The MCT hanging wall folded regionally to ESE plunging non-cylindrical pattern as the thrust plane swerved from the ENE to NE direction towards the eastern part. Therefore the thrust propagation in the area is rather towards ENE compared to SSE/SE propagation in the western Arunachal Himalaya (Singh 1990; Yin 2006). The migmatization indicates high temperature and partial melting in the core of the fold. In most cases, the host fabric element (HE) deflects near the cross-cutting veins/fractures. This indicates that the HE and CE were subject to the same ductile deformation. However, in few examples, the CE develop probably after the HE folded (Fig. 17.4a,b). Moreover, the small volume of leucosome confined within the pre-existing foliation has the diffused boundary with the host foliation. It may be due to the shallow level of deformation causing *in-situ* decompression melting. On the other hand, at the margin of the pre-existing foliation melt pockets of melanosomes also occur.

The foliation and extension crenulations cleavages, designated as S- and C'-, are at low angles (~10°; Fig. 17.4a) and C' trend ENE. We present a model for the S-fabric developments in the ductile shear zone at Tato (Fig. 17.8). The N–S or NNW–SSE compression at the MCT ductile zone at Tato produced a duplex type of structure by the subsequent formation of a brittle fault (base) near Pene, designated as the MCT-1 and the MCT-2 in Fig. 17.8a. However, during the ductile–brittle transition, the western part of the Siang dome was WNW–ESE compressed and the MCT swerved from ENE to NE. The ductile shear zone was extended possibly along ENE (Fig. 17.8b). Stretching lineations in Fig.17.3e parallel this direction. The folded Higher Himalayan slab in the region probably became non-cylindrical as the thrust transport direction took an easterly course from its earlier SSE or southward propagation.

17.6   CONCLUSIONS

The major geotectonic regimes affecting the Higher and Lesser Himalayan sequences in the Siyom valley can be divided in to two distinct types. The first one is the horizontal intense shear in the NNW–SSE direction that emplaced the Higher Himalayan sequences over the Lesser Himalayan sequences in terms of a mega duplex with a ductile zone WNW of MCT, marked as MCT-1. The second phase may be coeval with the development of a brittle fault zone at the immediate south of MCT (marked as MCT- 2) and N–S fold pattern in the Lesser Himalayan metasedimentary sequences. The subsequent deformations relate to the compression along ENE–WSW and created asymmetric folds in the MCT ductile zone and this compression relate to the Indo-Burmese collision swinging the MCT from ENE to NE. The asymmetric structures developed in the migmatitic

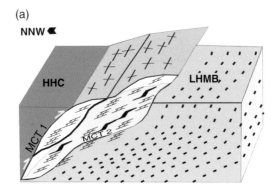

(a)

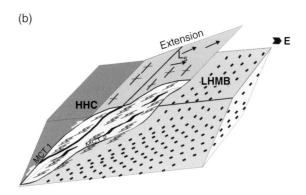

(b)

Fig. 17.8. (a) Block diagram represents cration of a ductile shear zone at Tato (MCT-1) followed by a brittle base at Pene (MCT-2). Within the Higher Himalayan Crystallines, thus a duplex structure is created. (b) The late WNW–ESE compression resulted in the extension of the folded Higher Himalayan slab to the ENE directions and the ESE to ENE plunging stretching lineation (Ls), asymmetric shear fracture boudins, S-C fabric, flanking folds are produced (MCT: Main Central Thrust, HHC: Higher Himalayan Crystallines, LHMB: Lesser Himalayan Metamorphic Belt).

gneisses in the Tato area are good shear sense indicators including flanking folds. We noted both normal- and reverse drags when shear senses were deduced. Vergence of the folds are considered. The positive slip, reverse drag and over roll are the most observed senses of the flanking folds in the migmatitic gneisses. The slip is positive towards ESE. This indicates that the MCT hanging wall folded and stretching lineations show low plunge towards ENE/ESE. Therefore, these stretching lineations indicate a late extension paralleling the thrust transport direction i.e. ESE. Thus, the ductile shear zone at Tato shows asymmetry in the migmatitic gneisses with right lateral shear. This matches well with the WNW–ESE compression that thrusted and extended the MCT hanging wall. This led to brittle fracturing, veins and flanking folds in the migmatitic gneisses. Stretching lineations at low-angle to the regional-scale F3 fold hinge reflect, along with simple shear, a component of wrenching might be there (Ridley 1986). However, if the entire fold hinge is considered, stretching lineations show obliquity varying to ~30° from Gapo to Pene.

ACKNOWLEDGMENTS

T.K.G. is grateful to Department of Science and Technology, New Delhi for providing financial assistance under the pilot project No.SR/S4/ES/531/2010 dated. 16.05.2012. The authors are thankful to Soumyajit Mukherjee for reviewing and editing this work which significantly helped to improve the manuscript.

REFERENCES

Ahmed T, Harris N, Bickle M, Chapman H, Prince C. 2000. Isotopic constraints on the structural relationships between lesser Himalayan series and higher Himalayan crystalline series, Garhwal Himalaya. Tectonophysics 112, 467–477.

Coelho S, Passchier C, Grasemann B. 2005. Geometric description of flanking structures. Journal of Structural Geology 27, 597–606.

Druguet E, Passchier C, Carreras J, Victor P, den Brok S. 1997. Analysis of a complex high strain zone at Cap de Creus, Spain. Tectonophysics 280, 31–45.

Exner, U, Mancktelow NS, Grasemann B. 2004. Progressive development of s-type flanking folds in simple shear. Journal of Structural Geology 26, 2191–2201.

Gayer RA, Powell DB, Rhodes S. 1978. Deformation against metadolerite dykes in the Caledonides of Finnmark, Norway. Tectonophysics 46, 99–115.

Grasemann B, Stuwe K, Vannay J-C. 2003. Sense and non- sense of shear in flanking structures. Journal of Structural Geology 25, 19–34.

Grasmann B, Exner U, Tschegg, C. 2011. Displacement length scaling of brittle faults in ductile shear. Journal of Structural Geology 33, 1650–1661.

Gururajan NS, Chowdhury BK. 2003. Geology and tectonic history of the Lohit Valley, Eastern Arunachal Pradesh, India. Journal of Asian Earth Science 21, 731–741.

Hamblin WK. 1965. Origin of 'reverse drag' on the down thrown side of normal fault. Geological Society of America Bulletin 76, 1145–1164.

Hanmer S, Passchier CW. 1991. Shear–sense indicators: a review. Geological Survey of Canada, paper 90–17, 72.

Hubert M, Emilien DYP, Nzenti JP, Jean B, Boniface K, Emmanual SC. 2011. Major structural features and the tectonic evolution of the Bossangoa-Bossembele Basement, Northwestern Central African Republic. The Open Geology Journal 5, 21–32.

Hudleston PJ. 1989. The association of folds and veins in shear zones. Journal of Structural Geology 11, 949–957.

Jain AK, Manickavasagam RM, Singh S, Mukherjee S. 2005. Himalayan collision zone: new perspectives-its tectonic evolution in a combined ductile shear zone and channel flow model. Himalayan Geology 26, 1–18.

Kumar G. 1997: Geology of Arunachal Pradesh. Geological Society of India Publication, Bangalore.

Lin S, Williams PF. 1992. The geometrical relationship between stretching lineation and movement direction in shear zones. Journal of Structural Geology 14, 491–497.

Mukherjee S. 2005. Channel flow, ductile extrusion and exhumation of lower mid-crust in continental collision zones. Current Science 89, 435–436.

Mukherjee S. 2007. Geodynamics, deformation and mathematical analysis of metamorphic belts of the NW Himalaya. Unpublished Ph.D. thesis. Indian Institute of Technology, Roorkee.

Mukherjee S. 2010a. Structures in Meso- and Micro-scales in the Sutlej section of the Higher Himalayan Shear Zone, Indian Himalaya. e-Terra 7, 1–27.

Mukherjee S. 2010b. Microstructures of the Zanskar shear zone. Earth Science India 3, 9–27.

Mukherjee S. 2011. Flanking microstructures from the Zanskar Shear Zone, NW Indian Himalaya. YES Bulletin 1, 21–29.

Mukherjee S. 2012a. Simple shear is not so simple! Kinematics and shear senses in Newtonian viscous simple shear zones. Geological Magazine 149, 819–826.

Mukherjee S. 2012b. Tectonic implications and morphology of trapezoidal mica grains from the Sutlej section of the Higher Himalayan Shear Zone, Indian Himalaya. The Journal of Geology 120, 575–590.

Mukherjee S. 2013a. Deformation Microstructures in Rocks. Springer, Heidelberg, pp. 55–71.

Mukherjee S. 2013b. Channel flow extrusion model to constrain dynamic viscosity and Prandtl number of the Higher Himalayan Shear Zone. International Journal of Earth Sciences 102, 1811–1835.

Mukherjee S. 2013c. Higher Himalayan in the Bhagirathi section (NW Himalaya, India): its structures, backthrusts and extrusion mechanism by both channel flow and critical taper mechanisms. International Journal of Earth Sciences 102, 1851–1870.

Mukherjee S. 2014a. Mica inclusions inside Host Mica Grains from the Sutlej section of the Higher Himalayan Crystallines, India-Morphology and Constrains in Genesis. Acta Geologica Sinica 88, 1729–1741.

Mukherjee S. 2014b. Review of flanking structures in meso-and micro-scales. Geological Magazine 151(6), 957–974.

Mukherjee S. 2015. A review of out-of-sequence deformation in the Himalaya. A review on out-of-sequence deformation in the Himalaya. In Tectonics of the Himalaya, edited by S. Mukherjee, R. Carosi, P. van der Beek, B.K. Mukherjee, and D.M. Robinson, Geological Society, London. Special Publication vol. 412. doi. org/10.1144/SP412.13.

Mukherjee S, Koyi HA. 2009. Flanking microstructures. Geological Magazine 146, 517–526.

Mukherjee S, Koyi HA. 2010a. Higher Himalayan Shear Zone, Sutlej section: structural geology and extrusion mechanism by a combination of simple shear, pure shear and channel flow in shifting modes. International Journal of Earth Sciences 99, 1267–1303.

Mukherjee S, Koyi HA. 2010b. Higher Himalayan Shear Zone, Zanskar Indian Himalaya – microstructural studies and extrusion mechanism by a combination of simple shear and channel flow. International Journal of Earth Sciences 99, 1083–1110.

Mukherjee S, Koyi HA, Talbot CJ. 2012. Implications of channel flow analogue models for extrusion of the Higher Himalayan Shear Zone with special reference to the out-of-sequence thrusting. International Journal of Earth Sciences 101, 253–272.

Mukherjee S, Biswas R. 2014. Kinematics of horizontal simple shear zones of concentric arcs (Taylor–Couette flow) with incompressible Newtonian rheology. International Journal of Earth Sciences 103, 597–602.

Mukherjee S, Biswas R. 2016. Biviscous horizontal simple shear zones of concentric arcs (Taylor Couette flow) with incompressible Newtonian rheology. In Ductile Shear Zones: From Micro- to Macro-scales, edited by S. Mukherjee and K.F. Mulchrone. John Wiley & Sons, Chichester.

Mukherjee S, Carosi R, van der Beek PA, Mukherjee BK, Robinson DM. (eds) 2015. Tectonics of the Himalaya: an introduction. Geological Society, London, Special Publications, 412, 1–3.

Mulchrone KF. 2007. Modeling flanking structures using deformable high axial ratio ellipses: Insights into finite geometries. Journal of Structural Geology 29, 1216–1228.

Passchier C. 2001. Flanking structures. Journal of Structural Geology 23, 951–962.

Passchier CW, Coelho S. 2006. An outline of shear-sense analysis in high-grade rocks. Gondwana Research 10, 66–76.

Ramsay JG. 1967. Folding and Fracturing of Rocks. McGraw-Hill, New York, pp. 103–109.

Ramsay JG. 1980. Shear zone geometry: a review. Journal of Structural Geology 2, 83–99.

Ridley J. 1986. Parallel stretching lineations and fold axes oblique to a shear displacement direction – a model and observation. Journal of Structural Geology 8, 647–653.

Saha D. 2011. Along strike variation in the Himalayan Orogen and its expression along major intracontinental thrusts – the case of MCT in Sikkim and Arunachal Pradesh, India, Cenozoic tectonics, seismology, and palaeobiology of the eastern Himalayas I and Indo-Myanmar range. Geological Society of India Memoir 77, 350.

Searly MP, Law RD, Godin L, et al. 2008. Defining the Himalayan Main Central thrust in Nepal. Journal of the Geological Society of London 165, 523–534.

Singh S, Chowdhury PK. 1990. An outline of the Geological framework of the Arunachal Himalaya. Journal of Himalayan Geology 1(2), 189–197.

Singh S, Malhotra G. 1980. Progress Report, Geological Survey of India for F.S. 1981–82, 1–58.

Singh RK Bikramaditya, Gururajan NS. 2011. Microstructures in quartz and feldspars of the Bomdila Gneiss from western Arunachal Himalaya, Northeast India: Implications for the geo-tectonic evolution of the Bomdila mylonitic zone. Journal of Asian Earth Sciences 42, 1163–1178.

Singh S. 1993. Geology and tectonics of the eastern syntaxial bend, Arunachal Himalaya. Journal of Himalayan Geology 4, 149–163.

Wiesmayr G, Grasemann B. 2005. Sense and non-sense of shear in flanking structures with layer-parallel shortening: implications for fault related folds. Journal of Structural Geology 27, 249–264.

Yin A. 2006. Cenozoic tectonic evolution of the Himalayan orogen as constrained by along-strike variation of structural geometry, exhumation history, and foreland sedimentation. Earth-Science Reviews 76, 1–131.

Index

Note: Numbers in **bold** refer to tables and figures

Ductile Shear Zones: From Micro- to Macro-scales, First Edition. Edited by Soumyajit Mukherjee
and Kieran F. Mulchrone.
© 2016 John Wiley & Sons, Ltd. Published 2016 by John Wiley & Sons, Ltd.